Achieving sustainable greenhouse cultivation

It is widely recognised that agriculture is a significant contributor to global warming and climate change. Agriculture needs to reduce its environmental impact and adapt to current climate change whilst still feeding a growing population, i.e. become more 'climate-smart'. Burleigh Dodds Science Publishing is playing its part in achieving this by bringing together key research on making the production of the world's most important crops and livestock products more sustainable.

Based on extensive research, our publications specifically target the challenge of climate-smart agriculture. In this way we are using 'smart publishing' to help achieve climate-smart agriculture.

Burleigh Dodds Science Publishing is an independent and innovative publisher delivering high quality customer-focused agricultural science content in both print and online formats for the academic and research communities. Our aim is to build a foundation of knowledge on which researchers can build to meet the challenge of climate-smart agriculture.

For more information about Burleigh Dodds Science Publishing simply call us on +44 (0) 1223 839365, email info@bdspublishing.com or alternatively please visit our website at www.bdspublishing.com.

Related titles:

Advances in crop modelling for a sustainable agriculture
Print (ISBN 978-1-78676-240-5); Online (ISBN 978-1-78676-242-9, 978-1-78676-243-6)

Precision agriculture for sustainability
Print (ISBN 978-1-78676-204-7); Online (ISBN 978-1-78676-206-1, 978-1-78676-207-8)

Chapters are available individually from our online bookshop:
https://shop.bdspublishing.com

climate•SMART•publishing
IN AGRICULTURAL SCIENCE

BURLEIGH DODDS SERIES IN AGRICULTURAL SCIENCE
NUMBER 63

Achieving sustainable greenhouse cultivation

Edited by Professor Leo F. M. Marcelis and
Dr Ep Heuvelink, Wageningen University, The Netherlands

burleigh dodds
SCIENCE PUBLISHING

Published by Burleigh Dodds Science Publishing Limited
82 High Street, Sawston, Cambridge CB22 3HJ, UK
www.bdspublishing.com

Burleigh Dodds Science Publishing, 1518 Walnut Street, Suite 900, Philadelphia, PA 19102-3406, USA

First published 2019 by Burleigh Dodds Science Publishing Limited

Library of Congress Control Number: 2019946947

British Library Cataloguing in Publication Data
A catalogue record for this book is available from the British Library

ISBN 978-1-78676-280-1 (Print)
ISBN 978-1-78676-283-2 (PDF)
ISBN 978-1-78676-282-5 (ePub)
ISSN 2059-6936 (print)
ISSN 2059-6944 (online)

DOI 10.19103/AS.2019.0052

Typeset by Deanta Global Publishing Services, Chennai, India

Contents

Series list

Title	Series number
Achieving sustainable cultivation of maize - Vol 1 From improved varieties to local applications Edited by: Dr Dave Watson, CGIAR Maize Research Program Manager, CIMMYT, Mexico	001
Achieving sustainable cultivation of maize - Vol 2 Cultivation techniques, pest and disease control Edited by: Dr Dave Watson, CGIAR Maize Research Program Manager, CIMMYT, Mexico	002
Achieving sustainable cultivation of rice - Vol 1 Breeding for higher yield and quality Edited by: Prof. Takuji Sasaki, Tokyo University of Agriculture, Japan	003
Achieving sustainable cultivation of rice - Vol 2 Cultivation, pest and disease management Edited by: Prof. Takuji Sasaki, Tokyo University of Agriculture, Japan	004
Achieving sustainable cultivation of wheat - Vol 1 Breeding, quality traits, pests and diseases Edited by: Prof. Peter Langridge, The University of Adelaide, Australia	005
Achieving sustainable cultivation of wheat - Vol 2 Cultivation techniques Edited by: Prof. Peter Langridge, The University of Adelaide, Australia	006
Achieving sustainable cultivation of tomatoes Edited by: Dr Autar Mattoo, USDA-ARS, USA & Prof. Avtar Handa, Purdue University, USA	007
Achieving sustainable production of milk - Vol 1 Milk composition, genetics and breeding Edited by: Dr Nico van Belzen, International Dairy Federation (IDF), Belgium	008
Achieving sustainable production of milk - Vol 2 Safety, quality and sustainability Edited by: Dr Nico van Belzen, International Dairy Federation (IDF), Belgium	009
Achieving sustainable production of milk - Vol 3 Dairy herd management and welfare Edited by: Prof. John Webster, University of Bristol, UK	010
Ensuring safety and quality in the production of beef - Vol 1 Safety Edited by: Prof. Gary Acuff, Texas A&M University, USA & Prof. James Dickson, Iowa State University, USA	011
Ensuring safety and quality in the production of beef - Vol 2 Quality Edited by: Prof. Michael Dikeman, Kansas State University, USA	012
Achieving sustainable production of poultry meat - Vol 1 Safety, quality and sustainability Edited by: Prof. Steven C. Ricke, University of Arkansas, USA	013
Achieving sustainable production of poultry meat - Vol 2 Breeding and nutrition Edited by: Prof. Todd Applegate, University of Georgia, USA	014

Chapter 1

Achieving sustainable greenhouse production: present status, recent advances and future developments

Leo F. M. Marcelis, Wageningen University, The Netherlands; Joaquim Miguel Costa, Universidade de Lisboa, Portugal; and Ep Heuvelink, Wageningen University, The Netherlands

1 Introduction

The current world population of 7.7 billion is expected to reach 9.7 billion in 2050 (United Nations, 2019) and eating patterns are changing. It is anticipated that the demand for plant food will increase by 60% by 2050 (Alexandratos and Bruinsma, 2012). Furthermore, strong urbanization has already been taking place for several decades, and more than half of the world population is now living in cities. At present, there are about 33 megacities with more than 10 million inhabitants each (United Nations, 2018). Consumers are getting more interested in the quality of their food. Quality includes a good taste, nice shape and colour, good shelf life, but also refers to food safety and healthy food. The society is increasingly demanding that plants are produced sustainably. The use of fossil fuel and pesticides should be substantially reduced, and ultimately, should be abandoned. Growers need to be restrictive when using natural resources such as phosphorous for plant nutrition. These resources are not endless and can pollute the environment. Soil and water pollution with nutrients or chemicals must be prevented. Furthermore, water saving and salinization (De Pascale, this book) are important issues. Emission

http://dx.doi.org/10.19103/AS.2019.0052.01

of light from supplemental lighting is considered as environmental pollution and in many places light emission is no longer accepted (e.g. the Netherlands). Also, good working conditions for employees are considered crucial. In many countries supermarkets dominate the supply chain and are requesting cheap products and are increasingly demanding guarantees on quantity, quality and sustainability of the products and the production process. Nevertheless, the risks remain largely at the side of growers, especially for those that are small and not associated.

This whole context sets demanding goals on production systems to sustainably produce crops with very high yield and quality on a limited land area, and preferably nearby urban areas. Greenhouse systems, when properly managed, improve growth conditions for plants in order to increase yield and quality and to extend the growing season. Protected cultivation, controlled environment agriculture (CEA) and greenhouse horticulture are often used as synonyms, all three terms referring to crop production systems that allow manipulation of the crop environment. Over the past 50 years the level of environment control in greenhouses has grown rapidly through increased use of technology and improved knowledge. The level of control has improved due to introduction of:

1 heating,
2 cooling or dehumidification by controlling the window opening, cooling by pad-and-fan or even mechanical cooling,
3 CO_2 supply,
4 fogging,
5 extending the photoperiod by low-intensity supplementary light or shortening by black-out screens,
6 lighting with high-intensity supplementary light to increase assimilation, initially only by HPS lamps with fixed spectrum, and nowadays also by LEDs where spectrum can be chosen,
7 biological and/or integrated pest and disease management,
8 soilless cultivation, crop growth on an artificial substrate, with a dripper for water and nutrients for each plant or plants can even be grown without any substrate.

In modern greenhouses, climate, irrigation and fertilization are often fully computer controlled. Meanwhile, automation is gradually improving labour conditions for workers and helps to reduce production costs, deal with shortage of skilled labour and so on (Van Henten, this book). The height of greenhouses has increased to create a more stable climate and to accommodate the use of technology, for example screens, lamps and hanging gutters with rooting medium, as well as to grow a taller canopy; nowadays the gutter height of newly built greenhouses is often about 6 to 7 m in the Netherlands. An increased level of technology implies higher investment costs. The economically optimal level of control depends on the type of product produced, the market segment aimed for, and also the climate and socioeconomic factors (e.g. labour skills, investment capacity of growers, policies and subsidies). This is called the adaptive greenhouse: greenhouse design based on local boundary conditions (Van Henten et al., 2006).

The effects of increasing the level of control of the production process are clearly revealed by an analysis of the statistical data on yield in the Netherlands. The annual production per unit greenhouse area (kg m^{-2}) for tomato, sweet pepper and cucumber has increased substantially over the past few decades (Fig. 1). For tomato and sweet pepper, the yield doubled in the last three decades and this increase is even more when the companies are included that use supplementary lighting. Besides the application of new technology (e.g. soilless systems, higher greenhouse transmissivity, CO_2 enrichment) and cultivation knowledge, improved cultivars (more productive, more resistant to diseases) also contributed to this yield increase. Yield of modern greenhouse

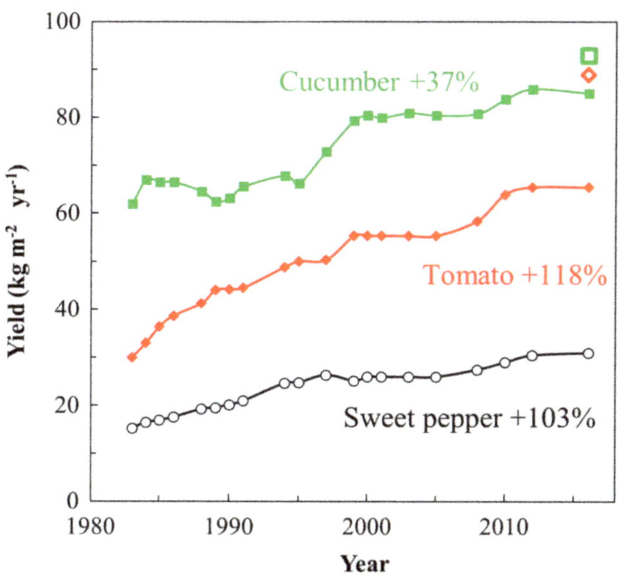

Figure 1 Average annual yields for cucumber (closed green squares), tomato harvested as individual fruits (closed red diamonds) and red sweet pepper (open black circles) in Dutch greenhouses without use of supplementary lighting. For cucumber (open green square) and tomato (large fruits on the vine; open red diamond) grown with supplementary light, the yield is also shown for the year 2016. Data were derived from annual statistical overviews as updated from De Gelder et al. (2012) with later data until 2016 (Vermeulen, 2017).

tomato cultivars (hybrids) was shown to be about 40% higher than that of old cultivars when cultivated under identical conditions (Higashide and Heuvelink, 2009). This increase was not related to a higher harvest index but rather due to increased light-use efficiency (LUE), which results in an increase in total dry matter production.

2 The state of current greenhouse horticulture

The worldwide area of greenhouses (Table 1), defined as permanent structures (so excluding simple low or high tunnels), is estimated at 473 000 ha (Hadley,

Table 1 Non-exhaustive overview of greenhouse vegetable production worldwide

Region/country	Vegetables production area (ha)	References
Asia + Oceania	*175 800*	Stanghellini et al. (2019)
China	82 000	Rabobank (2018)
South Korea	56 400	USDA (2018)
Japan	43 000[a]	MAFF (2018)
Europe	*178 000–210 000*	Stanghellini et al. (2019)/ Rabobank (2018)
Spain	70 000	Rabobank (2018)
Italy	42 800	Rabobank (2018)
France	11 500	Rabobank (2018)
The Netherlands	5 000	Rabobank (2018)
Poland	6 700	Rabobank (2018)
Américas	*16 500*	Stanghellini et al. (2019)
Mexico	16 500	Aurélio (2017)
Canada	1 560	Rabobank (2018)
USA	1 050	Hortidaily (2019)
Africa	*30 300*	Stanghellini et al. (2019)
Morocco	20 000–24 000	Freshplaza (2018), Rabobank (2018)
Near East	*74 800*	Stanghellini et al. (2019)
Turkey	41 400–45 500	Rabobank (2018), Tüzel and Öztekin (2016)
Israel	11 000	Rabobank (2018)

[a] Total greenhouse area including ornamentals and fruit crops.
Italics are aggregated values over several countries, while the non-italics are values for individual countries.
Greenhouses for ornamentals are not included, for lack of information (Hickman, 2018). However, in equatorial countries both in Africa (e.g. Kenia) and in America (e.g. Colombia) ornamentals are produced in greenhouse (usually at high elevation) for export, and the acreage can be several times over the acreage of vegetable production (Stanghellini et al., 2019).
Source: adapted from Stanghellini et al. (2019).

2017) or even at 500 000 ha (Rabobank, 2018). Europe has the largest area of greenhouses (210 000 ha), followed by Asia (180 000 ha), Africa (44 700 ha), the Middle East (27 000 ha), North-Central America (23 000 ha), South America (14 000 ha) and Oceania (1300 ha) (Rabobank, 2018). When other types of structures (e.g. low and high tunnels and Chinese solar greenhouses) are included, worldwide greenhouse area easily surpasses 3 million ha (Hadley, 2017). The expansion of global greenhouse horticulture is closely linked to the increasing use of plastic since the 1950s. On an area basis, about 90% of world's greenhouses are plastic-covered structures whereas about 10% consists of glasshouses (47 000 ha; Hadley, 2017). In China, the area covered by glasshouses is only about 1% of the total (Costa et al., 2004) but in countries such as the Netherlands it reaches 100% of the total area (9500 ha). Plastic greenhouses are the most common greenhouse type in Mediterranean countries, where greenhouse tomato is the most important crop. In many places the greenhouse area is (rapidly) increasing, be it for vegetables (e.g. Mexico, Turkey and Morocco, mainly tomato) or cut flower production (e.g. Ecuador and Colombia, Kenya and Ethiopia).

3 Sustainability

The term 'sustainability' can be defined in several ways, in essence the term points to long-term ability of systems and processes to sustain the well-being of future generations (UN World Commission on Environment and Development). Environmental impact categories typically used in a life cycle assessment (LCA) are energy use, abiotic depletion, acidification, eutrophication, global warming and photochemical oxidation (see Anton, this book). Human toxicity and aquatic toxicity are also relevant measures. Usage of resources per unit horticultural product, for example water, fertilizer, energy and land usage, can be used as indicators for sustainability.

Though there are some negative aspects (e.g. plastic waste, energy use, water and soil pollution), greenhouse horticulture can positively contribute to sustainability: higher yields per unit area which reduces land use, improved water-use efficiency, more and better options for biological control (given the closed environment) and hence less pesticide use. These aspects differ between soilless cultivation in greenhouses and soil-based greenhouse production systems. For example, water use in soilless cultivation (substrates like stone wool and NFT systems) is potentially much lower than in soil-based (conventional) systems. Whether this potential is realized depends on the irrigation strategy, the application of recirculation and the quality of the irrigation water. Soilless cultivation systems can in principle realize zero water losses, because the nutrient solutions can be recirculated (Beerling et al., 2014), although in practice 100% recirculation is often not achieved. Water and nutrient-use efficiency differs significantly between growers/greenhouses, even in the most advanced production systems. For example in the Netherlands, for the top 20% tomato growers, the discharge is almost zero, whereas for the bottom 20% tomato growers it is 746 m^3 ha^{-1} $year^{-1}$ which is about 10% of the annual overall nutrient solution used (Beerling et al., 2014). Comparing a closed versus an open soilless cultivation system, a reduction in fertilizer use per kg of product of 20% up to 78% has been observed in closed systems, without any reduction in yield or product quality (Pardossi et al., 2011; Giuffrida and Leonardi, 2012). When grown on substrates there are no specific requirements for soil quality since the growing system is separated from the soil. Hence, land unfit for soil-based production (poor soils and contaminated soils, for example because of high heavy metal or salinity levels), or land infected with soil-borne diseases can be used for production of vegetables and ornamentals on substrates.

4 Current and future trends

The beginning of this chapter describes the trend of increasing control over the production process in recent decades. We expect rapid further developments in even more precise control and automation of the production process. In this section we will briefly describe some of these developments.

There are a wide range of typologies of greenhouses, ranging from net and screen houses (Tanny, this book), solar greenhouses, low-tech greenhouses that are unheated and have simple climate control systems, to high-tech greenhouses with advanced and fully automatized climate control (Montero, this book). While in the past greenhouses were mainly found in cool climates, to protect plants from low temperatures, nowadays greenhouses are spread in all regions to protect plants from unfavourable conditions (wind, rain, hail, heat, cold). Water saving has become an important driver for expansion of the greenhouse industry, in particular in arid and semi-arid regions. To save water, collection and re-use of irrigation water is essential (De Pascale, this book), which can be most efficiently done in soilless culture, costs and the available knowhow are still a limitation for certain countries. Greenhouse production can be highly sustainable, but more investment and incentives are needed to improve sustainability and circularity (EIP-AGRI, 2019). LCA can help to guide these developments (Anton, this book).

In northern countries energy saving is still of crucial importance, even though energy efficiency has increased tremendously. In the Netherlands CO_2 emissions from greenhouse horticulture decreased from 8.1 Mton in 2010 to 5.7 Mton in 2015 (Van der Velden and Smit, 2017), and primary energy use per unit product in 2015 was 58% less than in 1990 (Van der Velden and Smit, 2017). The most important routes to save energy are (Hemming, this book):

1 maximizing the use of natural sunlight;
2 maximizing insulation;
3 efficient use of energy by different technologies and control strategies;
4 replacing fossil fuels by sustainable energy sources.

In many countries the main energy sources are fossil fuels such as natural gas and oil. Considering general changes in energy economy and policy, we expect a strong increase in the relative importance of electricity as an energy input in greenhouses. All-electric greenhouses will arise with consequences for the control of the production system.

We expect that application of new technologies will continue to rapidly expand in the next few years. The use of biological control methods (predators, parasites, antagonists etc.) is nowadays a common way to control most pests and a number of diseases in medium to high-tech greenhouses (Messelink, this book). The use of biostimulants is rapidly increasing (Dorais, this book; Mattson and Currey, this book; De Pascale, this book). In practice many positive results have been reported with respect to yield increase, disease suppression, water saving and so on, though there have also been a number of tests where the effects of biostimulants were disappointing (Dorais, this book). Our understanding of biostimulants is still in its infancy and urgently needs to be improved, in order to make full use of the potential benefits (Bulgari et al., 2015). Furthermore, more transparency is needed by companies in order to know precisely what growers are truly applying to crops.

Organic greenhouse production is still relatively small but it is an expanding segment (Dorais, this book). Organic food is gaining market share worldwide, for example in the EU (Rabobank, 2018). The share of organic fruit and vegetable sales (in total fresh fruit and vegetable sales) has already passed 10% in wealthy countries such as Switzerland, Sweden, Austria and Denmark (Fig. 2). Despite the fact that many people associate organic horticulture with a low level of technology, the opposite holds true as well. Successful organic horticulture goes often together with significant use of technology, though with regulatory limitations in some countries, for instance relating to prohibiting use of rooting media (instead of soil) or specific chemicals. In the light of major global challenges and trends, aquaponics, which combines fish production with plant production, is receiving attention within the greenhouse sector (Alsanius, this book). This parallels ongoing efforts to increase circularity.

In greenhouses crops are often grown in soilless conditions, using stone wool or peat- or coir-based substrates. Efforts are being undertaken to develop more environmentally friendly substrates (Zheng, this book). An integrated rootzone management (IRM) approach is necessary to achieve optimal

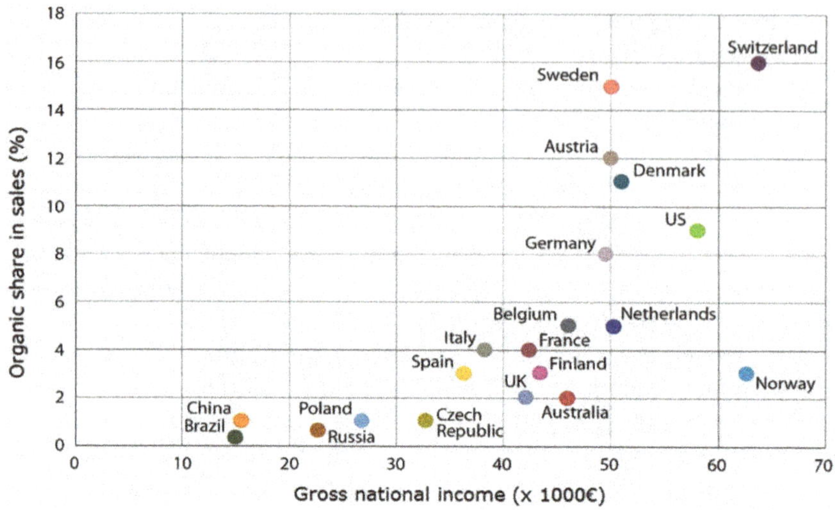

Figure 2 Share of organic fresh fruit and vegetable sales vs. gross national income in purchasing power parity, 2016 for 20 European countries. Source: reprinted from Rabobank (2018).

rootzone conditions, using in-situ rootzone sensors and artificial intelligence (Zheng, this book).

Nutrient management continues to move towards more precise nutrient applications to limit environmental nutrient loading and to enhance plant yield and quality at harvest and post-harvest stages (Mattson and Currey, this book). Precisely meeting plant nutritional needs requires an integrated approach relying on the adoption of several technologies. Greater ability to sense the nutrient solution composition, rootzone environment or plant status in real-time will allow for more agile decision making and correction of nutrient disorders before yield and quality are impaired (Mattson and Currey, this book).

In recent years there have been interesting developments with respect to light use in greenhouses (both natural and artificial light). Examples include introduction of diffuse greenhouse covers (glass, plastics) and screens which have resulted in yield increases of up to 10% (Hemming, this book). In shade-tolerant plants the use of a diffuse cover requires less shading which could increase yield by up to about 50% in anthurium (Li et al., 2014). Light transmission in greenhouses has been substantially improved over the last couple of decades (Hemming, this book). In order to improve yield and quality, and obtain year-round production, growers are making increasing use of supplemental light in greenhouses. High pressure sodium lamps are still the most commonly used light source, but rapid developments in LED technology have encouraged implementation of LED lighting in the greenhouse sector. Modern LEDs are a more energy efficient light source than high pressure sodium lamps and are an important next step in controlling growth and quality of plants. Though at present fixed light spectra are typically used, in the future dynamic illumination strategies may be adopted where the light spectrum is varied in relation to the development phase of the crop as well as in relation to other growth conditions. There are huge efforts worldwide in researching plant responses to LED lighting, and the expectation that the costs for this technology will further reduce will support a widespread implementation of LED lighting in horticulture. Recent reviews on supplementary light in greenhouse horticulture are provided by Kozai et al. (2016) and Stanghellini et al. (2019).

The availability of efficient LEDs has also boosted the development of vertical farms ('plant factories' with artificial light, PFAL) (Kozai, this book). The development of these high-tech indoor farms fits neatly into the general trend of ever-increasing control over the production process. Although there is still a lot to learn about producing crops in vertical farms, it offers a novel opportunity to produce high-value plant products. Many developments on vertical farming are expected in the coming years and differences between conventional horizontal greenhouse horticulture and vertical farming will tend to become smaller.

An integrated approach to crop growth and development and the technical aspects of greenhouse cultivation and climate management is provided by Stanghellini et al. (2019). Plant Empowerment is a new approach to growing crops in greenhouses, also known as 'Next Generation Growing' (Geelen et al., 2019). It is based on a combination of physical and physiological principles rather than practical experiences, human feelings and 'green' fingers. Plant Empowerment aims to optimize growth by supporting the plant in keeping its balances regarding energy, water and assimilates in equilibrium (Geelen et al., 2019).

Turning to our understanding of crop physiology, there is already a significant body of knowledge about plant responses to growth conditions. This knowledge often derives from single-factor experiments, where one factor is changed at a time, while other conditions are kept constant (Kubota, this book). Responses of plants to the environment are seldom linear, and show many interactions (Poorter et al., 2013). Therefore, research should not only study plant responses under constant conditions, but also analyse multiple interacting factors under fluctuating conditions (Kubota, this book; Gruda, this book; Marcelis et al., 2018). In addition, it is not only growth but other quality attributes that might also need to be improved, and which may require different optimum conditions and interact in different ways with the crop genotype (Gruda, this book). More research on optimal phenotyping of greenhouse crops will be also needed to optimize selection of the most competitive genotypes for increased resource use efficiency, yield and quality.

We expect many efforts to be made to develop autonomous greenhouses, in particular in the most technological advanced countries. In an autonomous greenhouse most of the manual work is replaced by robots and both the above-ground and below-ground climates are controlled without daily interference by the grower. Simulation models, sensors for plants and their environment (Körner, this book) will play a crucial role in autonomous greenhouses, especially by supporting more robust decision support systems (Gupta et al., 2010; Canãdas et al., 2017; Aiello et al., 2018). The whole greenhouse will be controlled from a distance, that is, the physical location of the grower controlling the greenhouse could now be far removed from the greenhouse. Despite the many challenges in replacing human labour by machines, more robots are gradually being introduced in greenhouses (Van Henten, this book). This implies machines adapted to the plants they are helping to grow, but it will also imply that plant type and also the cultivation system will be adapted to the limitations posed by the use of machines (Van Henten, this book). Artificial intelligence, machine learning and use of big data will change the way greenhouses are controlled. Low-cost sensors (e.g. for soil, plant and air) for more precise climate control will play a role as well. Imaging sensors and machine vision approaches are also expected to play an increasingly important role in supporting greenhouse climate control as well as pest and disease monitoring (Mahlein, 2016; Sun et al., 2016). An indication of the potential of artificial intelligence can be seen in the first 'Autonomous greenhouse challenge' organized by Wageningen UR Greenhouse Horticulture in 2018 where five teams each operated a cucumber greenhouse at a distance by using artificial intelligence. One of the teams managed to realize a higher net profit (income minus costs) than a manually grown system using current commercial best practice (Hemming et al., 2019).

Circularity and clustering are two issues that should be considered together. A circular economy consists in minimizing waste generation and maintaining the economic value of products, materials and resources as long as possible (http://www.circulary.eu/about/). Circular horticulture or agriculture has now become a 'hot topic' in Europe (EIP-AGRI, 2019). The sector needs to face the challenge of 'zero waste' and minimize its environmental impact. Clustering can be used to improve sustainability, resulting in large greenhouse areas, preferably located next to industries storing or producing CO_2 or waste heat (EIP-AGRI, 2019). Clusters involve two major attributes: (1) geographical/spatial distribution and (2) a sectoral dimension. In greenhouse horticulture several examples of clustering exist: Westland (the Netherlands), Almeria (Spain) (Aznar-Sánchez and Galdeano-Gómez, 2011) or the Lake Naivasha Cluster (Kenya) (Bolo, 2006). Clusters of small greenhouse growers are viable either by acting as third-party suppliers to existing large firms, and making use of latter's packing and distributional facilities, or by working within cooperatives and making use of shared packing and distribution infrastructure. This is the case in Almeria, which allows small producers to compete more effectively with large companies

(Aznar-Sánchez and Galdeano-Gómez, 2011; Hadley, 2017). Clustering may also result in gains in terms of efficiency and environmental protection by optimizing logistics and recycling procedures. Indeed, logistics costs are a major issue when talking about recycling (EIP-AGRI, 2019). Clusters can also favour innovation and cooperation between companies but the capacity and willingness of the industry to collaborate needs to improve, as well as in the sharing of big data between producers.

Traditionally, vegetables (including fruit vegetables), cut flowers and potted or bedding plants are grown in greenhouses. With the legalization of cannabis for medicinal use in a number of countries, and in some countries even for recreational use, the cultivation of this high-value crop is rapidly expanding. As consistency and reliability of the product is extremely important, this is a crop typically suited for greenhouses and vertical farms. Furthermore, greenhouses and vertical farms are in particular suited for producing crops for which a guaranteed content of specific metabolites is essential, such as for pharmaceuticals and cosmetics industries, which may become an important new market segment for the greenhouse and vertical farming sector.

5 Conclusion

Greenhouse horticulture is a sector rapidly adopting new technologies. Greenhouse production can now be characterized by its extremely high production rates of high-quality produce. On many sustainability measures, greenhouses perform very well. Nevertheless, further improvements are of utmost importance, in particular in reducing the use of fossil energy and water. The high technology nature of greenhouse production makes it one of the agricultural sectors most suited to apply rules of circularity for greater sustainability.

Modern greenhouses will gradually develop towards autonomous greenhouses with a high level of automation, use of sensors, big data analytics and artificial intelligence, where all growth factors are precisely measured and controlled in an integrated way. These developments are not just about technology, but about the interplay between technology and biology. Technology will help and advise growers who need a high level of education and technical skill to make good use of these developments.

6 Where to look for further information

Complementary information on recent and future developments can be found in the chapters of this book and in other recent and less recent references focused on the greenhouse horticulture sector worldwide (FAO, 2013; Cajamar 2016; Yang et al., 2017; Stanghellini et al., 2019).

7 References

Aiello, G., Giovino, I., Vallone, M., Catania, P. and Argento, A. 2018. A decision support system based on multisensor data fusion for sustainable greenhouse management. *Journal of Cleaner Production* 172, 4057–65. doi:10.1016/j.jclepro.2017.02.197.

Alexandratos, N. and Bruinsma, J. 2012. World agriculture towards 2030/2050: the 2012 revision. ESA working paper no. 12-03. FAO, Rome, 147pp.

Aurélio, B. T. 2017. Evolución y Situación Actual de la Agricultura Protegida en México. *Memorias VI Congresso Internacional de Investigation en Ciencias Básicas Y Agronómicas*, 21–22 Septiembre de 2017. Rev. Chapingo. Available at: http://dicea.chapingo.mx/wp-content/uploads/2018/05/MEMORIA_MESA_3_2_CONGRESO2017.pdf.

Aznar-Sánchez, J. and Galdeano-Gómez, E. 2011. Territory, cluster and competitiveness of the intensive horticulture in Almería (Spain). *The Open Geography Journal* 4, 103–14.

Beerling, E. A. M., Blok, C., van der Maas, A. A. and van Os, E. A. 2014. Closing the water and nutrient cycles in soilless cultivation systems. *Acta Horticulturae* 1034, 49–55. doi:10.17660/ActaHortic.2014.1034.4.

Bolo, M. 2006. *Knowledge, Technology and Growth: the Case Study of Lake Naivasha Cut Flower Cluster in Kenya*. World Bank Institute, Washington DC.

Bulgari, R., Cocetta, G., Trivellini, A., Vernieri, P. and Ferrante, A. 2015. Biostimulants and crop responses: a review. *Biological Agriculture and Horticulture* 31(1), 1–17. doi:10.1080/01448765.2014.964649.

Cajamar. 2016. *Greenhouse Agriculture in Almeria. A Comprehensive Techno-Economic Analysis*. March 2016, Cajamar Caja Rural. ISBN: 978-84-95531-75-9.

Canãdas, J., Sanchez-Molina, J. A., Rodrıguez, F. and Del Aguila, I. M. 2017. Improving automatic climate control with decision support techniques to minimize disease effects in greenhouse tomatoes. *Information Processing in Agriculture* 4(1), 50–63. doi:10.1016/j.inpa.2016.12.002.

Costa, J. M., Heuvelink, E. and Botden, N. (Eds). 2004. Greenhouse horticulture in China: situation and prospects. Report on a study tour 11 October – 2 November 2003. Wageningen University, 140p. Available at: http://edepot.wur.nl/331170.

De Gelder, A., Dieleman, J. A., Bot, G. P. A. and Marcelis, L. F. M. 2012. An overview of climate and crop yield in closed greenhouses. *The Journal of Horticultural Science and Biotechnology* 87, 193–202.

EIP-AGRI. 2019. EIP-AGRI focus group circular horticulture – final report. Available at: https://ec.europa.eu/eip/agriculture/sites/agri-eip/files/eip-agri_fg_circular_horticulture_final_report_2019_en.pdf.

FAO. 2013. Good Agricultural Practices for greenhouse vegetable crops Principles for Mediterranean climate areas - Principles for Mediterranean climate areas. FAO Plant Production and Protection paper 217. Available at: http://www.fao.org/3/a-i3284e.pdf.

Freshplaza. 2018. Available at: https://www.freshplaza.com/article/9052907/greenhouse-acreage-expanding-faster-in-morocco-than-in-spain/ (accessed on 8 July 2019).

Geelen, P. A. M., Voogt, J. O. and van Weel, P. A. 2019. *Plant Empowerment: the Basic Principles*. Letsgrow.com, Vlaardingen, 328p.

Giuffrida, F. and Leonardi, C. 2012. Nutrient solution concentration on pepper grown in a soilless closed system: yield, fruit quality, water and nutrient efficiency. *Acta*

Agriculturae Scandinavica, Section B - Soil and Plant Science 62(1), 1–6. doi:10.1080/09064710.2011.560123.

Gupta, M. K., Samuel, D. V. K. and Sirohi, N. P. S. 2010. Decision support system for greenhouse seedling production. *Computers and Electronics in Agriculture* 73(2), 133–45. doi:10.1016/j.compag.2010.05.009.

Hadley, D. 2017. Controlled environment horticulture industry potential in NSW. University of New England. Available at: https://www.une.edu.au/__data/assets/pdf_file/0010/174565/controlled-environment-horticulture-industry-potential-hadley.pdf.

Hemming, S., de Zwart, F., Elings, A., Righini, I. and Petropoulou, A. 2019. Remote control of greenhouse vegetable production with artificial intelligence–greenhouse climate, irrigation, and crop production. *Sensors* 19(8), 1807. doi:10.3390/s19081807.

Hickman, G. W. 2018. *International Greenhouse Vegetable Production Statistics*. Cuesta Roble Greenhouse Consultants, Mariposa, CA, 170p. Available at: http://www.cuestaroble.com/statistics.htm.

Higashide, T. and Heuvelink, E. 2009. Physiological and morphological changes over the past 50 years in yield components in tomato. *Journal of the American Society for Horticultural Science* 134(4), 460–5. doi:10.21273/JASHS.134.4.460.

Hortidaily. 2019. Available at: https://www.hortidaily.com/article/9093453/us-greenhouse-vegetable-acreage-up-15-percent-between-2012-and-2017/ (accessed on 8 July 2019).

Kozai, T., Fujiwara, K. and Runkle, E. S. 2016. *LED Lighting for Urban Agriculture*. Springer Singapore, 454p.

Li, T., Heuvelink, E., van Noort, F., Kromdijk, J. and Marcelis, L. F. M. 2014. Responses of two *Anthurium* cultivars to high daily integrals of diffuse light. *Scientia Horticulturae* 179, 306–13.

MAFF. 2018. Situation of greenhouse horticulture. Ministry of Agriculture, Forestry and Fisheries. Available at: http://www.maff.go.jp/e/policies/agri/attach/pdf/index-16.pdf.

Mahlein, A. K. 2016. Plant disease detection by imaging sensors - parallels and specific demands for precision agriculture and plant phenotyping. *Plant Disease* 100(2), 241–51. doi:10.1094/PDIS-03-15-0340-FE.

Marcelis, L. F. M., Kaiser, E., van Westreenen, A. and Heuvelink, E. 2018. Sustainable crop production in greenhouses based on understanding crop physiology. *Acta Horticulturae* 1227, 1–12. doi:10.17660/ActaHortic.2018.1227.1.

Pardossi, A., Carmassi, G., Diara, C., Incrocci, L., Maggini, R. and Massa, D. 2011. Fertigation and substrate management in closed soilless culture. Deliverable report no. 15, EUPHOROS project, 63pp. Available at: www.euphoros.wur.nl.

Poorter, H., Anten, N. P. R. and Marcelis, L. F. M. 2013. Physiological mechanisms in plant growth models: do we need a supra-cellular systems biology approach? *Plant, Cell and Environment* 36(9), 1673–90 doi:10.1111/pce.12123.

Rabobank. 2018. World vegetable map 2018. RaboResearch Food & Agribusiness. Available at: https://research.rabobank.com/far/en/sectors/regional-food-agri/world_vegetable_map_2018.html.

Stanghellini, C., Van't Ooster, B. and Heuvelink, E. 2019. *Greenhouse Horticulture: Technology for Optimal Production*. Wageningen Academic Publishers, Wageningen, The Netherlands, 311p.

Sun, G., Li, Y., Zhang, Y., Wanga, X., Chen, M., Li, X. and Yan, T. 2016. Nondestructive measurement method for greenhouse cucumber parameters based on machine vision. *Engineering in Agriculture, Environment and Food* 9(1), 70–8. doi:10.1016/j.eaef.2015.06.003.

Tüzel, Y. and Oztekin, G. B. 2016. Recent developments in protected cultivation of Turkey. *Acta Horticulturae* 1142, 435–42. doi:10.17660/ActaHortic.2016.1142.66.

United Nations. 2018. World urbanization prospects: the 2018 revision [Keyfacts]. Available at: https://population.un.org/wup/Publications/Files/WUP2018-KeyFacts.pdf.

United Nations, Department of Economic and Social Affairs, Population Division. 2019. World population prospects 2019: highlights (ST/ESA/SER.A/423). Available at: https://population.un.org/wpp/Publications/Files/WPP2019_Highlights.pdf.

USDA. 2018. Korea's controlled horticulture. GAIN report no. KS1820. Available at: https://gain.fas.usda.gov/Recent%20GAIN%20Publications/Korea's%20Controlled%20Horticulture_Seoul_Korea%20-%20Republic%20of_6-7-2018.pdf.

Van der Velden, N. and Smit, P. 2017. *Energiemonitor Glastuinbouw*. Wageningen Economic Research (Rapport / Wageningen Economic Research 2016-099), Wageningen. doi:10.18174/393130.

Van Henten, E. J., Bakker, J. C., Marcelis, L. F. M., Van't Ooster, A., Dekker, E., Stanghellini, C., Vanthoor, B., Van Randeraat, B. and Westra, J. 2006. The adaptive greenhouse - an integrated systems approach to developing protected cultivation systems. *Acta Horticulturae* 718, 399–406.

Vermeulen, P. C. M. 2017. Report GTB 5154 quantitative information on Dutch greenhouse horticulture 2016–2017. Report GTB 5154. Wageningen UR Greenhouse Horticulture, Wageningen, The Netherlands, 332p.

Yang, Q., Luo, W. and Jiang, W. 2017. International Symposium on New Technologies for Environment Control, Energy-Saving and Crop Production in Greenhouse and Plant Factory - GreenSys 2017, *Acta Horticulturae* 1227.

Part 1

Production systems

Chapter 2

Advances in greenhouse design

Juan I. Montero, formerly Institut de Recerca i Tecnología Agroalimentaries, Spain; and Yi Zhang, Qichang Yang and Xinglin Ke, Institute of Environment and Sustainable Development in Agriculture, China

1 Introduction

Greenhouses have been used for decades as a way of protecting crops from unfavourable weather conditions (low or excessive temperature, strong winds, heavy rains etc.). Other benefits are the increase in water use efficiency and the defence of crops against pest attacks. Greenhouses, if properly designed and managed, are efficient systems to produce healthy food. In most countries, greenhouse crops yield the highest agricultural production values with the lowest use of available surface for agricultural activities. Such efficiency in the use of natural resources (land, water, light) makes greenhouses a solid tool to confront the threatening problem of food security.

The greenhouse internal climate is the result of a balance between all energy inputs (solar radiation, heating etc.) and outputs (ventilation, losses by convection and thermal radiation etc.). During the day, greenhouses are solar heat collectors. They traditionally maintain a higher internal temperature compared with external ambient temperature, with solar radiation being the major input of energy in the greenhouse. At night, particularly in cold areas, artificial heating is needed to modulate the unfavourable external conditions and offer more reliability to the grower. Benefits associated with greenhouses can only be achieved by using technology, which ranks from very simple crop protection structures to very complex high-tech greenhouse systems. As a rule, increasing technology increases (in most cases) yield, security and stability in the greenhouse operation.

The question of how much technology is 'the best' for a specific location is open to debate. Vanthoor (2011) conducted simulation studies comparing

http://dx.doi.org/10.19103/AS.2019.0052.02

different production systems in different areas of the world. A general conclusion was that, in terms of economic profitability, technology pays back in the long term. So, for instance, a glass-covered greenhouse with minimal heating could be more profitable than simple greenhouse shelters, very common in mild climate areas such as the Mediterranean. This holds true if the market conditions are stable and growers can finance their investments. In practice, there is a variety of greenhouses and associated equipment, and this is due to the adaptation of greenhouses to climate, economic and social conditions.

This chapter examines recent developments in greenhouse designs. Among the various greenhouse typologies, this chapter focusses on three relevant production systems, namely low-tech greenhouses, high-tech greenhouses and the Chinese solar greenhouses (CSG). Other categories like net houses, shade houses, plant factories or vertical farms will be discussed in other chapters.

Low-tech greenhouses are relatively cheap, passive greenhouses with little climate control. They are mostly covered with plastic film. Plastic low-tech greenhouses prevail around the world. In the Mediterranean basin, out of an estimated 220 000 ha of greenhouses 90% are covered with plastic (FAO, 2013), and they also occur in arid areas as well as tropical, subtropical and equatorial areas of the world. Very often they suffer from excessive temperature in spring and summer and, in some areas, suboptimal winter temperature. Consequently, production may be stopped during some unfavourable months and yield is much affected by the outside weather conditions.

High-tech greenhouses are typical of temperate and colder regions, where heating and sometimes lighting is required to produce vegetables in winter and part of the spring and fall seasons. Northwestern Europe has the highest number of high-tech greenhouses. Glass is the prevailing cover of high-tech greenhouses. There are approximately 47 000 ha of glass houses in temperate to subarctic areas of Europe with different levels of technology (Stanghellini et al., 2019). High-tech greenhouses are efficient and reliable, but expensive in terms of investment and managing costs. Their major challenge is the reduction of their dependence on fossil fuel for heating.

The CSG is the most wide-spread greenhouse structure in the world, with over 1 Mha in China. The CSG is a plastic-covered passive greenhouse, but its technical and social peculiarities as well as its extent deserves a particular discussion separate from that of the low-tech greenhouses.

2 Advances in low-tech greenhouses

Low-tech greenhouses have driven the economic, social and demographic development of many rural areas of the world, such as the Almería province in Southern Spain. Design details of relevant low-tech greenhouses for mild climate areas can be found in Von Zabeltitz (2011). There is room for technological improvement of low-tech greenhouses, as proven by the intermediate production values. For instance, average tomato production (9–10 months cycle) in Almería's greenhouses is around 17 kg/m^2, with some growers reaching yields of about 21 kg/m^2 (Valera et al., 2016). These values are below the yield obtained in high-tech heated greenhouses of Northern Europe or North America (50–65 kg/m^2, and some Dutch growers can reach up to 70 kg/m^2).

Similarly, experimental data from the Canary Islands show that the improvement factor in terms of tomato yield of a well-designed greenhouse, compared with a traditional Canarian greenhouse, can be up to three times (Raya-Ramallo, 2014). Of course, the improved greenhouse requires adaptation of the crop management to the upgraded greenhouse structure.

Three major areas of interest have been identified for the development of better low-tech greenhouses.

- Ventilation is often poor. The ventilation surface, type of vents and greenhouse geometry are not sufficient to avoid overheating in hot periods.
- Light transmission is low, particularly in greenhouses with low roof slope.
- Night-time winter temperature is suboptimal. This slows crop growth and implies problems of excessive humidity.

2.1 Natural ventilation

Natural ventilation is the most essential tool that growers can use to maintain the greenhouse microclimate (temperature, humidity and CO_2) within the best ranges for the crop. The greenhouse is essentially a solar collector and only a tiny fraction of the solar energy entering the greenhouse is converted into dry matter by photosynthesis. Most of the energy becomes sensible or latent heat by means of plant transpiration. When the sensible heat load is too high, the greenhouse temperature becomes too high and must be decreased by ventilating (Vox et al., 2010; Giacomelli et al., 2012). When relative humidity (RH) becomes too high, ventilation can decrease it as external air is usually drier. Moreover, natural ventilation can avoid CO_2 depletion if the greenhouse CO_2 concentration is lower than that of external air.

Studies on greenhouse ventilation have taken place for the last 30 years approximately, but it has been more recently, from the beginning of the current century, when research in the subject has intensified and has produced significant achievements on the understanding of air exchange and greenhouse internal climate. Such achievements have been based on the applications of computational fluid dynamics (CFD) in the modelling of ventilation systems, as discussed later.

Bournet and Boulard (2010) did the most complete review of experimental and modelling ventilation studies, up to the date of publication. The review included the two major forces that induce ventilation: buoyancy-driven ventilation and wind-driven ventilation. Ventilation studies are based on the use of three methodologies: scale model studies (most of them in wind tunnels), *in situ* experimentation (tracer gas, anemometry etc.) and modelling (mostly CFD modelling).

Scale models in wind tunnels or water tunnels have been analysed by Sase et al. (1984), Okushima et al. (1989), Montero et al. (2001a), Kacira et al. (2008), among others. Proportional ratios ranged from 1:3 scale of a Venlo-type greenhouse (Bailey et al., 2004) to 1:100 scale replica of a multi-span arched-type greenhouse (Teitel et al., 2008). The main advantage of the wind tunnel is that it allows the control of the airflow regime without external disturbances. It is also a relatively cheap and fast way to conduct experimentation provided sound facilities are available. Strict scaling of wind and turbulence Reynolds number for the simulated flow is not possible, but still, wind tunnel experimentation is a powerful tool to understand the airflow pattern in and around greenhouses.

In situ measurements have also provided good insight to the ventilation problem. At the real scale, many ventilation studies measure the ventilation rate, that is, the number of air exchanges produced during a given time (decay-rate method and constant concentration method, used among others by Kittas et al. (1995) and Perez Parra et al. (2004)). The decay-rate method provides an overall estimation of ventilation but does not explain either the details of the air movement or the relevance of each ventilator on the air exchange or the influence of the greenhouse geometry on ventilation.

More detailed information is obtained from direct measurement of the temperature and velocity fields. Such technique requires a large number of sensors to map the climatic parameters inside the greenhouse, which are expensive and difficult to handle. Sonic anemometry has been used by Boulard et al. (1996) (1D sonic anemometry), Wang and Deltour (1999) (2D sonic anemometry), Boulard et al. (2000) and Bartzanas et al. (2007) (3D sonic anemometry) and Molina-Aiz et al. (2004) (hot ball anemometry).

CFD application in greenhouses was extensively reviewed by Norton et al. (2007). CFD allows detailed description of air movement in and around the greenhouse, as well as the determination of temperature field, humidity field and most relevant climatic variables for a wide range of greenhouse geometries and boundary conditions. The work by Bournet and Boulard (2010) discussed the basis of CFD modelling, the choice of turbulence models, radiation sub-model, crop sub model as well as the flow characteristics through insect-proof screens.

The aforementioned methodologies have been applied to the most common greenhouses typologies, viz. mono-span greenhouses (Mistriotis et al., 1997; Bartzanas et al., 2004), 'Mediterranean' multi-span greenhouses (Boulard and Fatnassi, 2005; Teitel et al., 2008), Venlo greenhouses (de Jong, 1990; Bailey et al., 2004), gothic greenhouses (Kacira et al., 2004), parral greenhouses (Perez-Parra et al., 2004; Molina-Aiz et al., 2004; Baeza, 2007) and Canarian greenhouses (Majdoubi et al., 2009). Most studies compare the greenhouse response to different ventilator configurations for a varied range of boundary conditions, typically wind speed and direction. From the abundant set of ventilation studies, some conclusions can be established.

Windward ventilation provides higher ventilation rates than leeward ventilation. For instance, Pérez-Parra et al. (2004) showed that with flap roof ventilators only, the normalized ventilation rate was reduced by a third when ventilators faced away from the wind (leeward ventilation), in comparison with ventilators facing the wind (windward ventilation). Nevertheless, climate uniformity under windward ventilation is usually lower than under leeward ventilation.

The combination of side wall ventilation and roof ventilation is very convenient, particularly for large greenhouses. Such combination increases ventilation rate as well as climate uniformity (Kacira, 2008). In small greenhouses, side ventilation produces even more air exchange than roof ventilation (Montero et al., 2001a; Bartzanas et al., 2004). As the greenhouse length increases, the importance of side ventilation decreases compared to roof ventilation (Kacira et al., 2004). When the ratio of roof area to sidewall area is higher than five (as in large greenhouses of more than 1 ha), little benefit is obtained from sidewall vents (Baeza, 2007).

Flap roof vents are nearly twice as efficient per unit opening area than rolling up roof vents, as reported by the work of Muñoz et al. (1999) in arched-type greenhouses and Pérez-Parra et al. (2004) in parral-type greenhouses.

Higher ventilation rates are obtained when vents are perpendicular to dominant wind than when parallel (Shklyar and Arbel, 2004). In larger-sized greenhouses, the ventilation rate with winds parallel to the side vents is 30–75% lower than with wind perpendicular to the side vents (Hong et al., 2008). Moreover, the flow and temperature patterns inside the greenhouse are very different for wind parallel or perpendicular to the vents.

The crop acts as a porous medium that slows down the airflow. Lee and Short (2000) reported a reduction of 12% in ventilation due to the drag effect of the crop compared to an empty greenhouse. Plant rows perpendicular to side ventilators reduce the air movement more than plant rows parallel to side ventilators. For instance, in a small greenhouse, Sase (1989) reported maximum internal speed of 36% of the external wind speed with the plant rows perpendicular to the side vents and 16% with plant rows parallel to the side vents.

Other aspects of the geometry such as greenhouse height also have an effect on ventilation. Fatnassi et al. (2017) compared greenhouses with 4, 5, 6 and 7 m gutter heights. Inside air temperature at 2 m height decreases by 2K when moving from 4 to 5 m height, but only by 0.2K when rising from 6 to 7 m height. This study demonstrated that increasing greenhouse height by more than 6 m was unnecessary. Nevertheless, the effect of height on greenhouse temperature may depend on the efficiency of the ventilation system, so poorly ventilated greenhouses can benefit more from the increase in height than well-ventilated greenhouses. More research is needed to analyse the combined effect of greenhouse height and ventilation rate on greenhouse temperature.

Concerning greenhouse roof slope, Baeza (2007) showed that increasing the roof slope from 10° to 12° (which is typical of parral greenhouses) to 25°–30° increases the ventilation rate by almost a factor of two and improves climate uniformity. Increasing the roof slope by more than 30° did not produce any significant benefit.

The greenhouse length and the number of spans affect the air exchange rate and flow pattern. Lee et al. (2000) and Kacira et al. (2004) showed that increasing the span number enhanced the development of circulation cells. For large span number greenhouses, two elliptical cells appear with opposite circulating directions. Where both cells meet, generally a hot temperature spot is created (Fig. 1). Therefore, it is recommended to use the combination of side ventilators and roof ventilators and to maintain the greenhouse length up to 50 m to avoid the occurrence of hot spots in central areas (Stanghellini and Montero, 2012).

An important aspect that can greatly affect the efficacy of greenhouse natural ventilation is the presence of neighbouring buildings or other obstacles to air movement. Such buildings may change the airflow pattern around the greenhouse, thereby reducing ventilation and modifying the internal air distribution (Gómez-Mataix et al., 2013). Ideally, greenhouses should be free of obstacles for at a least 20 m all around, a situation that is not practically feasible.

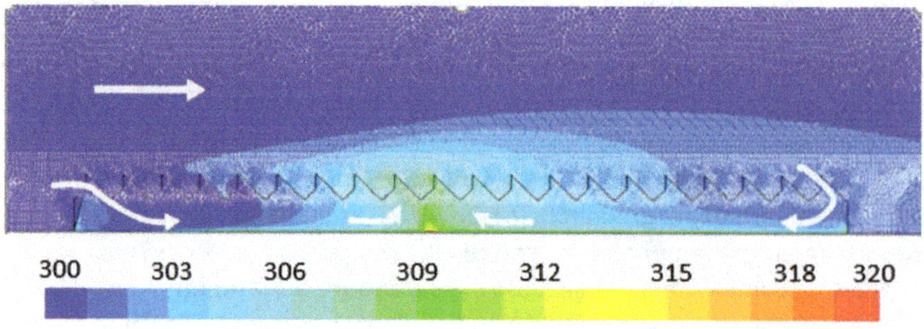

Figure 1 Map of temperature in a 100 m wide greenhouse. A counterclockwise circulation cell on the windward spans meets a clockwise circulation cell on the leeward spans. Both cells meet in the central spans and create a hot spot (yellowish colour). Source: Stanghellini and Montero (2012).

2.2 Light transmission of low-tech greenhouses

It is a common misconception that solar radiation exceeds crop requirement in countries, such as those of the Mediterranean basin, where solar radiation is abundant. Research has demonstrated that productivity is directly proportional to accumulated light intercepted by the crop (Heuvelink, 2005), provided other crop-related parameters (e.g. temperature, irrigation etc.) are under control. The efficiency with which available global radiation is transformed into vegetable production can be assessed by the radiation use efficiency (RUE) index. In a recent study on an urban rooftop greenhouse (RTG) in Barcelona (Spain), average RUE of three tomato crops in the RTG was 8 g/MJ of outside solar radiation accumulated from transplantation, while the average RUE for a conventional greenhouse in the same area was 11.56 g/MJ (Montero et al., 2017b). The major reason for such difference was that the average transmission of the RTG was 50% due to severe shade from construction elements and equipment, while that of the conventional greenhouse was 65% on the average. Higher greenhouse transmission meant higher RUE and thus higher yield.

While the transmission of (Dutch) glass-covered greenhouses has been thoroughly studied (Bot, 1983; Zwart, 1993; Heuvelink et al., 1995), less attention has been paid to the design of more transmissive Mediterranean greenhouses. The performance of traditional greenhouses with nearly horizontal flap roofs in terms of transmission is poor. For instance, it has been reported that parral-type greenhouses, which are typical from some areas in Southern Spain (Latitude 36.5°N), transmit about 55% in winter (Castilla, 2005, 2012), while well-designed greenhouses such as multi-tunnel-type greenhouses can transmit up to 70% (García et al., 2011).

Soriano et al. (2004) used an adaptation of Bot's model as well as measurements on scale models for conducting a parametric study for the optimization of transmission in plastic-covered greenhouses. Different symmetrical and asymmetrical greenhouse roof configurations were tested. Maximum differences were found around the winter solstice, when the roof with the greatest slope (45° south-facing slope, 27° north-facing slope) had the maximum transmission power (66%). Nonetheless, asymmetrical greenhouses are more costly than symmetrical greenhouses, and the increase in cost normally does not compensate for the little benefit in increasing the roof slope. So, Castilla et al. (2012) concluded that, for Mediterranean latitudes, a symmetrical flap roof with a 30° slope is a good compromise between transmission and construction costs.

Some studies have compared transmission of curved roof and flap roof greenhouses. Studies by Critten (1993) for Northern latitudes and Hernandez (2009) for Southern latitudes showed similar transmission for curved and flap roof shapes of the same average slope. A very interesting outcome of

Hernandez's work was that, when fully open, flap roof ventilators make the roof shape horizontal in the ventilator areas. Since horizontal surfaces reflect more solar radiation than sloped surfaces in winter, due to the unfavourable angle of incidence of solar radiation, this produces a reduction in light transmission of as much as 7.4% in the winter months. Perhaps the opening angles of flap ventilators in winter should be maintained at a minimum to combine sufficient ventilation with good transmission.

In terms of orientation, there is a general agreement that, in winter, when good transmission is necessary, an east-west (E-W) orientation is preferred to north-south (N-S) since it is more transmissive, although light uniformity is better in N-S oriented greenhouses (Bot, 1983; Castilla, 2005, 2012). As shown in Fig. 2a and b, for a roof slope of 10°, transmission increases along the season from winter to summer, but it is not very different for E-W and N-S oriented greenhouses. Orientation makes a big difference in winter transmission for greenhouses with roof slope of 30° being higher for the E-W orientation.

As mentioned earlier, light uniformity is better in N-S oriented greenhouses, but the N-S orientation does not guarantee good light uniformity. Teitel et al. (2012) measured light distribution in three N-S oriented greenhouses which were different in terms of the number of ventilators per span and span width. Experiments were done on clear sky days only. This study concluded that PAR distribution was not homogeneous. For instance, the daily loss in irradiance below the gutter (ventilators were located near the gutter) was as high as 27% relative to that of the span's central line. The authors concluded that the installation of ventilators and gutters is essential, but much more careful design of construction elements and ventilators can be done to enhance global transmission and light homogeneity.

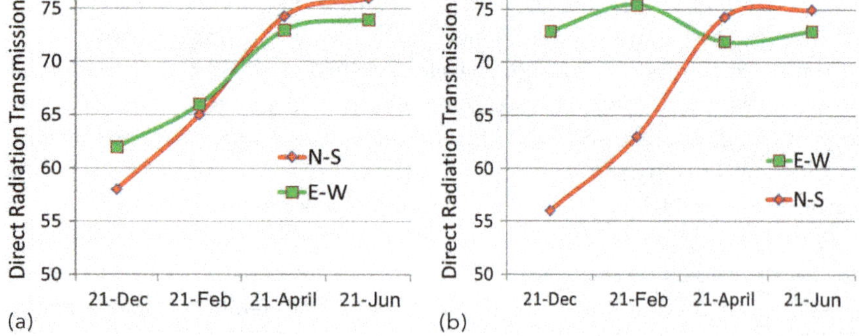

Figure 2 Direct radiation transmission for greenhouses with the main axis North-South and East-West. Latitude 37° North. (a) Roof slope 10°. (b) Roof slope 30°. Source: Adapted from Castilla (2005, 2012).

Insect screens reduce the light levels in greenhouses, particularly when they are dirty. Their transmission depends on their porosity (Möller et al., 2010). As the porosity decreases, dust accumulation increases and loss of light transmission increases. For instance, Montero et al. (2001b) reported a loss in transmission at normal incidence angle from 90% to 63% in an insect-proof screen after 1 year of exposure. Plastic films also lose transmission with dirt, but not as much as insect-proof screens. In a double-walled polycarbonate sheet, losses around 12–13% after 18 months of exposure have been reported, partially due to dust accumulation between the two polycarbonate sheets (Geoola et al., 1998). Cleaning of covering films restores transmission to original values. Cleaning with water jet proved to be more efficient than with air jet (Sangpradit, 2014).

Condensation also affects light transmission. The loss in light transmission varies with the drop size: bigger drops reduce the transmission minimally than smaller drops. Geoola and Peiper (1994) reported average values of transmission of a PE sheet to be 80.3% for the wet state and 88.7% for the dry state. Anti-drip films had nearly the same transmission for the wet state and the dry state. Pollet et al. (1999) conducted a laboratory study on the effect of condensation on the availability of solar energy in greenhouses. The study showed that drop-wise condensation affected the solar radiation transmission more in plastic-covered greenhouses than in glasshouses. It was also observed that film-wise condensation gave rise to a transmittance increase of up to 2% when compared with the phase of condensation without runoff. This is due to the more favourable refractive index of the water film with air.

Anti-drip additives are frequently added to agricultural films to produce a continuous layer of condensed water, but such additives migrate to the film surface and are washed out after some time. Recent multi-layer plastics use one of their central layers as a reservoir of anti-drip additives, so that they replace the additives lost by washing (FAO, 2013).

2.3 Improving the night-time climate in passive greenhouses

In many areas of the world with mild winter temperatures, active heating is not considered economically viable. So, most greenhouses are unheated (also called passive) greenhouses. The night-time climate of unheated greenhouses has been less studied than the day-time climate. It is commonly thought that excessive temperature and light levels are the most limiting factors in greenhouse production in mild climate areas. But the winter greenhouse climate in such areas is suboptimal (Bartzanas et al., 2005) and has a negative effect on yield and fruit quality (Lopez et al., 2008). Besides, low temperatures lead to high RH and to the development of fungal diseases (Baptista, 2007).

In a passive greenhouse, at night, most energy losses are due to far-infrared radiation (FIR) emissions (Baille et al., 2006). A CFD study conducted by Piscia et al. (2012) presented a temperature map of a passive greenhouse at night under clear sky conditions. The greenhouse cover was the coolest surface in the greenhouse (Fig. 3); it was a sink of energy since it emitted more FIR than it received from the clear sky. The major source of energy was the soil, since the energy stored during the day is released at night. Values around 20 W/m^2 of heat transferred from the soil to the greenhouse air have been reported (Baille et al., 2006). Since the greenhouse air is confined and is surrounded by the cold surface of the cover, the inside air temperature is frequently lower than the outside air, and thermal inversion occurs (Montero et al., 2013). Under overcast skies, the amount of FIR received is more than that received under clear skies, the cover is warmer and thermal inversion is unlikely to happen.

Therefore, to increase the greenhouse night-time temperature without artificial heating, two basic approaches can be used either separately or combined: reducing the FIR loses and/or increasing the greenhouse thermal inertia to enhance the heat transfer from the soil.

2.3.1 Reducing FIR losses

Thermal screens that reduce convective greenhouse losses in addition to reducing radiative energy transfer (FIR), particularly if they are aluminized screens, have been in use for years in heated greenhouses. Alternatively, in unheated greenhouses, fixed or movable screens as well as double-walled covers are mostly used to reduce FIR losses since convective losses are relatively small.

Figure 3 Map of temperature at night in an unheated greenhouse under clear sky conditions. Dark blue areas denote cooler surfaces and light blue areas denote warmer surfaces. Source: Adapted from Piscia et al. (2012).

Impermeable screens are probably the cheapest and the most popular screen material in Mediterranean areas. Hernandez et al. (2017) compared fixed and movable impermeable screened greenhouses and unscreened greenhouses. Screened greenhouses increased night-time temperature, reduced RH and condensation, and promoted early yield in a melon crop. Surprisingly, the comparison of movable (folded at night) versus fixed screens produced little difference in terms of climate and crop response. This can be attributed to the small differences in short wave radiation reaching the crop. The mechanism of folding and unfolding the screen reduced greenhouse transmission even when the screen was folded. Better engineering design is needed for an effective use of movable screens in many Mediterranean greenhouses.

External movable screens are generally used during the day for shading, but they can also be used at night as a barrier for thermal radiation emissions (Montero et al., 2013). As in the study by Piscia et al. (2012), the benefit of external and internal screens was better under clear sky conditions (an increase in air temperature up to 3°C referred to a single clad greenhouse).

Optical properties of covering materials in terms of emissivity, reflectivity and transmission to FIR have a clear effect on the FIR losses, therefore they play a significant role in the climate of heated and unheated greenhouses. Ultrathermic films with low FIR transmission and high emissivity have proven to be effective in saving 5–10% of energy in heated greenhouses (Espi et al., 2006), but their effect on unheated greenhouses has not been assessed. According to simulations (Piscia et al., 2013) a highly emissive covering material (such as ultra-thermic films) does not reduce FIR losses due to its high power of emission. Preferred materials are those with low emissivity and high reflectivity to FIR. The study by Piscia (Fig. 4) showed that under a clear sky the predicted temperature gain by using a high-reflectance material compared to a regular one on an unheated greenhouse was approximately 4°C. The RH in the same scenario was reduced by almost 10%.

While low-emissivity glasses are commercially used in greenhouse production (Kempkes et al., 2014), to the best of our knowledge low-emissivity and high-reflectivity plastic films or semi-rigid sheets are not available in the market as a greenhouse cladding. Solovyev et al. (2015) presented a study on PE films and PET sheets with low-emissivity–high-reflectivity properties. Such materials are made by deposition mostly performed by magnetron sputtering of multi-layer coating. At least one layer is a metallic layer with a thickness of about 8–12 nm (preferably silver, since silver has the lowest absorption coefficient compared to some other metals). The silver layer is usually covered by dielectric protective layers such as oxides of zinc, tin or titanium with a thickness below 40 nm with the purpose of stabilizing the silver layer, since moisture intensifies the migration of Ag atoms from the film. If such

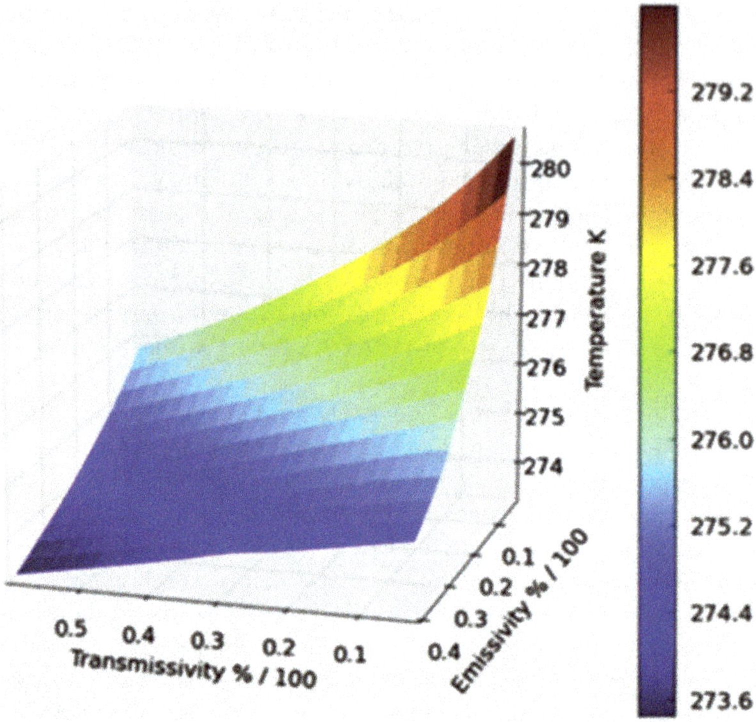

Figure 4 Effect of cover properties (FIR transmission and emissivity) on night-time greenhouse air temperature. Source: Adapted from Piscia et al. (2013).

low-emissivity materials become a commercial reality, relevant improvements in the night-time climate in passive greenhouses, as well as strong reduction in energy use in heated plastic greenhouses, would be achieved.

2.3.2 Increasing the greenhouse thermal inertia

Greenhouses are light structures with little thermal inertia. As soon as the sun sets and outside temperature drops, the greenhouse air temperature falls. Mulching has been used to increase soil temperature and night-time air temperature. Bonachela et al. (2012) compared black mulch, transparent mulch and no mulch treatments in unheated greenhouses. In late autumn and early winter, black mulch had positive effects on soil temperature and air temperature (average increase in daily air temperature 1.4°C compared to transparent mulch), but such enhancement of soil heat storage was reduced when the greenhouse had to be ventilated for suitable crop growth. The authors of this study recommend keeping minimal winter ventilation and maintaining a close soil-mulching contact to enhance soil heat storage.

Water-filled polyethylene sleeves are a passive technique proposed and used in the 1980s, particularly in Northern Greece (Von Zabeltitz, 1989). The water sleeves are placed on the soil between the plant rows for collecting solar energy during the day and releasing the same as heat at night. An increase in greenhouse air temperature up to 3°C has been reported, particularly after transplantation, when solar radiation reaches the sleeves directly and the crop does not shade them. Later, such technique was abandoned, probably due to some practical difficulties associated with the occupation of the sleeves on the soil and some labour operations. More recent unpublished studies in Northern Mediterranean areas have shown that the sleeves collected up to 75% of the incident radiation when they were not shaded by the crop. In a greenhouse with 25% of soil surface covered with water sleeves, the increase in night-time temperature was only 1°C. More positive results on the use of water sleeves have been reported by Ntinas et al. (2015), who showed an increase of 10.8% in tomato production.

As indicated in the text above, passive methods produce apparently modest effects on night-time air temperature, soil temperature and humidity regime. Nevertheless, under mild winter conditions, such modest improvements should produce strong benefits in terms of yield and quality. Since the storage of energy in passive systems is on a daily basis, passive methods are particularly beneficial on cold nights with sunny days, when outside conditions are colder than normal and when frost may occur. When using passive techniques, it is advisable to combine them to achieve a synergistic effect among the techniques. For instance, the heat released from water sleeves at night is more efficient in a better insulated screened greenhouse than in a single cover greenhouse. The combination of several passive techniques may reach optimal greenhouse climate for crops with null artificial heating and so, with minimal CO_2 emissions.

2.4 Future trends in research on low technology greenhouses

With regard to greenhouse ventilation, most previous studies provide information on the air exchange rate of a given greenhouse under a given set of boundary conditions. But maximizing the air exchange rate may not lead to optimizing the internal climate, particularly within the greenhouse canopy. Some ventilator configurations do not produce good air mixing and a lack of uniformity may appear. Teitel and Wenger (2014) discussed the concept of ventilation efficiency (defined as the ratio between the shortest possible time and the actual average time needed to replace the air in the room), and indicated it is a better indicator of ventilation performance than air exchange rate.

Studies based on CFD modelling are expected to continue providing an inside view of the internal climate. They present maps of temperature, velocity vectors as well as other relevant climatic parameters. Such maps may help to detect regions where unfavourable conditions for crop growth can develop. Further CFD analyses can help to eliminate those unfavourable areas by testing different ventilator configurations or by making changes on the greenhouse design. As indicated by Bournet and Boulard (2010), future CFD studies can help to develop necessary indicators of climate uniformity (e.g. coefficient of uniformity or the normalized standard deviation of temperature, wind speed etc.).

In terms of light transmission, CFD modelling is also being applied for the year-round evaluation of covering materials (Baxevanou et al., 2018). Using the method of finite volumes, this study assessed the distribution of radiation in four wavelength bands: UV, PAR, NIR and FIR. The seasonal effect of different covering materials on air velocity and greenhouse temperature distribution was also presented. From such parameters, crop response could also be assessed seasonally. The study by Baxevanou et al. (2018) showed the potential of CFD for the design of more efficient greenhouses, not only in terms of ventilation as discussed earlier, but also regarding radiation distribution and homogeneity (Fig. 5). As CFD models develop and computational calculation power is enlarged, the seasonal crop response in a given greenhouse structure under a given external climate could also be investigated.

Relevant trends in future greenhouse technology research are focussed on the development of new covering materials. The necessity of producing highly FIR reflective-low-emissivity films has been discussed earlier, which is

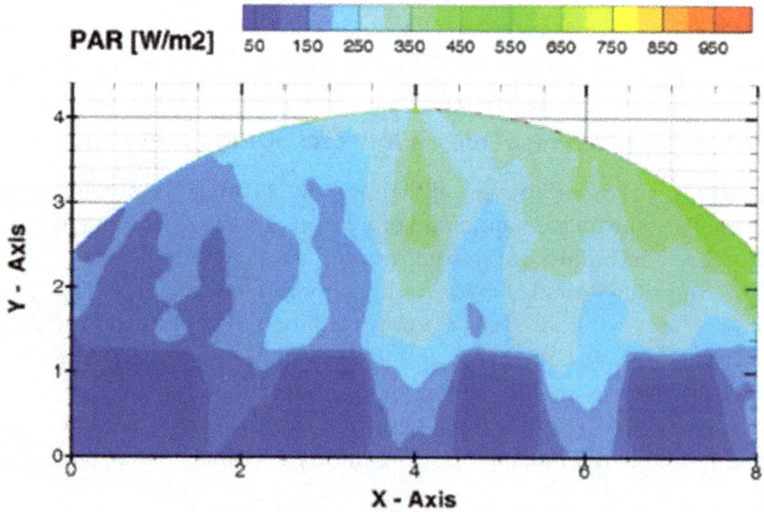

Figure 5 PAR distribution in a greenhouse covered with ethylene-vinyl acetate (EVA) film. Source: Adapted from Baxevanou et al. (2018).

still not available in the market. But it could be the case that the best optical properties for night-time conditions could have detrimental effects on the day-time climate. For instance, low-emissivity and high-reflectivity covering materials may reduce the cooling capacity of the greenhouse by radiation, thus overheating the greenhouse during sunny days.

3 High-tech greenhouses

High-tech greenhouses are common in cold climate regions, where technology is needed to overcome the unfavourable external conditions. Microclimate and all other growing factors (irrigation, nutrition, internal transportation etc.) are fully controlled in high-tech greenhouses. As a consequence, yields are high, and quality is assured, at the expense of high investment and high running costs.

3.1 Venlo-type glasshouses

Very probably the Venlo glasshouse is the most popular glasshouse in the world, which has been fully described, modelled and improved (Bot, 1983; Zwart, 1996). In the Netherlands, it is estimated that 80% of the energy consumed in the greenhouse is used for heating (Stanghellini et al., 2016), and 20% for humidity management if ventilation is combined with heating for humidity control (Zwart, 1996). Nevertheless, new equipment such as mechanically controlled ventilation can significantly reduce energy consumption for humidity management (Campen et al., 2009). Thus, most research and development efforts have been directed at reducing energy consumption for heating and dehumidification, and to achieve almost fossil energy-neutral glasshouse. We will concentrate here on some research results related to the Venlo structure and design, particularly concerning innovations in the roofing. Energy saving is addressed elsewhere in this book.

In this regard, efforts have been made to maximize the solar energy input. Only the topmost leaves of a crop canopy are light saturated, so productivity and solar radiation are closely related. To enhance solar radiation transmission, glasses with anti-reflective coatings have been investigated. For instance, Hemming et al. (2011) compared single glass and double glass with different anti-reflection coatings. The study showed that it was possible to produce double glass with higher light transmission than traditional single greenhouse glass (83–85% for hemispherical (diffuse) light, compared to 82–83% for traditional single glass). Furthermore, energy savings with the double roofing were considerable. The addition of a low-emissivity coating in an inner glass layer produced further energy savings, but at the cost of losing some light transmission (78.5% for hemispherical (diffuse)).

New prototypes of greenhouses are in development, like the 'Winter light greenhouse' in the Netherlands. Such a prototype has been previously modelled by making use of solar ray tracing models (Swinkels et al., 2017) capable of accounting for both the direct and diffuse fraction that will be available for the crop, and how this radiation is more or less efficiently scattered (i.e. more or less penetration in the canopy). The goals are achieving 10% more natural sunlight during the winter months (October–March) and more than 10% higher light use efficiency by the crop. This prototype also includes other construction details that help to maximize crop light interception, such as modified structural elements painted in white. A particular innovation is the highly transparent energy saving screen. It is not installed horizontally but at a certain angle (Fig. 6). This increases light transmission, compared to a horizontal screen, and allows closing the screen for longer periods, for better energy saving with little loss in light.

Using single or multiple thermal screens is a common practice in Dutch greenhouses (a subject to be discussed in a different book chapter), but focussing here on the greenhouse structure, new roofing concepts are being developed (Kempkes et al., 2017). They consist of a double covering material made of a combination of a high transparent glass panel and a high transparent film (F-Clean®). Compared to single glass and thermal screened greenhouses, significant advantages can be achieved in terms of energy saving and light transmission, because thermal screens are sometimes closed during day time for energy conservation, which results in significant light reduction. This is not the case with the new roofing concept.

Figure 6 The Winter light greenhouse. View of the v-shaped internal screen. Photo: Michiel van Spronsen.

Regarding the quality of the transmitted radiation, light diffusion has focussed much of the attention during the last decade. For decades, PE diffusive films have been widely used in passive Mediterranean greenhouses, but only recently some research works have demonstrated the benefits of diffuse materials.

The positive effect of diffuse light was shown in many simulation studies in the 1980s and earlier (e.g. Spitters, 1986; Gijzen, 1992), though not tested in experiments as there was no diffuse light glass material that had good light transmission. In practice, light diffusing materials have been widely used in passive greenhouses for decades (i.e. PE diffusive films in Mediterranean greenhouses). In glass-covered greenhouses, it has only been recently identified that different research works have helped to elucidate and quantify the real benefit of diffuse light for greenhouse crops (Hemming et al., 2006, 2008; Victoria-Garcia et al., 2012; Li et al., 2014a,b).

Diffuse materials are slightly less transmissive to solar radiation than clear materials. Such loss in transmission can be compensated by using anti-reflective coating and low FE content glass, so that the transmission is similar to that of standard glass. Diffuse light materials produce a more efficient use of light. With diffuse covering materials, light penetrates deeper in the crop, resulting in higher photosynthesis in the lower canopy. Dueck et al. (2012) reported an increase in tomato yield of 8% under Dutch conditions with diffusive glass, compared to standard glass. The most important reason for the increased production was an increase in individual fruit weight by 5–8 g. This has led glasshouse manufacturers to produce different kinds of diffusing glass to be used in commercial high-tech greenhouses.

3.2 Semi-closed greenhouses: the trade-off between temperature control and excessive humidity

Greenhouses with reduced or no (winter) ventilation were named as semi-closed or closed greenhouses. The concept of the closed greenhouse is that the greenhouse acts as a solar collector and the heat surplus is stored daily or on a seasonal basis; to store energy, ventilation is null (closed greenhouse) or strongly reduced (semi-closed greenhouse).

For daily energy storage, during the day, a low temperature air to water heat exchanger can transfer surplus heat from the greenhouse air to an external water tank (Fig. 7). At night, the water flow is reversed, and stored heat is transferred to the greenhouse air (Baeza et al., 2012). Fine wire heat exchangers for cooling purposes in semi-closed greenhouses have been studied by Bakker et al. (2006), among others.

In cold areas, seasonal storage is generally preferred. The heat surplus in summer is harvested, stored and reused in winter to heat the greenhouse.

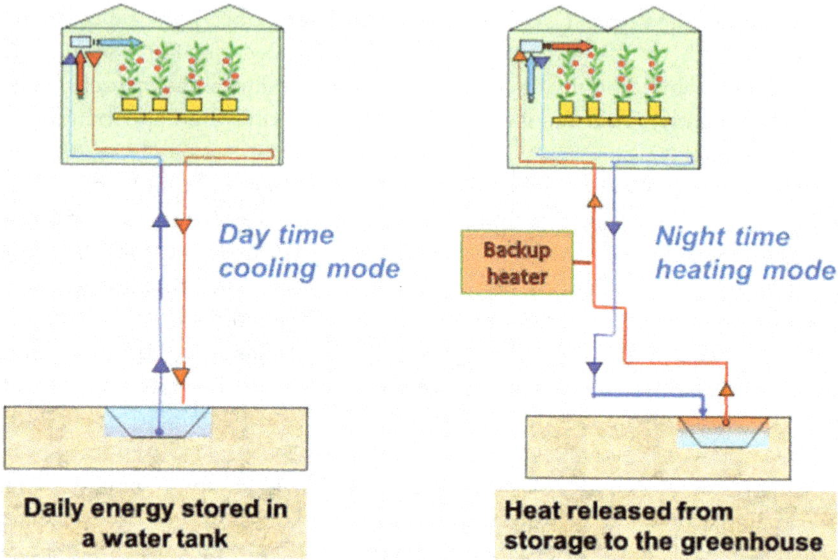

Figure 7 Operational scheme of a closed greenhouse. Daily energy storage operation. Source: Baeza et al. (2012), adapted from A. Grisey (CITFL, Balandran, France).

This requires a heat pump, underground seasonal energy storage (aquifer) as well as equipment for cooling, heating and dehumidification of air (Yildiz and Stombaugh, 2006).

Semi-closed greenhouses show different features in passive (no heating and little climate control) and active, high-tech greenhouses. For the passive greenhouse, the purpose is to reduce ventilation as much as possible to increase the average day-time temperature. However, minimal ventilation has two potential negative side effects: CO_2 depletion and excessive humidity. In terms of CO_2 depletion, Stanghellini et al. (2008) reported a loss in yield of 8% in a tomato crop in Almería (Spain) due to CO_2 shortage. Such loss can be compensated by applying CO_2 enrichment, at least up to the outside concentration, even when the greenhouse is ventilating.

Concerning excessive humidity, semi-closed unheated greenhouses are more humid than conventional greenhouses and fungal infections are a common factor (Baptista, 2007). Some innovative plastic greenhouses allow condensation collection and avoid water dripping on the crop, which helps to avoid fungal disease development. Nevertheless, in a semi-closed unheated greenhouse, even if water dripping is avoided, dew formation can occur on the fruits (and so fungal diseases) mostly in early morning hours. In this case, ventilation is needed to eliminate excessive humidity and the semi-closed greenhouse approach cannot be followed (Montero et al., 2017a).

In cold climates, semi-closed greenhouses were developed with an energy saving objective. Further benefits are associated with semi-closed greenhouses, namely the reduction of CO_2 emissions and the improvement of crop pest management. In Northern climates, reducing ventilation results in high CO_2 concentration (800–1000 ppm) throughout the year (Qian et al., 2012).

Energy saving of up to 35% and increase in yield of 20% have been reported in closed greenhouses in the Netherlands (Marcelis et al., 2009). An annual increase in production of 10–20% is realistic, with reduced amounts of supplied CO_2 (De Gelder et al., 2015). The yield increase is primarily obtained through increased rates of photosynthesis due to the higher CO_2 concentrations in closed greenhouses (Qian et al., 2012). Moreover, Katsoulas et al. (2015) reported that reducing ventilation increases water use efficiency. This can be achieved if water from crop transpiration is collected and reused after condensing on cold surfaces, such as the greenhouse roof or the condenser of the cooling equipment.

Vadiee and Martin (2012) reviewed the state of the art of the closed greenhouse integrated with thermal energy storage (TES) technology. They concluded that the closed greenhouse is technically feasible and shows good potential. Nevertheless, the authors pointed out the importance of using renewable energy for running the system to ameliorate its sustainability. Also, further technical developments of the TES systems are needed.

3.3 Research trends in high-tech greenhouses

As mentioned earlier, the Venlo greenhouse has been fully studied and improved throughout the years. Most recent research efforts for active greenhouses have focussed on reducing their dependency on fossil fuels to gain sustainability. The final goal is to build and operate climate-neutral greenhouses without the use of fossil energy. New energy-saving techniques, including double roofing, multiple thermal screens as well as new cultivation concepts (temperature integration depending on available solar radiation levels; higher humidity set points; intensified use of thermal screens), are under research and development. Forced ventilation of outside air (mechanical dehumidification) is also implemented (Campen et al., 2003) since between 10% and 20% of energy consumption is due to humidity control in Dutch greenhouses (Bakker et al., 2008).

The design of closed and semi-closed greenhouses still needs further refinement. For this purpose, Boulard et al. (2017) developed a numerical model of a closed Venlo greenhouse located in Southern France. Numerical models are very effective tools for the design and sizing of equipment such as the positioning of air ducts inside the closed greenhouse. A relevant novelty of this model is that it incorporates a sub-model of plant canopy photosynthesis, so that the crop response as a function of the closed greenhouse design and management can be assessed. The development and incorporation of crop physiological sub models (viz. transpiration, crop temperature and yield) into numerical models of the greenhouse climate is a research subject worthy of investigation, not only for the closed or semi-closed greenhouse but also for the other greenhouse typologies discussed in this chapter.

Ideally, in high-tech greenhouses, new covering materials should be 'smart', that is they should be able to switch their optical properties instantaneously according to crop requirements. Some of such smart materials are under investigation by a consortium led by Wageningen University and Research (the Netherlands), within the framework of the project PPS Smart materials, http://www.ppssmartmaterials.nl/. The project addresses innovations in high transparency–energy saving materials, changing diffusive light scattering as a function of specific triggers (temperature or light), and switchable modification of the solar spectrum (e.g. multi-layer plastics or liquid crystalline polymers capable of reversible reflection and transmission of incident NIR and outgoing thermal IR, modification of the solar spectrum to produce more photoactive radiation). It is an ongoing research project that shows promising but still unpublished results.

Another typology of high-tech greenhouses is the urban rooftop greenhouse (RTG). Urban agriculture is gaining relevance due to its potential for helping to solve the food security issue, increasing resource use efficiency for producing food and providing social services and environmental services. It is beyond the scope of this chapter to go deep into the urban agriculture subject. In terms of RTG technology, readers are advised to consult recent literature on the subject (e.g. Orsini et al., 2017; Nadal et al., 2017). Very briefly it is relevant to mention that benefits can be achieved if the greenhouse and building are integrated, so that they exchange and optimize energy flows, water flows and CO_2 flows The integration is made by the exchange of air and waste heat from the building to the greenhouse and vice versa: surplus heat in the greenhouse that is ventilated away can be used for heating the building. Moreover, harvested rainwater from the greenhouse roof can be used for irrigation, as well as properly treated grey water from the building. Additionally, air with high CO_2 concentration (if available in the building) can be conducted to the greenhouse for CO_2 enrichment.

RTGs show some peculiarities compared to commercial greenhouses in rural areas. RTG frames are usually stronger since they must meet construction codes, which are more demanding than the agricultural codes. The fact that RTG frames have more opaque elements than commercial greenhouses, in conjunction with the frequent existence of shadows cast from building elements or equipment, means that transmission to solar radiation of RTGs is generally lower than that of conventional on-soil greenhouses (Montero et al.,

2017b). Another peculiarity of RTGs is related to the wind regime in multi-storey buildings, which makes, on the one hand, ventilation better and, on the other hand, thermal losses by convection higher. Lastly, specific climate control equipment for RTGs needs to be developed, since RTGs are dependent on the building's climate and, very often must be subject to the building's climate requirement, which are prioritized over the crop requirements.

Research to analyse several RTG typologies and progress on technical solutions to solve the aforementioned problems are needed to develop sustainable urban agriculture in greenhouses.

4 Chinese solar greenhouse

4.1 CSG design

With over 1 million ha (Sheng et al., 2016), China has the largest solar greenhouse area in the world (Qichang, 2012). The Chinese solar greenhouse (CSG) is a special greenhouse design developed in China. The modern style of CSG (see Fig. 8) has an arc-shaped south-facing light-transparent roof, and an energy-storing north wall. It is 8–14 m wide with a 2.5–3.5 m high north wall. The arch is 3.8–5.8 m high. It fits the outdoor climate in north China between latitudes of 32°N and 43°N (core area) where the winter is cold (-20°C to 5°C). Due to the arc-shaped south surface of the CSG, more solar energy can be collected during the day, and be stored in the north wall and the soil. Without additional heating, temperature differences of 21–25°C between inside and outside are realized in winter (Gonghong et al., 2013).

The reasons why it can be applied in the extreme cold winter period without, or with low, energy consumption are the special functions of CSG including energy storage and insulation. Energy is stored in north wall during the day

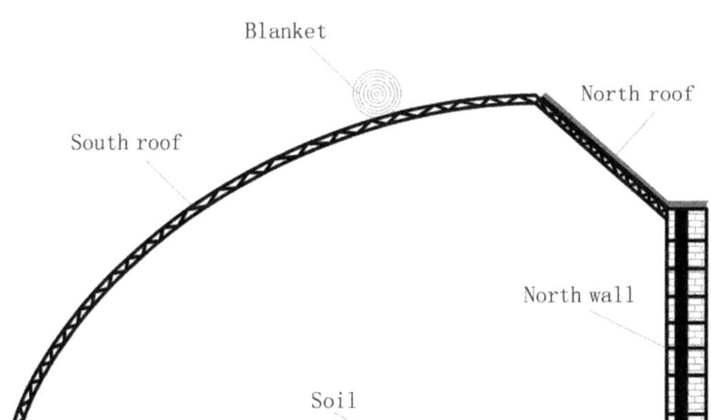

Figure 8 Sketch of modern CSG.

and this energy is released during night (Ming et al., 2016). High insulation and therefore low energy loss during the night is realized by the application of an external thermal blanket which is removed during day time (Chengwei et al., 2006). In addition to this, the policy of saving energy and environment protection also led to the development of CSG. Coal consumption will be controlled at 4.2 billion tons in 2020. Beijing phased out coal use in all villages in the plain areas at the end of 2018. Because of low construction cost and low energy consumption, a lot of farmers tend to use CSGs.

4.2 CSG typologies

In the early stages, due to the constraints of economic conditions and production levels, the pillars, beams, arches, tie bars and doors were constructed from crude bamboo and other available materials. The size was smaller and its span was generally 6.0–7.5 m (Lijuan et al., 2016; Shuo, 2018). With the improvement of economic and construction technology levels, brick–concrete structures and steel frame structures have been widely used. The growing space and environmental control capability have been increased as well.

CSGs can be divided into northeast type, northwest type and north type depending on the different climate and distribution location in China. The main distribution area of the northeast-type CSG lies at 38–53°N and 120–135°E. It is a temperate semi-humid continental climate. The annual temperature is between −29°C and 36°C. The average temperature is 8.3°C. This kind of CSG started from the mid-1980s. The representative design is named Liaoshen series CSG. The back wall of Liaoshen series CSG is made of light thermal insulation material, such as expanded polystyrene (see Fig. 9a) and brick instead of soil and clay.

In order to increase the land utilization rate, the arc-shaped south roof has been redesigned as a flap roof with the sides having different slopes. This allows more room for growing crops, compared to the arch-shaped roof, the roof slope is bigger near the south side (Fig. 9b). The average difference of inside–outdoors air temperature is nearly 24°C, and the range of difference in temperature is between 22°C and 31°C (Yikui et al., 2011). With the development

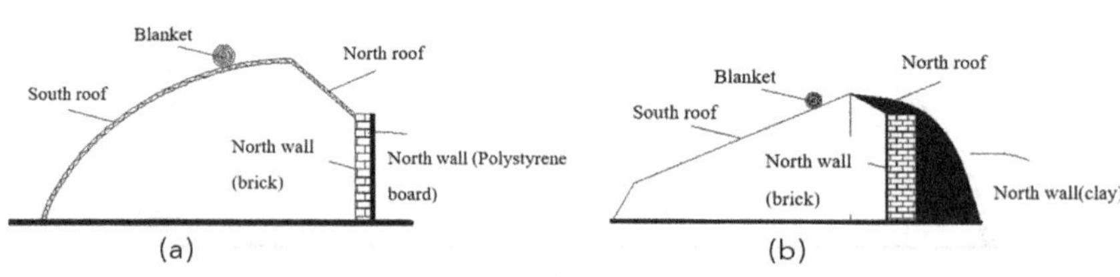

Figure 9 Sketch of a typical northeast-type CSG.

of northeast-type CSG, the span width has increased from 6 to 12 m and the thermal insulation material has increased as well by adding soil or clay (Fig. 9b).

The main distribution area of the north-type CSG lies at 32–42°N and 104–125°E. The area is close to the Bohai Sea and Yellow Sea and is mainly under temperate monsoon climate with four distinctive seasons. The annual average temperature in the north area is 13°C which is higher than that of the northeast area. Because of the lower latitudes, the north CSG type is generally flatter and shorter than the northeast type, since the sun elevation angle in winter is higher and the solar radiation reaches the CSG more perpendicularly than in more northern latitudes. So, the height of the north wall is lower, the front roof angle (the angle between the tangent line of the front roof and a horizontal line) is smaller, the span width is larger and the minimum height to span width ratio is 0.26.

Taking construction cost into consideration, in some regions (e.g. Shouguang city), farmers like to build north walls with clay. Figure 10a exhibits the Shouguang generation-5 type, whose north wall is 1.0–4.5 m thick. The typical design is with ground subsidence (ground sunk to a lower level). The major disadvantage of such a design is the low land utilization. Besides it is detrimental to soil for re-cultivation. In other regions such as Beijing, land is a valuable resource, so there are some simply assembled CSG with a thin wall. The wall is composed of two fibre cement boards and a polystyrene board in between (Fig. 10b). Some CSGs were dug around 50 cm in order to have better insulation effect (Fig. 10c). Steel frames are assembled on site to support the greenhouse and the thermal blanket is made of lightweight insulation materials which are rolled down at night. Compared with the traditional brick and clay thick north wall, land use of the new thin wall can decrease by 72% with better thermal insulation performance (Jun et al., 2003; Jiangjun et al., 2009).

The main distribution area of the northwest-type CSG lies at 37–50°N and 73–120°E. The area enjoys temperate continental climate with abundant sunlight, low winter snowfall and low night temperature. Basic snow pressure is around 20 kg/m^2, and basic wind pressure is less than 25 kg/m^2. Fully considering the climate features and weak economic foundations, the northwest type uses a thick earth wall or some other locally accessible materials, such as pebbles (Fig. 11a). Sometimes, the supporting frame of the southern side is made of

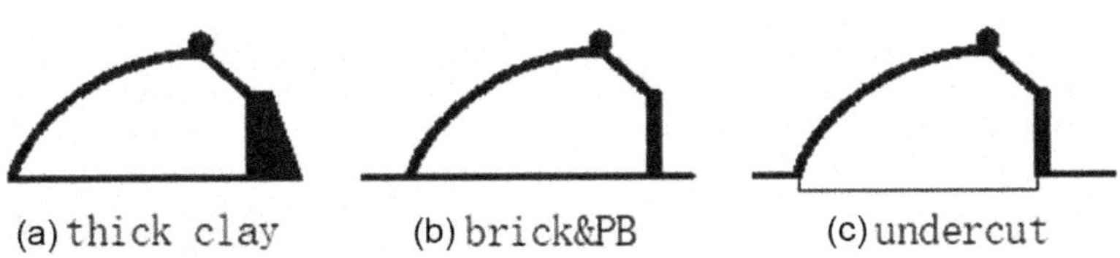

Figure 10 Sketch of typical north-type CSGs.

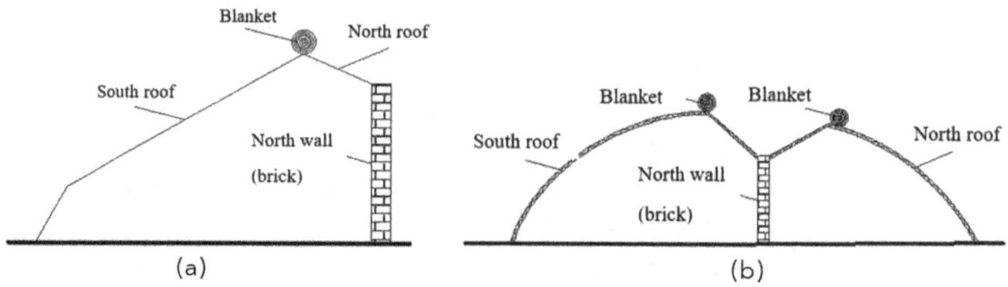

Figure 11 Sketch of typical northwest-type CSG.

tensioned wires. Such a greenhouse is referred to as a string CSG (Jun et al., 2003; Jiangjun et al., 2009).

Another CSG typology used in North Western China is the double-span CSG (Fig. 11b). Compared to the single-span CSG in northwest, with the double-span CSG the indoor temperature can be increased by 3–4°C, and the solar transmission can be increased by 20%. Besides, land utilization rate is improved by 65% approximately. The double-span greenhouse warms up quickly in the morning, warming up to a large extent, and slowly cooling down in the afternoon. In terms of RH, minor differences have been found between the single-span and double-span greenhouses in Shandong Shouguang (Jun et al., 2003; Jiangjun et al., 2009).

The main design parameters which include the front slope angle, back slope elevation angle, ridge-span ratio, back wall height (Fig. 12) determine the differences among CSG typologies (Yikui et al., 2011). Table 1 shows the main design parameters of the three CSG types described above.

In terms of climate control, only simple equipment is used in CSGs. The most significant environment factor that can be regulated is temperature, so most CSG facilities have a mat roller with heat preservation cover and film rollers for ventilation (Xiao et al., 2010). Because air temperature in a closed

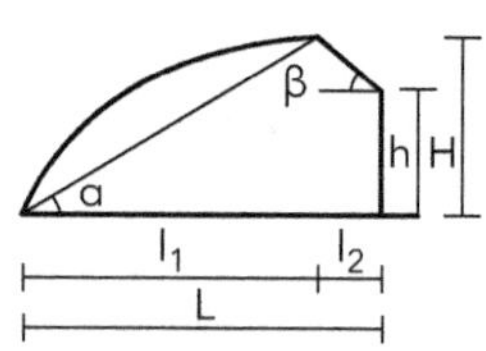

Figure 12 The parameter model diagram of CSG. Note: L: the net span (m) is the horizontal distance from the front foot of the skeleton's upper chord to the inner side of the northern wall; l1: the horizontal projection length of the front slope (m); l2: the horizontal projection length of the rear slope (m); H: the height of the ridge (m), it is the vertical distance from the highest point of the greenhouse to the ground; h: the height of the back wall refers to the vertical distance of the skeleton support from the ground (m); β: the elevation angle of the back slope (°); α: the front slope angle refers to the roof (°), it refers to angle of the line connecting the front foot between the horizontal plane.

Table 1 Main parameters of three CSG types

Type	Span (m)	Front slope angle (°)	Back slope elevation angle (°)	Back wall height (m)	Ridge height (m)	Wall thickness (m)
Northeast type	6.0-12.0	27.1-33.7	30-45	1.0-3.0	2.1-5.5	Clay wall: top width 0.9-4.5 bottom width 1.5-7.0 Brick wall: 0.5-0.7
North type	7.0-11.0	16.0-26.5	30-45	2.0-4.0	2.4-5.7	Clay wall: top width 0.4-3.0 bottom width 0.4-6.0 Brick wall: 0.4-0.6
Northwest type	7.0-7.5	16.5-27.4	33-55	2.0-3.0	3.2-3.6	Clay wall: top width 0.4-3.0 bottom width 0.4-6.0 Brick wall: 0.9-2.2

CSG at noon is high (above 30°C usually), ventilation is essential. So, CSGs have one rolling vent (E-W direction) on the top and another similar one at the bottom, 90 cm above the ground. Ventilation is done by running the film roller. The top vent will open in winter while the top and bottom vents will both open in summer. In some CSGs, east and west walls have a pad and fan cooling system with electric fans. Sometimes an external shading system is added for summer cooling.

Traditional insulation cover materials of CSG include grass carpet and straw mat, which are bulky and ageing. When it rains or snows, the heat preservation performance of soaked insulation cover is reduced. At present, insulation cover is made of industrial raw materials mainly acrylic staple fibre, needle felt and cotton felt. At the same time new insulation materials are waterproof on the outside surface, light weight, anti-ageing and have good insulation effect (Fangjun and Haizhen, 2010). The thermal insulation cover is installed as a whole, covering the full length of the CSG, and it is necessary to use the mat roller mechanism up and down for daily operations (Guoxiang et al., 2017).

Currently, the commonly used mat roller can be divided into two categories according to its principle of operation: rope pulling-type mat roller and reel-type mat roller. The drawstring type mat roller is usually fixed on the ground of the north wall or the side of the north wall. The insulation cover is connected with the motor by rope, pulley, winding shaft, coupling and reducer. The motor is placed on the ground. The speed of rotation is slowed down by the reducer, which drives the coil rope-shaft to rotate. The rope is wound around the shaft. When the motor is reversed, and the rope is relaxed, the insulation cover is lowered. The advantage is that when selecting a high-power motor, the heavy insulation cover can be rolled up. However, the disadvantage is that the installation is complex, and the amount of steel consumed is large. Usually hang motors are installed on the gable or on the centre of the front covers. The motor is directly connected to the shaft that rolls the thermal blanket up and down, the shaft motor rotation drives the shaft rotation. The advantage against the rope pulling mechanism is that the operation is not affected by high winds and the CSG structure remains safer. Also, the motor is easy to operate (Hong et al., 2001).

Many CSGs are not heated except when suffering from extreme cold weather. But with the development of the economy and science and technology, more and more facilities are used in CSG (Gao and Quangguo, 2009). Besides the mat roller with heat preservation cover, there is also other equipment, like heating systems, to increase temperature in CSG in winter. According to the energy source, the heating equipment can be divided into boiler by coal (only allowed in coal-free areas), heated wire by electricity (Yunlong et al., 2013), ground source heat pump by geothermal energy and active heat storage-release system by solar energy (Sun, 2014; Yi et al., 2012; Fang et al., 2015). Furthermore, there are also other environment control facilities such as integrated water and fertilization irrigation equipment, LED lighting system and so on.

4.3 Research trends in CSG

Although the modern CSG style has developed a lot, there are also some choke points and disadvantages. Firstly, to prevent shadowing from the north wall there must be sufficient distance between two neighbouring CSGs, leading to a low land use efficiency (Tianlai, 2005). Secondly, the curved south-facing roof limits the height for crops, especially in the most southern part. Moreover, the relatively small span width of the CSG limits the options for mechanization of labour which will be needed in the future. Thirdly, during winter a large part of the solar light can diffuse due to air pollution and low solar position (Li et al., 2014a,b). Low transmission for diffuse light due to the back roof and surrounding walls results in low irradiation levels inside the greenhouse. Fourthly, during the night with low outdoor temperatures, air temperature in CSG decreases to low levels of about 5–10°C (Lihong et al., 2010) and accordingly RH increases to over 90% (Bo et al., 2016), resulting in low crop production levels. Fifthly, the inside climate (RH, temperature and CO_2) (Weihua and Tianlai, 2003) cannot be controlled accurately because of the simple design of the climate control equipment, and its crude manual management, usually based on the grower's personal experience (Zhonghua et al., 2012; Zhongfu and Xuefen, 2004; Junwei et al., 2016). Last but not least, CSG has a specific and limited cultivation period around winter time, since in summer plant production must be stopped due to the combination of high irradiation and high outdoor air temperature (higher than 40°C) (Heuvelink, 2005). To extend the cultivation period, a greenhouse design considering both hot and cold periods is essential.

In order to solve these problems or improve the performance of CSG, there are several approaches identified for future application. Firstly, theoretical models of realizing the desired situation of CSG with high yield, high annual net financial return (NFR) and low energy consumption will be needed. Further modelling studies in CSG should link the greenhouse geometry, inside climate conditions and crop performance with yield, NFR and energy consumption.

Secondly, based on the results obtained from validated theoretical models, the structure design of CSG should be optimized, so that the design should provide optimal climate for crop growth as well as facilitate the installation process. It is relevant to simplify the design to economize manpower in the installation, and increase the manufacturing efficiency through industrialization and standardization of the greenhouse components.

The latter aim focusses on the optimization of the geometry (Yiming et al., 1992) and materials. Much has been done on this matter, but still there is room for optimization considering the shape of the south curve, function of north roof and walls, construction materials and thermal insulation materials. Lastly and importantly, the major feature of CSG, which is its good performance in terms of heat preservation and heat storage (Zhengwei et al., 2008), could be a solution for different greenhouse types in other areas of the world. For this purpose, CSG technology needs promotion and adaptation to different greenhouse production systems.

5 References

Baeza, E. (2007). Optimización del diseño de los sistemas de ventilación en invernaderos tipo parral. Tesis doctoral. Universidad de Almería, Spain.

Baeza, E. J., Pérez Parra, J., López, J. C., Gázquez, J. C., Meca, D. E., Stanghellini, C., Kempkes, F. L. K. and Montero, J. I. (2012). Performance of a day/night water heat storage system for heating and cooling of semi-closed greenhouses in mild winter climate. *Acta Horticulturae* 927, 535–42.

Bailey, B. J., Robertson, A. P. and Lockwood, A. G. (2004). The influence of wind direction on greenhouse ventilation. *Acta Horticulturae* 633, 197–202.

Baille, A., López, J. C., Bonachela, S., González-Real, M. M. and Montero, J. I. (2006). Night energy balance in a heated low-cost plastic greenhouse. *Agricultural and Forest Meteorology* 137, 107–18.

Bakker, J. C., De Zwart, H. F. and Campen, J. B. (2006). Greenhouse cooling and heat recovery using fine wire heat exchangers in a closed pot plant greenhouse: Design of an energy producing greenhouse. *Acta Horticulturae* 719, 263–70.

Bakker, J. C., Adams, S. R., Boulard, T. and Montero, J. I. (2008). Innovative technologies for an efficient use of energy. *Acta Horticulturae* 801, 49–62.

Baptista, F. (2007). Modelling the climate in unheated tomato greenhouses and predicting *Botrytis cinerea* infection. PhD thesis. Universidade de Évora, Portugal.

Bartzanas, T., Boulard, T. and Kittas, C. (2004). Effect of vent arrangement on windward ventilation of a tunnel greenhouse. *Biosystems Engineering* 88, 479–90.

Bartzanas, T., Tchamitchian, M. and Kittas, C. (2005). Influence of the heating method on greenhouse microclimate and energy consumption. *Biosystems Engineering* 91(4), 487–99.

Bartzanas, T., Kittas, C., Sapounas, A. A. and Nikita-Martzopoulou, C. (2007). Analysis of airflow through experimental rural buildings: Sensitivity to turbulence models. *Biosystems Engineering* 97, 229–39.

Baxevanou, C., Fidaros, D., Bartzanas, T. and Kittas, C. (2018). Yearly numerical evaluation of greenhouse cover materials. *Computers and Electronics in Agriculture* 149, 54–70.

Bo, Z., Yi, Z., Qichang, Q., Hui, F. and Wei, L. (2016). Dehumidification in a Chinese solar greenhouse using dry outdoor air heated by an active heat storage-release system. *Applied Engineering in Agriculture* 32(4), 447–56.

Bonachela, S., Granados, M. R., López, J. C., Hernández, J., Magán, J. J., Baeza, E. J. and Baille, A. (2012). How plastic mulches affect the thermal and radiative microclimate in an unheated low-cost greenhouse. *Agricultural and Forest Meteorology* 152, 65–72.

Bot, G. P. A. (1983). Greenhouse climate: From physical processes to dynamic model. PhD thesis. Agricultural University, Wageningen, The Netherlands.

Boulard, T. and Fatnassi, H. (2005). Greenhouse aeration and climate optimization based on CFD studies. *Plasticulture* 124, 38–57.

Boulard, T., Meneses, J. F., Mermier, M. and Papadakis, G. (1996). The mechanisms involved in the natural ventilation of greenhouses. *Agricultural and Forest Meteorology* 79(1–2), 61–77.

Boulard, T., Wang, S. and Haxaire, R. (2000). Mean and turbulent air flows and microclimatic patterns in an empty greenhouse tunnel. *Agricultural and Forest Meteorology* 100(2–3), 169–81.

Boulard, T., Roy, J. C., Pouillard, J. B. Fatnassi, H. and Grisey, A. (2017). Modelling of micrometeorology, canopy transpiration and photosynthesis in a closed greenhouse using computational fluid dynamics. *Biosystems Engineering* 158, 110–33.

Bournet, P. E. and Boulard, T. (2010). Effect of ventilator configuration on the distributed climate of greenhouses: A review of experimental and CFD studies. *Computers and Electronics in Agriculture* 74, 195–217.

Campen, J. B., Bot, G. P. A. and de Zwart, H. F. (2003). Dehumidification of greenhouses at northern latitudes. *Biosystems Engineering* 86, 487–93. doi:10.1016/S0021-8634(89)80081-2.

Campen, J. B., Kempkes, F. L. K. and Bot, G. P. A. (2009). Mechanically controlled moisture removal from greenhouses. *Biosystems Engineering* 102 (4), 424–32.

Castilla, N. (2005). *Invernaderos de plástico. Tecnología y manejo*. Ediciones Mundi Prensa, Madrid, 462pp.

Castilla, N., Baeza, E. J. and Papadopoulos, A. P. (2012). *Greenhouse Technology and Management*. 2nd ed. CAB International, Wallingford, UK, pp. 1–335.

Chengwei, M., Junfang, Z., Midao, Q. and Ruichun, L. (2006). Theoretical analysis using the method of energy equilibrium in covering layer and validation of the overall heat transfer coefficient of greenhouse covering materials. *Transactions of the Chinese Society of Agricultural Engineering* 22(4), 1–5.

Critten, D. L. (1993). A review of the light transmission into greenhouse crops. *Acta Horticulturae* 328, 9–31.

De Gelder, A., Dieleman, J. A., Bot, G. P. A. and Marcelis, L. F. M. (2015). An overview of climate and crop yield in closed greenhouses. *Journal of Horticultural Science and Biotechnology* 87(3), 193–202.

De Jong, T. (1990). Natural ventilation of large multi-span greenhouses. PhD thesis. University of Wageningen, Wageningen, The Netherlands, 116p.

Dueck, T. A., Janse, J., Li, T., Kempkes, F. L. K. and Eveleens-Clark, B. A. (2012). Influence of diffuse glass on the growth and production of tomato. *Acta Horticulturae* 956, 75–82.

Espí, E., Salmerón, A., Fontecha, A., García-Alonso, Y. and Real, A. I. (2006). New ultrathermic films for greenhouse covers. *Journal of Plastic Films and Sheeting* 22(1), 59–68.

Fang, H., Yi, Z., Qichang, Y., Wei, L., Bo, Z. and Sheng, Z. (2015). Performance testing on warming effect of heat storage-release metal film in Chinese solar greenhouse. *Transactions of the Chinese Society of Agricultural Engineering* 31(15), 177–182.

Fangjun, Z. and Haizhen, C. (2010). Development status and tendency of heat insulating covering for greenhouse. *China Textile Leader* 1, 73–6.

FAO. (2013). Good Agricultural Practices for greenhouse vegetable crops. Principles for Mediterranean climate areas. Food and Agriculture Organization of the United Nations, Rome, Italy. E-ISBN: 978-92-5-107650-7 (PDF).

Fatnassi, H., Boulard, T., Benamara, H., Roy, J. C., Suay, R. and Poncet, C. (2017). Increasing the height and multiplying the number of spans of greenhouse: How far can we go? *Acta Horticulturae* 1170, 137–43.

Gao, Y. and Quangguo, L. (2009). Categories and application effect of regulating and auxiliary equipment for Solar Greenhouse environment. *Journal of Agriculture Mechanization Research* 31(2), 244–246.

García, M. L., Medrano, E., Sánchez-Guerrero, M. C. and Lorenzo, P. (2011). Climatic effects of two cooling systems in greenhouses in the Mediterranean area: External mobile shading and fog system. *Biosystems Engineering* 108(2), 133–43. doi:10.1016/j.biosystemseng.2010.11.006.

Geoola, F. and Peiper, U. M. (1994). Outdoor testing of the condensation characteristics of plastic film covering materials using a model greenhouse. *Journal of Agricultural Engineering Research* 57(3), 167–72.

Geoola, F., Kashti, Y. and Peiper, U. M. (1998). A model greenhouse for testing the role of condensation, dust and dirt on the solar radiation transmission of greenhouse cladding materials. *Journal of Agricultural and Engineering Research* 71(4), 339–46.

Giacomelli, G. A., Sase, S., Cramer, R., Hoogeboom, J., MacKenzie, A., Parbst, K., Scarascia-Mugnozza, G., Selina, P., Sharp, D. A., Voogt, J. O., van Weel, P. A. and Mears, D. (2012). Greenhouse production systems for people. *Acta Horticulturae* 927, 23–38.

Gijzen, H. (1992). Simulation of photosynthesis and dry matter production of greenhouse crops. Simulation Report CABO-TT 28. CABODLO, Wageningen, The Netherlands, 69pp.

Gomez-Mataix, G., Montero, J. I., Raya, V. and Suay, R. (2013). Benchmark study of the distance between greenhouses and its effect on wind driven ventilation. *Acta Horticulturae* 1008, 207–12.

Gonghong, T., Christopher, D. M. and Tianlai, L. 2013. Passive solar energy utilization: A review of cross-section building parameter selection for Chinese solar greenhouses. *Renewable and Sustainable Energy Reviews* 26, 540–8.

Guoxiang, Z., Zetian, F., Lingxian, Z., Jin, Y., Biao, Z. and Xinxing, L. (2017). Development status and prospect of mechanical rolling shutter technology in solar greenhouse in China. *Transactions of the Chinese Society of Agricultural Engineering* 33(1), 1–10.

Hemming, S., van der Braak, N., Dueck, T., Elings, A. and Marissen, N. (2006). Filtering natural light by the greenhouse covering - More production and better plant quality by diffuse light? *Acta Horticulturae* 711, 105–10.

Hemming, S., Dueck, T. A., Janse, J. and Noort, F. R. van (2008). The effect of diffuse light on crops. *Acta Horticulturae* 801, 1293–300.

Hemming, S., Kempkes, F. L. K. and Mohammadkhani, V. (2011). New glass coatings for high insulating greenhouses without light losses - energy saving crop production and economic potentials. *Acta Horticulturae* 893, 217–26.

Hernández, G. (2009). Valoración radiativa. microclimática y productiva de dos invernaderos con diferente geometría de cubierta (hemicircular y lineal simétrica a dos aguas). Tesis doctoral. Universidad de Almería, Spain.

Hernandez, J., Bonachela, S., Granados, M. R., Lopez, J. C., Magán, J. and Montero, J. I. (2017). Microclimate and agronomical effects of internal impermeable screens in an unheated Mediterranean greenhouse. *Biosystems Engineering* 163, 66–77.

Heuvelink, E. (2005). *Tomatoes*. CABI, Wallingford, UK and Cambridge, MA.

Heuvelink, E., Batta, L. G. G. and Damen, T. H. J. (1995). Transmission of solar radiation by a multispan Venlo-type glasshouse: Validation of a model. *Agricultural and Forest Meteorology* 74(1–2), 41–59.

Hong, X., Jijun, L. and Xiaobo, Y. (2001). The mechanical technology of mechanical rolling shutter of China Solar Greenhouse is described. *Journal of Agriculture Mechanization Research* 4, 103–4.

Hong, S. W., Lee, I. B., Hwang, H. S., Seo, I. H., Bitog, J. P., Yoo, J. I., Kim, K. S., Lee, S. H., Kim, K. W. and Yoon, N. K. (2008). Numerical simulation of ventilation efficiencies of naturally ventilated multi-span greenhouses in Korea. *Transactions of the ASABE* 51(4), 1417–32.

Jiangjun, Y., Zhirong, Z., Zhi, Z., Yunbing, W., Zhixin, Z. and Fei, Y. (2009). Optimization of earth wall thickness and thermal insulation property of solar greenhouse in northwest China. *Transactions of the Chinese Society of Agricultural Engineering* 25(8), 180–5.

Jun, L., Zhirong, Z., Xu, Y. and Xiaohong, C. (2003). Analysis and discussion and application on solar greenhouse about azimuth angle and lighting surface angles. *Acta Agriculturae Boreali-occidentalis Sinica* 12(2), 105–8.

Junwei, W., Zhongju, H., Shirong, G., Shuhai, L., Jin, S., Jian, W. and Bin, W. (2016). Temperature humidity performance and application of solar greenhouse suiting for winter and summer. *Transactions of the Chinese Society of Agricultural Engineering* 32(23), 190–8.

Kacira, M., Sase, S. and Okushima, L. (2004). Effects of side vents and span numbers on wind-induced natural ventilation of a gothic multi-span greenhouse. *Japan Agricultural Research Quarterly* 38, 227–33.

Kacira, M., Sase, S., Ikeguchi, A., Ishii, M., Giacomelli, G. and Sabeh, N. (2008). Effect of vent configuration and wind speed on three-dimensional temperature distributions in a naturally ventilated multi-span greenhouse by wind tunnel experiments. *Acta Horticulturae* 801, 393–400.

Katsoulas, N., Sapounas, A., De Zwart, F., Dieleman A. and Stanghellini, C. (2015). Reducing ventilation requirements in semi-closed greenhouses increases water use efficiency. *Agricultural Water Management* 156, 90–9.

Kempkes, F., Janse, J. and Hemming, S. (2014). Greenhouse concept with high insulating double glass with coatings and new climate control strategies; from design to results from tomato experiments. *Acta Horticulturae* 1037, 83–92.

Kempkes, F., Janse, J. and Hemming, S. (2017). Greenhouse concept with high insulating cover by combination of glass and film: Design and first experimental results. *Acta Horticulturae* 1170, 469–76.

Kittas, C., Draoui, B. and Boulard, T. (1995). Quantification of the ventilation of a greenhouse with a roof opening. *Agricultural and Forest Meteorology* 77(1–2), 95–111.

Lee, I. B. and Short, T. H. (2000). Two-dimensional numerical simulation of natural ventilation in a multi-span greenhouse. *Transactions of the American Society of Agricultural Engineers* 43(3), 745–53.

Lee, I. B., Short, T. H., Sase, S. and Lee, S. K. (2000). Evaluation of computational fluid dynamics for analysis of aerodynamics in naturally ventilated multi-span greenhouse. *Agricultural and Biosystems Engineering* 2, 73–80.

Li, T., Heuvelink, E., Dueck, T. A., Janse, J., Gorg, G. and Marcelis, L. F. M. (2014a). Enhancement of crop photosynthesis by diffuse light: Quantifying the contributing factors. *Annals of Botany* 114(1), 145–56.

Li, T., Heuvelink, E. and van Noort, F. (2014b). Responses of two *Anthurium* cultivars to high daily integrals of diffuse light. *Scientia Horticulturae* 179, 306–13.

Lihong, G., Mei, Q. and Huazhong, R. (2010). Structure, function, application, and ecological benefit of a single-slope, energy-efficient solar greenhouse in China. *HortTechnology* 20(3), 626–31.

Lijuan, Q., Fujiang, Z., Xiaoqian, H., Jianhui, H. and Qingzhen, Y. (2016). Analysis of temperature performance of three-span sunlight greenhouse. *Journal of Hebei Agricultural Sciences* 20(3), 27–31.

López, J. C., Baille, A., Bonachela, S. and Pérez-Parra, J. J. (2008). Analysis and prediction of greenhouse green bean (*Phaseolus vulgaris* L.) production in a Mediterranean climate. *Biosystems Engineering* 100, 86–95.

Majdoubi, H., Boulard, T., Fatnassi, H. and Bouirden, L. (2009). Airflow and microclimate patterns in a one-hectare Canary type greenhouse: An experimental and CFD assisted study. *Agricultural and Forest Meteorology* 149(6–7), 1050–62.

Marcelis, L. F. M., Gelder, A. de, Dieleman, J. A., Qian, T. and Raaphorst, M. G. M. (2009). Crop management in semi-closed greenhouses. *International Symposium on High Technology for Greenhouse Systems: GreenSys2009*, 14–19 June 2009.

Ming, L., Changji, Z., Tao, Z., Yilei, Y., Jianlu, F., Zhiqiang, W. and Changhong, Q. (2016). Heat transfer process of soil wall in Chinese solar greenhouse and its theoretical simplification methods. *Transactions of the Chinese Society of Agricultural Engineering* 32(3), 175–81.

Mistriotis, A., Arcidiacono, C., Picuno, P., Bot, G. P. A. and Scarascia-Mugnozza, G. (1997). Computational analysis of ventilation in greenhouses at zero- and low-wind-speeds. *Agricultural and Forest Meteorology* 88(1–4), 121–35.

Molina-Aiz, F. D., Valera, D. L. and Alvarez, A. J. (2004). Measurement and simulation of climate inside Almeria-type greenhouses using computational fluid dynamics. *Agricultural and Forest Meteorology* 125, 33–51.

Möller, M., Cohen, S., Pirkner, M., Israeli, Y. and Tanny, J. (2010). Transmission of short-wave radiation by agricultural screens. *Biosystems Engineering* 107, 317–27.

Montero, J. I., Hunt, G. R., Kamaruddin, R., Anton, A. and Bailey, B. J. (2001a). Effect of ventilator configuration on wind driven ventilation in a crop protection structure for the tropics. *Journal of Agricultural Engineering Research* 80(1), 99–107.

Montero, J. I., Antón, A., Hernández, J. and Castilla, N. (2001b). Direct and diffuse light transmission of insect-proof screens and plastic films for cladding greenhouses. *Acta Horticulturae* 559, 203–9.

Montero, J. I., Muñoz, P., Sanchez-Guerrero, M. C., Medrano, E., Piscia, D. and Lorenzo, P. (2013). Shading screens for the improvement of the night time climate of unheated greenhouses. *Spanish Journal of Agricultural Research* 11(1), 32–46.

Montero, J. I., Munoz, P., Baeza, E. and Stanghellini, C. (2017a). Ongoing developments in greenhouse climate control. *Acta Horticulturae* 1182, 1–14.

Montero, J. I., Baeza, E., Heuvelink, E., Rieradevall, J., Muñoz, P., Ercilla, M. and Stanghellini, C. (2017b). Productivity of a building-integrated roof top greenhouse in a Mediterranean climate. *Agricultural Systems* 158, pp. 14–22.

Munoz, P., Montero, J. I., Anton, A. and Giuffrida, F. (1999). Effect of insect proof screens and roof openings on greenhouse ventilation. *Journal of Agricultural Engineering Research* 73, 171–8.

Nadal, A., Llorach-Massana, P., Cuerva, E., Lopez-Capel, E., Montero, J. I., Josa, A., Rieradevall, J. and Royapoor, M. (2017). Building-integrated rooftop greenhouses: An energy and environmental assessment in the mediterranean context. *Applied Energy* 187, 338–51.

Norton, T., Sun, D. W., Grant, J., Fallon, R. and Dodd, V. (2007). Applications of computational fluid dynamics CFD in the modelling and design of ventilation systems in the agricultural industry: A review. *Bioresource Technology* 98, 2386–414.

Ntinas, G. K., Koukounaras, A. and Kotsopoulos, T. (2015). Effect of energy saving solar sleeves on characteristics of hydroponic tomatoes grown in a greenhouse. *Scientia Horticulturae* 194, 126–33. http://dx.doi.org/10.1016/j.scienta.2015.08.013

Okushima, L., Sase, S. and Nara, M. (1989). A support system for natural ventilation design of greenhouse based on computational aerodynamics. *Acta Horticulturae* 248, 129–36.

Orsini, F., Dubbeling, M., de Zeeuw, H. and Gianquinto, G. (2017). *Rooftop Urban Agriculture*. Springer International Publishing AG, Switzerland. ISBN 978-3-319-57719-7.

Perez Parra, J., Baeza, E., Montero, J. I. and Bailey, B. J. (2004). Natural ventilation of parral greenhouses. *Biosystems Engineering* 873, 355–66.

Piscia, D., Montero, J. I., Baeza, E. and Bailey, B. J. (2012). A CFD greenhouse night-time condensation model. *Biosystems Engineering* 111(2), 141–54.

Piscia, D., Montero, J. I., Bailey, B., Muñoz, P. and Oliva, A. (2013). A new optimisation methodology used to study the effect of cover properties on night-time greenhouse climate. *Biosystems Engineering* 116(2), 130–43.

Pollet, I. V., Thoen, F. P. and Pieters, J. G. (1999). Solar energy availability in greenhouses as affected by condensation on cladding materials. *Renewable Energy* 16(1–4), 769–72.

Qian, T., Dieleman, A., Elings, A. and Marcelis, L. F. M. (2012). Leaf photosynthetic and morphological responses to elevated CO_2 concentration and altered fruit number in the semi-closed greenhouse. *Scientia Horticulturae* 145, 1–9.

Qichang, Y. (2012). Energy efficient heating technologies and models in lean-to Chinese solar greenhouses: A review. In *IV International Symposium on Models for Plant Growth, Environmental Control and Farm Management in Protected Cultivation. Acta Horticulturae* 957, 83–9.

Raya-Ramallo, V. (2014). Mejora de la productividad del cultivo de tomate para exportación en Canarias. Tesis doctoral. Universidad de la Laguna, Spain.

Sangpradit, K. (2014). Study of the solar transmission of plastic cladding materials and influence of dust and dirt on greenhouse cultivations. *Energy Procedia* 56, 566–73.

Sase, S. (1989). The effect of plant arrangement on airflow characteristics in a naturally ventilated glasshouse. *Acta Horticulturae* 245, 429–35.

Sase, S., Takakura, T. and Nara, M. (1984). Wind tunnel testing on airflow and temperature distribution of a naturally ventilated greenhouse. *Acta Horticulturae* 148, 329-36.

Sheng, Z., Yi, Z., Qichang, Q., Ruifeng, C., Hui, F., Xinglin, K., Wei, L. and Bo, Z. (2016). Performance of active heat storage-release unit assisted with a heat pump in a new type of Chinese solar greenhouse. *Applied Engineering in Agriculture* 32(5), 641–50.

Shklyar, A. and Arbel, A. (2004). Numerical model of the three-dimensional isothermal flow patterns and mass fluxes in a pitched-roof greenhouse. *Journal of Wind Engineering and Industrial Aerodynamics* 92, 1039–59.

Shuo, W. (2018). Investigation on types of solar greenhouses and main passage benefits of vegetables in Wuqing district, Tianjin City. *China Vegetables* 4, 93–5. (in Chinese).

Solovyev, A. A., Rabotkin, S. V. and Kovsharov, N. F. (2015). Polymer films with multilayer low-E coatings. *Materials Science in Semiconductor Processing* 38, 373–80.

Soriano, T., Montero, J. I., Sanchez-Guerrero, M. C., Medrano, E., Antón, A., Hernández, J., Morales, M. I. and Castilla, N. (2004). A study of direct solar radiation transmission in asymmetrical multispan greenhouses using scale models and simulation models. *Biosystems Engineering* 88(2), 243–53.

Spitters, C. J. T. (1986). Separating the diffuse and direct component of global radiation and its implications for modelling canopy photosynthesis: II. Calculation of canopy photosynthesis. *Agricultural Forest Meteorology* 38, 231–42.

Stanghellini, C. and Montero, J. I. (2012). Resource use efficiency in protected cultivation: Towards the greenhouse with zero emissions. *Acta Horticulturae* 927, 91–100.

Stanghellini, C., Incrocci, L., Gazquez, J. C. and Di Mauro, B. (2008). Carbon dioxide concentration in Mediterranean greenhouses: How much lost production? *Acta Horticulturae* 801 Part 2, 1541–9.

Stanghellini, C., Baptista, F., Eriksson, E., Gilli, C., Giuffrida, F., Kempkes, F., Muñoz, P., Stepowska, A. and Montero, J. I. (2016). Sensible use of primary energy in organic greenhouse production. BioGreenhouse COST Action FA 1105, www.biogreenhouse.org, ISBN: 978-94-6257-535-6. doi: http://dx.doi.org/10.18174/373582.

Stanghellini, C., Van't Ooster, B. and Heuvelink, E. (2019). *Greenhouse Horticulture. Technology for Optimal Crop Production*. Wageningen Academic Publisher, Wageningen, The Netherlands. https://doi.org/10.3920/978-90-8686-879-7

Swinkels, G. L. A. M., Kempkes, F. and Hemming, S. (2017). Using a ray-tracing model to design a greenhouse with maximized transmission during winter. *Acta Horticulturae* 1154, 179–83.

Teitel, M. and Wenger, E. (2014). Air exchange and ventilation efficiencies of a monospan greenhouse with one inflow and one outflow through longitudinal side openings. *Biosystems Engineering* 119, 98–107.

Teitel, M., Ziskind, G., Liran, O., Dubovsky, V. and Letan, R. (2008). Effect of wind direction on greenhouse ventilation rate, airflow patterns and temperature distributions. *Biosystems Engineering* 101(3), 351–69.

Teitel, M., Deriugin, M., Haslavsky, V. and Tanny, J. (2012). Light distribution in multispan gutter-connected greenhouses: Effects of gutters and roof openings. *Biosystems Engineering* 113(2), 120–8.

Tianlai, L. (2005). Current situation and prospects of green house industry development in China. *Journal of Shenyang Agricultural University*, 2, 001.

Vadiee, A. and Martin, V. 2012. Energy management in horticultural applications through the closed greenhouse concept, state of the art. *Renewable and Sustainable Energy Reviews* 16(7), 5087–100.

Valera, D., Belmonte, L., Molina-Aiz, F. D. and López, A. 2016. Greenhouse agriculture in Almería. A comprehensive techno-economic analysis. *Serie Economia*. Cajamar, Caja Rural, Spain, 408pp. http://www.publicacionescajamar.es/series-tematicas/economia/greenhouse-agriculture-in-almeria-a-comprehensive-techno-economic-analysis/

Vanthoor, B. (2011). A model-based greenhouse design method. Doctoral thesis. University of Wageningen, Wageningen, The Netherlands, 307pp. ISBN 978-90-8585-919-2.

Victoria-Garcia, N., Kempkes, F. L. K. and Van Weel, P. (2012). Effect of a diffuse glass greenhouse cover on rose production and quality. *Acta Horticulturae* 952, 241–8.

Von Zabeltitz, C. (1989). Passive solar heating of greenhouses with water-filled polyethylene tubes. Food and Agriculture Organization (FAO), Regional Office for Europe, 79pp. ISBN: 978-3926203045.

Von Zabeltitz, C. (2011). *Integrated Greenhouse Systems for Mild Climates: Climate Conditions, Design, Construction, Maintenance, Climate Control*. Springer, Berlin and Heidelberg, pp. 1–363.

Vox, G., Teitel, M., Pardossi, A., Minuto, A., Tinivella, F. and Schettini, E. 2010. Sustainable greenhouse systems. In: A. Salazar and I. Rios (Eds), *Sustainable Agriculture: Technology, Planning and Management*. Nova Science Publishers, Inc., New York, NY, pp. 1–80.

Wang, S. and Deltour, J. M. (1999). Lee-side ventilation-induced air movement in a large scale multi-span greenhouse. *Journal of Agricultural Engineering Research* 74, 103–10.

Weihua, G. and Tianlai, L. (2003). Effects of different mixed organic matter application on the CO_2 concentration in greenhouse and the physiological effects of tomato plants. *Acta Horticulturae Sinica* 30(5), 592–4.

Weituo, S. (2014). Design and optimization of active storage-release associated with heat pump heating system [D]. Masters Thesis. Chinese Academy of Agricultural Sciences, pp. 34–65. (in Chinese with English abstract).

Xiao, M., Fei, Q., Xiaoming, D., Shunshu, B. and Fen, H. (2010). Main achievements in China facility horticulture. *Journal of Agriculture Mechanization Research* 32(12), 227–31.

Yi, Z., Qichang, Y. and Huifang, F. (2012). Research on warming effect of water curtain system in Chinese solar greenhouse. *Transactions of the Chinese Society of Agricultural Engineering* 28(4), 188–93.

Yikui, B., Tianlai, L., Tieliang, L. and Wenhe, L. (2011). Progress of solar greenhouses construction in Liaoning Province. *Northern Horticulture* 2011(1), 62–7.

Yildiz, I. and Stombaugh, D. P. (2006). Simulated performances of a heat pump system for energy and water conservation in open and confined greenhouse systems. *Acta Horticulturae* 718, 341–50. http://dx.doi.org/10.17660/ActaHortic.2006.718.39.

Yiming, W., Yonghua, C., Zhongfu, S. and Youxiang, L. (1992). An analytical method of light transmission in greenhouse design-The first part of serial studies in simulation of light environment in protected cultivation. *Transactions of the Chinese Society of Agricultural Engineering* 8(3), 73–80.

Yunlong, Z., Xianchang, Y., Yansu, L., Chaoxing, H. and Yan, Y. (2013). Application of electric carbon crystal soil-warming system for tomato production in greenhouse. *Transactions of the Chinese Society of Agricultural Engineering* 29(7), 131–8.

Zhengwei, Q., Zhirong, Z. and Shuangxiao, Y. (2008), Heat preservation test of one kind of thermal insulation quilt for solar greenhouse. *Journal of Agricultural Mechanization Research* 6, 131–3.

Zhongfu, S. and Xuefen, Z. (2004). Greenhouse climate control techniques in china and the setup of the control system for a chamber. In *International Conference on Sustainable Greenhouse Systems-Greensys2004*, p. 691.

Zhonghua, B., Yu, W., Xiaohui, H., Zhirong, Z. and Jing, Z. (2012). Effects of outer type straw reactor on solar greenhouse environment and tomato photosynthetic performance. *Journal of Northwest A & F University (Natural Science Edition)*, 40(2), 136–42.

Zwart, H. F. de (1993). Determination of direct transmission of a multispan greenhouse using vector algebra. *Journal of Agricultural Engineering Research* 56, 39–49.

Zwart, H. F. de (1996). Analyzing energy-saving options in greenhouse cultivation using a simulation model. PhD thesis. Agricultural University, Wageningen, The Netherlands, 236p.

Chapter 3

Advances in screenhouse design and practice for protected cultivation

Josef Tanny, Agricultural Research Organization – Volcani Center, Israel

1 Introduction

The increasing world population, along with water and land shortages and climate change, led growers worldwide to protected cultivation systems. Such systems allow active manipulation of internal microclimate as well as recycling and saving of inputs. Hence, protected cultivation systems allow increasing the produce quantity and quality, in order to meet the increasing demand for high-quality food. Plastic and glass greenhouses induce crop microclimate, which is isolated from the outside. This allows using active climate control systems for heating, cooling, dehumidification, lighting and so on. Such systems increase the yield and its quality; however, the use of such high-energy consumption systems is undesirable due to their large environmental footprint. Unless these systems use renewable energy sources (e.g. solar, wind), their overall benefit is deteriorated due to their environmental impact.

In many regions of the world, cultivation of high-quality products is possible without active climate control, for example, regions with mild winter climates like the Mediterranean. In these regions, minor climate modifications are usually sufficient to facilitate economic crop production. Besides, in recent years, when food markets became global and highly dynamic and competitive, farmers could not always secure markets for their yields. Hence, in many regions farmers are not inclined towards high economic investments in production systems of

http://dx.doi.org/10.19103/AS.2019.0052.03

crops whose marketing potential may vary with time. Therefore, in such regions, farmers look for low-cost production systems, which provide acceptable conditions for profitable crop production, with a high level of flexibility.

The screenhouse is one example of such systems. It is built with a skeleton of basic construction elements enveloped by a porous screen, which may be either a shading or an insect-proof screen. Another type of structure is a screen cover without sidewalls, usually used for shading only. In both types, the screen cover provides a moderate modification of the microclimate, in particular reducing incoming solar radiation and wind speed. Hence, it allows crop production in regions where external conditions are otherwise unfavourable. In certain regions and crops, insect-proof screenhouses are popular. These screens consist of tiny holes, which avoid insect penetration, thus allowing growers to reduce pesticide application, which allows them better access to world markets. Screenhouses are low-cost structures, hence they are popular among growers in developing as well as developed countries. Their capability in saving irrigation water under certain conditions made them popular especially in arid and semi-arid regions where water is either scarce or expensive, or both.

Crops cultivated within screenhouses or under horizontal screens are in high interaction with the outside environment. This is due to the screen porosity, which allows the exchange of mass, momentum and energy between the air adjacent to the crop and the outside. Therefore, active manipulation of crop microclimate is not feasible and climate is passively controlled. This means that screen properties including mesh size and colour as well as structure geometry (height, horizontal extent) and plant traits all affect the microclimate in a rather complicated way.

In the rest of this chapter, Section 2 will describe in detail common types of structures and typical materials used by growers. Sections 3 and 4 will discuss, respectively, studies regarding microclimate and evapotranspiration of crops under screens. Section 5 will be focussed on the question whether it is possible to increase water-use efficiency in screenhouses, and Section 6 will review studies where product quality under screens was examined. The chapter will conclude with Section 7 that will depict knowledge gaps and suggest future research directions. The reader is also referred to two review papers on the effects of screens and screenhouses on crop microclimate, evapotranspiration and produce quality (Tanny, 2013; Mahmood et al., 2018).

2 Structures and screen materials

2.1 Structures

Three basic types of screen constructions are popular among growers: screen strips, a screen cover and a screenhouse. Screen strips are commonly used for moderate shading of deciduous and evergreen orchards. They are deployed horizontally above the hedgerows only (Tanny and Cohen, 2003) leaving the paths between trees unshaded. By covering the trees only, this system allows sufficient radiation for lower parts of the foliage, thus reduction of photosynthesis due to shading is limited. The second type of structure is a screen cover, where the roof is fully covered by a horizontal screen; the sides are usually open to facilitate ventilation (Kittas et al., 2012), but folded sidewalls are also popular. Horizontal screen covers are mainly used for shading against surplus radiation and for hail protection in regions of high hailstorms risk. Horizontal screens may cover various types of crops, from flowers and vegetables to deciduous fruit trees. In the latter case, for example when a screen is covering apple trees, screen deployment should be scheduled with fruit set which requires specific light and microclimate conditions (Tanny et al., 2009). Hence, for such applications the screens are not permanent, but deployed only during the warm season, when the risk of fruit sunburn is high and there is high potential for saving irrigation water through shading.

The third type of structure is the screenhouse which is totally enveloped by a screen on both its roof and sidewalls (Tanny et al., 2003). Such structures are covered by either shading or insect-proof screens. Shading screenhouses provide both protection from high radiation (like horizontal covers) and high wind speed. The sidewalls serve as artificial wind breaks (Wilson and Flesch, 2003; Brandle et al., 2004) that reduce wind speed and hence possible damage to leaves and fruits and potential water saving (Tanny et al., 2006). Such technology allows farmers to cultivate in regions of high wind speed that are unsuitable for farming of unprotected crops.

Insect-proof screenhouses have become very popular in recent years, mainly because they allow growers to reduce pesticide application, which makes their products healthier and more environmentally friendly. This also provides growers with access to markets which strictly limit the levels of pesticide residues on the products. In such structures the screen is deployed throughout the production season to avoid the invasion of insects at all stages of plant growth. To maintain high protection, the screenhouse should be closed, and usually a double door entrance is available to avoid insect invasion during door opening.

2.2 Types, colours and textures of screens

Most commercially available screens are made of a plastic material like polyethylene (PE). This proved to be a durable and economically feasible material for agricultural production. Screens differ by the colour and thickness of the thread, the distance between threads (mesh) and the texture (Möller et al., 2010). The type of screen should be chosen according to its purpose, as well as the crop type and regional climatic conditions. However, it should be recalled that whatever the purpose, covering a crop by any type of screen would always reduce radiation load and wind speed near the crop. This should be considered, especially when the main purpose of screen deployment is insect exclusion or hail protection. For this reason, research was focussed on investigating microclimate under insect-proof screens (Desmarais et al., 1999; Tanny et al., 2003; Seginer et al., 2017), as lower radiation and ventilation might deteriorate photosynthesis and production. Shading screens are either knitted or woven (Pirkner et al., 2014b), whereas insect-proof screens are always woven which allows the fine mesh texture. Knitted screens provide a complex texture and are usually more flexible mechanically compared to woven screens that are stiffer.

Insect-proof screenhouses of different mesh are used to block different insects. In the context of insect-proof screens deployed in greenhouse openings, Teitel (2007, Table 1) compiled information from several sources regarding the size of common pests (e.g. thrips, whiteflies, aphids, miners) and the maximum hole size, or mesh, in a screen that can block the invasion of each insect into the house. For example, the smallest insects, thrips, would be blocked by fine mesh screens with mesh size of about 190 μm, and mesh would increase with insect size up to 610 μm. Using screens with larger mesh is desired to facilitate adequate ventilation and CO_2 supply to the plants. Although the data is compiled with relevance to screened greenhouse openings, it appears to be relevant for insect-proof screenhouses as well.

Besides mesh size, screens differ in colour, and there is a variety of colours used for different purposes. Black screens induce intense shading in cases where radiation and heat load can be harmful for plants and fruits. For example, a common practice for pepper cultivation during the summer in the western Negev region in Israel is growing within a black 30% shading screenhouse throughout the season (Moeller and Assouline, 2007). On the other hand, pepper cultivated in greenhouses in the Jordan Valley region of Israel during the winter is covered by a 40% shading screen only during the warm fall and spring seasons, at the beginning and end of the cultivation period, respectively (see Fig. 1). During winter the black shading screen is removed to provide sufficient radiation to the plants.

Shading is also achieved using white or clear screens; however, with these screens the shading level is usually moderate. Another advantage of such screens (as well as other coloured screens) is that the light intercepted by the screen threads is partially transmitted into the house as diffuse radiation (Romero-Gámez et al., 2012). Diffuse radiation reaches the foliage at low levels, that is usually shaded from direct solar radiation, and hence can contribute to increased productivity (Gu et al., 2002; Gu et al., 2003). As shown by Li et al.

Figure 1 High-tunnel greenhouses covered by 40% shading screens during the initial period of pepper cultivation in the Jordan Valley region of eastern Israel. Plantation was in mid-August and the photo taken on 15 September 2016. Towards the winter, when radiation load reduces, the black shading screens are removed. They are deployed again towards the warm spring. Source: J. Tanny.

(2014), photosynthesis of greenhouse tomatoes increased under diffused light conditions. This advantage is absent in black shading screens where the threads absorb most of the solar radiation and only direct radiation is transmitted through the screen holes.

Besides black or white screens, in recent years coloured, photo-selective screens have become available on the market. Such screens demonstrate several effects related to plant growth and yield (Shahak, 2014). A variety of field experiments were conducted to investigate the effects of different screen colours (blue, red, grey, pearl and yellow) on crop development and productivity. For example (Shahak, 2014), for ornamental crops, experiments showed stimulated vegetative vigour (under the red and yellow nets), dwarfing (blue), enhanced branching (grey, pearl), as well as various effects on leaf size, variegation, time-to-flowering and flower quality. In vegetables (bell peppers, tomatoes), the red, yellow and pearl shade nets were found to markedly increase productivity, compared with the common-practice covers.

The large variety of screen constructions and screen materials, textures and colours provide growers with a wide choice for protecting their crops with screens, according to their needs. However, most knowledge on the effects of screens on crops is nowadays available from growers' experience. Hence more research is needed in order to provide a solid knowledge base about these effects for the benefit of growers and consumers.

3 Microclimate under screens

Screens provide a physical barrier between the crop and the atmosphere. Therefore screens modify the exchange processes of mass, momentum and energy between the crop and the external environment. Being a porous material the modification is generally moderate and depends on the screen properties. The modified transport through screens, as compared to undisturbed transport, causes variations in crop microclimate which affects the meteorological variables near the crop, mainly radiation, air velocity, temperature and humidity. The variations of these properties might, in turn, affect the transport processes in the soil-plant-atmosphere continuum, which then interact with the modified properties. Hence covering a crop may induce an interactive response between the crop and its immediate environment. In this section, we will highlight selected results on the effects of screens on the above-mentioned four microclimate variables.

3.1 Radiation

Solar radiation is the major energy source for crop production. The radiation spectrum relevant to agricultural applications can be roughly divided into two wave bands: (1) short-wave (SW, 400–700 nm), which includes the photosynthetically active radiation (PAR); and (2) long wave (LW) which is mostly thermal and includes the near-infrared (NIR). PAR is needed for photosynthesis and hence its absorption by the plants is essential for production. The LW radiation is thermal; in cool climates, it may save artificial heating requirements, but under warm climates, excess LW may limit production and hence should be avoided.

Radiation that reaches the screen would be either transmitted towards the crop through the screen holes and threads, absorbed by the screen material or reflected back. Screen transmittance dictates the amount of radiation reaching the plants; hence, it is a basic parameter for the screen's characterization. Usually, manufacturers provide nominal transmittance (or shading level) of a screen, however, due to dust accumulation or other environmental effects this can vary with time. For example, radiation measurements by Möller et al. (2010), under a shading screen covering a large banana plantation, showed a steady reduction rate of 0.1% per day during the rainless period in Israel from May to October 2005. During the winter following this period, transmittance was resumed almost to its nominal value due to washing of the screen by rain (Fig. 2). Haijun et al. (2015) measured radiation inside and outside of a banana screenhouse during the periods between May and October of 2005 and 2006. Regressions between inside (y-axis) and outside (x-axis) data showed slopes of 0.80 and 0.77 (with R^2 = 0.93 and 0.89) for solar and net radiation,

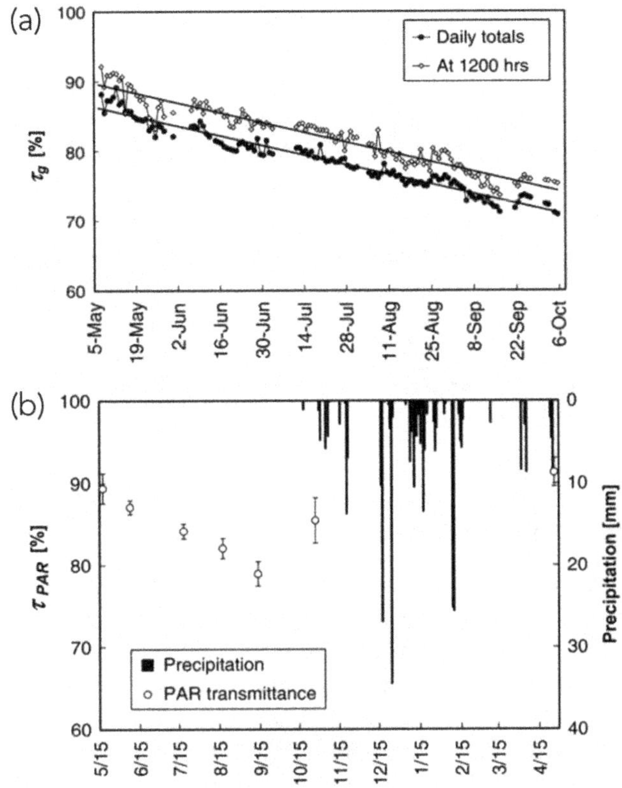

Figure 2 (a) Seasonal short-wave transmittance τ_g of the shading screen, measured in a banana screenhouse from 6 May to 5 October 2005. (b) Midday PAR-transmittance τ_{PAR} of the shading screen of the banana screenhouse, measured for 7 days from 17 May 2005 to 26 April 2006. Vertical bars represent two standard errors of the mean. Daily rainfall measured at the nearby station is indicated on the secondary y-axis. More details can be found in Fig. 10 of Möller et al. (2010).

respectively. Thus, a reduction of about 20% in radiation was realized under these screens.

In contrast to shading screens that have relatively large holes, insect-proof screens are much denser with very tiny holes. Therefore transmittance through such screens is lower. Measurements in an insect-proof screenhouse in which pepper was grown (Möller et al., 2003) showed that an average 52% of the incoming short-wave radiation (RG) was transmitted through the screen ($RG_{in} = 0.52\ RG_{out}$; $R^2 = 0.98$). However, the short-wave screen transmissivity was not independent of the solar elevation and the maximum daytime values - those typically reached at noon - were within the range 0.57–0.59. An approximately similar result (transmittance of 0.5) was obtained by Al-Mulla et al. (2017), who measured microclimate in an insect-proof screenhouse in Oman. Somewhat higher transmittance values were obtained by Kitta et al. (2014a) who studied three different types of screenhouses: (1) a pearl insect-proof screen (IP-78), (2) a white insect-proof screen (IP-59) and (3) a green-shade screen (GS-62) with nominal values of the transmittance to PAR of 78%, 59% and 62%, respectively. According to their measurements, transmittance to global radiation of the above-mentioned three screens was 0.81, 0.66 and 0.70, respectively.

As mentioned above, an important advantage of porous screens is the high transmission of diffuse radiation. Montero et al. (2001) measured transmission of direct and diffuse radiation through several types of screens and plastic films. Their results showed that the insect-proof screen with hole dimensions of 1.4 × 1.8 mm, provided highest transmission of diffuse radiation (81%) as compared to the denser insect-proof screen (with holes of 0.4 × 0.8 mm, 75%) and the plastic films (59–77.5%).

3.2 Air velocity and turbulence

Screen threads induce a drag force resistance on the flow thus reducing the internal air velocity in screenhouses and under screen covers, as compared to the external undisturbed wind speed. Estimating the effect of horizontal screen covers or screenhouses on wind speed is challenging due to the natural variation of wind speed with height above the ground. In the atmospheric boundary layer, wind speed above a uniform horizontal surface increases logarithmically with height. This region is usually denoted as the surface layer. The increase of wind speed with height implies that to study the effect of screens on wind speed, measurements with and without screens should be done at the same height. This task is not always feasible, because similar neighbouring crops with and without screens are not available in all cases.

The relation between inside air velocity and outside wind speed was explained by Waggoner et al. (1959) using principles of fluid mechanics. They suggested that as external wind increases from zero up to a certain threshold, the drag by the screen threads increases rapidly and inside velocity remains nearly zero. As outside wind further increases to a second threshold, a measurable but small increase in inside velocity can be detected. An increase in external wind speed above this second threshold will induce a constant drag, and hence an increase in internal air velocity linearly related to the outside wind speed. For a horizontal screen cover without a crop, Waggoner et al. (1959) obtained the relation: $u_{in} = 0.67(u_{ex} - 0.833)$ (m/s), where u_{in} and u_{ex} are internal air velocity and external wind speed, respectively, measured 1 m above the ground. Using the logarithmic wind profile model, Tanny et al. (2006) used external wind speed measured at 18 m height to simulate air velocity over an open banana plantation. The relation between inside and outside velocity at 5 m height for a banana screenhouse was obtained as: $u_{in} = 0.6\,(u_{ex} - 0.18)$ (m/s) (Tanny, 2013).

The use of the logarithmic wind profile to estimate wind speed at various heights was also examined by Tanny et al. (2009) in a field experiment where air velocity was measured at 3.2 m height within a screen-covered apple orchard, and outside wind speed was measured at 6 m by a nearby external meteorological station. The measurement at 6 m was used to predict air velocity under the screen, assuming a constant friction velocity. The analysis showed that air velocity calculated at 3.4 m was in best agreement with that measured at 3.2 m, see Fig. 3. The small difference between this height (3.4 m) and the actual measurement height of 3.2 m indicated the approximate validity of the logarithmic wind profile model under the screens.

The drag reduction by the screen is implemented through absorption of momentum; hence, momentum flux, or shear stress, above and below the screen should be different. This phenomenon was observed by Tanny et al. (2018), who measured vertical profiles of the friction velocity above and below a shading screen covering a large banana plantation. Results clearly showed (see Fig. 4) the reduction in friction velocity below the screen as compared to its value above the screen.

For the particular case of the white woven 17 mesh 'crystal' screen with 9% nominal shading, the mean ratio between friction velocity below and above the screen was 0.54 ± 0.035 (Tanny et al., 2018).

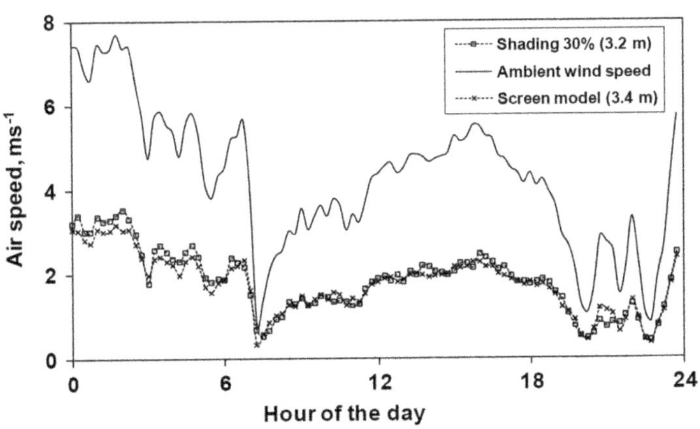

Figure 3 Diurnal curves of wind speed and air velocity on a typical day (17 July 2004). Solid line - external wind speed at 6 m height; squares - air velocity measured at 3.2 m under the 30% shading screen covering an apple orchard; asterisks - air velocity over a shaded orchard at 3.4 m, calculated from the external wind speed and logarithmic wind profile model for screenhouses. Source: adapted from Tanny et al. (2009).

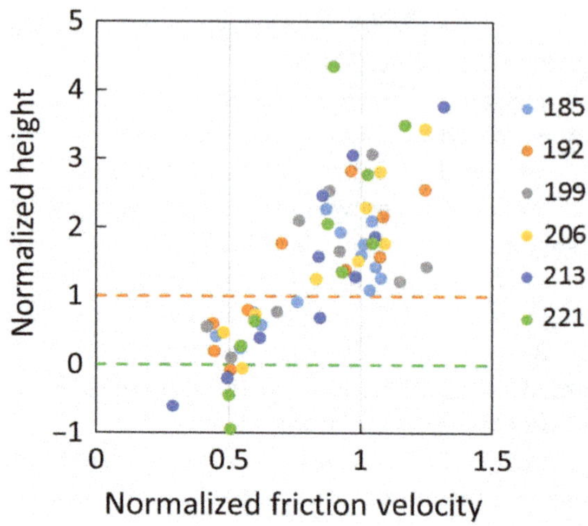

Figure 4 Vertical profiles of the friction velocity, u_*, normalized by the mean friction velocity above the screen. The vertical axis is the normalized height, expressed as $Z_{cs} = (z-h_c)/(h_s-h_c)$, where z, h_c and h_s are measurement, canopy and screen heights, respectively. Each symbol colour represents a profile measured on a different day of the year as indicated in the legend. The screen is at $Z_{cs} = 1$ (orange dashed line) and canopy top is at $Z_{cs} = 0$ (green dashed line). For more details the reader is referred to Tanny et al. (2018).

Another turbulent property of interest is the decay rate of turbulent energy spectrum with frequency. This rate indicates to what extent the flow under study resembles the flow of undisturbed boundary layers. In theory, for a steady state turbulent boundary layer, the power relating energy spectrum and frequency, or the decay rate, is −5/3 (Tennekes and Lumley, 1972). Using high-frequency measurements, Tanny et al. (2018) calculated the energy spectrum below and above the screen. Their results (see Fig. 5) showed that the decay rate deviated from the theoretical −5/3; however, the deviation was highest within the canopy and below the screen. Above the screen, the deviation decreased with height, suggesting that the disturbance of the flow by the screenhouse decreased with height above the screen, as expected.

Besides the magnitude of air velocity and its turbulent characteristics, the air flow direction is also of interest, with relation to ventilation and insect invasion. Two contrasting results illustrate the effect of screen porosity on airflow direction in screenhouses. In a 50-mesh insect-proof screenhouse, in which pepper was grown, Moeller et al. (2003) have shown (see their Figs. 10 and 11) that over part of the screenhouse, the internal air flow was in opposite direction to the external wind speed. This observation suggested that internal airflow pattern is of a complex nature with a counter-flow in the windward region of the screenhouse. On the other hand, in a banana screenhouse, Tanny

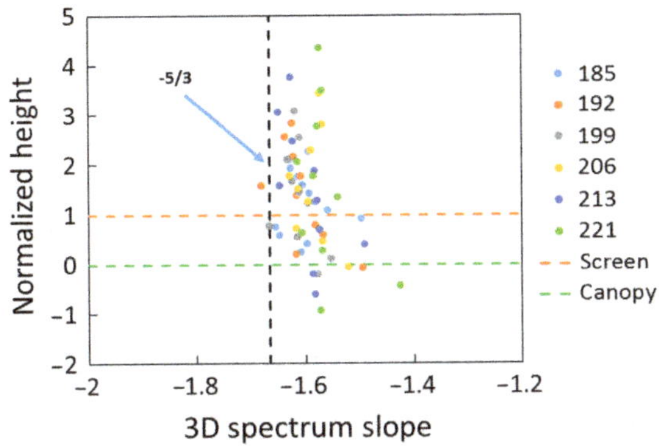

Figure 5 Variation with height of energy spectra slopes measured in the different profiles. The vertical axis is the normalized height, defined as in Fig. 4. The vertical black dashed line represents the −5/3 (−1.67) slope typical of the inertial sub-range in steady-state turbulent boundary layers. Each symbol colour represents a profile measured on a different day as indicated in the legend. The screen is at Z_{cs} = 1 (orange dashed line) and canopy top is at Z_{cs} = 0 (green dashed line). Source: adapted from Tanny et al. (2018).

et al. (2006) have shown (see their Figs. 3 and 4) that for wind speed higher than 3 m/s, internal airflow was in the same direction as the external wind. However, at lower wind speeds there were some cases of opposite airflow directions, probably due to streamline curvature around the structure.

3.3 Air temperature and humidity

In contrast to the clear effects of screens in reducing the radiation and wind speed, the effects on temperature and humidity are not straightforward.

Measurements within the foliage of shaded apple trees (Tanny et al., 2009) showed a decrease of about 1.5°C in air temperature within the canopy and about 300 Pa in vapour pressure deficit (VPD) under shaded treatments as compared to unshaded control. The reduced VPD was explained by the lower air velocity under the screen which inhibited the ventilation and the removal of water vapour away from the canopy. Kitta et al. (2014a) studied the effect of three different types of screenhouses on sweet pepper microclimate and yield. In their experiment they also conducted measurements in an uncovered pepper crop, which was similar to the crop within the screenhouses. They obtained approximately similar air temperature and VPD at 1.5 m above the ground, for the four treatments examined: three screenhouses and open field.

Kittas et al. (2012) studied the effects of four shade nets with different colour transmission properties and shading levels (two black nets with shading levels of 40% and 49%, a green net with a shading of 34%, and a mixed green and black shade net with a shading of 40%) on the screenhouse microclimate, and on the growth and development of a tomato crop. Their results showed that air temperature and VPD values were similar under all four shade nets and without shading. However, canopy temperature and, accordingly, the canopy-to-air VPD were significantly lower under shading than without shading. In particular, canopy temperature under the screens was about 5°C lower than in the open, which resulted in a reduction of about 50% in canopy-to-air VPD under the screens. Kittas et al. (2012) explained that since the structure consisted of a screened roof only, without sidewalls, and screen area was relatively small, high ventilation of air near the canopy eliminated the difference in air temperature between shaded and unshaded treatments. On the other hand, the reduced radiation under the shade caused a lower canopy temperature compared to the unshaded control.

The above results suggest that in highly ventilated screenhouses air temperature under the shade is not much different from that of an unshaded canopy. This is due to the high air exchange rate in such screenhouses which diminishes differences between inside and outside conditions. Such situations may occur either when the screen is deployed only on the roof and there are no sidewalls, or when external wind speed is high and efficiently penetrates into the screenhouse. However, shading affects leaf temperature, as can be concluded from the measurements by Kittas et al. (2012) and Tanny et al. (2009). We recall that in the latter study air temperature was measured within the canopy, hence was probably influenced by the reduced temperature of the shaded leaves.

4 Crop evapotranspiration

Screens modify internal microclimate; hence they affect the evaporative demand in screenhouses and under screens. Due to the reduction in radiation, wind speed and VPD, the evaporative demand usually decreases under screens. One of the purposes of cultivating under screens is the possibility of water saving, or, increasing the water-use efficiency (WUE), namely the yield obtained per unit of irrigation water. Hence, in recent years much research was focussed on measurements and modelling of evapotranspiration (ET) of screenhouse crops, as tools for improved and more accurate irrigation management.

In an early research on this topic, Moeller et al. (2004) studied a 50-mesh insect-proof screenhouse in which pepper was grown. They determined crop water use by sap flow measurements using the heat-pulse technique, and by an eddy-covariance system that was operated for 2 days only. Agreement between the two methods for these two days was very good (R^2 = 0.95). Estimations of soil evaporation showed that in this environment, of mature pepper plants and a dense insect-proof screen, soil evaporation could be neglected.

In addition to ET measurements, Moeller et al. (2004) derived a modified Penman–Monteith model that was adapted to the screenhouse environment.

This was done by including an additional boundary layer resistance which accounts for the air layer between the crop and the horizontal screen on the roof. In addition to the screenhouse model, a greenhouse Penman-Monteith model that was earlier derived by Fuchs (1993) was examined. Half hourly predictions of both models yielded good agreement with measured sap flow (R^2 = 0.94–0.95), as shown in Fig. 6 for 2 days. Daily crop water use predicted with the greenhouse model (PM-gr) was in better agreement with the measurements than that predicted by the screenhouse model (PM-sc). Measured and predicted transpiration rates inside the screenhouse were approximately 1.8–2.1 mm/day during the most active stage of growth, while those simulated for a similar crop grown outside would be 4.5–5.3 mm/day on average. The large difference between internal and external ET illustrated the potential water saving in cultivating pepper within screenhouses.

Following this research, Tanny and co-authors conducted several studies, measuring and modelling ET in banana and table grapes screenhouses, for example, Tanny et al. (2006), Dicken et al. (2013a,b), Pirkner et al. (2014a) and Tanny et al. (2018). In these studies ET was measured by the eddy-covariance technique and various properties of the turbulent flow which are relevant for the performance of the eddy-covariance method were explored. In addition, several Penman-Monteith approaches for modelling screenhouse ET were examined. For example, Dicken et al. (2013a) measured ET of banana plants during their growth and compared the data with Penman-Monteith predictions using the reference evapotranspiration model (Allen et al., 1998) and the screenhouse model (Moeller et al., 2004). Results showed (see Fig. 7) that the models, predictions were in better agreement with measurements of larger than

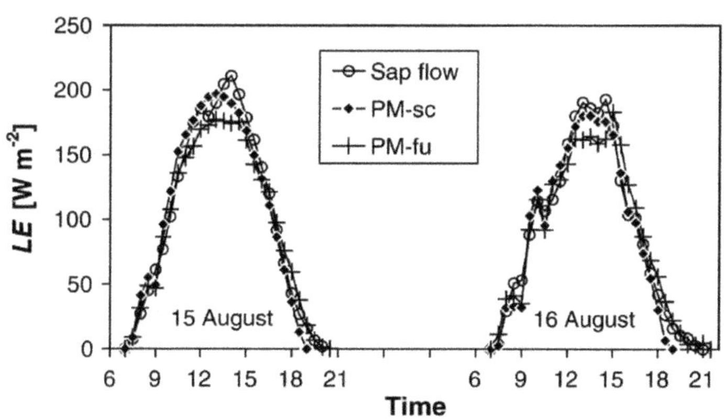

Figure 6 Diurnal variation of transpiration, LE (Wm^{-2}) measured by sap flow, and calculated from the meteorological models for screenhouses (PM-sc) and greenhouses (PM-fu) on 15 and 16 August 2001; 30 min average data (Moeller et al., 2004). The vertical axis presents the latent heat flux ($W\ m^{-2}$) required for water transpiration.

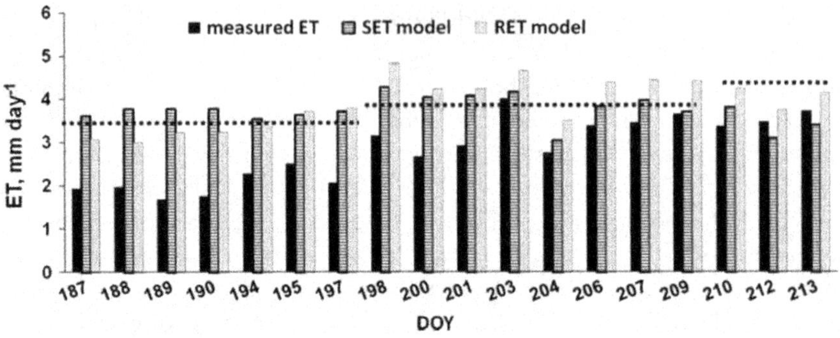

Figure 7 Total daily evapotranspiration as measured by the EC system and estimated by the two models during DOY 187-213. SET – screenhouse evapotranspiration; RET – reference evapotranspiration. The dashed line is the irrigation given by the grower. DOY 187-200: smaller plants; DOY 201-213: larger plants. Source: adapted from Dicken et al. (2013a).

smaller plants. This was explained by the fact that the Penman–Monteith model assumes full canopy cover, which is more valid for the larger than the smaller plants. In addition, results showed that the screenhouse model provided more accurate predictions than the reference evapotranspiration model. This result is rather expected, since reference ET represents a short grass crop whereas the screenhouse model described actual banana plants.

In the above studies, model calculations were mostly based on internal microclimate conditions. However, Kitta et al. (2014b) adopted a somewhat different approach. They measured both internal and external climate conditions and used these data to derive two screenhouse parameters: one for radiation and the other for advection. Using these parameters they predicted reference evapotranspiration based on external conditions only. Their results showed that the predictive performance of the model was fair (RMSE = 0.11 mm/day). They suggested that the proposed model, based only on outside climate data and the knowledge of two screenhouse-related parameters, provides a straightforward way to estimate screenhouse reference evapotranspiration (ETo), a necessary step towards the determination of screenhouse–crops water requirements.

Real-time knowledge of evapotranspiration of screenhouse crops can be of assistance for growers in irrigation management. Therefore in recent years research was focussed on examining simple methods for real-time ET estimation. One of these methods is surface renewal (SR). This technique utilizes high-frequency readings of air temperature near or above the vegetation by low-cost fine-wire thermocouples, to indirectly estimate the sensible heat flux from which the latent heat flux (evapotranspiration) can be estimated assuming energy conservation. For example, Mekhmandarov et al. (2015) studied the SR method in two types of screenhouses: a shading screenhouse covering banana plants and an insect-proof screenhouse in which pepper was grown. In both campaigns, the SR technique was used by installing several miniature thermocouples at different heights. Reference measurements of sensible heat flux were conducted using the eddy-covariance (EC) technique. Results showed a better performance of the SR technique in the shading than the insect-proof screenhouse. Turbulence characteristics were observed to be affected by the porosity of the screen, which is conjectured to be the main cause for reduced performance of the SR technique in the insect-proof than the shading screenhouse.

Another simple method examined was the flux-variance (FV) technique, also based on high-frequency measurements of air temperature using a miniature thermocouple, and the calculation of its standard deviation. This method was examined in three types of structures: the above-mentioned two screenhouses and a naturally ventilated greenhouse (Ahiman et al., 2018), and data was validated against direct EC measurements. In the greenhouse, due to the impermeable plastic roof, the EC results were not sufficiently reliable to validate those by the FV. On the other hand, results showed that the FV technique was suitable for reliable estimates of ET in shading and insect-proof screenhouses with R^2 of the regressions between FV latent heat flux and latent heat flux deduced from energy balance closure of 0.99 and 0.92, respectively.

5 Water-use efficiency

WUE is generally defined as the ratio between crop production and the water used by the plants. Hence, there are several ways to define WUE. One definition is at the plant level, that is, the flux WUE, namely, the ratio between net rates of CO_2 assimilation and transpiration. Another definition considers the whole canopy, and defines the WUE as the ratio between yield and irrigation. While the first is more important to understand, and perhaps control, plant physiological processes, the second is of more practical significance for the grower.

Since the screenhouse environment reduces the evaporative demand, it is anticipated that water saving will be feasible without yield reduction, hence, the WUE would increase. Nevertheless, very few studies are available in the literature which actually estimated WUE under screens. To examine this possibility, an irrigation experiment was carried out in a banana screenhouse in the Jordan Valley region of northern Israel (Tanny et al., 2015). Four irrigation treatments were applied: 100%, 85%, 70% and 55%, where 100% was the amount of irrigation (2200 mm/y) provided in open banana plantations in this region. Results obtained during 4 years (see Fig. 8) showed that at 85% irrigation the yield did not decrease as compared to the 100% level.

Irrigation at 70% reduced the yield slightly but this reduction was not statistically significant. The lowest irrigation level of 55% did cause a significant

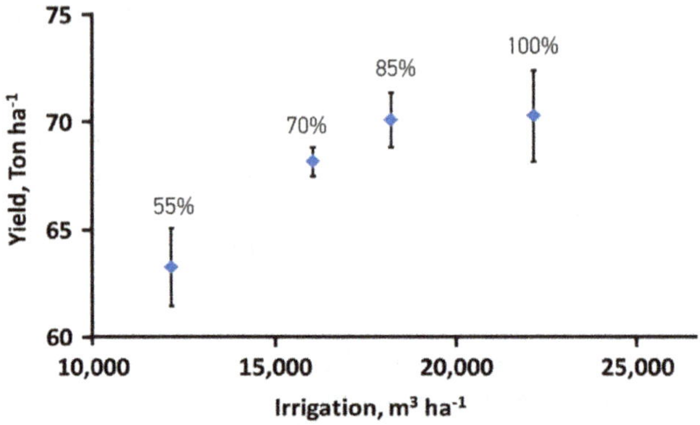

Figure 8 Mean annual banana yield as a function of the mean annual irrigation, for the years 2004–07. Percentage indicates the irrigation amount relative to open banana plantations in this region. Source: adapted from Tanny et al. (2015).

reduction in yield. Hence, WUE can be increased by about 20–30% by reducing irrigation to 70% of its recommended value when growing bananas in screenhouses in this region. In practice, banana growers in this region reduce irrigation by about 30% and obtain the same yield.

In a study where turbulent fluxes of water vapour and CO_2 were measured in a banana screenhouse, Dicken et al. (2013a) estimated the flux WUE for smaller and larger banana plants with leaf area index (LAI) of 0.67 and 1.47, respectively. Results showed that water vapour and CO_2 fluxes per unit leaf area were independent of plant size and averaged at 51 and 0.29 g/m/day, respectively. Consequently, WUE, defined as the ratio between net vertical fluxes of CO_2 and water vapour, was nearly constant during plants' growth.

Moeller and Assouline (2007) investigated microclimate, irrigation, yield and WUE in a pepper crop covered by a black shading screen. They compared WUE obtained in their screenhouse, in units of kg produce per m^3 of irrigated water, to data from open field reported by various literature studies. The results showed that in the first year of their screenhouse study, yield was in the upper range of what was reported for the open field, and in the second year screenhouse yield was significantly higher than all available evidence for sweet pepper grown in the open field. The amount of irrigation water in the screenhouse was lower than outside, and in both years, screenhouse WUE was significantly higher than that of the reported outside crops.

6 Product quality

Screens physically protect the fruits from external weather hazards like high wind speed, supra-optimal radiation and hail. Hence, it is generally anticipated that product quality of screenhouse crops would be higher. On the other hand, the screens reduce radiation levels, may change the spectrum of light reaching the plants and thus may modify produce quality and colour. Studies on the effects of screens on produce quality examined both shading and insect-proof screens, as well as photo-selective nets.

A recent paper by Ilić and Fallik (2017) reviewed recent studies demonstrating the effects of screens of different type and colour on produce quality at harvest and postharvest. Their Table 3 provides a comprehensive review on the effects of screens on different produce attributes. For example, they indicate that shade nets protected fruits from sunburn, thus increasing marketable yield, and resulted in high yields of good quality tomatoes (Ilić et al., 2015). Ambrózy et al. (2016) studied the effects of various coloured nets (CN red, yellow, green and white) all with 40% shading, on yield, fruit quality and pigments of sweet pepper. Higher yield was obtained under the yellow and red nets as compared to unshaded control. Sunscald was reduced under the nets and fruits from first harvest had higher levels of red than yellow pigments.

Screens also affect postharvest attributes of fruits. Selahle et al. (2015) examined the postharvest response of red and yellow sweet pepper, to three types of photo-selective nets and a standard black net. Results showed that red and yellow peppers grown under pearl and yellow nets resulted in a higher percentage of marketable fruit, after storage. Red pepper grown under the yellow net showed a higher number of odour active aroma compounds in the fruit, while black nets significantly affected the synthesis of odour active aroma compounds during storage. They reported that sensory analysis indicated a preference for red pepper fruits after storage from plants grown under pearl nets.

The above review suggests that nets of different colours and photo-selective properties would affect fruit quality in diverse ways. Ilić and Fallik (2017) indicated in their review that during production, the light quality under the photo-selective shade nets exerts a positive effect on the yield, quality parameters and phytochemical contents of commonly consumed vegetables such as tomatoes, sweet peppers, lettuce and aromatic herbs at harvest and after storage. This technology has the ability to extend the shelf life of produce, thereby lowering postharvest losses.

Besides the above effects, screens protect the crop from high wind speed which improves fruit cleanliness and reduces fruit waste. Also, screenhouse cultivation facilitates a more uniform distribution of harvest during the year, thus allowing higher market prices, as reported for banana under shading screenhouses in Israel (Israeli et al., 2009). Another observation regarding banana cultivation in screenhouses was that although plants were not fully protected from frost, they recovered from frost events faster than uncovered banana plants (Israeli et al., 2009).

7 Future trends and conclusion

Cultivation under screens and in screenhouses provides several benefits for the growers, which makes this technology economical and profitable. In recent years the production under screens increased, mainly because of the improved fruit quality, the possibility for water saving and the decrease in pesticide application when insect-proof screens are used. Hence this technology provides advantages for both growers and consumers, and facilitates a more environmentally friendly production.

In recent years, studies on screens and screenhouses focussed mainly on microclimate, evapotranspiration, WUE and produce quality. Still, research has several challenges in each of these aspects. Selected microclimate variables are routinely measured in almost all field experiments on screenhouse cultivation. However, many of these studies are done in relatively small structures, where edge effects on internal conditions are significant. Therefore, these measurements represent the specific structure under study but cannot be generalized to a large area covered by a screen. To overcome this limitation, the challenge is to perform experiments in large areas or large enough plots, where edge effects could be neglected.

Evapotranspiration measurements are a significant challenge, for both open field and protected crops. Methods like lysimeter or heat pulse, measure a single plant and extrapolation is required for whole field data. To overcome this limitation, recent studies used micrometeorological methods like the eddy covariance, Bowen ratio (Dicken et al., 2013b), SR and flux variance, which measure whole canopy fluxes. These methods are based on the turbulent properties of the atmospheric boundary layer; they were derived for flux measurements in open fields, and the challenge here is using these techniques under screens which somewhat modify turbulence properties.

To be an economic technology for farmers, the cost invested in screenhouse construction should be returned by increased profits, as compared to open field cultivation. WUE is an important parameter with high economic significance mainly in arid regions where irrigation water is scarce and expensive. Research should focus on investigating WUE of screenhouse crops for the benefit of farmers and consumers in such regions.

The effect of screens on yield and quality in harvest and postharvest is another aspect of the economics of screenhouse cultivation. Several studies have shown the advantages of this cultivation technology; however, due to the wide variety of growing conditions, including soil and water properties, fertilization and microclimate, such studies are needed in almost any region where this type of cultivation is considered by farmers.

In summary, although the use of screenhouses and screen-covered cultivation systems is increasing worldwide, there are still significant knowledge gaps that require research at various levels: from the single plant to the whole canopy. It is preferable that such studies would be conducted in cooperation with extension service experts so that the knowledge gained will be disseminated for the benefit of the society.

8 References

Ahiman, O., Mekhmandarov, Y., Pirkner, M. and Tanny, J. 2018. Application of the flux-variance technique for evapotranspiration estimates in three types of agricultural structures. *International Journal of Agronomy* 2018, 1–13. doi:10.1155/2018/7935140.

Al-Mulla, Y. A., Al-Balushi, M. I., Al-Busaidi, H. A., Al-Mahdouri, A. A., Kittas, C. and Katsoulas, N. 2017. Analysis of microclimate and cucumber fruit yield in a screenhouse and an evaporatively cooled greenhouse in a semi-arid location. *Transactions of the ASABE* 61(2), 619–29. doi:10.13031/trans.12144.

Allen, R. G., Pereira, L. S., Raes, D. and Smith, M. 1998. Crop evapotranspiration - guidelines for computing crop water requirements. FAO Irrigation and drainage paper 56. FAO, Rome.

Ambrózy, Z., Daood, H., Nagy, Z., Darázsi Ledó, H. and Helyes, L. 2016. Effect of net shading technology and harvest times on yield and fruit quality of sweet pepper. *Applied Ecology and Environmental Research* 14(1), 99–109. doi:10.15666/aeer/1401_099109.

Brandle, J. R., Hodges, L. and Zhou, X. H. 2004. *Windbreaks in North American Agricultural Systems*. Springer, Dordrecht, pp. 65–78. doi:10.1007/978-94-017-2424-1_5.

Desmarais, G., Ratti, C. and Raghavan, G. S. V. 1999. Heat transfer modelling of screenhouses. *Solar Energy* 65(5), 271–84. doi:10.1016/S0038-092X(99)00002-X.

Dicken, U., Cohen, S. and Tanny, J. 2013a. Effect of plant development on turbulent fluxes of a screenhouse banana plantation. *Irrigation Science* 31(4), 701–13. doi:10.1007/s00271-012-0346-0.

Dicken, U., Cohen, S. and Tanny, J. 2013b. Examination of the Bowen ratio energy balance technique for evapotranspiration estimates in screenhouses. *Biosystems Engineering* 114(4), 397–405. doi:10.1016/j.biosystemseng.2012.11.001.

Fuchs, M. 1993. Transpiration and foliage temperature in a greenhouse. In: *International Workshop on Cooling Systems for Greenhouses*. Agritech, Tel-Aviv, pp. 55–68.

Gu, L., Baldocchi, D., Verma, S. B., Black, T. A., Vesala, T., Falge, E. M. and Dowty, P. R. 2002. Advantages of diffuse radiation for terrestrial ecosystem productivity. *Journal of Geophysical Research: Atmospheres* 107, ACL 2-1–ACL 2-23. doi:10.1029/2001JD001242.

Gu, L., Baldocchi, D. D., Wofsy, S. C., Munger, J. W., Michalsky, J. J., Urbanski, S. P. and Boden, T. A. 2003. Response of a deciduous forest to the mount Pinatubo eruption: enhanced photosynthesis. *Science* 299(5615), 2035–8. doi:10.1126/science.1078366.

Haijun, L., Cohen, S., Lemcoff, J. H., Israeli, Y. and Tanny, J. 2015. Sap flow, canopy conductance and microclimate in a banana screenhouse. *Agricultural and Forest Meteorology* 201, 165–75. doi:10.1016/j.agrformet.2014.11.009.

Ilić, Z. S. and Fallik, E. 2017. Light quality manipulation improves vegetable quality at harvest and postharvest: a review. *Environmental and Experimental Botany* 139, 79–90. doi:10.1016/J.ENVEXPBOT.2017.04.006.

Ilić, Z. S., Milenković, L., Šunić, L. and Fallik, E. 2015. Effect of coloured shade-nets on plant leaf parameters and tomato fruit quality. *Journal of the Science of Food and Agriculture* 95(13), 2660–7. doi:10.1002/jsfa.7000.

Israeli, Y., Hudi, G., Arzi, A., Nameri, N., Shapira, O., Levi, Y., Tanny, J., Cohen, S. and Schwartz, A. 2009. Saving irrigation water in a banana plantation in the Jordan Valley using shading screens. *Alon Hanotea* 63(3), 20–1.

Kitta, E., Baille, A. D., Katsoulas, N., Rigakis, N. and González-Real, M. M. 2014a. Effects of cover optical properties on screenhouse radiative environment and sweet pepper productivity. *Biosystems Engineering* 122, 115–26.

Kitta, E., Baille, A., Katsoulas, N. and Rigakis, N. 2014b. Predicting reference evapotranspiration for screenhouse-grown crops. *Agricultural Water Management* 143(0), 122–30. doi:10.1016/j.agwat.2014.07.006.

Kittas, C., Katsoulas, N., Rigakis, V., Bartzanas, T. and Kitta, E. 2012. Effects on microclimate, crop production and quality of a tomato crop grown under shade nets. *Journal of Horticultural Science and Biotechnology* 87(1), 7–12. doi:10.1080/14620316.2012.11512822.

Li, T., Heuvelink, E., Dueck, T. A., Janse, J., Gort, G. and Marcelis, L. F. M. 2014. Enhancement of crop photosynthesis by diffuse light: quantifying the contributing factors. *Annals of Botany* 114(1), 145–56. doi:10.1093/aob/mcu071.

Mahmood, A., Hu, Y., Tanny, J. and Asante, E. A. 2018. Effects of shading and insect-proof screens on crop microclimate and production: a review of recent advances. *Scientia Horticulturae* 241, 241–51. doi:10.1016/J.SCIENTA.2018.06.078.

Mekhmandarov, Y., Pirkner, M., Achiman, O. and Tanny, J. 2015. Application of the surface renewal technique in two types of screenhouses: sensible heat flux estimates and turbulence characteristics. *Agricultural and Forest Meteorology* 203, 229–42. doi:10.1016/j.agrformet.2015.01.010.

Möller, M. and Assouline, S. 2007. Effects of a shading screen on microclimate and crop water requirements. *Irrigation Science* 25(2), 171–81.

Möller, M., Tanny, J., Cohen, S. and Teitel, M. 2003. Micrometeorological characterization in a screenhouse. *Acta Horticulturae* 614, 445–51.

Möller, M., Tanny, J., Li, Y. and Cohen, S. 2004. Measuring and predicting evapotranspiration in an insect-proof screenhouse. *Agricultural and Forest Meteorology* 127(1–2), 35–51. doi:10.1016/j.agrformet.2004.08.002.

Möller, M., Cohen, S., Pirkner, M., Israeli, Y. and Tanny, J. 2010. Transmission of short-wave radiation by agricultural screens. *Biosystems Engineering* 107(4), 317–27. doi:10.1016/j.biosystemseng.2010.09.005.

Montero, J. I., Anton, A., Hernandez, J. and Castilla, N. 2001. Direct and diffuse light transmission of insect-proof screens and plastic films for cladding greenhouses. *Acta Horticulturae* 559(559), 203–10. doi:10.17660/ActaHortic.2001.559.29.

Pirkner, M., Dicken, U. and Tanny, J. 2014a. Penman–Monteith approaches for estimating crop evapotranspiration in screenhouses–a case study with table-grape. *International Journal of Biometeorology* 58(5), 725–37. doi:10.1007/s00484-013-0653-z.

Pirkner, M., Tanny, J., Shapira, O., Teitel, M., Cohen, S., Shahak, Y. and Israeli, Y. 2014b. The effect of screen type on crop micro-climate, reference evapotranspiration and yield of a screenhouse banana plantation. *Scientia Horticulturae* 180, 32–9. doi:10.1016/j.scienta.2014.09.050.

Romero-Gámez, M., Suárez-Rey, E. M., Castilla, N. and Soriano, T. 2012. Evaluation of global, photosynthetically active radiation and diffuse radiation transmission of agricultural screens. *Spanish Journal of Agricultural Research* 10(2), 306. doi:10.5424/sjar/2012102-260-11.

Seginer, I., Pirkner, M., Achiman, O., Mekhmandarov, Y. and Tanny, J. 2017. Aerodynamic characterisation of two large screenhouse roofs. *Acta Horticulturae* 1170(1170), 329–36. doi:10.17660/ActaHortic.2017.1170.40.

Selahle, K. M., Sivakumar, D., Jifon, J. and Soundy, P. 2015. Postharvest responses of red and yellow sweet peppers grown under photo-selective nets. *Food Chemistry* 173, 951–6. doi:10.1016/J.FOODCHEM.2014.10.034.

Shahak, Y. 2014. Photoselective netting: an overview of the concept, research and development and practical implementation in agriculture. *Acta Horticulturae* 1015(1015), 155–62. doi:10.17660/ActaHortic.2014.1015.17.

Tanny, J. 2013. Microclimate and evapotranspiration of crops covered by agricultural screens: a review. *Biosystems Engineering* 114(1), 26–43. doi:10.1016/j.biosystemseng.2012.10.008.

Tanny, J. and Cohen, S. 2003. The effect of a small shade net on the properties of wind and selected boundary layer parameters above and within a citrus orchard. *Biosystems Engineering* 84(1), 57–67. doi:10.1016/S1537-5110(02)00233-7.

Tanny, J., Cohen, S. and Teitel, M. 2003. Screenhouse microclimate and ventilation: an experimental study. *Biosystems Engineering* 84(3), 331–41. doi:10.1016/S1537-5110(02)00288-X.

Tanny, J., Haijun, L. and Cohen, S. 2006. Airflow characteristics, energy balance and eddy covariance measurements in a banana screenhouse. *Agricultural and Forest Meteorology* 139(1–2), 105–18. doi:10.1016/j.agrformet.2006.06.004.

Tanny, J., Cohen, S., Grava, A., Naor, A. and Lukyanov, V. 2009. The effect of shading screens on microclimate of apple orchards. *Acta Horticulturae* 807(807), 103–8. doi:10.17660/ActaHortic.2009.807.11.

Tanny, J., Cohen, S. and Israeli, Y. 2015. Covering agricultural crops with shade nets to increase water use efficiency – the case of a banana plantation. *Israel Agriculture* 2015, 8–9.

Tanny, J., Lukyanov, V., Neiman, M., Cohen, S., Teitel, M. and Seginer, I. 2018. Energy balance and partitioning and vertical profiles of turbulence characteristics during initial growth of a banana plantation in a screenhouse. *Agricultural and Forest Meteorology* 256–257, 53–60. doi:10.1016/j.agrformet.2018.02.028.

Teitel, M. 2007. The effect of screened openings on greenhouse microclimate. *Agricultural and Forest Meteorology* 143(3–4), 159–75. doi:10.1016/j.agrformet.2007.01.005.

Tennekes, H. and Lumley, J. 1972. *A First Course in Turbulence*. MIT Press, Cambridge, MA.

Waggoner, P. E., Boyd-Pack, A. and Reifsnyder, W. E. 1959. The climate of shade. *The Connecticut Agricultural and Experiment Station* 626, 5–39.

Wilson, J. D. and Flesch, T. K. 2003. Wind measurements in a square plot enclosed by a shelter fence. *Boundary-Layer Meteorology* 109(2), 191–224. doi:10.1023/A:1025414805676.

Chapter 4

Sustainable systems for integrated fish and vegetable production: new perspectives on aquaponics

B. W. Alsanius and S. Khalil, Swedish University of Agricultural Sciences (SLU), Sweden; A. Tadesse, Debre Berhan University, Ethiopia; A. K. Rosberg, K. J. Bergstrand, R. Hartmann, L. Mogren, M. Alam, M. Grudén and T. Naznin, Swedish University of Agricultural Sciences (SLU), Sweden; and A. Getahun, Addis Ababa University, Ethiopia

1 Introduction

According to the FAO, 'food security exists when all people, at all times, have physical, social and economic access to sufficient, safe and nutritious food which meets their dietary needs and food preferences for an active and healthy life' (FAO, 2009). Global food security is characterized by four major features, namely food availability, physical and economic access to food, food utilization and food stability (FAO, IFAD and WFP, 2013). As not only food quantity, but also the nutritional value of food is essential to public health, high-yielding production systems encouraging production of both animal protein and fish lipids and high-value horticultural produce high in fibres, minerals and bioactive compounds are highly interesting. One potential system is *aquaponics* (AP), integrating fish and seafood rearing (*aquaculture*) with soilless production of horticultural produce (*hydroponics*). Such engineered systems also meet the demands posed by some of the grand global challenges, such as global population growth, urbanization,

http://dx.doi.org/10.19103/AS.2019.0052.04

global climate change, limited access to resources for agricultural activities (e.g. water and land use) and sustainability.

The concept of multitrophic water-based culture systems, such as aquaponic systems, is ancient and has long been used for integrated production of fish and rice (Kangmin, 1988) and of shellfish and mangroves. Such systems are now attracting increasing attention, both economically and from a scientific point of view, in the case of mangrove systems for use in coastal areas (Chakravartty et al., 2017). Departing from a general perception of the aquaponic crop production component as wetland and the horticultural crop produced as a by-product, there is growing awareness of fish and plant products as equitable production lines. Aquaponic production systems are multitrophic (Fig. 1). Poorly planned implementation of hydroponics and aquaculture into aquaponics leads to suboptimal solutions for integrated land use and food production systems, an issue illustrated by the recently released definition by Palm et al. (2018). Irrespective of the use of low- or high-tech approaches, aquaponic culture systems are highly engineered. Optimization of the output from aquaponic systems, with respect to fish and plant produce quantity and quality and mitigated environmental impacts from production, requires a structured approach. This is incorporated by Palm et al. (2018), who define aquaponics as 'a production system of aquatic organisms and plants where the majority (>50%) of nutrients sustaining the optimal plant growth derives from waste originating from feeding the aquatic organism' (p. 818). In this chapter, flows of resources and loops of connected parameters within aquaponic systems are considered as the basis for a more structured course of action. This is a novel approach and creates an innovative platform to optimize

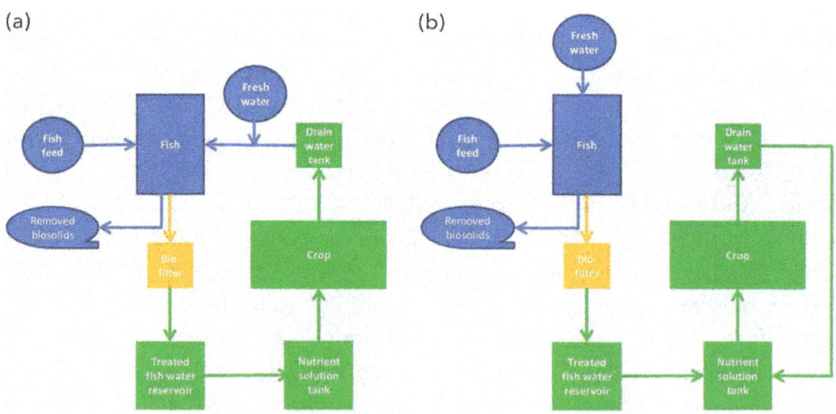

Figure 1 Simplistic design of coupled one-loop (a) and decoupled (b) aquaponic systems consisting of the fish rearing unit (blue), the water treatment unit (nitrogen conversion) (yellow) and the crop production unit (green). Illustration: B. Alsanius.

aquaponics from a fish and plant crop perspective. A general systems design of aquaponic systems, as featured in the majority of existing publications, is presented in Fig. 1. Multitrophic water-based culture systems have a high level of complexity and, apart from fish or seafood and plants, can encompass units for alternative fish feed production. For simplicity, the analysis provided in this chapter does not include the fish feed part, but is based on three distinct units: (1) fish rearing, (2) water treatment and (3) crop production.

2 Terminology of aquaponics systems

The term '*culture system*' encompasses the entire aquaponic system. '*Co-culture*' considers production of different trophic groups of organisms within the same culture system, whereas the production of two or more species within the same trophic group in the same culture system is called '*polyculture*'.

Based on systems design, Palm et al. (2018) categorize four types of aquaponic systems, namely (1) open aquaponics, entailing pond aquaponics, (2) domestic aquaponics, including mini-, hobby and backyard systems, (3) demonstration aquaponics, serving educational purposes, for example container solutions and living walls and (4) commercial aquaponics, with small-scale and semi-commercial aquaponics, urban gardening, rooftop aquaponics, large-scale aquaponics and aquaponic farming. Within the framework of this categorization, the soilless plant production unit is divided into two: (1) hydroponics in a narrow sense (*sensu strictu, ss*) using liquid soilless culture methods without growth medium, such as the nutrient film technique (NFT) or deep flow technique (DFT) and (2) hydroponics in a wider sense (*sensu latu, sl*) employing solid soilless culture with growing medium. Aquaponics can be based on one single loop (Fig. 1a), referred to as *one-loop aquaponics*, *coupled aquaponics* or *classical aquaponics* (Monsees et al., 2017), or separate loops for fish and crop production, called *decoupled aquaponics* (Fig. 1b).

Aquaponics have been studied by experts from different disciplines, such as aquaculturists, ichthyologists, horticulturists, plant scientists and ecologists, and the terminology differs between disciplines. In this introductory section, terms used in this chapter are briefly explained. The fish section (fish rearing unit) includes the fish rearing section (fish tank) and all input and output means (e.g. freshwater supply, fish feed, additives, bio-solids removal). In some publications, this section is called the aquaculture part (Rakocy, 2012) or rearing tank (Monsees et al., 2017). To distinguish between open or closed hydroponic systems, the section in this chapter discussing horticultural plant production is entitled the crop production section/unit. The term '*biofilter*' is limited to the unit converting and mineralizing organically bound nutrients to nutrients readily available for plants and microorganisms, while filter systems used for disinfection of the nutrient-enriched water, such as slow filters and sedimentation filters, are called the *disinfection unit*. This term also encompasses devices based on techniques other than filtration. Nutrition supplied to fish is named fish feed, whereas the term 'nutrients' refers exclusively to plant nutrients. Water is a major resource in aquaponics. To underline its potential for the entire production system, we call the nutrient-enriched water leaving the fish section (sometimes called waste water) '*fish tank effluent*' and, once it enters the plant section, '*nutrient solution*'. Likewise, the term '*drainage water*' is used to characterize nutrient-enriched water effluent from the plant section, but not sediment (Rakocy, 2012).

3 Aquaponics system design

3.1 Fish compartment

Aquaculture systems can be land-based, freshwater-based or marine-based and, depending on the nature of the site, either flow through or recirculating. As aquaponics is the subject of this chapter, the focus is on five land-based, recirculating systems, namely pond, tank, raceway, cage and pen fish culture systems. Systems design and choice of system are dependent on production intensity (extensive/traditional, semi-intensive, intensive) and on the environmental (water quality, temperature, oxygen) and trophic (carnivorous, omnivorous, herbivorous) requirements of the stocked fish (Edwards, 2013). The system can be open, semi-closed or closed (Tidwell, 2012b) (Table 1). Aquaculture systems have recently been comprehensively described by various authors (Tidwell, 2012a,b; Masser, 2012a,b; Mims and Onders, 2012; Fornshell et al., 2012; Tucker and Hargreaves, 2012; Ebeling and Timmons, 2012). These five systems differ in the level of habitat modification. In pond culture, fish grow in large water bodies formed by cut or modified land masses (earthen, concrete or lined with plastic sheeting) with groundwater, surface water or wastewater as a water source (Tidwell, 2012a). Fish reared in cage culture are contained in small mesh cages (all sides) of 1–4 m^3 volume suspended or submersed in an open water body (large pond, lake, ocean) (Tidwell, 2012b). In pen culture systems, the fish are held in 40 000–60 000 m^3 pens attached to the lake/ocean floor (Tidwell, 2012a). Raceway systems are defined as shallow, narrow and long canals (troughs) (30:3:1; L:W:D), with a flow rate which enables removal of solid waste (Stickney, 2017). In tank-based recirculating aquaculture systems, 90% of the water is reused (Stickney, 2017). Of these five systems, the tank-based approach is the most intensive fish rearing system.

The productivity in pond fish production systems is low compared with that of raceway and tank systems. The area of ponds commonly used in fish production industries ranges from 0.5 to 10 ha and pond depth is 1.5 to 1.75 m (Tucker and Hargreaves, 2012). The average production of fish per year with continuous supplemental feeding and aeration is 10 000 kg ha^{-1}, with 3–10 m^3 of

Table 1 Characteristics of various aquaculture systems of importance for integrated fish-plant (aquaponic) production

	Hydraulic properties				Solids		
Aquaculture system	Water use efficiency ($m^3\ kg^{-1}$)	Flow rate ($m^3\ h^{-1}$)* alt. mixing rate (% per day)**	Stocking density ($kg\ m^{-3}$)	Feed source	Accumulation	Removal	References
Pond culture	1.2-4.6	10-20*	2-10	Natural[a] complementary (floating, sinking)[b]	High accumulation from siltation, faeces, uneaten diet and sedimentation	Settling basin, clarifier, sand filter	Tucker and Hargreaves (2012)
Cage and pen culture	100-1000	<250*	3.5-13.5	Natural[c] complementary (floating, sinking)[d]	Medium accumulation from uneaten diet and faeces	Movement between cage water and open water column	Masser (2012a)
Raceway culture	98	15-36**	32	Artificial diet (sinking, pelleted)	Low waste accumulation from siltation, faeces and uneaten feed due to the low hydraulic retention time or residence time	Clarifier, settling basin, protein skimmer	Stickney (2017)
Recirculating aquaculture	1.4	2-21**	100-300	Artificial (floating pellet) Biofloc developed in the system	High accumulation from faeces and uneaten feed	Clarifier, settling basin, protein skimmer, rotating biological contactor, filter membrane	Edwards (2013), Masser (2012a), Pillay and Kutty (2005)

[a] Natural feed includes bacteria, phytoplankton, zooplankton and invertebrates in freshwaters.
[b] For intensive production, complementary artificial feed and fertilization of pond will be provided.
[c] Natural feed includes bacteria, phytoplankton, zooplankton and invertebrates in freshwaters and marine waters.
[d] For intensive production, complementary artificial feed, either floating or sinking pellets, will be provided.

water exchange per kg fish produced (Ebeling and Timmons, 2012). However, raceway systems (commonly used for salmonids) can produce 270 000 kg ha^{-1} (27 kg m^{-3}) of fish, with 98 m^3 water exchange per kg fish produced (Tucker and Hargreaves, 2012). Well-managed tank production (flow through or recirculating) can achieve output of 1 200 000 kg ha^{-1} (120 kg m^{-3}) of fish, with only 0.1 m^3 water use per kg fish produced.

The nutrient load in fish wastewater varies with carrying capacity (stoking density), feeding rate and water exchange rate. Raceway, pond and recirculating systems produce 9.6, 0.2 and 0.01 L fish tank effluent per hour. As recirculating aquaculture systems have the highest carrying capacity (stocking density) and employ biofilters for conversion of organic nitrogen, the effluent water is more suitable for replacing the hydroponic nutrient solution for soilless plant production than the effluent water from pond and raceway aquaculture systems. In intensive recirculating aquaculture systems, 5–10% of the total water volume needs to be exchanged daily due to high nutrient load (Ebeling and Timmons, 2012), unless a biological treatment unit (biofilter) for nitrogen and phosphorus removal is included.

3.2 Water treatment compartment

To reduce the load of particulate matter and decrease high levels of ammonia in intensive aquaculture, closed systems need to be equipped with a solid removal unit (clarifier) and an ammonia oxidizer unit.

In aquaponics, the solid (particulate) wastes mainly consists of fish faeces and uneaten feed (Rakocy, 2012). Waste solids can be classified into four categories: (1) settleable, (2) suspended, (3) fine floatable and (4) dissolved solids. In aquaponic systems, the first two are the primary concern, but dissolved organic solids can become a problem in systems when the flow rate is low (Ebeling and Timmons, 2012). Despite some positive attributes, where for example particulates act as an attachment site for bacteria and as a source for biofloc fish feed generation, large amounts of particulate matter cause increased biological oxygen demand and ammonia concentration in the system.

Settleable solids are generally the easiest to deal with and should be removed from the tank and filtration components as rapidly as possible using a sedimentation tank, swirl separator (clarifier) and/or mechanical filter (granular or microscreen). Unlike settleable solids, suspended solids are difficult to remove by sedimentation due to a density gradient, and therefore microscreen and fine granular media filters (sand filter) are required (Ebeling and Timmons, 2012). Fine suspended solids and dissolved solids are the most difficult particulate wastes to deal with in aquaponics. They constitute a large proportion of total particulate waste and have negative impacts on fish (gill irritation), biofilters (oxygen competition) and plants. Fine and dissolved solids cannot be easily removed by sedimentation or mechanical filtration technology, and instead protein skimming or foam fractionation needs to be employed. Particulate matter removal thus demands a combination of different technologies.

Through biofiltration, high levels of ammonia are converted biologically in a sequential two-step process (nitrification) to nitrate by chemoautotrophic bacteria (*Nitrosomonas*, *Nitrobacter*). During the first step, *Nitrosomonas* convert ammonia to nitrite, which is only slightly toxic to scaled fish and moderately toxic to scaleless fish. In the second step, *Nitrobacter* convert nitrite to nitrate, which is a major plant nutrient required at a concentration of 50–280 mg L^{-1} in the form of nitrate (Rakocy, 2012). This process can be mediated by nitrifiers either attached to biofilm or suspended in the water column. Biofilm-based systems are regarded as more stable and efficient than systems where nitrifiers are suspended in the water column (Gutierrez-Wing and Malone, 2006). Biofilm-based biofilters employ carrier material in either fixed (settled) or moving beds. Biofilm-based biofiltration systems integrated into aquaculture systems have recently been reviewed by Drennan II et al. (2006), Rusten et al. (2006) and Suhr and Pedersen (2010). Total ammonia nitrogen removal is a function of filter medium volume, filter-specific surface area, biofilm diffusion and the oxygen level, trophic level, and physical and chemical properties of the effluent water (temperature, salinity, pH, alkalinity).

3.3 Crop production compartment

In conventional hydroponic systems, plants are grown in water enriched with water-soluble fertilizers. The level of nutrients, pH and electrical conductivity are continuously adjusted according to crop demands. Efficient use of water and nutrients and high crop yield are decisive for such systems. These cropping systems are characterized by high yield per unit area (ten-fold the yield of field cropping systems) and by high produce quality (Gruda, 2009). From a systems design perspective, two general approaches can be distinguished: (1) solid hydroponics and (2) liquid hydroponics. In solid hydroponics, the soil is replaced with a limited volume of growing medium (*e.g.*, peat, pumice, perlite or rockwool). In liquid hydroponics, there is no growth medium. The plants are anchored mechanically and the roots grow into, and maintain continuous contact with, the nutrient solution. Hydroponic systems are applied in greenhouse horticulture, but also increasingly in the field, using gutters, cultivation beds or potting containers. They can be organized horizontally on the greenhouse floor or on the ground, but also as hanging gutters and beds, or on tables. Vertical and three-dimensional (3D) organization offer more efficient use of space, but require different solutions to improve light and climate conditions in the crop cultivation unit. Hydroponic systems are engineered systems with different requirements on power supply, monitoring and maintenance (low-tech, high-tech).

Distribution systems for water and nutrients differ between these basic approaches. Plants are irrigated either by drip, capillary and subirrigation, by shallow and continuous water flow (NFT), or by larger, continuously moving water volumes (DFT). The highest form of specialization is aeroponic systems, where the plant root is suspended in a root chamber without growing medium or a steady supply of water and nutrients. Instead, water and nutrients are sprayed in pulses on the root and excess nutrient solution is drained from the root chamber. Table 2 shows the technical features of different hydroponic systems. The choice of systems design is dictated by, among other factors, the crop grown. In general, all crops can be grown hydroponically. The most common hydroponically grown food crops are tomato, cucumber, sweet pepper, eggplant and different types of leafy vegetables such as lettuce, rocket, Swiss chard as well as herbs (e.g. basil, parsley, chive, coriander, dill). However, hydroponic systems are also used for other vegetable plants, such as French beans, radish, kohlrabi and berries (e.g. strawberry, blueberry, blackberry, raspberry) grown under controlled conditions as early vegetables or off-season, if economically competitive. Some ornamentals, including nursery stock plants produced in commercial settings, are also grown in hydroponics. Urban horticulture solutions (green walls, extensive green roofs, roof gardening) are based on hydroponic principles. In intensive horticulture with vegetables, the single water/nutrient loop considers only one crop species, but crop diversification within the same greenhouse can be achieved by using several parallel water/nutrient loops. In intensive production of ornamentals and herbs, and in urban settings, several crop species are served by the same water/nutrient loop.

Hydroponic systems are organized in either open systems or closed systems with recirculation of the nutrient solution. Closing the nutrient and water loop prevents run-off of nutrient-rich irrigation water into the soil and catchment. However, a risk of dispersal of root diseases is a concern in closed systems (Stanghellini and Rasmussen, 1994). Systems design therefore needs to consider the crop grown, its disease profile, its economic value and the potential economic impact of crop losses. During the past 20 years, solutions to mitigate the risk of dispersal of plant pathogens by the recirculating nutrient solution have been developed. Two principles are involved: (1) prevention of pathogen transmission by the recirculating nutrient solution and (2) prevention of pathogen transmission between plant roots. In spot disinfection, measures that either kill or remove the target pathogen are applied at one specific place within the system. They can comprise thermal treatment (pasteurization), irradiation (UV treatment), oxidation (O_3, H_2O_2, photocatalysis), use of halogens (Cl, B, I), use of heavy metals (Cu, Ag) and filtration (slow filtration, sedimentation

Table 2 Technical features of different system approaches in hydroponics. Suggestions for suitable spatial organization, that is horizontal, vertical or three-dimensional are presented

System	Slope (%)	Flow rate (*L m^{-2} h^{-1} for solid hydroponics alt. **L min^{-1} for liquid hydroponics)	Water volume (L $plant^{-1}$ day^{-1})	Irrigation events (# day^{-1})	Power dependence	Plants (m^{-2})	Suitable organization (H, horizontal; V, vertical; 3D, three-dimensional)*
Solid hydroponics							
Gravel culture	2.5–5 (Resh, 2013)	NA[a,b]	Depending on crop and growing stage	Winter: 3–4 (from 8 am to 8 pm)[b,c] Summer: At least 12 (from 6 am to 7 pm)[b,c]	High	Depending on crop and growing stage	H
Gutter[d]	0.2-0.5 (Herbold, 1994; Lieth and Oki, 2008)		Depending on light intensity and leaf/transpiration area (Lieth and Oki, 2008) 1.09 (tomato, closed system) (Herbold, 1994)	Depending on light intensity and leaf/transpiration area (Lieth and Oki, 2008) Last irrigation: 2.5-3 h before sunset (Resh, 2013)	High	2-5 (Van Os et al., 2008)	H, 3D
Subirrigation	0.2–0.5 (Herbold, 1994) 0.25–1 (Mackroth, 1998)	Depending on gutter width, slope, distance between pots (longitudinal distribution) 2.0 (Mackroth, 1998; Montesano et al., 2015)	NIA[e]	Depending on the growing medium's hydraulic conductivity (Göhler and Molitor, 2002) and presence of capillary mat (Mackroth, 1998) Without capillary mat: 0.5–1 (Löfkvist et al., 2009) With capillary mat: 0.25 (Löfkvist et al., 2009)	High	Potted plants: 30–45 (Löfkvist et al., 2009)	H

(Continued)

Table 2 (*Continued*)

System	Slope (%)	Flow rate (*L m^{-2} h^{-1} for solid hydroponics alt. **L min^{-1} for liquid hydroponics)	Water volume (L $plant^{-1}$ day^{-1})	Irrigation events (# day^{-1})	Power dependence	Plants (m^{-2})	Suitable organization (H, horizontal; V, vertical; 3D, three-dimensional)*
Ebb and flow (bench basin)	Sloping profiled surface within the tray 0.2–0.4 (Van Os et al., 2008)	See specifications below[f,g]	See specifications below[h]	2	High	Depending on crop, production goal and growing stage	H
Tank culture	NA	2–3*[i]	Depending on crop and growing stage	Continuous	High	Depending on crop and growing stage	H
Liquid hydroponics							
Nutrient film technique	1–4[j]	Depending on gutter length; increased flow rate with longer gutters Gutter length 12 m: 2* (Cooper, 1979) Gutter length 45 m: 6–8* (Göhler and Molitor, 2002)	1.6	Continuous	Very high	Flexible[k]	H, V, 3D

Deep flow system	2 (Blok et al., 2017)	10–14* (Alsanius, 2011)	11	Continuous	Very high	Tomato: 2.3 (Gertsson et al., 1994)	H
Aeroponic	NA	0.65* (Göhler and Molitor, 2002) 2* (Van Os et al., 2008)	NA	Depending on the position of the nozzles (container bottom, root zone) 60 h^{-1} (Tan et al., 2002) 20–30 h^{-1} (Maloupa, 2002)	Very high	Depending on crop, production goal and growing stage	H, V, 3D

[a] Not applicable.
[b] Total duration of filling and drainage cycle: 20-30 min (Resh, 2013).
[c] Depending on aggregate particle size and surface, crop, crop size, climatic factors, time of day (Resh, 2013).
[d] Using rockwool (rockwool slab sizes (L*W*H): 0.9 m*0.15-0.2 m*0.075 m, tomato, sweet pepper; 0.9 m*0.2-0.3 m*0.075, cucumber) (Göhler and Molitor, 2002).
[e] No information available.
[f] Duration of flooding 10-30 min given that the growing medium has a high unsaturated hydraulic conductivity (Van Os et al., 2008).
[g] Recommended duration for filling, flooding and drainage of the basin: 5 min, 5-10 min and 10 min respectively (Mackroth, 1998).
[h] 2-5 cm of the pot bottom submerged during flooding.
[i] For a basin volume of 3.6 m^3 (L*W*D: 30 m*0.6 m*0.2 m); exchange rate: 1 day^{-1} (Resh, 2013).
[j] No upper limit for slope (Cooper, 1979).
[k] Depending gutter width, systems design (single or multiple rows), crop.

filtration, membrane filtration) (Hong et al., 2014). These measures may be target-specific or affect the entire microbiota inhabiting the nutrient solution. Dispersal of root pathogens between plants that share the same volume of growth medium can be counteracted by using biological control agents (BCA), surfactants (including biosurfactants) and ClO_2.

3.4 Combined systems for fish and horticultural crop production

As multitrophic culture systems, aquaponic systems are highly affected by synergies or conflicts between different system compartments (units) and parameters (water quality, nutrient, temperature, oxygen), and by the product (nutrient) or function (biological, chemical, physical) of the other system. Production goals (fish species, crop species, yield, quality and timing) have to be considered when choosing management practices and production intensity. These factors dictate the choice of production system, level of complexity and technology. The ratio between crop and fish is another key factor when integrating aquaculture and hydroponic systems.

Cultivation in aquaponic systems is performed using coupled, one-loop systems in which water is directly transferred from the fish rearing unit to the crop production unit, and the same water quality is shared by these two units (Monsees et al., 2017). However, this poses challenges regarding water quality and nutrient supply to the plants, and thereby to the use of aquaponics on a commercial basis. Decoupled systems (Fig. 1b) have therefore been developed as a step towards commercial application with efficient use of water and nutrients (Monsees et al., 2017). Decoupled systems use an approach where water is circulated between the aquaponics units using two separate loops, leading to efficient use and control of water and nutrients. Goddek and Keesman (2018) recommend use of desalination technology in decoupled systems, as this could improve the nutrient balance, leading to increased productivity.

The choice of hydroponic unit and constituents used in aquaponic systems are critical for plant growth and yield. For example, the biomass of mint and lettuce is promoted when they are grown in crumbled stone and crumbled stone supplemented with coco fibre as the growth medium, compared with liquid hydroponics or river stone (Jordan et al., 2018; Shete et al., 2017). On the other hand, the type of hydroponic method, for example NFT, drip irrigation or floating raft culture, is reported to have no effect on the productivity and quality of tomato (Schmautz et al., 2016).

The amount of feed needed is related to the amount and quality of the fish feed and is affected by the number of fish and plants as well as the nitrogen content and assimilation in the system. Lennard and Lennard (2006) review approaches used to calculate feeding rate ratio related to plant numbers and growing area and calculated 1 kg of fish food per day to grow 1500 heads of lettuce. The Rakocy feeding rate ratio is based on the amount of fish feed per m^2 of crop production area and day (Rakocy, 2007), which ranges between 60 and 100 g m^{-2} day^{-1}. In this ratio, the feed is supplied to an amount that meets plant requirements, leading to an excess of nitrogen in the form of ammonia in the system. The Lennard approach is based on the nitrogen balance in the system, where the amount of nitrogen in the feed is equal to the amount of nitrogen in the form of nitrate used by the plants – a so-called offline approach where feed rate is based on active and regular nitrate concentration determination. Recently, Petrea et al. (2013a,b) showed that high plant densities and good light intensity are needed for effective nitrogen assimilation in aquaponic systems, and at the same time plant density would have an effect on plant growth parameters, plant length and leaf area.

System size is discussed by Lennard and Lennard (2006) that might be based on (1) the ratio between the volume of water in the fish tank and the volume of the nutrient solution or (2) the ratio between the fish tank volume and the growing medium volume used in the crop production unit. Feed frequency and photoperiod are other factors that affect fish and plant growth in aquaponic systems (Liang and Chien, 2013). Good water quality and growth can be achieved by higher feeding frequency in the fish rearing unit and longer photoperiod in the crop production unit.

For water quality, temperature, dissolved oxygen, carbon dioxide, pH, water flow rate and nitrogen wastes are of great importance (Lennard and Lennard, 2006). Temperature in the cultivation system not only affects growth parameters, but also the efficiency of the biofilter and feed conversion efficiency. In addition, the high pH value in the system is also a crucial parameter that affects nutrient uptake in the plants. As discussed by Masser et al. (1999), this parameter is not crucial in systems rearing warm water fish species but is a major consideration in aquaponic systems designed for cold water fish species. Furthermore, careful maintenance of pH is required in aquaponic systems where different pH ranges are present, for example for fish 6–9.5 (Lennard and Lennard, 2006), for the biofilter 7–8 (Tokuyama et al., 2004) and for plants 5.5–6 (Göhler and Molitor, 2002). A value below six reduces the efficiency of biofilters due to acid formation. In addition, high water flow rate in aquaponic systems can have a positive effect on water quality parameters and nutrient removal (Endut et al., 2010).

Aquaponic systems are variable with regard to size (small- and large-scale) and technological input. As aquaponics is still in the infancy as regards commercial implementation, the level of technological input is limited and offers considerable potential for development. System carrying capacity is determined by the processing efficiency of each of the three distinct units, which in turn is a function of the design and management. Systems with low carrying capacity (<45 kg fish m^{-3}) and with conventional water treatment (such as clarifiers and biological contactors) units are categorized as low-tech aquaponics and systems with high carrying capacity (>45 kg fish m^{-3} fish) and innovative technological additives (such as liquid oxygenation, UV treatment, protein skimmer, low head oxygenation) as high-tech aquaponics.

3.4.1 Low-tech aquaponics

Essentially, low-tech systems focus on optimized production of fish and plant biomass based on fish feed ratio. Using small-scale systems with low technological input increases the possibility of using locally produced food (Pade and Nelson, 2007). Low-tech systems are based on a simple design with basic, locally available and cheap system components, mechanical system solutions for water flow and aeration, and low demand for electricity. In general, the fish stocking density is low. The low-tech aquaponic setup integrates a fish rearing unit with biofilter and clarifier attached. The biofilter is connected to the crop production unit and recycles drained nutrient solution from that unit back to the fish rearing unit (Rakocy, 2012). Systems design varies with respect to the level or outlet of the crop production unit and fish tank, and the site of pump attachment. Low-tech aquaponics often makes use of single-loop system (Fig. 4) using ebb and flow, NFT, DFT or floating raft systems in the hydroponic unit. However, productivity is limited. Effluents are not efficiently controlled and removal of solids is a challenge. Some effluents may be released to the environment. Recently, Palm et al. (2018) characterized aquaponic systems as domestic, backyard, small-scale systems. Within that characterization, low-tech aquaponics can include domestic systems and backyard aquaponic systems.

3.4.2 High-tech aquaponics

The highest, most sustainable and most predictable production can be expected to be achieved by aquaponics with maximum advanced technological integrity. In high-tech systems which also include decoupled systems, the fish rearing unit serves as a site for both fish rearing and solid waste removal (Cornel dual drain system; removal of solid waste 90%; water exchange 10%) (Ebeling and Timmons, 2012). The efficiency may be further improved by integrating additional trophic groups using the removed solids (sludge concentrate) for biofloc for in-site fish feed production. The water treatment unit integrates the biochemical conversion of fish waste to plant nutrients and physio-chemical changes of minerals from fish feed. Hence, the nutrient budget of the system is expected to have a net balance of zero (Yogev et al., 2016). To maximize the treatment efficiency, a moving bed bioreactor, rotary screen, sand filter and membrane filter can be used for solid and soluble waste removal (Ebeling and Timmons, 2012). Drainage water released from the crop production unit may be further improved chemically (ion exchange) or biologically (UV radiation) (Gendel and Lahav, 2012) to match the fish demand (Ghafari et al., 2008). Integration of technology for automated regulation of biological, chemical and physical factors allows optimization of input factors, prognosis and timing of production.

4 Flows in aquaponics systems

4.1 Water loop

Water is the medium connecting the fish rearing and crop production units in aquaponics. However, water is considered differently in these two parts of the system. In the fish rearing section, water is a habitat, carrier of fish feed and additives, and matrix for discharge of animal faeces. In the crop production section, water is an input and precondition for plant growth and development (irrigation water) and a solvent for plant nutrients and plant protection products (PPP). Water used in the fish and crop units needs to be of high quality for production of healthy animal- and plant-based foodstuffs. However, the definition of water quality is not consistent for high-quality fish farming and plant production, but is based on the demands of the species grown. Due to different disciplinary traditions and information value, water quality in aquaculture and hydroponics and in environmental sciences is assessed by different tools, which complicates system comparison. Among other factors, the form and concentration of nitrogen, especially ammonia, are important for water quality (Klinger-Bowen et al., 2011). Ammonia in water occurs in either un-ionized (NH_3 or UIA) or ionized (NH_4^+) form and is referred to as total ammonia nitrogen (TAN; unit: mg L^{-1} or ppm). Ammonia must be converted to nitrate (NO_3^- ; mg L^{-1} or ppm) to become available to plants.

Tolerance limits in fish production are <1 mg L^{-1} of NH_4^+-N, <0.2 mg L^{-1} of NO_2^--N and <150 mg L^{-1} of NO_3^--N, while pH should be kept within the range of 7–8. Data from various experiments show that these limits are met in aquaculture of tilapia and perch for most parameters except nitrite and oxygen content, although some parameters may be close to the threshold value (Graber and Junge, 2009).

Water quality is not as critical for the crop production unit in aquaponic systems, which frequently takes the form of wetland to bioremediate water quality rather than producing high-quality horticultural produce. The crop grown dictates chemical and physical water quality parameters. Plants can cope with pH levels between 4 and 8, but to ensure optimal nutrient uptake and optimal content of dissolved nutrients, pH > 6 should be avoided in hydroponics (Göhler and Molitor, 2002). Macro- and microelements in the fish tank effluent serving as plant macro- and micronutrients must be accounted for when adjusting the plant nutrient solution to optimal conditions. However, some elements, for example sodium and chloride, are detrimental to some horticultural crops when present in elevated concentrations or to all crops on exceeding threshold values (Sonneveld, 2000). From the perspective of crop production, nutrient imbalance rather than nutrient load is the main problem with respect to chemical water quality.

4.1.1 Temperature

An adequate temperature regime in the water recirculating between the fish and plant culturing units is essential for high-quality food production as temperature affects physical factors in both the root environment and the corresponding plant response. Electrical conductivity increases and oxygen availability decreases with increasing temperature in the nutrient solution. Plants respond with depressed growth to both suboptimal and supraoptimal temperatures. Temperature optima are crop dependent and must be adjusted to light intensity. This might be a reason for the conflicting results presented in the literature which make it difficult to draw general conclusions. Plant growth can be impaired at root zone temperatures <15°C (Martin and Wilcox, 1963; Wilcox et al., 1962) and promoted by higher temperatures. For example, a root zone temperature of 24°C leads to higher biomass production by butterhead lettuce (*Lactuca sativa* L. cv. Ostinata) compared with 17°C and 31°C, although the air temperature regime needs to be synchronized (Thompson et al., 1998). In the root zone of hydroponic cropping systems, temperatures between 24°C and 30°C increase plant growth, leaf development and carbon assimilation in cucumber and tomato (Hurewitz and Janes, 1983; Gosselin and Trudel, 1985), but decrease yield in cucumber (*Cucumis sativa* cv. Toska) and tomatillo (*Physalis ixocarpa* Brot. ex Hornem) (Gosselin and Trudel, 1985; Diáz-Pérez et al., 2005). Root zone temperature also affects crop tolerance to pathogens, for example the fungi-like root pathogen *Pythium aphanidermatum* is tolerated by tomato plants at 20°C, but not at higher temperatures (Panova et al., 2004), while disease development increases in pepper at temperatures >23°C (Sopher, 2012). Fungal root pathogens also respond to elevated temperatures (O'Brien and van Bruggen, 1993). Root zone cooling reduces plant disease attack (Jensen and Malter, 1995). The temperature in the root environment must be optimized with respect to plant yield.

Water temperature strongly influences feeding and growth of cultured fish (Stickney, 2017). The water temperature preferences of most cultured fish are well known and can be maintained year-round in recirculating aquaculture systems. In general, cultured fish species can be classified as either cold water species that prefer temperatures of 10–18°C (trout), cool water species that prefer temperatures of 18–25°C (yellow perch) and warm water species that prefer temperatures of 25–32°C (channel catfish) (Stickney, 2017). Water temperature also influences the water quality processes occurring in recirculation systems. For example, the optimum water temperature for bacterial nitrification activity is about 29°C, which may or may not be optimal for the cultivated fish species (Goddek et al., 2015).

Energy conservation is one of the major advantages of recirculating aquaculture systems. Once the tank water is heated to the optimal temperature for fish growth, only a small amount of heat energy is required to maintain the temperature due to limited heat loss. However, heat losses from the building and from the water via evaporation, splash out or wastewater must be minimized to ensure economic success. Heat losses through convection, conduction and heat radiation are minimal in a well-insulated building. Heat can be provided by heating the air in the building or by directly heating the water (Goddek et al., 2015).

4.1.2 Oxygen

Differences in terminology and available methods for measurement of oxygen content limit the generalization of results in different studies, and thereby conclusions on optimization of oxygen conditions in aquaponics. Oxygen content, expressed as mg O_2 L^{-1}, denotes the amount of dissolved oxygen (DO), whereas oxygen saturation (%) is a relative measure indicating the upper limit of oxygen that can be dissolved under given conditions (e.g. temperature). The biochemical oxygen demand (BOD; mg O_2 L^{-1}) is defined as 'the amount of oxygen required by bacteria while stabilizing decomposable organic matter under aerobic conditions' (Lester and Birckett, 1999). It is determined after either five (BOD_5) or seven (BOD_7) days of incubation. The chemical oxygen demand (COD; mg O_2 L^{-1}) describes the 'total quantity of oxygen required to oxidize substances to carbon dioxide and water' (Lester and Birckett, 1999). Oxygen content in the aquaponics water loop is a function of physical factors, such as temperature; chemical factors, such as the load of organic compounds; and the associated macro- and microorganisms. In the fish unit, oxygen content must be >6 mg O_2 L^{-1}. Oxygen is a limiting factor for ammonification and nitrification (Alsanius and Jung, 2018). Göhler and Molitor (2002) identify a BOD level of 10 mg O_2 L^{-1} and a COD level of 60 mg O_2 L^{-1} as crucial threshold parameters for water quality in hydroponics. In the root zone, there is constant competition between the plant root and root-associated microbiota (Alsanius and Wohanka, 2018; Alsanius et al., 2014). Aerobic conditions need to be maintained adequately for both water and nutrient uptake processes (Flannery and Lieth, 2006) and for microbial organic carbon utilization and degradation (see Sections 4.2.2 and 4.3). However, depending on the root density and external factors in the root zone, small, confined oxygen-depleted sites (*microloci*) may develop (Schroeder and Engwicht, 2005). Oxygen-depleted sites not only affect plant physiological parameters, but also favour certain plant pathogens (Chérif et al., 1997). In hydroponic systems, a good oxygen regime can be maintained by appropriate systems design. Further studies are needed to determine whether mechanical solutions are satisfactory in aquaponics, or whether there is a need for additional aeration in the water loop.

Apart from the impact on fish and plant well-being, dissolved oxygen concentration and aeration are crucial for nitrification and denitrification. It has been shown that changes in aeration patterns (50% lower O_2 flux compared with control and intermittent O_2 supply) reduce gaseous nitrogen emissions by approximately 30% and 50% from the fish rearing unit and by approximately 50% and 70% from the crop production unit compared with ambient air flux (0.8 L h^{-2}, continuous flow) (Fang et al., 2017). Thus aeration must be optimized with respect to product outputs and gaseous emissions.

4.1.3 Electrical conductivity

Electrical conductivity (EC; mS cm^{-1}) indicates the concentration of dissolved charged ions that are able to conduct electric current and is thus a measure of the osmotic load. EC is affected by temperature. In the fish rearing unit, the EC level should not exceed 1.2 mS cm^{-1}. In the crop production unit, the EC is dependent on the crop grown, plant developmental stage and product quality (lettuce: 2.3 mS cm^{-1}; tomato: 2.2 mS cm^{-1}; cucumber: 2.2 mS cm^{-1}; roses: 2.2–2.4 mS cm^{-1}; gerbera: 1.7 mS cm^{-1}) (Sonneveld and de Kreij, 1994; Göhler and Molitor, 2002). Elevated EC values are also sometimes used to increase quality in fruit-bearing crops (Sonneveld and Welles, 1988; Hao et al., 1998). Non-desired elements may also affect the EC. For example, an increase in sodium (Na^+) and chloride (Cl^-) in the nutrient solution also leads to increased EC. Salinity may cause crop losses as a consequence of osmotic stress, nutrient imbalance and cation antagonism.

4.2 Trophic loops

4.2.1 Nutrients

The aquaponic nutrient pool is characterized by nutrient sources (fish feed, plant fertilizers, freshwater) and nutrient converters (fish, target crop and other autotrophic heterotrophic microbiota associated with biofilters) (Schneider et al., 2005). The desired quality and quantity of the main nutrient input (fish feed) depends on the fish type, size and age, but all fish have a high protein requirement during early growth stages. Herbivorous and planktivorous fish (e.g. grass carp) have a protein requirement (% of dietary protein) of 18–23%, omnivorous fish (e.g. Nile tilapia) have a protein requirement of 24–33% and carnivorous fish (e.g. trout) have a protein requirement of 35–50% (Cho et al.,

1999). Excretion of nitrogenous waste depends on fish feeding behaviour and feed quality, so carnivorous, omnivorous and herbivorous fish have high, medium and low nitrogen excretion ability, respectively (Cho et al., 1999).

The protein requirements of fish decrease with age. For example, the protein requirements of larval, juvenile and adult tilapia are reported to range between 35–40%, 30–40% and 20–30%, respectively. African catfish fry and fingerlings require 35–40% protein in their diet, while grown catfish require 30–35% protein. The estimated protein requirement for catfish fry, grow-out and broodstock is 35–40%, 25–35% and 28–32%, respectively (Cho et al., 1999).

Fish species, feed integrity and quality, especially with respect to energy density, amount of fish feed added in relation to stocking density, feeding level, fish age and feed conversion ratio (FCR = kg feed/kg fish meat gain) are decisive for biomass conversion and nutrient excretion by reared fish. As summarized by Schneider et al. (2005), FCR can vary between 0.71 and 3.00. Excretion of nitrogen and phosphorus is estimated to range between 14 and 48 g N kg^{-1} feed and 3 and 5 g P kg^{-1} feed, respectively, depending on fish species, feeding regime and feed qualities (Schneider et al., 2005). To meet crop demands and to optimize crop growth and development, crop yield, quality and timing of the crop, the ambient nutrient content in the water after passage through the biofilter needs to be supplemented with readily available nutrients.

Besides water, light and carbon dioxide, plants need nutrients. Nutrients are used as building blocks in organic matter production during photosynthesis, such as formation of cell walls and storage compounds, are constituents in enzymes needed for plant functions and are involved in electron transfer (Taiz and Zeiger, 2006). Nutrients are primarily taken up via roots, as ions are dissolved in water. Thirteen elements are generally considered essential and are grouped into macronutrients (N, P, K, S, Ca, Mg) and micronutrients (B, Cl, Cu, Fe, Mn, Mo and Zn) (Marschner, 1995). In addition to these elements, there are elements considered beneficial.

Plant nutrient requirements vary between plant species and phenological stages. In general, the N:K ratio is higher during vegetative growth of the crop and decreases during the generative stage (fruit set and onwards). Conventional nutrient solutions for hydroponic systems are prepared from pure salts, for example KNO_3, NH_4NO_3, $MgSO_4$ and so on, or from acids (H_2PO_4, H_2SO_4). Micronutrients, such as Fe, are often added in chelated form (Göhler and Molitor, 2002). In this way, the nutrient solution of a hydroponic system can easily be adjusted to the changing crop demand and environmental conditions (temperature, light). Plant performance is not only a question of plant-available nutrient load, but also of the pH level. Optimal pH level varies for different nutrients (Marschner, 1995). The nutrient solution in hydroponic systems is adjusted to a pH of 5.5–6, depending on plant species and developmental stage. The nutrient solution produced from the fish tank effluent normally does not meet the demands of the crop with respect to pH or to the concentrations of different ions of interest from a plant nutrition point of view. Some nutrients such as K, Fe, Ca and micronutrients are insufficient in fish feed and need to be added to the crop production unit in aquaponic systems.

Most studies reported in the literature focus on nitrogen, which is the dominant element released by the fish and a macronutrient for plants, but high nitrogen levels do not necessarily ensure high biomass production (Dediu et al., 2012). Nutrient output from the fish tanks may vary with stocking density (Rakocy et al., 2006), while crop requirements may vary with developmental stage (vegetative, generative) (Roosta and Hamidpour, 2011). Supply of potassium is a major problem, as it is not needed or excreted by fish and is normally not administered through the fish feed (Graber and Junge, 2009). It therefore needs to be added to the nutrient solution. Microelements such as Zn, Cu, B and Mn are likely to be present in too low amounts in aquaponic systems, thus inducing deficiencies in crop plants and resulting in reduced yield and product quality. Accumulation of nutrients to levels toxic to fish or plants is another problem in aquaponic systems (Treadwell et al., 2010). Moreover, high concentrations of ammonium decrease uptake of K, Ca and Mg (Rakocy, 1997) and the high pH in the solution, while favouring ammonium and nitrate uptake, limits the uptake of other elements, such as the micronutrients Zn, Mn and Cu.

As discussed by Ru et al. (2017), fish feed has a low content of P, K, Fe, Mn and S, and thus it may be necessary to add macro- and micronutrients to aquaponic systems in order to compensate for nutrient imbalances and to improve growth and yield. Besides active supply of nutrients, plant nutrient availability may be enhanced through management strategies. For example, P availability increases as a consequence of bacterization (*Bacillus* spp.) (da Silva Cerozi and Fitzsimmons, 2016). Intercropping systems can also have positive effects on plant nutrient uptake and thus growth and quality of crops (Maucieri et al., 2017). Intercropping of lettuce with red chicory increases lettuce sugar content and reduces its acid content.

4.2.2 Nitrogen

Fish assimilate nitrogen in the form of protein with the feed. Part of this nitrogen is released through the gills as ammonia (NH_3), as a result of metabolic activity (Tyson et al., 2011). According to Ebeling and Timmons (2012), the amount of metabolized ammonia can be estimated as:

$$P = F * PC * 0.092 \tag{1}$$

where P is the ammonia production rate per day (kg * day^{-1}), F is the daily feed level (kg * day^{-1}) and PC is the protein content in the feed (percentage * 100^{-1}).

In water, ammonia occurs in two forms, which establish an equilibrium: $NH_4^+ \leftrightarrow NH_3 + H^+$. This equilibrium is influenced by water temperature and pH (Tyson et al., 2011). Ammonia is toxic to fish and therefore its removal is one of the main objectives in aquaponic systems. This is achieved by converting ammonia into nitrate, which can be assimilated by plants. This nitrification process takes place in a biofilter housing different bacterial species and comprises two main steps, ammonia oxidation and nitrite oxidation.

Nitrogen is the nutrient needed in the highest amount by plants (Taiz and Zeiger, 2006) and, after carbon, is the most important nutrient for microbial growth (Madigan et al., 2009). In aquaponics, there are three primary sources of nitrogen: (1) fish feed, (2) nitrogen fertilizers added to optimize plant growth and (3) autochthonous nitrogen. From the nitrogen pool, nitrogen may be consumed by the cultured fish and crop and taken up by biofilms and free-living microorganisms. Nitrogen may leave the aquaponic system as gaseous losses to the atmosphere or, depending on the system design, as solids. Organic and inorganic nitrogen released by the fish, residual and mineralized fish feed, and organic debris from dead and dying organisms and nitrogenous root exudates are fed back to the nitrogen pool through autochthonous nitrogen and may be utilized and fixed in planktonic biofilm or macrophyte biomass (Fig. 2). Ammonia is produced by fish as a waste product, which is excreted through the gills. In aquaculture and aquaponics, accumulation of ammonia is a major concern, since it can quickly reach levels that are toxic to the fish (Rakocy, 2012). Toxic levels can be avoided through nitrification, an aerobic oxidation process which converts NH_4^+ to NO_2^- and then to NO_3^-. Nitrification is facilitated by nitrifying bacteria and one species of *Archaea*, which is widely distributed in soil and water. The first step in nitrification, where ammonia is converted to nitrite, is carried out by a group of bacteria called nitrosifiers, of which *Nitrosomonas* is an important genus, and by the *Archaea* species *Nitrosopumilus*. The second step, where nitrite is oxidized to nitrate, is driven by another group of nitrifying bacteria, in which *Nitrospira* and *Nitrobacter* are important genera (Madigan et al., 2009). Nitrifying bacteria grow as a biofilm on inert material and on organic particles. A biofilter provides large surface areas for such biofilm formation. Biological nitrogen conversion is not restricted to the biofilter, but occurs at all places where recirculating water comes into contact with biofilms of nitrifying bacteria, for example inside pipes and tank walls and on plant roots (Rakocy et al., 2006). However, the conditions for optimum nitrogen conversion are in conflict with the growing conditions in soilless systems, due to the higher pH optimum required for nitrification (7.0–9.0). Rakocy (2012) suggests compromising on the pH level for plant production (pH 7). However, early cucumber yield is negatively affected when the crop is exposed to increased pH level (Tyson et al., 2008). Furthermore, pH (and temperature) is important for the relative concentrations of un-ionized

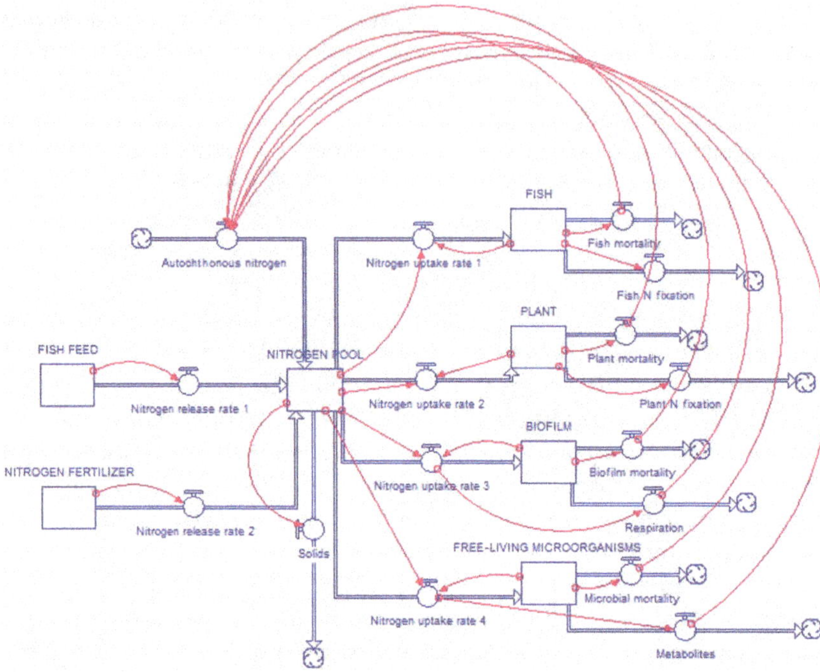

Figure 2 Simulation model conceptualized in Stella (isee systems inc.) for the functional organization of nitrogen flows in closed aquaponic systems. The boxes represent stocks of nitrogen and the blue arrows represent the flows. The red arrows are connectors which illustrate the factors regulating the flows. Inputs of autochthonous and allochthonous organic and inorganic nitrogen to the general nitrogen pool are provided by fish feed and additional nitrogen fertilizers. Nitrogen inputs support the growth of fish, plants, biofilm and microbiota. In return, the mortality of fish, plants and biofilm are large contributors to the autochthonous nitrogen pool. Illustration: A.K. Rosberg.

ammonia and ionized ammonium (NH_4^+). With increasing temperature or pH, the proportion of ammonia increases (Ebeling and Timmons, 2012). Due to the high toxicity of ammonia to the fish and plants and plant inability to take up ammonia, it is of the utmost importance to maintain an appropriate pH.

The amount of nitrogen in aquaponics can be directly related to the density of fish and the amount and protein content of the fish feed (Endut et al., 2010; Petrea et al., 2013a; Seawright et al., 1998). Approximately 10% of the protein in fish feed is converted to ammonia (Tyson et al., 2011). Seawright et al. (1998) noted that at higher fish densities nitrification, but also denitrification, occur, leading to nitrogen losses. Denitrification is an anaerobic process that takes place in solid particles, such as fish faeces. Unless solid particles are removed, there will be an increase in gaseous losses of nitrogen and more supplementary nutrients will be needed for plant growth.

4.3 Carbon loop

As is the case for oxygen, both the terminology and measures to estimate organic carbon content in natural and artificial water-based production systems differ. The total organic carbon content comprises both autochthonous and allochthonous carbon sources. These may be grouped based on their size into particulate organic matter (POM) and dissolved organic matter (DOM). As colloidal and small particles may pass through filters with pore size 0.20 and 0.45 μM, this grouping is of an operational, but not of a functional, nature (Münster, 1993). Coarse POM (cPOM) comprises organic matter >1 mm, whereas fine POM (fPOM) comprises particles between 0.20 μm and 1.00 mm in diameter. DOM, which is measured as dissolved organic carbon (DOC), represents organic carbon compounds smaller than 0.20 μm. They may consist of either living or dead material with differing resistance to degradation. An additional carbon source is volatile organic carbon. The organic carbon fractions in aquatic systems may also be grouped based on their bioavailability, into biodegradable DOC, semi-labile DOC, which is resistant to microbial attack, and recalcitrant DOC. With respect to molecular weight, DOC may be separated into low (<1000 Da) and high (>1000 Da) molecular weight fractions. Some researchers use indirect methods, such as COD and BOD measurement, for estimation of organic carbon content.

The carbon loop in hydroponic cropping systems has recently been described in detail by Alsanius et al. (2014). However, the carbon loop in aquaponics is more complex (Fig. 3). The organic carbon loop is closely related to the level of dissolved oxygen and the microbial loop.

The carbon pool in aquaponic systems has seven major sources, namely the macrophyte, phytoplankton, fish, fish feed, autochthonous carbon, additives to ensure plant or fish health (e.g. PPP; antibiotic compounds) and biofilm. Carbon may be fed into the system in dissolved or in particulate form. DOC may be released as deposits from plant roots (Rosberg, 2014), as metabolic waste products from fish (Seawight et al., 1998; Tyson et al., 2011), as non-utilized degraded fish feed or as metabolic phytoplanktonic compounds. Autochthonous organic carbon may originate from the raw water, the plant growing medium and the equipment (Alsanius et al., 2014). Chemical agents promoting fish and plant health contribute to the pool of DOC in aquaponic systems, but also affect carbon uptake ratio and mortality ratio in the system. Particulate organic carbon consists of plant debris, organic plant growing medium, phytoplankton, fish debris and non-consumed fish feed.

To the best of our knowledge, the organic carbon pool and its turnover in aquaponic systems have not yet been determined. However, for the crop production part of such systems, results from root exudation research may give some guidance. Approximately 10% of the carbon assimilated through

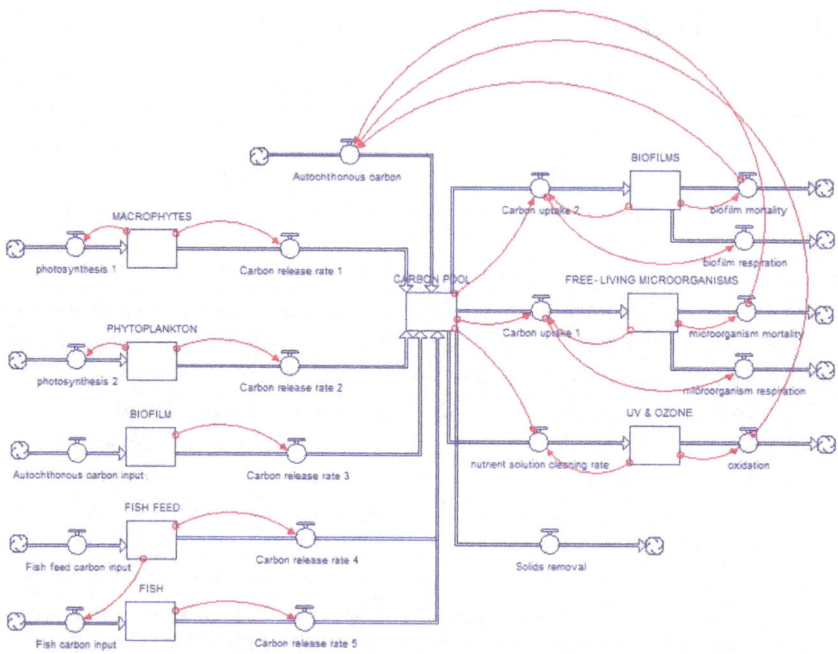

Figure 3 Simulation model conceptualized in Stella (isee systems inc.) for the functional organization of carbon flows in closed aquaponic systems. The boxes represent stocks of carbon and the blue arrows represent the flows. The red arrows are connectors which illustrate the factors regulating the flows. Inputs of autochthonous and allochthonous organic carbon to the general carbon pool are provided by fish and fish feed, macrophytes, phytoplankton, carbon sources related to the growing system, and biofilms associated with the growing system. Carbon inputs support the growth of microbial decomposers and opportunists while mortality of biofilms and free-living microorganisms contribute to the autochthnonous carbon pool. Illustration: A.K. Rosberg.

photosynthesis is released as root debris and root exudates (Uren, 2001). Apart from this active allocation of organic carbon by the plant, plant disease attack may add to passive release of organic carbon, in both dissolved and particulate form. The favourable conditions in the plant environment may affect degradation of organic material, even including the breakdown of organic plant growing media (e.g. peat, composts, cocofibre, wood fibre). Measurements of total and DOC in the nutrient solution have revealed levels between 4 and 60 mg L^{-1} (Alsanius and Bergstrand, 2007; Rosberg et al., 2014) or 3 and 39 mg L^{-1} (Waechter-Kristensen et al., 1997; Rosberg et al., 2014). PPP add to the organic carbon pool. In this context their chemical form, application frequency and mode of application are important parameters.

Organic carbon in aquaponics may be consumed by free-living or biofilm-bound microorganisms. Carbon uptake rate, on the one hand, and microbial and biofilm respiration and mortality, on the other, determine the amount of autochthonous carbon re-allocated to the carbon pool. Furthermore, devices to improve the water quality and reduce dispersal of plant pathogens often use non-selective measures, such as oxidation (e.g. ozone, hydrogen peroxide) or UV irradiation, which degrade particulate organic carbon and DOC in the nutrient solution. Degradation of organic carbon by these means is dependent on the concentration and application rate used, but also on exposure time and flow rate.

Readily degradable organic carbon in the recirculating water leads to oxygen consumption. Due to the competition for oxygen by macro- and microorganisms in the fish rearing and crop production sections and in the water treatment unit, knowledge on the organic carbon pool size and contribution by different sources is important to optimize aquaponic systems. Impaired oxygen content is deleterious to the target macroorganisms (see Section 4.1.2) in aquaponic systems and to the microbial turnover process, as described in Section 4.3. Besides pool size, the quality of organic compounds is an essential issue for optimizing aquaponics from a biological and microbiological point of view. For example, for optimizing biological control of plant pathogens by the indigenous microbiota, precursor compounds and other physio-chemical conditions must be present, as reviewed by Alsanius et al. (2007).

4.4 Microbial loops

In general, the microbiota associated to aquaponics can be categorized based upon their impact on the target organism (fish, plant) into the resident commensal microbiota, deleterious microbiota (microorganisms pathogenic to fish and/or plants roots) and beneficial microbiota (microorganisms improving fish health and/or stimulatory to plant growth and development and improving root health). Microbial communities associated with aquaponic systems may also be viewed from the perspective of their function in cycling energy. As fish and crops cultured in aquaponics eventually end up as human food, loops of human pathogens within the aquaponic system need to be considered.

Information on microbial loops in the scientific literature is scarce, but the issue has received more attention during the past two years. In addition to these data, some assumptions can be made from the existing literature on the microbial ecology of hydroponic systems. However, there are substantial differences between these two ecosystems and the microbial dynamics in the aquatic phase of co-cultured fish and plants are more complex. The fish and crop production units both consist of pelagic and benthic habitats. The microbial community structure in aquaponics is affected by available sources of energy, by external factors (light, temperature, replenishment with freshwater) and by competition. Based upon organic carbon levels and nutrient concentrations, the trophic status of the fish rearing compartment may vary, depending on the reared fish species, from oligotrophic to eutrophic (Malone and DeLosReyes, 1997), whereas the crop production compartment is considered oligo- to mesotrophic considering the concentration of oxygen and total organic carbon. However, with respect to nutrient levels, the trophic status in the nutrient solution is regarded as eutrophic.

The two principal habitats within aquaponics are dominated by different types of macroorganisms (heterotrophic and autotrophic) and the distinct microbial community structures in the biofilm associated with these macroorganisms. On phylum level, the microbial community structure of fish faeces is dominated by four phyla (Fusobacteria, Bacteriodetes, Firmicutes and Proteobacteria) and differs substantially from that of the plant roots (Schmautz et al., 2017). There are also considerable differences in biodiversity (Shannon index), with higher values for plant roots than fish faeces.

Another major difference between the two habitats is the presence/absence of light. The aquatic phase associated with the crop production unit (root zone, nutrient solution and water tanks) is shielded from light. Exclusion of light excludes proliferation of photoautotrophic organisms in this compartment and decreases competition for crop nutrients. In contrast, photoautotrophic organisms are part of the fish rearing compartment, provisioning the system with additional nutrients (nitrogen, carbon). Due to high fish stocking densities in aquaponics and the need for feeding, phytoplankton in the fish section may not be exposed to high predation pressure, which leads to uncontrolled growth (Sigee, 2005). The steady replenishment of fish tank water with freshwater as a consequence of water depletion due to plant water uptake is expected to have a regulating impact on the phytoplankton community.

Knowledge of microbial community structure and the relative abundance of different taxa in the different aquaponic habitats is crucial for understanding microbial loops in aquaponic systems, but its significance in the present form is limited. A deeper analysis is imperative in order to relate spatial and temporal distribution of microbes to function, rather than general taxonomic information. Furthermore, in order to use metagenomics data for operational decisions and microbial optimization of aquaponics, a gold standard needs to be developed with regard to choice of site, sample size, number of replicates, change of dynamics over time and sampling practice (*e.g.*, What can small root hair samples from the hydroponic part reveal about the microbial community structure of roots grown in aquaponics?). The wealth of insights available from microbiology studies on hydroponic systems using culture-dependent and culture-independent methods may be transferrable to aquaponic systems (see Alsanius and Wohanka, 2018, and references therein). Apart from the obvious sampling target sites in aquaponics, very few studies shed light on the microbial community structure in biofilms in aquaponics and their dynamics. Some conclusions may be drawn from distribution systems for irrigation water and the impact of management factors. For example, flow rate and the kind of water distribution system (continuous, intermittent water flow) and the presence of stagnant water may affect biofilm dynamics in terms of retention and release of microorganisms in pipelines (Pachepsky et al., 2011; Shelton et al., 2012). However, this approach is still in its infancy and needs considerable support from in-depth studies on microbial processes and mechanisms of relevance in aquaponic systems and their different habitats.

4.4.1 Fish pathogens

Fish pathogens are commonly classified into microparasites and macroparasites. The microparasites include viral, bacterial, fungal and protozoan parasites, while the macroparasites belong to the groups Platyhelminthes, Acanthocephala, Nematoda, Hirudinea and Crustacea (Graham, 2008).

Bacterial taxa are largely associated with fish diseases and are the primary cause of losses in fish in aquaculture systems. They may be primary pathogens (*Aeromonas salmonicida, Flavobacterium psychrophilum* and *Renibacterium salmoninarum*), but are more often opportunistic pathogens (*Flavobacterium columnare*), which colonize and cause disease in already damaged hosts. The initial weakening processes may involve pollution or a physiological state that results from unsuitable temperature, pH or O_2 levels in the system. However, it is important to note that the disease is not necessarily caused by a single bacterial taxon, but that there may well be synergistic interactions between two or more taxa. Therefore, the disease is usually the outcome of an interaction between the host (fish), the disease-causing situation (pathogen) and external stressor(s) (unsuitable changes in the environment; poor hygiene; stress) (Austin and Austion, 2007). Many of the bacterial pathogens are members of the normal water and fish microbiota. Others have been associated only with clinically diseased or covertly infected (asymptomatic) fish.

Fungal growth on the surface of fish and shellfish eggs and larvae can cause extensive direct mortalities. Fungi also occur as common secondary invaders in wounds, lesions or abrasions caused by bacterial pathogens, parasitic organisms, abusive handling or unfavourable environmental conditions. Some fungal agents are primary pathogens, especially in salmon, trout and catfish, and they could be a cause of mortality in rearing units (Meyer, 1991).

Protozoan and other macroparasites are one of the most important groups of pathogens which negatively affect the health state of cultured fish. Besides direct losses caused by mortality, parasites may have considerable impacts on growth and behaviour of fish, their resistance to other stressing factors and their susceptibility to predation. Their presence may also reduce the marketability of fish (Scolz, 1999).

Viral diseases have been studied for trout and salmon in conventional aquaculture systems. Most are a serious threat to the survival of fry and fingerlings and can cause major losses among young-of-the-year fish (Meyer, 1991). Infectious haematopoietic necrosis, infectious pancreatic necrosis and viral haemorrhagic septicaemia are the major viral diseases of trout and salmon.

Human infections caused by pathogens transmitted from fish or the aquatic environment are quite common. These pathogens are often bacterial species that are facultatively pathogenic to both fish and humans (Novotny et al., 2004). There is still a high demand for improved methods to control fish parasites. Available effective drugs may conflict with parasite resistance, toxicity of chemicals and persistence of chemical residues. Preventive measures in fish parasite control, including effective quarantine, are often ignored, resulting in much higher costs for pathogen elimination once outbreaks have occurred. Imports of infected fish and introduction of parasites by humans into new areas also cause serious problems.

4.4.2 Plant pathogens

A wide range of viruses, bacteria, fungi-like organisms and fungi provoke root diseases in some major hydroponically grown crops (Hong et al., 2014). Root disease affects root development, water and nutrient uptake, and hence canopy development, biomass production and yield. Plant pathogens adapted to an aquatic life style, such as oomycetes (*Pythium*, *Phytophthora*) and zoospore-forming fungi (*Olpidium*), are of major importance (Ivors and Moorman, 2014), causing devastating crop losses. However, oomycete pathogens relevant for plants do not overlap with those of importance for fish (e.g. *Pythium* sp., *Aphanomyces* sp., *Saprolegnia* sp.) (Mastan et al., 2012; Diéguez-Uribeondo et al., 2009; Phillips et al., 2008). *Olpidium* (Phycomycetes) not only acts as a pathogen, but also as a vector for plant viruses. For sustainable production in aquaponic systems, plant disease-suppressive and -conducive conditions need to be considered. Dispersal of root pathogens by the recirculating nutrient solution can be prevented by spot disinfection (Ehret et al., 2001). These measures may be target-specific or broad spectrum treatments. Choice of disinfection method has to match the disease profile of the crop grown and a worst-case scenario regarding technological features of the equipment (peak water volume, peak pathogen concentrations in the nutrient solution). Direct treatments are based on chemical and biological mechanisms, using pesticides and antagonistic microorganisms, respectively. In some countries, administration of pesticides is limited in vegetable crops with continuous harvest, such as tomato, cucumber and sweet pepper. To comply with waiting times, limitations may also occur if crops at different phenological stages are grown in the same water-nutrient loop. When administering pesticides to the nutrient solution in order to treat root diseases, their potential impact on the stocked fish must be considered. Low microbial degradation of pesticides in the nutrient solution (Alsanius et al., 2013; Alsanius and Bergstrand, 2014) and in surface water (Abbate et al., 2007) has been observed in recent studies. Suppressive effects of decisive fungal-like and fungal plant pathogens have been reported upon administration of fish tank effluents, and there are indications that these effects may be of microbial origin (Gravel et al., 2015). However, the efficacy of control of root pathogens differs between different types of aquaponic system. These encouraging results need to be followed up by more detailed analysis to understand underlying mechanisms and modes of action.

4.4.3 Human pathogens

Bacterial organisms pathogenic to humans are also associated with fish. In an aquaculture situation, this might be of low interest as long as workers use appropriate protective clothing and consider hygiene rules, and the fish-based food is properly prepared and not consumed raw. However, in an aquaponics situation, their dispersal to the crop unit needs specific attention as these crops are mostly consumed raw (see Section 5). Plants are also reservoirs for opportunistic human pathogens (Berg et al., 2005), such as *Stenotrophomonas maltophilia*, *Bacillus cereus*, *Burkholderia cepacia*, *Pseudomonas aeruginosa*, *Salmonella* Typhimurium, *Enterobacter cloacae* and *Klebsiella pneumoniae*. Among the isolates recovered from the nutrient solution of a closed hydroponic system with tomato, Berkelmann (1992) listed several species belonging to biosafety level 2 (*Pseudomonas aeruginosa*, *Stenotrophomonas maltophilia*, *Yersinia* spp.). Comparable results have been found in a study isolating bacterial strains from the schmutzdecke of a slow filter integrated into a closed cropping system with tomato (Brand, 2001). Some of these may also be pathogenic to fish (e.g. *Aeromonas*, *Pseudomonas*, *Streptococcus*). Ports for entry of human pathogens into aquaponic systems are discussed in detail in Section 6 of this chapter. However, it needs to be emphasized that human pathogenic bacteria transported in the water loop of aquaponic systems may not only colonize the plant root, but may also be internalized and translocated within the plant to the stem, leaves and fruit. To date, these aspects have received very little attention, especially for human pathogens associated to fish or to aquaponic systems.

4.4.4 Beneficial microorganisms

High production intensity can compromise fish health and plant health. To create resilient production systems, taxonomic and functional biodiversity are important factors. Beneficial organisms may be a natural part of the microbial communities of fish (Sirakov et al., 2016) and hydroponic systems (Rosberg, 2014; Berkelmann, 1992; Postma et al., 2000, 2005; Postma, 2004). However, external factors and spatial and dose-related parameters determine whether disease in the production system can be mitigated. Fish and plant disease may be treated by chemical means (antibiotics, PPP). However, in closed multitrophic systems such treatments are questionable with regard to system resilience. Therefore, administration of beneficial organisms may be considered. For plant root pathogens transmitted by irrigation water, mechanisms such as competition for space and nutrients, production of antagonistic metabolites and induced resistance mediate disease suppressiveness (Postma et al., 2000, 2005; Berkelmann, 1992; Hultberg et al., 2008). Experiences from hydroponics show that root disease and dispersal of root-borne pathogens may also be counteracted by introduction of microorganisms with antagonistic activities. Some fungi (*Trichoderma harzianum*, *Gliocladium*) and bacteria (*Streptomyces griseo-viridis*, *Bacillus subtilis*, *Lysobacter enzymogenes*) have been commercialized and are registered for use in soilless greenhouse cultivation systems in some areas or for some crops. Many bacterial strains, including *Pseudomonas fluorescens*, *Pseudomonas koreensis* and *Enterobacter cloacae*, have shown good antagonistic effects in laboratory studies and hydroponic small-scale systems (Gupta et al., 2002). However, biocontrol success is a function of the pathosystem, timing of addition of BCA, external management factors and microbial competition with non-target organisms (Khalil and Olsson, 2012; Khalil, 2011; Khalil and Alsanius, 2011). In aquaponics, choice of BCA also needs to consider side effects on the reared fish and food safety. Suppressive effects of decisive fungal-like and fungal plant pathogens have been reported upon administration of fish tank effluents and there are indications that these effects may be of microbial origin (Gravel et al., 2015). However, the efficacy of control of root pathogens differs between different types of aquaponic systems. These encouraging results need to be followed up by more detailed analysis to understand underlying mechanisms and modes of action.

5 Assessing aquaponics systems for food safety

Food safety is a key issue for both animal and food crop production and a challenge in multitrophic systems. According to the Codex Alimentarius (CAC, 1997), a food safety hazard is defined as a 'biological, chemical or physical agent in, or condition of food with the potential to cause an adverse health effect', where physical hazards are posed by foreign bodies (e.g. soil, stones, pieces of glass, metal, plastic or wood), chemical hazards may occur through allergens, mycotoxins, antibiotics, persistent organic pollutants or heavy metals, and microbiological (biological) hazards may be posed by viruses, bacteria, fungi, protozoa, helminths, prions and antibiotic resistance genes

(Wallace et al., 2011). In general, food safety issues in aquaponics have not received sufficient attention in the scientific literature and, when studied, the crop production unit and the plant-based produce are not considered. In an aquaponic context, physical food hazards may be low. Estimation of food safety hazards in aquaponic systems must be based on an ecosystem approach and consider the dynamic nature of the system. The feedback loop of resources, chemicals and microbes adds a high level of complexity to the approach. Chemical and biological hazards in aquaponic systems are summarized in Table 3.

5.1 Chemical food hazards

Few studies consider chemical food hazards associated with aquaponic production systems. One particular feature of aquaponics is the expected high nitrate load in the recirculating water. This issue is often considered a positive plant nutrition factor in aquaponics and is also a decisive parameter for the choice of crop. Besides fish excreta and fish feed, nitrate sources in aquaponics are the raw water, the growth medium and added plant fertilizers (see Table 3). High levels of nitrate promote biomass production during the vegetative plant stage and should be matched with plants with high nitrogen demand, such as leafy vegetables and herbs. Plant density is negatively correlated with the nitrate concentration of spinach leaves grown in an aquaponic system (Petrea et al., 2013b). Nitrate assimilation by plants is influenced by external factors, such as light intensity and temperature. Imbalanced nitrate supply in terms of supply and demand impairs the uptake and content of other plant nutrients and leads to nitrate accumulation in the vegetative parts. This will subsequently be transformed to nitrite and nitrosamine, both of which are a concern to human health with proven links to methaemoglobinaemia and cancer (Umar and Iqbal, 2007; Iammarino et al., 2014). The European Union has set maximum nitrate levels for plant-based food (European Union, 2011). To cope with maximum nitrate levels in the final product and their suitability to different groups of consumers (infants, young children, adults), the nitrate supply to crops grown in aquaponics must correspond to the demand of the crop grown. This again calls for a shift in perspective from aquaponic plants as phytobioremediators of inferior fish water quality to a multitrophic production system for high-quality fish and plant food.

Antibiotic compounds are supplied to cure bacterial disease in fish in aquaculture. These compounds are directly suspended in the pond or tank water or administered with the fish feed. Among the compounds used are quinolones, tetracyclines, phenicols and sulphonamides (Miranda et al., 2013; Romero et al., 2012). Many of these compounds are also used to treat bacterial diseases in humans. Public health concerns posed by antibiotic resistance are acknowledged by a recent FAO report (Romero et al., 2012). As various hotspots for antibiotic resistance genes (e.g. fish colon, aquatic environment, rhizosphere) are present in multitrophic systems such as aquaponics, this issue needs specific attention in future studies. Until the threat of antibiotic resistance has been ruled out, the use of antibiotic compounds in aquaponics is questionable.

Residues of pharmaceutical and personal care products (PPCP) may be introduced into aquaponic systems, depending on the quality of the water source and use of pharmaceuticals other than antibiotics for fish farming. PPCP residues may accumulate in fish meat, but plant uptake has also been demonstrated for various compounds (Redshaw et al., 2008; Herklotz et al., 2010; Winker et al., 2010; Shenker et al., 2011; Karnjanapiboonwong et al., 2011). However, these studies do not deal with factors which affect persistence that may be relevant for greenhouse horticulture or aquaponic crop production which affect persistence, nor do they consider the fact that PPCP may be present in combination and interact. To ensure the food safety of both fish and plant products from aquaponics, high water quality and low administration of pharmaceutical products are important. The same applies to PPP, residues of which may originate from compounds directly administered to the recirculating water. Moreover, they can be detected after foliar administration, due to run-off from the canopy directly to the water loop or due to percolation through the growth medium (Vermeulen et al., 2010).

The use of polluted water in aquaponic systems or of metal-contaminated mineral fertilizers in the crop production unit may give rise to heavy metal accumulation in the fish and plant-based food. In fish, bioaccumulated heavy metals are predominantly found in the kidneys and liver, and in lower concentrations in the muscles (Dallinger et al., 1987). Bioaccumulation of heavy metals in plants may affect roots and aerial parts (January et al., 2008).

5.2 Biological food hazards

Biological food safety hazards occurring in aquaculture have been covered well by different literature reviews (Howgate, 1998; Reilly and Käferstein, 2008; Chalmers, 2004). However, few studies consider the biological food safety hazards in aquaponics (Chalmers, 2004; Fox et al., 2012; Alanis and Fitzsimmons, 2007). Occurrence of relevant zoonotic bacterial pathogens depends on the properties of the aquatic environment (marine, brackish, freshwater, water quality). Fish may or may not display symptoms in response to zoonotic parasites. Gastroenteritis (*Vibrio cholerae*, *V. vulnificus*, *V. fluvialis*, *V. alginolyticus*, *Edwardsiella tarda*, *Aeromonas* spp.), listeriosis (*Listeria monocytogenes*), botulism (*Clostridium botulinum*), meningoencephalitis (*Edwardsiella tarda*) and sepsis (*Vibrio cholerae*, *V. vulnificus*, *V. fluvialis*,

Table 3 Summary of potential chemical and biological food safety hazards at different points within aquaponic systems

	Potential critical input factors and measures adventuring chemical and biological food safety						
Hazard	Water	Fish	Fish feed	Fish farming measure	Crop	Growing medium	Crop management measure
Chemical hazards							
Nitrates	•	•	•			•	•
Antibiotic compounds	•		•	•			(•)
Other pharmaceutical residues	•		•	•			
Plant protection products (PPP)	•		•		•	•	•
Persistent organic pollutants	•		•				
Mycotoxins							
Heavy metals	•		•	•		•	•
Biological hazards							
Virus	•	•	•	•			
Bacteria	•	•	•	•			
Antibiotic resistance genes	•	•	•	•	•	•	
Fungi	•	•	•				
Protozoa	•	•	•		•	•	
Helminths	•	•	•			•	

Aeromonas spp.) are among the diseases that may be transmitted by fish (Hoffmann, 2005). Enteric bacteria can survive in water for long periods, as recently reviewed by Alsanius (2014). While aquaponic crops were previously conceived as dead ends (Chalmers, 2004), the role of aquaponic-grown plants with respect to biological food safety hazards needs to be reconsidered in light of evidence that plants act as vectors of zoonotic bacteria (Matthews et al., 2014) and pose a risk to the food safety of vegetables and fruit (Matthews et al., 2014). This is of special interest as vegetables, herbs and berries grown aquaponically are most probably consumed raw or with minimal processing. To assess the biological food hazards in aquaponics, barriers (e.g. pH, temperature, osmotic stress, nutrient availability, disinfection measures) imposed within the different aquaponic habitats and potentially impeding their dispersal need to be studied. Survival in a viable but non-culturable state under harsh conditions in non-host environments, including aquatic environments, has been shown for *Salmonella*, shigatoxin-producing *Escherichia coli* and *Vibrio cholerae* (Roszak et al., 1984; Wang and Doyle, 1998; Asakura et al., 2007). *Vibrio cholerae* can also persist as conditionally viable environmental cells (CVEC) (Faruque et al., 2006). Quiescent lifestyles of bacteria in response to exposure to harsh conditions may lead to false negative results when screening different aquaponic habitats using culture-dependent methods and to underestimation of risks. To avoid cross-contamination from the fish rearing tank to the crop production unit through irrigation water, decontamination using UV irradiation has been tested. Disinfection by UV irradiation has been tested for different water purposes, for example potable water and irrigation water (nutrient solution) in closed hydroponic greenhouse systems. However, the studies performed in an aquaponics context are not conclusive regarding whether UV light is a suitable prevention practice (Elumalai et al., 2017; Moriarty et al., 2018). Apart from diametrically different results, these studies highlight the need for scientific discussion on adequate experimental set up and analyses. Membrane filtration has also been tested as a method to prevent dispersal of microorganisms from the fish rearing to the crop production unit, and can achieve improvements in water quality parameters (Wang et al., 2016). However, these positive results should be viewed in the light of experiences from membrane filtration in hydroponic contexts, where this measure has been abolished due to clogging and leakage of the membrane and the relatively high investment costs.

6 Assessing aquaponics systems for sustainability

Aquaponic systems are engineered and site-independent systems. Over the years, most scientific papers have stated that aquaponics are sustainable food production systems due to their contribution to food provisioning and nutrient recovery from fish effluents, and thus reduced eutrophication potential. However, it is only recently that the first analyses on sustainability have been published and a more distinct picture of aquaponics had emerged. These first sustainability assessments are based on commercial, experimental or virtual aquaponic systems.

Aquaponic systems may be located in urban, periurban or rural contexts. They are flexible not only regarding their placement, but also land use, as they can be integrated into buildings using 'grey' areas for fish rearing and rooftops or walls for crop production (Alsanius et al., 2017). However, their land use impact is still considerable, as production of protein-rich fish feed, especially plant-based feedstuffs, requires land. Boxman et al. (2017) estimated that 98% of land use competition is due to land-based fish feed production. Therefore alternatives, such as microalgae-based feed production, expanding these dual-purpose systems to tertiary systems should be considered.

Access to water and guaranteed power supply are two decisive determinants to be taken into account in site selection and in assessment of ecosystem services and environmental footprint. Boxman et al. (2017) identified electricity as a crucial hotspot of environmental impact for a system without a biofilter, that is requiring aeration, with aeration of the fish rearing tank and crop production unit and degassing of the fish tank being the main electricity-consuming processes. The use of NFT for the crop production unit is also an energy-intensive solution. However, in a study on a virtual scenario, cumulative energy demands were high irrespective of the choice of aquaponic system (Forchino et al., 2017). It is worth noting that the Boxman study considered Virgin Island conditions, and that energy consumption, and thus the impact of electricity, may change dramatically when aquaponic systems at higher latitude are considered. To reduce the electricity-based environmental impact of aquaponics, the form of power supply (Goddek et al., 2015) and the systems design should be reconsidered.

Aeration of aquaponic systems not only supports the life of fish and plants, but also assists in nitrogen conversion. However, aeration also contributes to greenhouse gas emissions (Fang et al., 2017), which are considerably higher in the crop production unit than in the fish rearing unit. The choice of aeration strategy (continuous, intermittent) plays an important role for N_2O emissions and goal conflicts regarding biomass production, and greenhouse gas emissions need to be considered when optimizing aquaponic systems.

7 Future trends and conclusion

Aquaponic systems are often suggested as a solution to meet the grand global challenges such as food security, food safety and sustainability. Their flexibility regarding land use, location and system size offers potential for food provisioning close to the consumer. However, success in production and provisioning is dependent not only on the site of production, but also on the resources, system design and timing used. System optimization is essential for

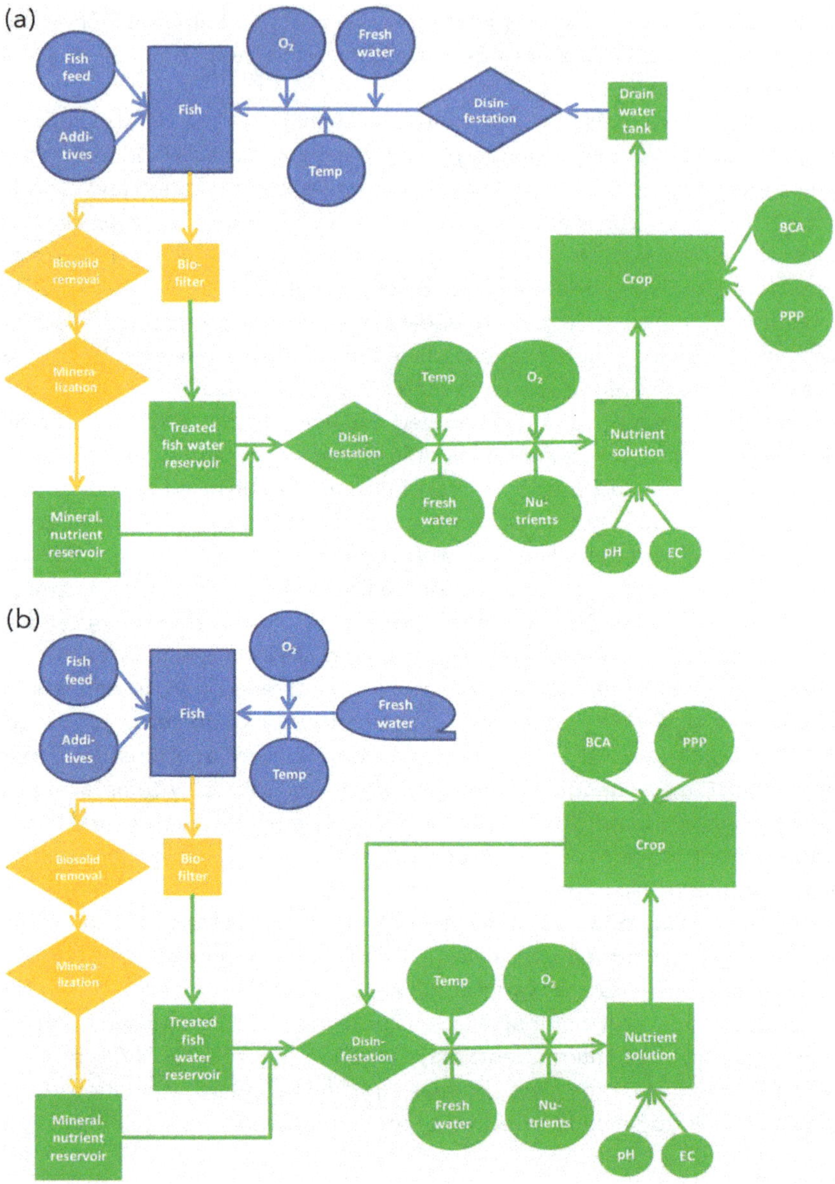

Figure 4 Optimized aquaponic systems based on either a coupled one-loop (a) or a decoupled (b) approach. The different basic units are shown in different colours (blue: fish rearing; yellow: water treatment (nutrient conversion); green: crop production). BCA = supply of biological control agents; EC = unit for control of electrical conductivity; O_2 = O_2 level control and supply; pH = pH level control and adjustment unit; PPP = supply of plant protection products; Temp = temperature control and adjustment unit. Illustration: B. Alsanius.

these functions. The circular approach, based on system-dependent, global flows and loops, used in this chapter is novel to evaluations of aquaponic systems and offers the possibility to pinpoint weak links and to identify gaps in knowledge and possible solutions.

A precondition for optimization of aquaponic production is a shift in paradigm and attitude as regards: (1) the function of the crop production unit and the quality and use of the final plant product, (2) surveillance and maintenance, irrespective of system scale and level of technology, (3) identification of goal conflicts and (4) improvements with respect to production, product quality, health and welfare of the reared animals and crops and the consumers, environmental impact and economic feasibility.

In general, aquaponic systems benefit from a holistic approach and good understanding of both horticultural and aquacultural terms and constraints. There is an extensive body of knowledge from both disciplines, but they are not always acknowledged in experimental studies. Future coupled and decoupled aquaponic systems should include the achievements that both domains have set into practice. Optimized coupled one-loop and decoupled aquaponics are displayed in Fig. 4.

Essential knowledge gaps need to be addressed, for example concerning microbial loops and food safety aspects. Initial steps have been taken to shed light on these issues and some findings have been translated into DNA-based methods. To use their full potential, fundamental issues need to be resolved and a gold standard needs to be put in place. Food safety assessment should also entail assessment of risks and definition of good production practices.

The increasing interest in the sustainability and economic feasibility of aquaponic systems is an exciting development. The studies performed to date point to the need for structured approaches in research in order to optimize aquaponic system development and management strategies and practices, thereby enabling accurate prediction, decision making and policies.

8 References

Abbate, C., Borzi, D., Caboni, P. and Baglieri, A. 2007. Behavior of fenhexamid in soil and water. *Journal of Environmental Science and Health Part B*, 42, 843–9.

Alanis, P. G. and Fitzsimmons, K. M. 2007. *Food Safety Study of Leafy Greens Irrigated with Tilapia Farm Effluents*. Tamaulipas: Universidad Autónoma de Tamaulipas.

Alsanius, B. W. 2011. Att förbereda för recirkulering av näringslösning i växthus. *Landskap Trädgård Jordbruk Rapportserie*, 39, 1–30.

Alsanius, B. W. 2014. *Hygien och bevattning*. Alnarp: Swedish University of Agricultural Sciences.

Alsanius, B. W. and Bergstrand, K.-J. 2007. *Make the Difference – om att sluta systemet*. Slu (Ed.). Alnarp: Department of Horticulture.

Alsanius, B. W. and Bergstrand, K.-J. 2014. Degradation of pesticides in nutrient solution from closed hydroponic systems. *Acta Horticulturae*, 1034, 33-9.

Alsanius, B. W. and Jung, V. 2018. Oxygen, not phenolic acids, is the primary driver for nitrification in soilless systems. *Acta Horticulturae* (in press).

Alsanius, B. W. and Wohanka, W. 2018. Root zone biology of soilless cropping systems. In: Raviv, M. and Lieth, H.-J. (Eds), *Soilless Cropping Systems*. Amsterdam: Elsevier.

Alsanius, B. W., Jung, V. and Brand, T. 2007. Enhanced antimicrobial metabolite formation by the microflora inhabiting nutrient solution in closed growing systems. *Acta Horticulturae*, 819, 173-80.

Alsanius, B. W., Bergstrand, K. J., Burleigh, S., Gruyer, N. and Rosberg, A. K. 2013. Persistence of fenhexamid in the nutrient solution of a closed cropping system. *Agricultural Water Management*, 127, 25-30.

Alsanius, B. W., Rosberg, A. K., Hultberg, M., Khalil, S. and Jung, V. 2014. Understanding and utilizing naturally occurring microbes against plant pathogens in irrigation reservoirs. In: Hong, C., Moorman, G. W., Wohanka, W. and Büttner, C. (Eds), *Waterborne Plant Pathogens*. St Paul, MN: American Phytopathological Society.

Alsanius, B. W., Khalil, S. and Morgenstern, R. 2017. Rooftop aquaponics. In: Orsini, F., Dubbeling, M. and Gianquinto, G. (Eds), *Rooftop Urban Agriculture*. New York, NY: Springer.

Asakura, H., Ishiwa, A., Arakawa, E., Makino, S.-I., Okada, Y., Yamamoto, S. and Igimi, S. 2007. Gene expression profile of *Vibrio cholerae* in the cold stress-induced viable but non-culturable state. *Environmental Microbiology*, 9, 869-79.

Austin, B. and Austin, D. A. 2007. *Bacterial Fish Pathogens: Diseases of Farmed and Wildfish*. Chichester: Praxis Publishing Ltd.

Berg, G., Eberl, L. and Hartmann, A. 2005. The rhizosphere as a reservoir for opportunistic human pathogenic bacteria. *Environmental Microbiology*, 7(11), 1673-85.

Berkelmann, B. (Ed.). 1992. *Charakterisierung der Bakterienflora und des antagonistischen Potentials in der zirkulierenden Nährlösung einer Tomatenkultur (Lycopersicon esculentum MILL.) in Steinwolle*. Geisenheim.

Blok, C., Jackson, B. E., Guo, X., De Visser, P. H. B. and Marcelis, L. M. F. 2017. Maximum plant uptakes for water, nutrients, and oxygen are not always met by irrigation rate and distribution in water-based cultivation systems. *Frontiers Plant Science*, 8, 562.

Boxman, S. E., Zhang, Q., Bailey, D. S. and Trotz, M. A. 2017. Life cycle assessment of a commercial-scale freshwater aquaponic system. *Environmental Engineering Science*, 34(5).

Brand, T. (Ed.). 2001. *Beurteilung der Bakterienflora von Langsamfiltern als biologische Filtrationskomponente in geschlossenen Bewässerungssystemen des Gartenbaus*. Geisenheim.

CAC. 1997. Principles and guidelines for the establishment and application of microbiological criteria related to foods. In: Codex Alimentarius Commission, C. (Ed.), *GL21*.

Chakravartty, D., Mondal, A., Raychowdhury, P., Bhattacharya, S. B. and Mitra, A. 2017. Role of aquaponics in the sustenance of coastal India – Aquaponics is a solution for modern agriculture in ecologically sensitive Indian mangrove Sundarbans: A review. *International Journal of Fisheries and Aquatic Studies*, 5, 441-8.

Chalmers, G. A. 2004. *Aquaponics and Food Safety*. Lethbridge-Alberta, Canada: DVM.

Chérif, M., Tirilly, Y. and Bélanger, R. R. 1997. Effect of oxygen concentration on plant growth, lipidperoxidation, and receptivity of tomato roots to Pythium F under hydroponic conditions. *European Journal of Plant Pathology*, 103, 255-64.

Cho, Y., Cowey, C., Dabrowski, K., Hughes, S., Lall, S., Lovell, R., Murai, T. and Wilson, R. 1999. *Nutrient Requirements of Fish. National Research Council (NRC).* Washington DC: National Academy Press.

Cooper, A. 1979. *The ABC of NFT*. London, UK: Grower Books.

da Silva Cerozi, B. and Fitzsimmons, K. 2016. Use of *Bacillus* spp. to enhance phosphorus availability and serve as a plant growth promoter in aquaponics system. *Scientia Horticulturae*, 211, 277-82.

Dallinger, R., Prosi, F., Segner, H. and Back, H. 1987. Contaminated food and uptake of heavy metals by fish: A review and a proposal for further research. *Oecologia*, 73, 91-8.

Dediu, L., Cristea, V. and Zhang, X. S. 2012. Waste production and valorization in an integrated aquaponic system with bester and lettuce. *African Journal of Biotechnology*, 11, 2349-58.

Diáz-Pérez, J. C., Phatak, S. C., Giddings, D., Bertrand, D. and Mills, H. 2005. Root zone temperature, plant growth and fruit yield of tomatillo as affected by plastic film mulch. *HortScience*, 40, 1312-19.

Diéguez-Uribeondo, J., García, M. A., Cerenius, L., Kozubíková, E., Ballesteros, I., Windels, C., Weiland, J., Kator, H., Söderhäll, K. and Martín, M. P. 2009. Phylogenetic relationships among plant and animal parasites, and saprotrophs in Aphanomyces (*Oomycetes*). *Fungal Genetics and Biology*, 46, 365-76.

Drennan II, D. G., Hosler, K. C., Francis, M., Weaver, D., Aneshansley, E., Beckman, G., Johnson, C. H. and Cristina, C. M. 2006. Standardized evaluation and rating of biofilters II. Manufacturer's and user's perspective. *Aquacultural Engineering*, 34, 403-16.

Ebeling, J. M. and Timmons, M. B. 2012. Recirculating aquaculture systems. In: Tidwell, J. H. (Ed.), *Aquaculture Production Systems*. Ames, IA: John Wiley & Sons, Inc.

Edwards, P. 2013. Philosophy, principles and concepts of integrated agri-aquaculture systems. In: Gooley, G. J. and Gavine, F. M. (Eds), *Integrated Agri-Aquaculture System*. Canberra, Australia: Rural Industries Research and Development Corporation.

Ehret, D. L., Alsanius, B. W., Wohanka, W., Menzies, J. M. and Utkhede, R. S. 2001. Disinfestation of recirculating nutrient systems. *Agronomie*, 21, 323-39.

Elumalai, S. D., Shaw, A. M., Pattillo, D. A., Currey, C. J., Rosentrater, K. A. and Xie, K. 2017. Influence of UV treatment on the food safety status of a model aquaponic system. *Water*, 9, 27.

Endut, A., Jusoh, A., Ali, N., Nik, W. B. W. and Hassan, A. 2010. A study on the optimal hydraulic loading rate and plant ratios in recirculation aquaponic system. *Bioresource Technology*, 101, 1511-17.

European Union. 2011. COMMISSION REGULATION (EU) No 1258/2011 of amending Regulation (EC) No 1881/2006 as regards maximum levels for nitrates in foodstuffs. In: Union, E. (Ed.), *L 320/15*. Brussels: Official Journal of the European Union.

Fang, Y., Hu, Z., Zou, Y., Fan, J., Wang, Q. and Zhu, Z. 2017. Increasing economic and environmental benefits of media-based aquaponics through optimizing aeration pattern. *Journal of Cleaner Production*, 162, 111-17.

FAO. 2009. *Reform of the Committee on World Food Security: Final Version*. In: Security, C. O. W. F. (Ed.). Rome: FAO.

FAO, IFAD and WFP. 2013. *The State of Food Insecurity in the World - The Multiple Dimensions of Food Security*. Rome: FAO.

Faruque, S. M., Biswas, K., Udden, S. M. N., Ahmad, Q. S., Sack, D. A., Nair, G. B. and Mekalanos, J. J. 2006. Transmissibility of cholera: *In vivo*-formed biofilms and their relationship to infectivity and persistence in the environment. *Proceedings of the National Academy of Sciences of the United States of America*, 103, 6350–5.

Flannery, R. J. and Lieth, J. H. 2006. Effect of plant root oxygen deprivation in relation to water and nitrate uptake for rose. *Acta Horticulturae*, 766, 53–8.

Forchino, A. A., Lourguioui, H., Brigolin, D. and Pastres, R. 2017. Aquaponics and sustainability: The comparison of two different aquaponic techniques using the Life Cycle Assessment (LCA). *Aquacultural Engineering*, 77, 80–8.

Fornshell, G., Hinshwaw, J. and Tidwell, J. H. 2012. Flow-through raceways. In: Tidwell, J. H. (Ed.), *Aquaculture Production Systems*. Ames, Chichester and Oxford: Wiley-Blackwell.

Fox, B. K., Tamaru, C. S., Hollyer, J., Castro, L. F., Fonseca, J. M., Jay-Russell, M. and Low, T. 2012. A preliminary study of microbial water quality related to food safety in recirculating aquaponic fish and vegetable production systems. In: Resources, C. O. T. A. A. H. (Ed.), *Food Safety and Technology*. Manao: University of Hawaii.

Gendel, Y. and Lahav, O. 2012. A novel approach for ammonia removal from fresh-water recirculated aquaculture systems, comprising ion exchange and electrochemical regeneration. *Aquacultural Engineering*, 52, 27–38.

Gertsson, U. E., Hansson, I., Waechter-Kristensen, B., Lundqvist, S., Svedelius, G. and Weich, R. G. 1994. Tomato grown in circulating nutrient solution using rockwool and as hydroponics. *Acta Horticulturae*, 361, 237–44.

Ghafari, S., Hasan, M. and Aroua, M. K. 2008. Bio-electrochemical removal of nitrate from water and wastewater – a review. *Bioresource Technology*, 99, 3965–74.

Goddek, S. and Keesman, K. J. 2018. The necessity of desalination technology for designing and sizing multi-loop aquaponics systems. *Desalination*, 428, 76–85.

Goddek, S., Delaide, B., Mankasingh, U., Ragnarsdottir, K. V., Jijakli, H. and Thorarinsdottir, R. 2015. Challenges of sustainable and commercial aquaponics. *Sustainability*, 7, 4199–224.

Göhler, F. and Molitor, H.-D. 2002. *Erdelose Kulturverfahren im Gartenbau*. Stuttgart: Ulmer.

Gosselin, A. and Trudel, M. 1985. Influence of root-zone temperature on growth, development and yield of cucumber plants cv. Toska. *Plant and Soil*, 85, 327–36.

Graber, A. and Junge, R. 2009. Aquaponic systems: Nutrient recycling from fish wastewater by vegetable production. *Desalination*, 246, 147–56.

Graham, A. L. 2008. Ecological rules governing helminth–microparasite coinfection. *Proceedings of the National Academy of Sciences of the United States of America*, 105, 566–70.

Gravel, V., Dorias, M., Dey, D. and Vandenberg, G. 2015. Fish effluents promote root growth and suppress fungal diseases in tomato transplants. *Canadian Journal of Plant Science*, 95, 427–36.

Gruda, N. 2009. Do soilless culture systems have an influence on product quality of vegetables? *Journal of Applied Botany and Food Quality*, 82, 141–5.

Gupta, C. D., Dubey, R. C. and Maheshwari, D. K. 2002. Plant growth enhancement and suppression of *Macrophomina phaseolina* causing charcoal rot of peanut by fluorescent Pseudomonas. *Biology and Fertility of Soils*, 35, 399–405.

Gutierrez-Wing, M. T. and Malone, R. F. 2006. Biological filters in aquaculture: Trends and research directions for freshwater and marine applications. *Aquacultural Engineering*, 34, 163–71.

Hao, X., Papadopoulos, A., Dorais, M., Ehret, D., Turcotte, G. and Gosselin, A. 1998. Improving tomato fruit quality by raising the EC of NFT nutrient solutions and calcium spraying: Effects on growth, photosynthesis, yield and quality. *XXV International Horticultural Congress, Part 1: Culture Techniques with Special Emphasis on Environmental Implications*- 511, pp. 213–24.

Herbold, J. 1994. Bodenunabhängige Kulturverfahren im Gemüsebau. PhD, Hohenheim University.

Herklotz, P. A., Gurung, P., Vanden Heuvel, B. and Kinney, C. A. 2010. Uptake of human pharmaceuticals by plants grown under hydroponic conditions. *Chemosphere*, 78, 1416–21.

Hoffmann, R. W. 2005. *Fischkrankheiten*. Stuttgart: Ulmer UTB.

Hong, C., Büttner, C., Mooreman, G. and Wohanka, W. 2014. *Biology, Detection, and Management of Plant Pathogens in Irrigation Water*. St Paul, MN: American Phytopathological Society.

Howgate, P. 1998. Review of the public health safety of products from aquaculture. *International Journal of Food Science and Technology*, 33, 99–125.

Hultberg, M., Bergstrand, K.-J., Khalil, S. and Alsanius, B. W. 2008. Production of biosurfactants and antibiotics by fluorescent pseudomonads isolated from a closed hydroponic system equipped with a slow filter. *Antonie van Leeuwenhoek*, 93, 373–80.

Hurewitz, J. and Janes, H. W. 1983. Effect of altering the root temperature on growth, translocation, carbon exchange rate and leaf starch accumulation in tomato. *Plant Physiology*, 73, 46–50.

Iammarino, M., Di Taranto, A. and Cristino, M. 2014. Monitoring of nitrites and nitrates levels in leafy vegetables (spinach and lettuce): A contribution to risk assessment. *Journal of the Science of Food and Agriculture*, 94, 773–8.

Ivors, K. L. and Moorman, G. W. 2014. Oomycete plant pathogens in irrigation water. In: Hong, C., Moorman, G. W., Wohanka, W. and Büttner, C. (Eds), *Biology, Detection and Management of Plant Pathogens in Irrigation Water*. St Paul, MN: American Physiological Society.

January, M. C., Cutright, T. J., Van Keulen, H. and Wei, R. 2008. Hydroponic phytoremediation of Cd, Cr, Ni, As, and Fe: Can Helianthus annuus hyperaccumulate multiple heavy metals? *Chemosphere*, 70, 531–7.

Jensen, M. H. and Malter, A. J. 1995. *Protected Agriculture: A Global Review*. Washington DC: The World Bank.

Jordan, R. A., Geisenhoff, L. O., De Oliveira, F. C., Santos, R. C. and Martines, E. A. S. 2018. Yield of lettuce grown in aquaponic system using different substrates. *Revista Brasileira de Engenharia Agricola e Ambiental*, 22, 27–31.

Kangmin, L. 1988. Rice-fish culture in China: A review. *Aquaculture*, 71, 173–86.

Karnjanapiboonwong, A., Chase, D. A., Cañas, J. E., Jackson, W. A., Maul, J. D., Morse, A. N. and Anderson, T. D. 2011. Uptake of 17a-ethynylestradiol and triclosan in pinto bean, Phaseolus vulgaris. *Ecotoxicology and Environmental Safety*, 74, 1336–42.

Khalil, S. 2011. Influence of electrical conductivity on biological activity of *Pythium ultimum* and Binab T in a closed soilless system. *Journal of Plant Diseases and Protection*, 118, 102–8

Khalil, S. and Alsanius, B. W. 2011. Effect of growing medium water content on the biological control of root pathogens in a closed soilless system. *Journal of Horticultural Science & Biotechnology*, 86, 298–304.

Khalil, S. and Olsson, O. 2012. Influence of micronutrients on the biological control of *Pythium ultimum* in a closed soilless system. *Canadian Journal of Plant Protection*, 1, 134–41.

Klinger-Bowen, R. C., Tamaru, C. S., Fox, B. K., McGovern-Hopkins, K. and Howerton, R. 2011. *Testing Your Aquaponic System Water: A Comparison of Commercial Water Chemistry Methods*. Waimanalo, HI: Center for Tropical and Subtropical Aquaculture.

Lennard, W. A. and Lennard, B. V. 2006. A comparison of three different hydroponic sub-systems (gravel bed, floating and nutrient film technique) in an aquaponic test system. *Aquaculture International*, 14, 539–50.

Lester, J. N. and Birckett, J. W. 1999. *Microbiology and Chemistry for Environmental Scientists and Engineers*. New York, NY: E. & FN. Spon.

Liang, J. N. and Chien, Y. 2013. Effects of feeding frequency and photoperiod on water quality and crop production in a tilapia-water spinach raft aquaponics system. *International Biodeterioration & Biodegradation*, 85, 693–700.

Lieth, J. H. and Oki, L. R. 2008. Irrigation in soilless production. In: Raviv, M. and Lieth, J. H. (Eds), *Soilless Culture – Theory and Practice*. Amsterdam: Elsevier.

Löfkvist, K., Englund, J.-E., Larsen, R. U. and Alsanius, B. W. 2009. Light integral as an indicator of water use in commercial greenhouse nurseries. *Acta Agriculturae Scandinavica, Section B – Plant Soil Science*, 59, 326–34.

Mackroth, K. 1998. Technik für geschlossene Kulturverfahren für Topfpflanzen (und Dünnschichtkulturen). In: Jennerich, L. and Molitor, H.-D. (Eds), *Geschlossene Bewässerungsverfahren*. Braunschweig: Thalacker-Medien.

Madigan, M. T., Martinko, J. M., Dunlap, P. V. and Clark, D. P. (Eds). 2009. *Brock Biology of Microorganisms*. San Francisco, CA: Pearson Education Inc.

Malone, R. F. and Delosreyes, A. A. 1997. Categories of recirculating aquaculture systems. *Aquaculture Engineering Society Proceedings*, 9–12 November 1997, Orlando, FL. Aquaculture Engineering Society, pp. 197–208.

Maloupa, E. 2002. Hydroponic systems. In: Savvas, D. and Passam, H. (Eds), *Hydroponic Production of Vegetables and Ornamentals*. Athens: Embryo Publications.

Marschner, H. 1995. *Mineral Nutrition of Higher Plants*. London and San Diego: Academic Press.

Martin, G. C. and Wilcox, G. E. 1963. Critical soil temperature for tomato plant growth. *Science Society of America Journal*, 27, 565–7.

Masser, M. P. 2012a. Cage culture in freshwater and protected marine areas. In: Tidwell, J. H. (Ed.), *Aquaculture Production Systems*. Ames, Chichester and Oxford: Wiley-Blackwell.

Masser, M. P. 2012b. In-pond raceways. In: Tidwell, J. H. (Ed.), *Aquaculture Production Systems*. Ames, Chichester and Oxford: Wiley-Blackwell.

Masser, M., Rakocy, J. and Losordo, T. 1999. Recirculating aquaculture tank production systems: Management of recirculating systems. Southern Regional Aquaculture Center Publication and Southern Regional Aquaculture Center.

Mastan, S. A., Radha Krishna Reddy, M. and Sri Lakshmi, D. 2012. Oomycete infections in freshwater fish. *International Journal of Fisheries and Aquaculture*, 4, 186–90.

Matthews, K. R., Sapers, G. M. and Gerba, C. P. 2014. *The Produce Contamination Problem: Causes and Solutions*. San Diego, CA: Academic Press.

Maucieri, C., Nicoletto, C., Schmautz, Z., Sambo, P., Komives, T., Borin, M. and Junge, R. 2017. Vegetable intercropping in a small-scale aquaponic system. *Agronomy*, 7, 63.

Meyer, F. P. 1991. Aquaculture disease and health management. *Journal of Animal Science*, 69, 4201–8.

Mims, S. D. and Onders, R. J. 2012. Reservoir ranching. In: Tidwell, J. H. (Ed.), *Aquaculture Production Systems*. Ames, Chichester and Oxford: Wiley-Blackwell.

Miranda, C., D., Tello, A. and Keen, P. L. 2013. Mechanisms of antimicrobial resistance in finfish aquaculture environment. *Frontiers in Microbiology*, 4, 1–6.

Monsees, H., Kloas, W. and Wuertz, S. 2017. Decoupled systems on trial: Eliminating bottlenecks to improve aquaponic processes. *PLoS ONE*, 12, e0183056.

Montesano, F. F., Serio, F., Mininni, C., Signore, A., Parente, A. and Santamaria, P. 2015. Tensiometer-based irrigation management of subirrigated soilless tomato: Effects of substrate matric potential control on crop performance. *Frontiers in Plant Science*, 23, 1–11.

Moriarty, M. J., Semmens, K., Bissonnette, G. K. and Jaczynski, J. 2018. Inactivation with UV-radiation and internalization assessment of coliforms and Escherichia coli in aquaponically grown lettuce. *LWT – Food Science and Technology*, 89, 624–30.

Münster, U. 1993. Concentrations and fluxes of organic carbon substrates in the aquatic environment. *Antonie van Leeuwenhoek*, 63, 243–74.

Novotny, L., Dvorska, L., Lorencova, A., Beran, V. and Pavlik, I. 2004. Fish: A potential source of bacterial pathogens for human beings. *Veterinární medicína-Czech*, 49(9), 343–58.

O'Brien, R. D. and Van Bruggen, A. H. C. 1993. Effect of temperature on corky root disease of lettuce and growth of the pathogen Rhizomonas suberifaciens. *Canadian Journal of Plant Pathology*, 15, 85–90.

Pachepsky, Y., Morrow, J., Guber, A., Shelton, D., Rowland, R. and Davies, G. 2011. Effect of biofilm in irrigation pipes on microbial quality of irrigation water. *Letters in Applied Microbiology*, 54, 217–24.

Pade, J. S. and Nelson, R. L. 2007. Village aquaponics. *Acta Horticulturae*, 742, 197–9.

Palm, H. W., Knaus, U., Appelbaum, S., Goddek, S., Strauch, S. M., Vermeulen, T., Haïssam Jijakli, M. and Kotzen, B. 2018. Towards commercial aquaponics: A review of systems, designs, scales and nomenclature. *Aquaculture International*, 26, 813–42.

Panova, G. G., Grote, D. and Kläring, H.-P. 2004. Population dynamics of Pythium aphanidermatum and response of tomato plants as affected by root-zone temperature. *Journal of Plant Disease and Protection*, 111, 52–63.

Petrea, S. M., Cristea, V., Dediu, L., Contoman, M., Ion, S., Mocanu, M. and Antache, A. 2013a. A study of nitrogen cycle in an integrated aquaponic system with different plant densities. *Bulletin of University of Agricultural Sciences and Veterinary Medicine Cluj-Napoca. Animal Science and Biotechnologies*, 70, 55–64.

Petrea, S. M., Cristea, V., Dediu, L., Contoman, M., Lupoae, P., Placinta, S. I., Coada, M. T. and Enache, M. 2013b. A comparison of nitrate level in spinach grown both under different densities in aquaponic system and under natural growth conditions. *The Annals of the University Dunarea de Jos of Galati, Fascicle Vi – Food Technology*, 37, 47–58.

Phillips, A. J., Anderson, V. L., Robertson, E. J., Secombes, C. J. and Van West, P. 2008. New insights into animal pathogenic oomycetes. *Trends in Microbiology*, 16, 13–19.

Pillay, T. and Kutty, M. 2005. *Aquaculture Principles and Practices*. Oxford, UK: Blackwell.

Postma, J. 2004. Suppressiveness of root pathogens in closed cultivation systems. *Acta Horticulturae*, 644, 503–15.

Postma, J., Willemsen-De Klein, M. J. E. I. M. and Van Elsas, J. D. 2000. Effect of the indigenous microflora on the development of root and crown rot caused by Pythium aphanidermatum in cucumber grown on rockwool. *Phytopathology*, 90, 125–33.

Postma, J., Geraats, B. P. J., Pastoor, R. and Van Elsas, J. D. 2005. Characterization of the microbial community involved in the suppression of Pythium aphanidermatum in cucumber grown on rockwool. *Phytopathology*, 95, 808–18.

Rakocy, J. 1997. Integrating Tilapia culture with vegetable hydroponics in recirculating systems. *Tilapia Aquaculture in the Americas*. Baton Rouge, LA: World Aquaculture Society.

Rakocy, J. 2007. Ten guidelines in aquaponic system. *Aquaponic Journal*, 46, 14–17.

Rakocy, J. 2012. *Aquaponics - Integrating Fish and Plant Culture*. Ames, Chichester and London: Wiley-Blackwell.

Rakocy, J. E., Masser, M. P. and Losordo, T. M. 2006. Recirculating aquaculture tank production systems: Aquaponics-integrating fish and plant culture. *SRAC Publication - Southern Regional Aquaculture Center*, p. 454.

Redshaw, C. H., Wootton, W. G. and Rowland, S. J. 2008. Uptake of the pharmaceutical Fluoxetine Hydrochloride from growth medium by *Brassicaceae*. *Phytochemistry*, 69, 2510–16.

Reilly, A. and Käferstein, F. 2008. Food safety hazards and the application of the principles of the hazard analysis and critical control point (HACCP) system for their control in aquaculture production. *Aquaculture Research*, 28, 735–52.

Resh, H. M. 2013. *Hydroponic Food Production*. Boca Raton, FL: CRC Press.

Romero, J., Feijoó, C. G. and Navarrete, P. 2012. Antibiotics in aquaculture – use, abuse and alternatives. In: Carvalho, E. (Ed.), *Health and Environment in Aquaculture*. Rijeka: Intech open.

Roosta, H. R. and Hamidpour, M. 2011. Effects of foliar application of some macro- and micro-nutrients on tomato plants in aquaponic and hydroponic systems. *Scientia Horticulturae*, 129, 396–402.

Rosberg, A. K. 2014. Dynamics of root microorganisms in closed hydroponic cropping systems. PhD, Swedish University of Agricultural Sciences.

Rosberg, A. K., Wohanka, W., Hultberg, M. and Alsanius, B. W. 2014. Organic carbon and micronutrients have an effect on microbial dynamics in nutrient solutions of diseased and non-diseased hydroponic systems. *Acta Universitatis Agriculturae Sueciae*, 51, 1–10.

Roszak, D. B., Grimes, D. J. and Colwell, R. R. 1984. Viable but nonrecoverable stage of Salmonella enteritidis in aquatic systems. *Canadian Journal of Microbiology*, 30, 334–8.

Ru, D., Liu, J., Hu, Z., Jiang, L., Cheng, X. and Lv, Z. 2017. Improvement of aquaponic performance through micro- and macro- nutrient addition. *Environmental Science and Pollution Research*, 24, 16328–35.

Rusten, B., Eikebrokk, B., Ulgenes, Y. and Lygren, E. 2006. Design and operations of the Kaldnes moving bed biofilm reactors. *Aquacultural Engineering*, 34, 322–31.

Schmautz, Z., Loeu, F., Liebisch, F., Graber, A., Mathis, A., Griessler Bulc, T. and Junge, R. 2016. Tomato productivity and quality in aquaponics: Comparison of three hydroponic methods. *Water*, 8, 533.

Schmautz, Z., Graber, A., Jaenicke, S., Goesmann, A., Junge, R. and Smits, T. H. M. 2017. Microbial diversity in different compartments of an aquaponics system. *Archives in Microbiology*, 199, 613–20.

Schneider, O., Sereti, V., Eding, E. H. and Verreth, J. A. J. 2005. Analysis of nutrient flows in integrated intensive aquaculture systems. *Aquacultural Engineering*, 32, 379–401.

Schroeder, F. G. and Engwicht, K. 2005. Gas concentration in the root zone of *Rosa hybrida* L. grown in different growing media. *Acta Horticulturae*, 697, 49–55.

Scholz, T. 1999. Parasites in cultured and feral fish. *Veterinary Parasitology*, 84, 317–35.

Seawight, D. E., Stickney, R. R. and Walker, R. B. 1998. Nutrient dynamics in integrated aquaculture - hydroponics systems. *Aquaculture*, 160, 215–37.

Shelton, D. R., Kiefer, L. A., Pachepsky, Y. A., Blaustein, R. A. and Martinez, G. 2012. Coliform retention and release in biofilms formed on new and weathered irrigation pipes. *Irrigation Science*, 31(5), 971–81. doi:10.1007/s00271-012-0373-x.

Shenker, M., Harush, D., Ben-Ari, J. and Chefetz, B. 2011. Uptake of carbamazepine by cucumber plants - A case study related to irrigation with reclaimed wastewater. *Chemosphere*, 82, 905–10.

Shete, A. D., Verma, A. K., Chadha, N. K., Prakash, C., Chandrakant, M. H. and Nuwansi, K. K. T. 2017. Evaluation of different hydroponic media for mint (*Mentha arvensis*) with common carp (*Cyprinus carpio*) juveniles in an aquaponic system. *Aquaculture International*, 25, 1291–301.

Sigee, D. C. 2005. *Freshwater Microbiology*. Chichester, UK: Wiley.

Sirakov, I. N., Lutz, M., Graber, A., Mathis, A., Staykov, Y., Smits, T. H. M. and Junge, R. 2016. Potential for combined biocontrol activity against fungal fish and plant pathogens by bacterial isolates from a model aquaponic system. *Water*, 8, 518.

Sonneveld, C. 2000. Effects of salinity on substrate grown vegetables and ornamentals in greenhouse horticulture. PhD, Wageningen University.

Sonneveld, C. and De Kreij, C. 1994. Voedingsoplossingen voor groenten en bloemen geteeld in water of substraten. In: Proefstation Voor Tuinbouw Onder Glas in Naaldwijk (Ed.), *Voedingsoplossingen glastuinbouw.* 10th ed. Naaldwijk, Aalsmeer: Proefstation voor Tuinbouw onder Glas in Naaldwijk.

Sonneveld, C. and Welles, G. 1988. Yield and quality of rockwool-grown tomatoes as affected by variations in EC-value and climatic conditions. *Plant and Soil*, 111, 37–42.

Sopher, C. 2012. High temperature predisposition of sweet pepper to Pythium root rot and its remediation by Pseudomonas chlororaphis. PhD, University of Guelph.

Stanghellini, M. E. and Rasmussen, S. L. 1994. Hydroponics - a solution for zoosporic pathogens. *Plant Disease*, 78, 1128–38.

Stickney, R. 2017. *Aquaculture: An Introductory Text*. Oxfordshire, UK: CABI Publishing.

Suhr, K. I. and Pedersen, P. B. 2010. Nitrification in moving bed and fixed bed biofilters treating effluent water from a large commercial outdoor rainbow trout RAS. *Aquacultural Engineering*, 42, 31–7.

Taiz, L. and Zeiger, E. 2006. *Plant Physiology*. Sunderland, UK: Sinauer Associates, Inc.

Tan, L. P., He, J. and Lee, S. K. 2002. Effects of root zone temperature on the root development and nutrient uptake of *Lactuca sativa* L. 'Panama' grown in an aeroponic system in the tropics. *Journal of Plant Nutrition*, 25, 297–314.

Thompson, H. C., Langhans, R. W., Both, A.-J. and Albright, L. D. 1998. Shoot and root temperature effects on lettuce growth in a floating hydroponic system. *Journal of the American Society of Horticultural Science*, 123, 361–4.

Tidwell, J. H. 2012a. Characterization and categories of aquaculture production systems. In: Tidwell, J. H. (Ed.), *Aquaculture Production Systems*. Ames, Chichester and Oxford: Wiley-Blackwell.

Tidwell, J. H. 2012b. Function and characteristics of all aquaculture systems. In: Tidwell, J. H. (Ed.), *Aquaculture Production Systems*. Ames, Chichester and Oxford: Wiley-Blackwell.

Tokuyama, T., Mine, A., Kamiyama, K., Yabe, R., Satoh, K., Matsumoto, H., Takahashi, R. and Itonaga, K. 2004. Nitrosomonas communis strain YNSRA, an ammonia-oxidizing bacterium, isolated from the reed rhizoplane in an aquaponics plant. *Journal of Bioscience and Bioengineering*, 98, 309–12.

Treadwell, D., Taber, S., Tyson, R. and Simonne, E. 2010. Foliar-applied micronutrients in aquaponics: A guide to use and sourcing. *IFAS Extension*.

Tucker, C. and Hargreaves, J. 2012. Ponds. In: Tidwell, J. H. (Ed.), *Aquaculture Production Systems*. Ames, Chichester and Oxford: Wiley-Blackwell.

Tyson, R. V., Simonne, E. H., Treadwell, D. D., Davis, M. and White, J. M. 2008. Effect of water pH on yield and nutritional status of greenhouse cucumber grown in recirculating hydroponics. *Journal of Plant Nutrition*, 31, 2018–30.

Tyson, R. V., Treadwell, D. D. and Simonne, E. H. 2011. Opportunities and challenges to sustainability in aquaponic systems. *HortTechnology*, 21, 6–13.

Umar, A. S. and Iqbal, M. 2007. Nitrate accumulation in plants, factors affecting the process, and human health implications. A review. *Agronomy for Sustainable Development*, 27, 45–57.

Uren, C. 2001. Types, amounts, and possible functions of compounds released into the rhizosphere by soil-grown plants. In: Pinton, R., Varanini, Z. and Nannipieri, P. (Eds), *The Rhizosphere – Biochemistry and Organic Substances at the Soil-Plant Interface*. New York: Marcel Dekker Inc.

Van Os, E. A., Gieling, T. H. and Lieth, H.-J. 2008. Technical equipment in soilless production systems. In: Raviv, M. and Lieth, H.-J. (Eds), *Soilless Culture – Theory and Practice*. Amsterdam: Elsevier.

Vermeulen, T., Van Der Linden, A. M. A. and Van Os, E. A. 2010. Emissions of plant protection products from glasshouses to surface water in the Netherlands. *Wageningen UR Greenhouse Horticulture, GTB-1002*. Wageningen.

Waechter-Kristensen, B., Caspersen, S., Adalsteinsson, S., Sundin, P. and Jensén, P. 1997. Organic compounds and microorganisms in closed hydronic culture: Occurrence and effects on plant growth and mineral nutrition. *Acta Horticulturae*, 481, 197–204.

Wallace, C., Sperber, W. H. and Mortimore, S. E. 2011. *Food Safety for the 21st Century: Managing HACCP and Food Safety throughout the Global Supply Chain*. Oxford and Ames: Wiley-Blackwell.

Wang, G. and Doyle, M. P. 1998. Survival of enterohemorrhagic *Escherichia coli* O157:H7 in water. *Journal of Food Protection*, 61, 662–7.

Wang, C.-Y., Chang, C.-Y., Chien, Y.-H. and Lai, H.-T. 2016. The performance of coupling membrane filtration in recirculating aquaponic system for tilapia culture. *International Biodeterioration & Biodegradation*, 107, 21–30.

Wilcox, G. E., Martin, G. C. and Langston, R. 1962. Root zone temperature and phosphorus treatment effects on tomato seedling growth in soil and nutrient solution. *Journal of the American Society of Horticultural Science*, 80, 522–9.

Winker, M., Clemens, J., Reich, M., Gulyas, H. and Otterpohl, R. 2010. Ryegrass uptake of carbamazepine and ibuprofen applied by urine fertilization. *Science of the Total Environment*, 408, 1902–8.

Yogev, U., Barnes, A. and Gross, A. 2016. Nutrients and energy balance analysis for a conceptual model of a three loops off grid, aquaponics. *Water*, 8, 589.

Chapter 5

Advances in organic greenhouse cultivation

Martine Dorais, Université Laval, Canada

1 Introduction

While the origins of protected cultivation go back to Roman, Greek and even Babylonian antiquity, the benefits of using green manure in agriculture date back from the Chou dynasty (1100 BCE) in China (Korcak, 1992). However, it was only in 1940 that the term 'organic farming' was first used in England by W. Northbourne in his book *Look to the Land* (Paull, 2011), based on the Steiner's biodynamic concept adopted in 1924 by a group of European farmers in Germany. Biodynamic farming, also known today as Demeter, is believed to be a holistic approach encompassing soil, plants and animals in a self-sustaining system with astronomical influences. Thus, the farm is considered

http://dx.doi.org/10.19103/AS.2019.0052.05

as an organism in which plants, animals and human beings are integrated together (www.demeter.net). The founder of the modern organic movement has, however, been considered by many to be the English mycologist Sir Albert Howard (1873–1947) (Hershey, 1992). Nonetheless, it was only ten years after the publication of *Silent Spring* by Rachel Carson (1962), that the International Federation of Organic Agricultural Movements (IFOAM, www.ifoam.org) was founded (1972) in order to promote the 'worldwide adoption of ecologically, socially and economically sound systems that are based on the principles of organic agriculture'. Therefore, organic farming, also known as biological or ecological agriculture, is more than producing without synthetic fertilisers and pesticides, sewage sludge, growth hormones, antibiotics, nanomaterials and genetically modified varieties or their derivatives (i.e. farm inputs). Organic farming is a holistic production system based on the four IFOAM principles (health, ecology, fairness and care) and defined in 2008 as:

> *'a production system that sustains the health of soils, ecosystems and people. It relies on ecological processes, biodiversity and cycles adapted to local conditions, rather than the use of inputs with adverse effects. Organic agriculture combines tradition, innovation and science to benefit the shared environment and promote fair relationships and a good quality of life for all involved'.*

Ergo, organic farming (1) benefits the quality of ecosystems, (2) uses production methods that conserve and improve soil, maintain water quality and use water efficiently and responsibly, (3) prevents risks by adopting appropriate technologies and rejecting unpredictable ones (e.g. GMO, nanomaterials), (4) sustains and prevents degradation of biotic and abiotic resources, (5) selects species and varieties tolerant to pests (i.e. insects, diseases and weeds) and adapted to the local soil and climate conditions, (6) uses good practices to control pests, (7) adopts measures to prevent from contamination (e.g. GMO, prohibited substances) and (8) adopts packaging that has minimal impacts on the product quality and the environment (IFOAM, 2014). Crops have to be grown in living soils, providing support for a diversity of species. Soil fertility is maintained by returning microbial, plant and animal materials to the soil, with nutrient recycling or other practices mitigating nutrient losses and promoting soil biological activity. For protected crops, the energy used for climate control should be renewable, considering that appropriate technologies are incorporated to limit energy consumption. Other specific requirements are defined in the following sections.

Nowadays, 87 countries have defined their own organic standards, which have been adopted by 172 countries to regulate the use of the term 'organic', protect consumers and facilitate international trade. However, only 43 countries have standards endorsed as organic by IFOAM-Organic International and 18 countries are in the process of defining their legislation (Willer and Lernoud,

2018). The guidelines of the Codex Alimentarius Commission, updated in 2013, provide guidance and substance lists to governments on developing national regulations for organic food. Although organic regulations are country-specific and standard interpretation may vary from one country to the other, a prohibited substance should not have been used during the last 36 months before harvest. The conversion period for horticulture crops shall be 12 months for annuals and 18 months for perennials. Furthermore, the general principles previously stated should be respected. Although organic producers should provide consumers with nutritious and high-quality organic products, it is noteworthy that organic regulation is not a statement about food safety or nutritional value of fruits and vegetables. Examples of organic regulations include the European Commission regulation (EC 834/2007; EC 889/2008; Annexes 1 and 2) that was revised and adopted in 2018 and shall enter into force on 1 January 2021, the U.S. Department of Agriculture's National Organic Program (USDA-NOP), Canada's National Standards for Organic Agriculture (CAN/CGSB-32.310-2015, CAN/CGSB 32.311-2015), Australia's National Standard for Organic and Bio-Dynamic Produce, the Japanese Agricultural Standard for Organic Agricultural Products, South Korea's Organic Regulations and New Zealand's National Organic Standard. Additional guidelines have been issued by organic grower associations such as Bioland (www.bioland.de) and Demeter (www.demeter.de). Products that do not meet organic standards may not be referred to as organic or bear an organic logo. Bilateral equivalence agreements exist between several countries (e.g. the EU with the United States, Canada, Chile and others listed on Third Country list; the United States with Canada, Japan, South Korea and Switzerland; Canada with the EU, Costa Rica, Japan and Switzerland) to facilitate commercial trade, with some exemptions requiring extra verification (e.g. sodium nitrate, hydroponics). For small-scale and/or national markets, participatory guarantee systems (PGS) were also established in 66 countries in order to provide an affordable alternative to organic certification, by using locally focussed quality assurance systems (Willer and Lernoud, 2018).

1.1 Specifics of organic greenhouse cultivation

In organic greenhouse cultivation, restrictions on production growing systems, energy use, artificial lighting, CO_2 enrichment, water and waste management, fertilisers and permitted substances (e.g. biopesticides, soil stimulators, plant-growth promoters and substrates) are set by each country and certifying body. In accordance with IFOAM standards and the revised European organic regulation, organic greenhouse crops must be grown in soil-bound systems (European Council, 2018). Existing areas of confined beds that existed before 28 June 2017 are to be phased out by 2030. Indeed, the European commission (EC) confirms that the link with the soil ecosystem and the plant constitutes a basic principle of organic farming, except for the production of transplants and plants that are grown and sold in pots to the consumer. However, in China, North America, Latin America, Australia, New Zealand, Turkey, the Gulf States and a few other countries, greenhouse fruiting vegetables may be grown in demarcated soil beds or containers (Raviv, 2017; Olivares et al., 2015; Tuzel et al., 2015; Bhat et al., 2013). For organic greenhouse production systems out of the native soil (soilless), organic living substrates such as compost-based growing media must provide the major part of the nutrients to the plants, as compelling under the Canadian organic regulation. When greenhouse production takes place in soilless growing media, no conversion period is required and drained water is usually recycled. Hydroponics, where nutrients are mostly provided under the liquid form via the irrigation system, is generally prohibited, except in the United States (NOP USDA, 2018), which is common. Other examples of producing area specificities are: (1) heating is not allowed in Italy, whereas Demeter (Germany) allows heating up to 15°C, (2) 80% of the energy used should be renewable in Sweden, (3) CO_2 greenhouse enrichment for organic greenhouse crops is not allowed in Spain, while there is no restriction on its use in most EU countries or in North America, (4) CO_2 should come from renewable sources in Ireland, whereas in the Netherlands, CO_2 should be a by-product of heating (e.g. natural gas) and (5) artificial lighting can be used in North America in complement to sunlight (Dorais et al., 2017b), but it is limited to plant propagation in the EU (maximum of 16 h) (van der Lans et al., 2011).

The European Commission Nitrates Directive (EEC, 1991) specifies a maximum of 170 kg N ha^{-1} $year^{-1}$ from manure, a level well below the nitrogen requirement of greenhouse fruiting vegetables. Liquid fertilisers can be used in most countries as complementary sources of nutrients. Organic aquaponics (Chapter 4, Alsanius et al., 2019; Nichols and Savidov, 2012; Pantanella et al., 2010), which combines fish farming with growing systems for plants such as greenhouse lettuces, is under study in several countries that are debating its potential incorporation or not into organic regulations.

2 Expansion of organic protected cultivation around the world

Organic is one of the most dynamic sectors of European (11% increase in 2016 over 2015) and North American (8% increase in 2016 over 2015 in the United States) countries, reaching a global food and drink market of almost US$90 billion in 2016, with 47% of the global market located in the United States and 37% in the EU (Willer and Lernoud, 2018). In fact, organic farming is gaining substantial recognition from consumers, civil society and governments due to its environmental, social and economic benefits resulting from the adoption of sustainable agricultural practices. In order to partly fulfil this fastest growing segment in the global food market, the production of organic vegetables increased by fourfold since 2004, reaching 437 000 ha in 2016 (Willer and Lernoud, 2018). For greenhouse organic crops, the total area in Europe was estimated at around 5236 ha (2000 ha Italy; 2000 ha Spain; 600 ha France; 263 ha Germany; 125 ha The Netherlands; 80 ha the United Kingdom; 57 ha Switzerland; 32 ha Belgium; 30 ha Austria; 25 ha Denmark; 12 ha Sweden; 9 ha Finland; a total of 3 ha in Hungary, Iceland and Estonia), which is almost entirely used for fruit vegetables and lettuce. In non-EU countries, organic greenhouses cover 500 ha in Israel, 230 ha in Morocco, 31 ha in Turkey, 30 ha in Egypt and more than 2275 ha in North America, mainly located in Mexico (Dorais and Cull, 2017). Mexico is the largest organic greenhouse producing area in North America with more than 1875 ha, including more than 116 ha of high technology greenhouses. In the United States, organic protected vegetables cover 186 ha representing 20% of the total vegetable protected cultivated land, while in Canada 87 ha of heated vegetable greenhouses are organically certified, representing 5% of the total greenhouse area in 2017. Additional cultivated land worldwide for greenhouse vegetable production is also managed organically but not certified, owing to the substantial costs involved. Although it is relatively easier to control pests and growing conditions under greenhouse cultivation compared with field production, organic greenhouse cultivation represents less than 2% of the total greenhouse area, considering a total area of 497 815 ha (Hickman, 2018).

3 Types of greenhouse used for organic cultivation

Just like conventional greenhouse cultivation (see Chapters 2, 9, 10 and 11), the level of technology needed by organic greenhouse producers to reach adequate growing environment varies according to the cultivated species, production schedule, markets, local climatic conditions such as temperature, solar radiation, humidity, water availability and quality, as well as renewable and fossil energy cost. For example, light transmission combined with minimal heat loss is a key issue under Northern latitudes, while shading and/or cooling is more important for Southern countries. Furthermore, it also depends on the capital investment, technical expertise and the socio-economic environment of producers (FAO, 2017). Thus, the types of greenhouse used for organic greenhouse cultivation range from high-tech heated and cooled greenhouses generally found in Northern European and American countries to medium-tech greenhouses with relatively low heating capacity and low-tech and passive greenhouses mainly found in Southern countries (e.g. Mediterranean, Mexico) and Asia, which produce vegetables in unheated plastic high tunnels or shadehouses (also called screenhouses or net house), that are naturally ventilated with no climate control. Table 1 describes the main specificities of each type of greenhouse structure generally found for organic greenhouse farming.

Table 1 Characteristics of the three main organic greenhouse cultivation systems

	High-tech heated/ cooled greenhouses	Medium-tech greenhouses	Low-tech greenhouses
Investment	High	Moderate	Low
Structure	Permanent structure made from galvanised steel and aluminium, with main path concrete floors	Permanent structure made with lighter structure	High tunnel, shadehouses and none permanent structure
Covering materials	Mainly glass Plastic in North America is common	Plastic	Plastic or netting
Energy consumption	High to maintain optimal temperature Powerful heating and/or cooling systems Use thermal screens	Low heating capacity to maintain a minimum temperature (e.g. 5–10°C) and to control humidity	Low energy use No heating or active ventilation system No humidity control
Climate control	Sophisticated climate control/semi-closed greenhouses/mechanic or natural ventilation Cooling systems	Basic control system Natural ventilation In some cases, fog or mist system	No climatic control Natural ventilation
Irrigation system	Computerised advanced control system with drip water and nutrient delivery system	Adequate control system	Basic control system
CO_2 enrichment	From natural gas, liquid CO_2 or renewable sources (biogas, waste by-product of cogeneration) When ventilation is closed to maintain 800–1000 $\mu L\ L^{-1}$	CO_2 from the heating system	No CO_2 enrichment
Artificial lighting	Specific to the country organic regulation	No	No
Crop rotation	Limited to few species with similar pest susceptibility (e.g. tomato, sweet pepper, aubergine, cucumber) Biofumigation between consecutive crops	Sometime part of multi-crop growing systems (leafy vegetables, cauliflowers, endives etc.) Biofumigation or biosolarisation	Crop rotation with a large number of species (tomato, cucumber, sweet pepper, eggplant, squash, melon, leafy vegetables, strawberry, beans, cruciferous etc.) Leafy vegetable during winter Biofumigation or biosolarisation

	High-tech heated/ cooled greenhouses	Medium-tech greenhouses	Low-tech greenhouses
Green manure	No crop break with green manure Green manure is added from external input	Seldom crop break with green manure Green manure is added from external input	Part of the crop rotation
Growing period	Intensive and year-round production systems	Less intensive production systems To extend the production season One or two production cycle(s) per year according to the local climate	Low-intensity production systems To extend the production season
Annual yield	High yield	Moderate yield	Low yield
Main growing areas	Northern countries	Central and Northern Europe, Mediterranean, North America	Mediterranean, Asia, Southern countries and some Northern countries

4 Productivity and profitability of organic greenhouse cultivation

Although crop productivity represents a central aspect of farm profitability and food security, it should be viewed as only part of a range of ecological, social and economic factors contributing to benefits delivered by organic agriculture. Hence, a key challenge for sustainable organic greenhouse producers is to achieve yield and product quality as high as conventional farming with no or minimal loss of nutrients and greenhouse gas emission (GHG), while limiting the use of external farm inputs and maintaining an optimal biodiversity of this intensive production agroecosystem. Sale prices of these products should also be affordable for a whole range of the population when ensuring a fair profitability to producers. In addition, the nutritive value of organic greenhouse fruits and vegetables should contribute to the health of the population, and minimise risks associated with contaminants coming from cultural inputs and practices.

4.1 Yield gap

Yields of organic agriculture are usually reported to be lower than in conventional farming. For example, Mäder et al. (2002) reported a yield reduction of around 20% for long-term biodynamic and bio-organic production systems compared with conventional systems. A meta-analysis using organic-conventional comparative yield data from 362 crop yield observations (1984/85 to 2010) pointed out that organic yields were in average 80% of the conventional yields

(De Ponti et al., 2012). Yields for fruits and vegetables were 72% and 80% of conventional yields, respectively, but they vary widely by region (70–73% in Northern and Western Europe; 84% in North America; 88% in Central Europe; 89% in Asia) and crop (e.g. 59% for strawberries; 105% for root and tuber crops, excluding potato). Another meta-analysis conducted by Seufert et al. (2012) using 316 yield observations and 34 different crop species reported a 25% lower average yield for organic compared with conventional farming, which can be partly explained by differences in the amount of nitrogen input received by the two systems. This yield gap was 13% when best organic practices were used. However, they observed no significant yield difference between both systems for fruits, but a yield reduction of 33% when vegetables were grown under organic agriculture. Afterwards, using a larger dataset from 115 studies containing more than 1071 yield observations from 38 countries and 52 crop species over a span of 35 years, Ponisio et al. (2014) found a yield reduction of 19.2% for organic compared with conventional agriculture, with a 95% credible interval ranging from 15.5% to 22.9%. Multi-cropping practices of organic systems, however, reduced the yield gap to 9%, while crop rotations decreased to 8%. They found that polyculture and crop rotations increased yields in both organic and conventional cropping systems. Differences between reported yield gaps from meta-analyses can be explained by the data quality collected from published studies and the use of different analytical methods, which resulted in less or more robust estimates of the gap between both systems. In addition, the over-representation of specific practices or crops in the datasets may greatly influence the estimate of the yield gap (Ponisio et al., 2014). For different horticultural crops commonly grown under protected cultivation, Table 2 gives examples of the relative yield range observed by several studies (Dorais and Alsanius, 2015; De Ponti et al., 2012). It shall be notified that in general, inputs of fertilisers, pesticides and energy are generally lower in the organic growing systems than in the conventional ones.

When comparing the yield gap between organic and conventional agriculture, management systems (different nutrient inputs, cover crops, crop rotations, pest control etc.) shall be considered and not only individual management practices where the same cultivars, cover crops, crop rotation and so on, are compared, which will result in an input replacement comparison instead of a system comparison. Thus, a valid organic system comparison involves the use of different nutrient inputs, more diverse crop rotations, multi-cropping and the use of leguminous crops for nutrient and pest management (Seufert et al., 2012). Consequently, it does not request a nutrient supply equivalence or a non-limiting nutrient supply in both systems. However, it was pointed out that other criteria such as the hierarchy of data and the sampling variance within studies (Ponisio et al., 2014), soil history prior to organic farming, soil physical properties and fertility status, rates of nutrients supplied, periods

Table 2 Relative yield of organic-to-conventional agriculture

Crops	Relative yield (%)
Filed crops	
Bean	70-129
Blackcurrant	85
Blueberry	66
Broccoli	98
Cabbage	96
Carrot	75-112 (89[a])
Cauliflower	53-133
Lettuce	71-101 (86[a])
Melon	95-164
Muskmelon	96-110
Onion	49-83
Pepper	75-150
Savoy cabbage	122
Strawberry	20-150 (59[a])
Tomato	21-175 (81[a])
Watermelon	181
Zucchini	97
Protected crops	
Pepper	88
Raspberry	109
Strawberry	102
Tomato	69-119

[a] Average values according to de Ponti et al. (2012).
Data was obtained by dividing the average organic yield by its paired conventional yield.
Data from Dorais and Alsanius (2015) and de Ponti et al. (2012).

of cover crops and legumes, and even the off-farm inputs and amount of manure transferred from conventional agriculture (Kirchmann et al., 2016) are important for a more global yield gap estimation between different systems.

Nonetheless, under good system design, knowledge and management practices, organic farming can match and even exceed conventional yields. In most organic greenhouse cultivation systems, yield-limiting factors are generally defined by nutrient limitations, specifically N-limitation, due to high nutrient requirements of greenhouse crops such as fruiting vegetables (Section 8), and by a limited pest and disease control (Section 11). Several organic growers experimented years of high yield followed by years of lower yield with limited recovering control tools, making organic cultivation riskier than conventional production.

4.2 Profitability

Profitability of organic greenhouse cultivation varies between locations, years, selected species, targeted markets (e.g. exports, wholesalers, proximity markets, box schemes) and consumer demand. It also depends on cost related to greenhouse climate management (energy demand for heating and/or cooling), yield, total production cost (including labour), competition, price premiums, regulations and government support (e.g. transition period). Analysis of organic wholesale prices in Canada suggests that market conditions are favourable for organic greenhouse crops, as a result of significant price premiums over conventional products resulting in 27–119% higher returns, on a per-weight basis. Nonetheless, the level of profitability of organic greenhouse cultivation largely varies with the cultivated species and achieved yield as shown in Table 3. Likewise, a study conducted in the United Kingdom with strawberry, tomato and cucumber grown under polytunnels or unheated greenhouses showed that organic protected cultivation can be very profitable on a gross margin basis, although it depends on variable market prices and achieved yields (Schmutz et al., 2011). The labour cost (33–42% of the total producing cost) has the highest impact on organic crop profitability. Fertilisers and crop protection account for 5–13% of costs, while other inputs represent 17–23% of the total producing costs. Other important costs are packaging (7–12%) and transport (5–9%), while the structure accounts for only 2–7% of the total costs, assuming that the structure will last for 15 years. For organic greenhouse cucumber grown in Turkey, net return of organic farming was 2–3 times higher than conventional management (US$1.12–1.19 m^{-2} vs US$0.55 m^{-2}), which was associated with a price premium and lower production costs (Tuzel et al., 2005). However, price premium for organic products varies according to the product, region and climatic events. Generally, higher values are observed when the offer is limited or when the prices of conventional products are low.

In a meta-analysis combining results for 55 crops grown on five continents over 40 years, Crowder and Reganold (2015) found that organic agriculture is more profitable than conventional agriculture, requiring break-even premiums of only 5–7% to match conventional profits. They reported that total, variable and fixed costs did not differ between organic and conventional crops or growing systems. However, labour costs were higher for organic crops (13%) and growing systems (7%), while the use of non-renewable resources and inputs (e.g. fertilisers, pesticides) was lower in organic agriculture than in conventional agriculture. The use of longer or more diverse rotations resulted in higher organic-to-conventional financial performance. When the ecosystem services (e.g. biological pest control, soil formation and mineralisation of plant nutrients) provided by the production systems is considered, the total economic value of organic farming is around 1.6 times higher than for conventional agriculture

Table 3 Comparison of wholesale prices and profitability (CAN$) for main organic greenhouse crops under market conditions in Canada

	Organic price premium		Conv. price	Organic price
Category	Lowest (%)	Highest (%)	Range (min-max)	Range (min-max)
English cucumber	157	215	0.79-1.26 $ piece^{-1}	1.71-1.98 $ piece^{-1}
Mini cucumber	131	157	2.23-2.80 $ kg^{-1}	3.48-3.66 $ kg^{-1}
Red sweet pepper	128	150	3.75-5.58 $ kg^{-1}	5.62-7.14 $ kg^{-1}
Orange sweet pepper	129	162	4.10-5.89 $ kg^{-1}	6.66-7.61 $ kg^{-1}
Yellow sweet pepper	127	164	3.75-5.62 $ kg^{-1}	6.17-7.14 $ kg^{-1}
Red round tomato	171	219	1,70-2.69 $ kg^{-1}	3.70-4.59 $ kg^{-1}
Vine tomato	153	233	1.98-3.73$ kg^{-1}	4.61-5.71 $ kg^{-1}
Roma tomato	135	190	1.46-2.38 $ kg^{-1}	2.76-3.22 $ kg^{-1}
Cherry tomato	118	150	3.53-4.54 $ kg^{-1}	5.29-5.36 $ kg^{-1}
Grape tomato	105	138	3.09-4.45 $ kg^{-1}	4.25-4.67 $ kg^{-1}
Iceberg lettuce	143	219	0.83-1.35 $ piece^{-1}	1.82-1.93 $ piece^{-1}
Romaine lettuce	119	158	0.82-1.29 $ piece^{-1}	1.30-1.54 $ piece^{-1}
Red leafy lettuce	122	163	0.66-1.09 $ piece^{-1}	1.08-1.33 $ piece^{-1}
Green leafy lettuce	109	148	0.73-1.12 $ piece^{-1}	1.09-1.21 $ piece^{-1}
Strawberry[a]	158	145	5.24-6.61 $ kg^{-1}	8.27-9.65 $ kg^{-1}

(Continued)

Table 3 (*Continued*)

Difference between org and conv revenue (% return) according to different yield achievements[b]					
Yield assumptions (org/conv)	60%	70%	80%	90%	100%
Round tomato (52.6 kg m^{-2})[c]	+3% to +31%	+20% to +53%	+37% to +75%	+54% to +97%	+71% to +119%
English cucumber (164 fruits m^{-2})	+3% to +31%	+20% to +53%	+37% to +75%	+54% to +97%	+71% to +119%
Yellow Pepper (27 kg m^{-2})	−24% to −1.3%	−11% to +15%	+2% to +32%	+14% to +48%	+27% to +64%
Strawberry (1.5 kg m^{-2})[d]	−12%	+3%	+18%	+32%	+47%

[a] Average of weekly price for 2017.

[b] Differences in organic revenues were calculated by multiplying the yield of conventional and organic commodities per cultivated m^2 with their respective min and max wholesale prices, and then by subtracting organic sale income from the conventional sale income. Note that the largest percentage of return is observed with the minimum wholesale prices.

[c] Data within brackets is the conventional yield (100% yield assumption; 0% yield gap between organic and conventional systems).

[d] Yields are based on Canadian field production. Yield of greenhouse cultivation can reach 10 kg m^{-2} $year^{-1}$ (4.5 kg m^{-2} in winter and 5.5 kg m^{-2} in summer-fall).

Data are from February 2016 to November 2017, corresponding to the harvesting period. Organic price premium is the ratio of organic versus conventional price.

Source: Data from Stat. Canada CANSIM Tableau 001-0006 (https://infohort.agr.gc.ca).

(Sandhu et al., 2010). Organic agriculture can become even more profitable when a cost to the environmental burdens related to farming is considered (Crowder and Reganold, 2015; Section 13).

5 Species and variety selection

The main organic greenhouse cultivated crops are tomato (*Solanum lycopersicum* L.), pepper (*Capsicum annuum* L.), eggplant (*Solanum melongena* L.), melon (*Cucumis melo* L.), watermelon (*Citrullus lanatus* Thunb), zucchini (*Cucurbita pepo* L.), cucumber (*Cucumis sativus* L.), leafy vegetables such as lettuce (*Lactuca sativa* L.), beans (*Phaseolus* spp.), cauliflowers and berries such as strawberry (*Fragaria* × *ananassa* Duch.) (Tittarelli et al., 2016a). Rotation between species of different families is highly desirable to reduce problems related to soil-borne diseases. Although GMO is prohibited, the use of hybrids is accepted in organic horticulture farming. Most of the greenhouse growers use modern and high-yielding varieties selected for conventional and high-input growing systems. Only a few modern varieties have been developed to produce high yields under organic conditions (Lammerts van Bueren et al., 2011; Murphy et al., 2007). In fact, more than 95% of the crop varieties used in organic farming were bred to be cultivated with conventional high-input inorganic fertilisers or pesticides (Lammerts van Bueren et al., 2011). Depending on the market scheme (e.g. food baskets and local farmers' markets), heirloom varieties are becoming more important due to their taste and appearance. Besides flavour, consumers are paying increasing attention to characteristics related to the nutritive value and human health components (Orsini et al., 2016; Dorais and Alsanius, 2015; Maggio et al., 2013; Oliveira et al., 2013; Hallmann and Rembiałkowska, 2012; Smith-Spangler et al., 2012; Brandt et al., 2011; Dorais et al., 2008). In order to increase plant vigour and tolerance to major diseases and pests as well as to abiotic stresses (temperature, drought, flooding, salinity, soil pH, contaminants etc.), grafting selected varieties onto appropriate rootstocks are common (e.g. tomato, eggplant, cucumber, melon, pepper; Kleinhenz, 2015; Barrett et al., 2012; Janse and van der Wurff, 2011; Colla et al., 2010; Louws et al., 2010; Savvas et al., 2010; Schwarz et al., 2010).

6 Seeds and seedlings

All seeds, seedlings and rootstocks must come from organic production, and no genetically modified organisms are allowed. Selection of organic seeds and vegetative propagation materials are proposed by some organisations (www.organicxseeds.com). Seeds used by organic growers should be untreated or treated with methods approved by organic standards. Seed testing against diseases and alternative treatment techniques such as treatments with acidified nitrite in citric acid, copper-based compounds, vinegar, plant and compost extracts, mustard seed extract, chitin, essential oils, lime and milk powder, combined or not with beneficial microorganisms, along with thermal treatment may be used to avoid fungi and bacterial diseases (Koller et al., 2016; Kasselaki et al., 2011, 2008). However, phytotoxicity effects, and moderate or low efficacy, can be observed. Seedlings are generally made with a mixture of peat, or any local materials, and mature compost amended with commercial organic fertilisers (e.g. vermicompost, alfalfa meal, vinasse, potato protein malt, animal-based nitrogen fertilisers, rock phosphate, potassium sulphate) and limestone to reach a pH value of 6.0–6.5 (Jack et al., 2011; Koller et al., 2004). Liquid organic fertilisers are often provided as complements during the final weeks in the nursery.

7 Soil and soilless organic greenhouse growing systems

The organic greenhouse production system is a highly intensive production system with yields up to ten times higher than those produced in open fields (Dorais and Alsanius, 2015; Voogt et al., 2014). Even under Northern growing conditions, greenhouse cultivation continues year-round to fulfil demand for organic fruits and vegetables outside the field growing season. To achieve high productivity of these high-value crops, soil health that relies on building organic matter through cover crops, crop rotation, manure and compost is the cornerstone of organic growing systems. However, due to the cost of the infrastructure, particularly for high-tech greenhouses (Dorais, 2018), the use of agroecological service crops (e.g. cover crops, green manure and intercrops), crop rotation and limited use of off-farm inputs, as often seen for organic field crops, are unusual for organic greenhouse production systems (Tittarelli et al., 2016a). Indeed, minimisation of off-farm inputs is difficult, given that up to 85% of the fresh-crop biomass is exported off-farm. Consequently, organic greenhouse production systems may be perceived by some as a 'conventionalised' organic farming system, where off-farm synthetic inputs are replaced by off-farm organic inputs that are permitted under country-specific organic standards (Constance et al., 2015; Goldberger, 2011; Darnhofer et al., 2010). On the other hand, these intensive production systems offer great opportunity in terms of land, water and energy use, contributing to the global challenges of food and nutrition security under climate change, human population pressure and urbanisation.

7.1 Soil growing systems

7.1.1 Organic matter

Too much or too little organic matter can result in nutrient imbalances (excess or deficiency) and nutrient leaching to groundwater. A 3.4% organic matter content in the top soil (0–30 cm) was determined by the EC as the minimum threshold for maintaining a healthy soil (Morra et al., 2010), although that threshold is soil and site dependent. Nevertheless, low values (e.g. 1%) are often observed in some Mediterranean areas such as Almeria, because of the limitation related to the cost and availability of high-quality composts and manure products (Tittarelli et al., 2016a). In temperate regions, organic matter contents of 3–5% are recommended in intensive greenhouse cultivation (van der Wurff et al., 2016). Higher contents (11–33% after 15–20 years of cultivation) can also be observed under Northern growing condition following several years of important compost amendment (Pepin et al., 2008). On the whole, soil organic matter content of organic greenhouse crops is around 5–7%, owing to the yearly amendments of composts, crop residues and manure (Tittarelli et al., 2016a). Cultural practices that maintain or increase organic matter in the soil and promote soil disease suppressiveness shall be adopted by producers.

7.1.2 Soil properties

High soil biological activity that ensures soil and plant health along with optimal nutrient flows to the crop depends on the soil physicochemical properties. First, high-performing soil must have appropriate texture and structure (e.g. bulk density, air-filled porosity, gas diffusion, water-holding capacity, easily available water) that create an appropriate balance of air and water at the root level. Secondly, effective soil must have suitable pH, electrical conductivity, cation exchange capacity and right balance between nutrients. It should also provide enough carbon and nutrients to microorganisms ensuring a high and diverse biological activity, without a negative nutrient immobilisation (e.g. nitrogen, phosphate) that might affect the crop productivity. In order to promote the microorganism population and gradual and stable nutrient release, without nitrogen immobilisation, a C:N ratio of 25–30 was proposed (Müller et al., 2006). As a matter of fact, organic greenhouse soils generally have a larger pool of organic carbon and microbial biomass and higher enzyme activity compared with soils in conventional farming. Finally, the soil should be free of pathogens and pests, and full of beneficial microorganisms and their derived compounds, ensuring plant resilience to abiotic and biotic stresses and contributing to cropping performance, fruit quality and crop sustainability.

7.1.3 Agroecological service crops

The inclusion of agroecological service crops within an organic cultivation system is highly beneficial because they (1) are an excellent and economical way to add considerable amounts of organic matter and nutrients to the soil (Campiglia et al., 2011; Partey et al., 2011; Wang et al., 2009; Clark, 2007;

Sullivan, 2003), (2) have a sustainable soil-building role that reduces soil erosion, rebuilds soil texture and water retention, (3) aid in loosening the soil, (4) reduce the potential for N leaching losses and C losses and (5) suppress weeds, nematodes and diseases (Tittarelli et al., 2016a; Ciaccia et al., 2015). However, their use for organic greenhouse production face a lot of economic constraints (e.g. reduction of the cash crop production period) and depends on the farm structure (Tittarelli et al., 2016a; Voogt, 2014). For example, short-term green manure crops (e.g. *Leguminosae*, *Brassicaceae*, *Poaceae*; 5–7 weeks) during the summer season are common in less-intensive greenhouse growing systems (high tunnels, cold greenhouses), while for high-tech greenhouses their use is mostly for pest control (biofumigation). As shown in Table 4, growing legumes as cover crops (low C:N ratio) is an economical way to provide available N to the plant within a short time, while non-legume species (high C:N ratio) provide substantial organic matter but may lead to N immobilisation. It was earlier reported that mixing cereals and legumes, which is often recommended, increases the effectiveness of N sequestration and nutrient-use efficiency via the formation of hyphal bridges between the roots of the two species, transferring legume N to the cereals (Ranells and Wagger, 1997).

7.1.4 Crop rotation

The implementation of appropriate agroecological service crops combined with crop rotation (e.g. 3-year crop rotation) is a sustainable system growing approach that maintains soil and plant healthy. The design of a suitable rotation depends on disease and pest cycles, the requested amount of nutrients by the crops and the needs of maintaining or improving the soil structure. However, in practice, crop rotation in greenhouses is often limited to cash crops of similar families due to the high investment and operation cost related to these high-value products. Examples are given by Tittarelli et al. (2016a) and Dorais (2017).

7.2 Soilless growing systems

7.2.1 Advantages

Soilless organic greenhouse production, also known as raised or demarcated beds or bag/container growing system, constitutes sustainable organic growing systems that adhere to the organic principles of limited farm inputs when waste biomass and crop effluents are recycled to the production system and organic solid amendments provide the major part of nutrients via a high soil biological activity. Indeed, growing out of the soil that is connected to the earth's surface offer the advantages of (1) cultivating in area that would normally

Table 4 Examples of agroecological service crops used in organic greenhouse systems

Family	Species	Nutrients (kg ha^{-1})			Biomass	Functions[a]
		N	K	P	(t ha^{-1} dw)	
Poaceae						• Fast-growing crops • Improvement of soil structure • Build soil organic matter • Provide nutrients • Weed suppression • Beneficial to insects • Banker plants for biological control • Suppressive effect
Fall	*Secale cereale*	54-159 (78%)[b]	121	19	3.36–11.55	
	Avena sativa				2.24–11.20	
	Triticum spp.				3.36–8.96	
	L. multiflorum	33			2.24–10.08	
Summer	Sorghum spp.	30–110 (30–50%)	64–188	19-57	3.30–12.00	
	Pennisetum spp.	65			6.66	
Fabaceae						• N fixation • Provide nutrients • Improve soil organic matter
Fall	*Vicia* spp.	111–236 (70–92%)	149	20	2.58–8.29	
	Trifolium spp.	78–246[c]	160	18	2.24–11.20	
	Pisum spp.	111–168 (77%)[d]	178[d]	21[d]	4.48–5.60[d]	
Summer	*Vigna* spp.	75–243 (90%)[e]	111–124	15	2.80–11.70	

(*Continued*)

Table 4 *(Continued)*

Family	Species	Nutrients (kg ha^{-1}) N	K	P	Biomass (t ha^{-1} dw)	Functions[a]
Brassicaceae						• Fast-growing crops • Improvement of soil structure • Provide nutrients • Suppressive effects • Soil-borne pathogens and pests • Weed suppression
Fall	*Raphanus* spp.	56–224			4.48–7.84	
	Brassica spp.	43–224			2.24–10.37	
Summer	*Brassica* spp.	43–224			2.24–10.37	
Polygonaceae	*Fagopyrum esculentum*	48			2.24–7.41	• Build soil organic matter • Help to suppress soil pests • Weed suppression

[a] Adapted from Tittarelli et al. (2016).
[b] Percentage of N release within the growing season. Note: These values vary depending on climate, soil, seeding rate and cropping system; potential rate of N decreases with plant maturity. Sources: Sullivan (2003), Wang et al. (2009), Clark (2007), Campiglia et al. (2011) and Partey et al. (2011).
[c] *T. pratense, T. incarnatum, T. subterraneum, T. alexandrinum, T. repens.*
[d] *P. sativum.*
[e] *V. unguiculata.*

be unproductive due to poor soil texture, structure and fertility, persistent soil pests or contaminated soils (e.g. by salinity, alkalinity, pesticides, heavy metals, pharmaceutical and personal care products); (2) reducing risks related to infection by soil-borne diseases, nematodes, weeds and other soil-borne pests; (3) reducing nutrient run-off to groundwater by collecting crop effluent and reusing it on the farm; (4) achieving optimal crop-nutrient management by fine-tuning the water and fertilisation regime; (5) achieving similar yields to conventional crops, by producing fruits and vegetables with high-quality attributes; and (6) facilitating grower transition from conventional to organic greenhouse farming as no waiting conversion period is requested, resulting in higher farm profitability. In addition, these organic growing systems may be used for urban vertical or rooftop farming systems where space is limited at ground level, contributing to food security in urban areas and to food safety by stringently controlling physical, chemical and biological hazards to the produce.

7.2.2 Growing systems

The size of the containers or demarcated beds used by organic producers largely varies (e.g. from 11 L per container to over 180 L m^{-2} of demarcated beds) according to the growing systems and country regulation. For example, for staked indeterminate plants such as tomato, cucumber and sweet pepper, a minimum soil/growing media volume of 60 L m^{-2} is requested by the Canadian organic regulation (CAN/CGSB-32.310-2015), while no minimal volume is requested by the NOP-USDA. Whereas small container volume provides limited capacity for water and nutrient storage, larger soil volume offers a buffer against suboptimal water and fertilisation management, and generally promotes the microfauna. The growing media used are generally a mix of several components such as coconut coir, perlite, peat, compost and other authorised local substrates (e.g. pumice, volcanic rock/tuff, bark, sawdust, wood fibre, biochar, marine components, organic waste components of the agri-food industry). Depending on the organic regulation, a minimal content of minerals (e.g. clay, sand) shall be added. As for conventional production systems, soilless growing media should have a high macro-porosity, ensuring good drainage, gas diffusion and oxygen supply to the roots between irrigation events, which drive root water and nutrient uptake, as well as appropriate micro porosity for an optimal retention of water (Barrett et al., 2016). An optimal balance between air and water porosity is particularly important for organic crops because microorganisms compete with the plants for the oxygen. Furthermore, their physicochemical and biological properties as well as their lifetime (related to the microbial decomposition, shrinkage, loss of air-filled porosity, compaction), cost and sustainability largely vary according to the growing media (Barrett et al., 2016; Castilla, 2013; Urbain and Urbain, 2010; Raviv and Lieth, 2008). In all cases, the growing medium must provide an optimal biological activity and chemical environment to protect the plants against plant pathogens and insure an optimal plant-nutrient efficiency. It must also be affordable, easy to obtain and manageable (Barrett et al., 2016).

To fulfil plant nutrient requirements, organic solid fertilisers are provided at 1- to 3-week interval. Complementary liquid fertiliser is supplied (e.g. potassium sulphate), although clogging (biofilm development) may occur when organic liquid fertiliser is used regularly. The use of compost tea and biofertilisation with plant growth-promoting agents (e.g. PGPR, PGPF, seaweed extracts, horsetail extracts and humic acids; see Section 9) is a common strategy promoting the biological activity and plant growth, and suppressing plant pathogens. After the growing season, the growing medium is reused (e.g. large containers or demarcated beds), or replaced, depending on soil properties (e.g. soil physicochemical properties, diseases and pests).

7.3 Hydroponics and aquaponics

Organic hydroponics, seen as an oxymoron, is defined as a production system that provides nutrients exclusively via an aqueous nutrient solution with little involvement of the biological activity of the soil. Plants are grown in small containers, bags or slabs placed on soil, tables or gutters, as well as on raft growing systems (deep water) and nutrient film technique (NFT) where no soil or growing media is used. Organic aeroponics, used by some urban vertical farmers, is a sophisticated hydroponic system where roots are exposed to an air and mist environment. These hydroponic liquid nutrients contain assimilable inorganic and small organic molecules as well as beneficial microorganisms that promote plant growth, productivity and protect the crop from pathogens (Vallance et al., 2011). Organic liquid fertilisers used are commercial or on-farm products such as fermented grains, digestates from the anaerobic digestion of vegetable and animal household waste, fish extracts, fish effluents or any other liquid agricultural by-products approved by the organic standards. However, organic hydroponic production systems often have issues with clogging in the irrigation system due to biofilm formation, caused by the living organisms in organic liquid fertilisers. Commercial liquid fertilisers and soluble organic fertilisers are generally more expensive than solid organic amendments and fertilisers. Even though organic hydroponics and aeroponics may constitute sustainable production systems, it is prohibited in most countries, although being permitted in countries such as the United States.

7.4 Aquaponics

Aquaponics origins from ancient techniques used by the South American and Asian population using floating rafts or dike-pond growing systems. Nowadays, organic aquaponics is defined as a symbiotic integrated production system combining organic aquaculture and hydroponics. In aquaponic systems, the fish waste provides nutrients to the plant and the plant acts as a biofilter for the water, removing toxic fish compounds such as ammonia (Somerville et al., 2014). This organic production tandem involving nutrient cycling and water conservation, although not recognised as organic in European countries, is allowed in several countries such as the United States, Australia, China and it is awaiting approval/consideration in Canada. Studies found that, for some species (e.g. leafy vegetables and herbs), similar productivity levels to conventional growing systems may be achieved with a high fish density under suitable environmental conditions (Nichols and Savidov, 2012; Pantanella et al., 2010). Because some species such as tomato are sensitive to ammonia, which may cause physiological disorders, the ammonia in the fish waste should be converted to nitrate by bacteria via a biofilter and then supplied to the tomato plant rhizosphere in a closed recirculating system. It was also shown that the use of crude fish effluent reduced tomato root colonisation by *Pythium ultimum* and *Fusarium oxysporum*, and had a stimulating effect on tomato plant height, leaf area and root dry weight (Gravel et al., 2014). Conventional aquaponic system is covered in detail elsewhere in this book.

8 Fertilisation management

Soil fertility is the core of organic farming. The soil should provide the required nutrients and environment to the plant via its biological activity (living soil), which ensures plant health and optimal productivity. Unlike conventional production systems, fertilisation decisions in soil- or substrate-based organic growing systems are complex and involve considering short- to long-term nutrient storage, nutrient-use efficiency, nutrient cycling, disease resistance and the well-being of beneficial soil microorganisms. In general, the organic standard regulations make no difference between organic fertilisation management of greenhouse and field crops. Some additional regulations, however, may create important constraints such as the European Nitrate Directive where the application of animal manure is limited to the equivalent of 170 kg N ha^{-1} $year^{-1}$.

8.1 Soil nutrient balance

The main challenge for organic greenhouse fertilisation management is to synchronise well-balanced soil nutrient release (e.g. N, K, P, Mg and Ca; with limited undesirable Na, Cl and SO_4) with plant nutrient requirements and uptake, without or with a minimum amount of losses via denitrification, leaching or emissions into the environment. As previously mentioned, greenhouse production is an intensive and long season (up to year-round) growing system characterised by a high productivity, essential to recover investment costs, and consequently requires continuously high inputs of nutrients as compared with field production (Voogt et al., 2011). Although N immobilisation by microorganisms (e.g. high C:N ratio inputs) should be avoided or limited, too rapid N mineralisation (e.g. low C:N ratio) may result in too vegetative crop development and N leaching losses. Ratios between individual macro elements should be well balanced according to the crop requirements to avoid plant imbalances and environmental burdens. However, most of the organic amendments are unbalanced in terms of N, P and K. For example, the mutual mass ratio of crop N, P and K demand of greenhouse fruiting vegetables is respectively around 37%, 6% and 57% of the total N-P-K demand, while the averaged organic amendment generally supplied by producers is around 43%, 11% and 46% (Voogt et al., 2011). Consequently, it was reported that the total supply of P can be 1.8 times of the crop removal resulting in a significant P accumulation, while N losses by denitrification might be as high as 25% (De Visser et al., 2006).

8.2 Nutrient-crop demand

The nutrient demand is crop specific and varies according to the growing conditions (e.g. light, temperature, CO_2, water). For organic greenhouse crops, the nutrient demand by the plant is generally expressed by yield expectation as a close linear relationship between the crop yield and the uptake of nutrients (e.g. N, P and K) is observed (Voogt, 2014; Sonneveld and Voogt, 2009). For example, the annual nutrient demand of organic greenhouse tomato yielding 50 kg m^{-2} is established at 1019 kg N ha^{-1} (~150 kg N ha^{-1} $month^{-1}$ between the 10th and 42nd weeks of cultivation), 1745 kg K ha^{-1}, 282 kg P ha^{-1} and 172 kg Mg ha^{-1}, while for sweet pepper and cucumber yielding 23 and 60 kg m^{-2}, respectively, the nutrient demand is 901 and 995 kg N ha^{-1}, 1405 kg K ha^{-1}, 152 kg P ha^{-1} and 126 kg Mg ha^{-1} (Voogt, 2014). Nutrient requirements of the most common greenhouse cultivated crops are shown in Table 5.

8.3 Nutrient supply

In order to fulfil plant-nutrient demand achieving optimal yield, nutrient management must take into account the mineralisation rate of the organic amendments and fertilisers given as a base (before planting) and top dressing (during the production season) as well as the loss by denitrification and leaching. Over-fertilisation should be avoided by fine-tuning nutrient supply

Table 5 Examples of nutrient requirements of protected crops per kg of yield

	g kg^{-1} of expected yield			
Crop	N	P	K	Mg
Tomato	2.0-3.5	0.4-0.6	3.5-5.2	0.35-0.5
Pepper	3.9-5.4	0.5-0.7	4.2-6.1	0.5
Eggplant	5.0-5.2	0.6	5.0-5.2	0.5
Cucumber	1.4-1.7	0.25-0.5	2.2-2.9	0.2
Zucchini	3.8-4.5	0.7-1.3	7.5-7.9	0.2
Watermelon	1.7-2.3	0.6	2.3-2.9	
Melon	3.0-4.4	0.6-0.7	4.2-4.8	
Lamb's lettuce	4.0	0.7	6.0	0.3
Lettuce	2.0-3.2	0.3-0.5	3.1-5.0	0.3
Parsley	3.2-5.4	0.7	3.8-9.4	0.4
Beans	4.0-9.3	0.8-2.0	3.6-6.9	0.6
Strawberry	3.5-4.7	0.9-1.1	5.4-7.4	
Radish	3.0-3.9	0.4-0.5	2.5-4.4	0.2

Source: Adapted from Koller et al. (2016) and Voogt (2014).

with crop demand, and by using advanced irrigation schedules to limit any nutrient leaching (Dorais et al., 2016a; Voogt, 2014). Soil fertility management also varies according to the type of the cropping systems used as presented in Table 6. For all cases, however, the current strategy of nutrient supply of organic greenhouse crops is based on soil analyses (Kjeldahl, Mehlich and SSE analyses at soil depths of 0–25 cm and 25–50 cm) to determine the plant-available minerals. As the growing period for greenhouse fruiting vegetables is long, only part of the nutrient can be applied as base dressing to avoid excessive levels of nutrients and salinity during the first months where the plant is under development (low fruit load), and then create a lack of nutrients during the active period of production (high fruit load).

Based on the soil-nutrient analysis, 50–90% of the total N input and up to 75% of the K supply are provided during the production period (top dressing) in order to avoid excessive soil salinity values and unbalanced plants (e.g. generative to vegetative ratio), impairing root and crop development, productivity, product quality and soil microbial activity. Model-based approaches, according to nutrient supply, nutrient mineralisation rates (nutrient availability) and plant nutrient requirements, should be used to optimise crop performance and limit any nutrient emissions to the environment and increase the sustainability of organic greenhouse crops. A stepwise approach using a combination of simple calculations and model-based inputs was proposed for organic greenhouse vegetables (Voogt, 2014). Recently, based on the nitrogen mineralisation rate of different organic amendments during 1 year (Fig. 1) and

Table 6 Characteristics of soil fertility management according to different types of the cropping systems

	Northern greenhouses		Southern greenhouses	
	High-intensive system	Less-intensive system	High-intensive system	Less-intensive system
Producing period	8–12 months	Extended the production season	Extended winter season – off-season production	Short term – winter season
Crops	High-value crop (*Solanaceae, Curcurbitaceae, Asteraceae*)	Diverse	High-value crop (*Solanaceae, Curcurbitaceae, Asteraceae*)	Diverse
Yield per year	High	Low to moderate	Moderate	Low
Soil organic matter	5–7% up to >20%		Low (~1%)	Low
Compost and manures	Compost and 170 kg ha^{-1} animal manure	Compost and manure	1.0–1.5 t ha^{-1} $year^{-1}$ 170 kg ha^{-1} animal manure (goat, sheep)	1 to 30–50 t ha^{-1} $year^{-1}$
Fertilisers	No green manure	Green manure is not common	Legume or green manure Crop residues (50% of N, P and K)	Legume and green manure
Top dressing	Common for N and K Up to 50–90% of N input	Low level to commonly used (up to 50–60% of N input)	Common for N, K and S P when needed	Common for N and K Up to 30–40% N input
Fertigation	Widely used for K and Mg	Sometimes for fruiting vegetables	Mainly for N, K and Ca Amino acids, residues of sugar beet processing, humic acids, potassium sulphate	Mainly for N, K and Ca
Input-output nutrient budget	Oversupply of P	Total N and P supply are much higher than the crop N removal		
Micronutrients	Rare		Chelated forms when needed	Chelated forms when needed

the NLOS model (Bittman et al., 2001), a new user-friendly model was proposed for organic greenhouse producers growing in soil or in peat-based containers (Dion et al., 2019a,b).

8.4 Soil organic amendments and fertilisers

8.4.1 Compost

The ecological and agricultural benefits of using compost were clearly demonstrated by extensive studies. For organic farming, compost, as a product recycling process, constitutes the basement of a sustainable soil management by providing energy sources and nutrients to preserve soil biodiversity (macrofauna and microflora) and soil functionalities. For organic greenhouse horticulture, compost is used as an organic matter input to (1) improve the physicochemical properties of the soil (decrease the soil bulk density; improve soil aggregation, water transport, water-holding capacity, cation exchange capacity, pH stabilisation); (2) promote biological quality attributes of the soil contributing to soil disease suppressiveness (van der Wurff et al., 2016; Raviv, 2010); and (3) provide nutrients (van der Wurff et al., 2016; Reddy and Crohn, 2012). The methods, types, processes, nutrient values and quality of compost for organic greenhouse horticulture are well described by van der Wurff et al.

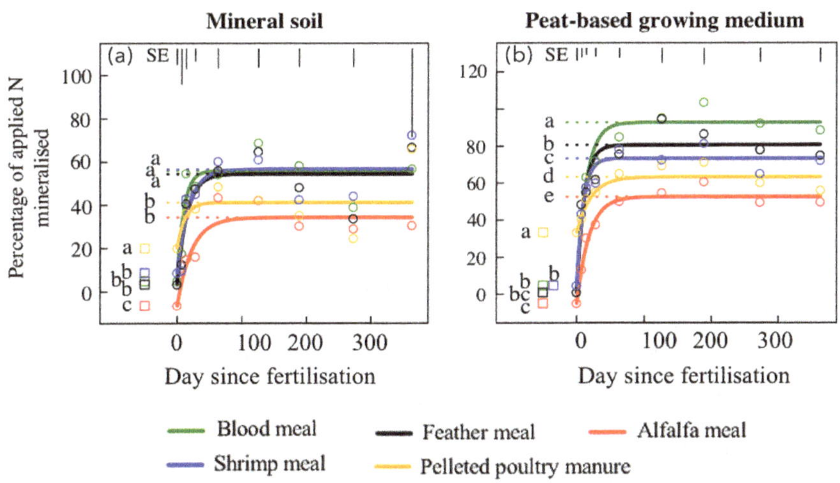

Figure 1 Nitrogen mineralisation curves of different organic amendments supplied to (a) mineral soil (loam) and (b) peat-based growing media over a 1-year period. The points represent the fraction of N mineralised after 18 weeks. Points with a same letter are not significantly different according to LSD test ($\alpha = 0.05$). Equations (not shown) follow the form $fN_{min} = y_0 + a\left(1 - e^{-bt}\right)$, where fN_{min} is the percentage of applied N mineralised after t days. Source: Dion et al. (2017a).

(2016). The decomposition rate of compost varies with the soil structure and texture, pH and temperature; in organic greenhouse soils of temperate regions, around 3–4% is decomposed annually. An annual application of compost is around 30 tonnes ha^{-1} for soil-bound organic greenhouses located under temperate climate, although higher amounts might be provided by some growers, while much smaller quantity is provided under Southern regions such as Almeria (Dorais and Schwarz, 2018; Tittarelli et al., 2016a; van der Wurff et al., 2016).

Composts and compost teas added to the soil were shown to increase disease suppressiveness against several species of *Fusarium*, *Verticillium*, *Pythium* and *Phytophthora*, as well as bacterial spot (*Xanthomonas vesicatoria*), early blight (*Alternaria solani*), bacterial canker (*Clavibacter michiganensis*), corky root rot (*Pyrenochaeta lycopersici*) and grey mould (*Botrytis cinerea*) (Dorais, 2017; van der Wurff et al., 2016; Dorais and Alsanius, 2015). Microbial content, nutrients and organic molecules such as humic or phenolic compounds in compost and compost tea were the main factors in the inhibition or prevention of plant disease (van der Wurff et al., 2016; Dorais and Alsanius, 2015). Microorganisms may act through direct pathogen antagonism (parasitism, antibiosis, competition, predation) or through plant resistance induction. Nutrients and organic compounds in compost may induce systemic resistance and/or may have a direct toxic effect on the pathogen.

The nutrient value of compost can vary by three to four times according to the input materials (van der Wurff et al., 2016; Reddy and Crohn, 2012). Depending on compost maturity only 0–10% of the N is available in the first year. Ammonium and ammonia are present in young composts, while nitrate and nitrite are present at low levels in more mature composts (van der Wurff et al., 2016). Nonetheless, composts with high C/N rarely provide enough N to meet greenhouse crop requirements (Burnett et al., 2016). To avoid risk of N immobilisation, the NO_3-N/N_{min} ratio should be >0.8 (van der Wurff et al., 2016). Excessive salinity and P levels are issues often related to the use of high quantity of compost. Indeed, EC of soil-compost mixtures can be as high as 10.6 mS cm^{-1} (Reddy and Crohn, 2012). Composts used for greenhouse production is usually a mix of manure and plant residues. Examples of nutritive values of compost are given in Table 7, and additional values for horticultural crops are given by van der Wurff et al. (2016) and Reddy and Crohn (2012). The pH value of compost depends on the input material and composting process, and often ranged from 7.2 to 8.5. The quantity of heavy metals in the compost depends on the input material used and should not be higher than the maximum-allowed contents, which is country specific. In Europe, the maximum content is in g t^{-1} DM: 45 Pb, 0.7 Cd, 70 Cu, 25 Ni, 0.4 Hg, 200 Zn, 70 Cr. Compost should be free of plant pathogens and weeds, which is normally achieved when the temperature during the composting process reached >55°C for 3 weeks or >65°C for 1 week (van der Wurff et al., 2016).

8.4.2 Biochar amendment

Biochar, defined as a charcoal made from the pyrolysis of plant biomass, is used as a soil amendment in cultivated soils to improve the agronomic efficiency of cropping systems and reduce the loss of nutrients into the environment (Shaaban et al., 2018). Indeed, numerous studies have demonstrated that biochar amendments increase soil mineralisation and nutrient retention by (1) stimulating mycorrhizae, symbiosis and the activity of microorganisms in the soil; (2) enhancing the enzymatic activity and the abundance of soil microorganisms and biofilms; (3) improving soil cation and anion exchange capacity; (4) reducing fertilisation needed and nutrient leaching; and (5) reducing the emissions of greenhouse gases of agricultural origin, hence contributing to the environmental performance of biochar-amended cropping systems (Lévesque et al., 2019, 2018; Kalakodio et al., 2018; Shaaban et al., 2018; De Tender et al., 2016; Lehmann et al., 2011; Thies and Rillig, 2009). Such improvements in soil's biological and physicochemical properties are partly attributable to the high specific surface area and large number of macro- and microscopic pores found in biochar fragments as well as its labile organic compounds, which stimulate microbial activity, promote biological immobilisation of N and limit N leaching (Vaccari et al., 2015). Biochar also has a high capacity to adsorb soluble organic matter, gases and inorganic nutrients and constitutes as such a suitable habitat for microbial populations to develop and reproduce. Despite these multiple benefits, its effect is highly variable according to the biomass source, pyrolysis temperature, physicochemical properties, application rate and economic feasibility (Kalakodio et al., 2018; Kavitha et al., 2018). For greenhouse crops, biochar was proposed to partly replace unsustainable growing media components such as peat and perlite (Lévesque et al., 2018; Dorais et al., 2017a; De Tender et al., 2016; Nemati et al., 2015). Adding biochar to the growing media increased greenhouse tomato and pepper yields (Lévesque et al., 2019; Kolton et al., 2011; Graber et al., 2010). For organic greenhouse tomato grown in different types of soil and growing media, biochar amendment (10–20% v/v) enhanced the soil biological activity by up to 83%, increased soil mineral content (by up to 47% NO_3, 17–25% P, up to 42% K, up to 14% Mg, 50–139% Cu, 35–98% Mn, 24–78% Fe) and decreased N losses by leaching by 30–61%, but had no significant effect on crop productivity (Gagnon et al., 2016; Laurin-Lanctôt, 2015). However, replacing 50% of the soil (sandy loam) or peat-based growing media by biochar decreased organically grown tomato seedling height and total dry mass, while no significant effect was observed for sweet pepper, basil and geranium. A proportion of up to 30% of biochar

Table 7 Examples of nutrient values of composts and organic fertilisers commonly used in organic greenhouse horticulture

Source	N (g kg^{-1} DM)	P (g kg^{-1} DM)	K (g kg^{-1} DM)	Ca (g kg^{-1} DM)	Mg (g kg^{-1} DM)	OM (% DM)	DM (% FW)	C:N	Salt content (mg KCl/100 g FW)
Greenhouse compost	8.6-25.2	2.1-8.8	5.5-27.8	29.5-69.4	4.4-13.3	20.9-52.8	32.2-64.5		328-1539
					(% DM)				
Digestates[a]	5.29-16.3	0.71-2.21	1.31-12.8	0.20-2.81	0.21-0.96	56.5-72.1	3.01-8.97	2.0-6.7	15.8-65.4
(g kg^{-1} FW)	N	P_2O_5[b]	K_2O	Ca	Mg	OM[c]	DM	C:N	% N release[d]
Animal-based composts									
Dairy manure	12-15	25-69	11-22					10.55	
Cattle manure	8	5-9	5-14			210	330		20-30
Sheep manure	11.5	7	23			260	360		20
Pig manure	7-11	10-43	3-21				310-343		
Poultry manure	13-50	24-69	7-35			479-577	637-780	5.7	50
Turkey manure	22-28	31-40	25-32			454-530	687-767		50
Plant-based composts									
Wood waste	4-15	1-4	15-25	14-71	0.6-4.8			25	
Green/yard waste	5-16	3-7	2-17	5-29	1.2-2.4	450	500	11-25	
Crop residues	12	5	17					9.3	
Fruit & vegetable waste	14	6	14						
Brewery waste solids	13-18	0.5	1.6-2.2						
Animal-based fertilisers									
Bone meal	20-114	150-160	0			558-934	940-950	4.2-4.7	55-66
Blood meal	111-158	0-7	0-8	31	1.4			3.7	
Feather meal	101-142	0-13	0-2.7	10	0.5	880	940	3.3-4.8	64-82
Fish meal	60-110	40-90	20				920-980		
Meat meal	81	93	6			667	940	4.8	70

Cattle manure	5.7 (31% NH_4)	3.6	5.3				210	16.6	45
Dairy manure	18	11	22	20	10			8.2	
Chicken manure	25-28 (21% NH_4)	23-44	18-39	140	7			11.6-14.5	70
Pelleted chicken manure	20-52	15-119	15-31	68	4			4.5-6.3	
Pig manure	2.7-6.0 (71% NH_4)	1.5-11.5	1.6-6.0				32	3.3	65
Seabird & Guano	90-160	30-202	10-30				840		82-93
Marine-based fertilisers									
Liquid fish residues	40	20	20						
Fish powder	137	14	13						
Crab waste	82	7	2-4						
Shrimp meal	57	109	1.2	118	9			4.7	
Algae products	5-14	1-2.5	3-217	3-11	1-6	200	150-200		
Plant-based fertilisers									
Alfalfa meal	26-40	3-5	19-32	16	3			15.9	
Castor cake	57	30	20			824	810	8.4	55-66
Peanut meal	70	15	12	3	1.8				
Soybean meal	70	7-20	10-24	3	1.8				70
Wheat bran	29	14	13						
Molasses	4-5	0.3	6						40-50

[a] Digestate feed stocks are household bio-waste, clover grass, energy crops, liquid cow manure, liquid pig manure, pig and poultry manure, catering and other food processing bio-waste materials, potato starch processing wastewater.
[b] In general, 65-80% of P and 80-100% of K are available during the growing season.
[c] OM: organic matter; DM: dry matter; FW: fresh weight; DW: dry weight.
[d] Within 1 year. Note: Heavy metal analyses (Cd, Cr, Cu, Hg, Ni, Pb, Se, Zn, As, Mo) should be performed to comply with organic production regulations.
Sources: Zublena et al. (1997), Greer and Diver (2000), Hartz et al. (2000), Hartz and Johnstone (2006), Gaskell et al. (2007), Ozores-Hampton et al. (2012), Laber (2003), Möller and Schultheiß (2014), Ahmad et al. (2016), van der Wurff et al. (2016), Dion et al. (2017a,b).

was proposed in replacement of peat for organic potted plants without any negative effect on plant growth and disease sensitivity (Dorais et al., 2016c).

8.4.3 Nutrient amendments

While compost is used as base dressing, rapid N mineralising fertilisers are applied throughout the growing season as top dressing in intensive vegetable growing systems and as a corrective fertilisation. Indeed, fertilisation management based solely on compost and solid animal manure (amendments with high C/N ratios) can result in a much lower nitrogen-use efficiency than a fertilisation including, in combination to compost, complementary organic fertilisers with lower C/N ratio (Zikeli et al., 2017). Examples are dried and/or composted animal manure (e.g. poultry pellets), dried slaughterhouse waste (e.g. blood meal, feather meal), dried plant-based fertilisers (e.g. alfalfa, soybean meal, maize meal, brewery waste, vinasse, hydrolysed plant material, fermented grains, seed meal etc.), products derived from the ocean and fish industry (e.g. shrimp meal, crab meal, fishmeal, extracts or emulsion, seaweed extracts), products derived from digesters and inorganic fertilisers (e.g. Epsom salts, potassium sulphate, limestone, gypsum) (Dorais, 2017; Dorais and Schwarz, 2018). Table 7 presents examples of the nutritive value of some organic fertilisers. Gypsum is a source of calcium that dissolves slowly but also adds SO_4, contributing to increased salinity, while limestone is mainly used for pH correction. Calcium chloride and calcium silicate from non-synthetic sources may be used in some countries to add additional calcium, although Ca is often sufficient when the animal source of amendments and compost are supplied. Enough available Ca is important to prevent physiological disorders such as tipburn in lettuce and blossom end rot in fruiting vegetables (e.g. tomato, pepper, eggplant, cucumber, squash). Authorised fertiliser products should be listed on the country-specific permitted substance lists (NOP, CAN/CGSB-32.311-2015). Crop residuals (stems, leaves, side shoots) can also be composted and reused in the cropping systems, contributing for fruiting vegetables to the recycling of 35–45% of N, P and K and 60% of Mg (Voogt et al., 2014). In general, animal fertilisers are richer in P than plant demand, while composts may contain high K levels and high salinity. In general, the use of diverse organic amendments provides sufficient micronutrients. Fe or Mn deficiency is generally associated with too high soil pH.

The mineralisation rate of organic fertilisers is variable and not always easy to predict and control as it varies according to their physicochemical and biological properties, the biodiversity, structure and functionality of the soil microbial communities, the microfauna and the soil temperature, O_2 and water content (micro- and macro-porosity). Based on their N mineralisation rate, they are usually divided into rapid, moderate and slow nutrient-release fertilisers as presented in Table 8. Nevertheless, for five commonly used organic

Table 8 Characteristics on the main organic amendments and fertilisers used for greenhouse crop production

	High N	Moderate N	Low N	High P	Moderate P	Low P	High K	Moderate K	Low K	High S	High Na
Rapid N mineralisation rate	Blood meal					Blood meal			Blood meal		
	Keratins[a]	Hydrolysed proteins[b]				Keratins			Keratins		
	Meat meal	Molasses[c]					Molasses			Molasses	Molasses
	Potato protein[d]					Molasses	Potato protein			Potato protein	
		Digestates[e]				Potato protein			Digestates		
Moderate N mineralisation rate		Poultry manure[f,g]		Poultry manure				Poultry manure			Poultry manure
		Shrimp meal[h]			Shrimp meal				Pulse crop seeds		
		Pulse crop seeds			Pulse crop seeds						
Slow N mineralisation rate		Maize residues[g,i]	Alfalfa and clover[j]		Maize residues	Alfalfa		Alfalfa	Maize residues	Maize residues	
			Mushroom culture waste	Mushroom culture waste					Mushroom culture waste	Mushroom culture waste	
None N source				Bonemeal[g,k]							

[a] Horn; hooves, feather meal, wool meal, hair meal.
[b] N content varies according to the product.
[c] Residues of sugar beet processing.
[d] Favourable nutrient spectrum in terms of N, P and K content (balanced NPK).
[e] Nutrient content depends on feedstock used; animal and/or plant origins; its use can be limited by the high salinity (NaCl).
[f] Composted or dehydrated poultry manure; high risk for NH_3 volatilisation and risk of crop damage; tomato is highly sensitive to NH_3.
[g] Unfavourable nutrient spectrum.
[h] Rich in calcium.
[i] Residues from the starch industry.
[j] Favourable nutrient spectrum in terms of N, P, K and S content (balanced NPKS).
[k] Low plant P availability under neutral and alkaline soil conditions; not suitable as complementary fertiliser to compost and digestate.

greenhouse fertilisers, an incubation soil trial performed with mineral soil and peat-based growing medium showed that the N mineralisation reached a plateau after only 9 weeks, with 40% of applied N mineralised for alfalfa meal, 40–70% for shrimp meal and poultry meal and 60–90% for feather and blood meals (Fig. 1, Dion et al., 2017a). Differences between the studied soils were partly related to their soluble organic carbon content. In both types of soil, alfalfa caused a net N immobilisation in the first week and achieved the lowest total N mineralisation of all fertilisers, but provided a higher soluble organic content which sustained microbial biomass and activity. In contrast, poultry meal provided a flush of mineral N at application time. Moreover, organic fertilisers supported contrasting bacterial communities, closely linked to soil biochemical properties such as mineral N, K, Ca, pH and soluble C. Alfalfa meal promoted the highest bacterial biodiversity in the mineral soil, whereas shrimp meal and poultry meal promoted it in the peat-based growing medium. On the opposite, blood meal and feather meal provided conditions that are less than optimal for bacteria (Dion et al., 2019a). Indeed, these N-rich organic fertilisers are poor in P and K, making them unbalanced fertilisers and unsuitable for sustaining the stoichiometry homeostasis of microbial communities. On the other hand, shrimp meal and poultry meal have a more balanced mineral input by providing P, Ca and Mg, along with a higher C/N, supporting a more diverse and abundant bacterial community (Dion et al., 2019a; Xu et al., 2015).

8.4.4 Organic N in the soil solution and its plant uptake for organic crops

Although inorganic N (i.e. NO_3 and NH_4) is the preferential form of plant N uptake over organic N (Nacry et al., 2013), there is compelling evidence that organic nitrogen (e.g. amino acids, small peptides) can be a non-negligible source of N for plants (Franklin et al., 2017; Czaban et al., 2016; Gruffman et al., 2014; Gioseffi et al., 2012; Paungfoo-Lonhienne et al., 2012; Ge et al., 2009; Näsholm et al., 1998). This could be particularly important for organic greenhouse crops where large amount of organic N fertilisers and protein hydrolysates from diverse sources are regularly provided to the plants. Organic N molecules have the advantage of providing reduced N along with a C source, which could improve the C budget and N-use efficiency of the plant by minimising the energy required for N reduction and assimilation (Franklin et al., 2017; Ganeteg et al., 2017). However, very few studies were published on the uptake of organic N molecules in organic horticulture (Song et al., 2009). Nevertheless, Dion et al. (2018) recently showed that cucumber plants exposed to a low inorganic:organic N ratio (with alanine as the organic N source), typical of conditions usually prevailing under organic greenhouse cultivation, acquired up to 6.84% (10:1 IN:ON) of their N from alanine, compared with 0.94% at a higher ratio (500:1), and do assimilate this alanine in the amino acids metabolism. The plants were able to increase their total organic N uptake when exposed to a higher total N concentration and alanine concentration (0.5, 5 and 15 mM N) within each inorganic:organic ratio (500:1, 100:1 and 10:1) in the presence of inorganic N, but its organic N uptake was highest at the smallest ratio (10:1). However, at total N concentration of 15 mM N, only 0.4–0.9% of the total N, that was taken up by seedlings, came from alanine. Consequently, under intensive organic production systems providing large amount of nitrogen, intact amino acid uptake seems to play a minor role in the total crop nitrogen budget. Nevertheless, the total soluble organic C and N pools in the growing medium do influence the xylem sap content of cucumber plants (Dion et al., 2017b). For lettuce, supplying the amino acid glycine instead of nitrate as a source of N promoted the accumulation of glycosylated flavonoids, ascorbic acid and most amino acids, while reducing their phenolic acids content (Yang et al., 2017).

8.4.5 pH

Organic crops usually perform better on weak-acidic to weak-alkaline soils (i.e. pH between 5.5 and 8.0) as under strongly alkaline and acidic conditions, P is less readily available to plants as it forms insoluble phosphates. Appropriate pH is also needed to maintain high N mineralisation as well as appropriate micronutrient plant availability. Most of the organic amendments and fertilisers have a neutral or basic pH, resulting in an increase in soil pH over time. Indeed, the imbalanced cation and anion content in plant biomass and organic fertilisers (i.e. surplus of cations) are mostly balanced by organic acids and carbonates. As the decomposition of these organic compounds is a process that utilises protons and releases CO_2 to the atmosphere, the soil becomes more alkaline (Yan et al., 1996). The composting process and storage conditions also contribute to the liming properties of composts because they reduce the availability of compounds providing acidity by ammonia volatilisation, denitrification and desulfurication (Zikeli et al., 2017). Peat can be used as conditioner to provide organic matter and decrease the soil pH, although it is not considered as a sustainable avenue. Elemental sulphur or sulphur pellets can be added to the soil to control the soil pH increase, despite the fact that it contributes to intensifying the soil SO_4 content and salinity. Other compounds can be used, such as citric acid, but have a weak control effect.

8.4.6 Salinity management

Salinity management is a major issue for organic greenhouse horticulture because most of the organic amendments and fertilisers are not well balanced with the crop-nutrient demand (e.g. N, P, K, Ca and Mg) and are rich in Na or SO_4 (Table 7). Adjusting the irrigation management to crop evapotranspiration in order to limit nutrient and water losses results in the accumulation of undesirable ions such as Na, SO_4 and Cl. Unbalanced soil nutrients might then result in physiological disorders (e.g. tipburn, blossom end rot, ripening disorders) and impair crop growth and productivity. To control soil salinity, growers often over-irrigate their soil, causing N leaching and environmental pollution. Poor water quality is also related to the increase of soil salinity (Dorais et al., 2016a). To avoid salinity build-up and the need to over-irrigate to control soil salinity, the input of residual salt should be limited as much as possible (Voogt et al., 2011). Plant grafting is a sustainable approach to mitigate high soil salinity, while the selection of well-balanced sources of organic fertilisers is the key issue for controlling the build-up of salinity. Next to negative effects on plant growth and development, salt accumulation in soil negatively affects the C and N mineralisation, enzymatic activity and CO_2 respiration, which is related to osmotic stress of microorganisms, as well as with plant roots (Pathak and Rao, 1998). Despite this, soil microorganisms have the ability to adapt to or tolerate osmotic stress caused by drought or salinity, especially when regularly confronted with such conditions (Wichern et al., 2006). Consequently, salinity build-up in organic greenhouses should be avoided because the complex system of soil fertility and root activity relies on a constant and highly active soil food web.

9 Biostimulants

In addition to the beneficial fungi (e.g. mycorrhizal fungi, *Trichoderma* spp.) and bacteria (Colla et al., 2015; Du Jardin, 2015; López-Bucio et al., 2015), biostimulants include humic substances (Canellas et al., 2015), hormone-containing products (Shekhar Sharma et al., 2014), amino acid–containing products (Colla et al., 2015; Pichyangkura and Chadchawan, 2015) and inorganic compounds such as silicon (Savvas and Ntatsi, 2015), phosphites (Gómez-Merinoa and Trejo-Téllez, 2015), Al, Co and Se (Du Jardin, 2015). For organic greenhouse horticulture, biostimulants are used to improve the suppressiveness of a soil and the plant resilience to abiotic and biotic stresses (Bulgari et al., 2015; Du Jardin, 2015). Their physiological functions or 'modes of action' are diverse, and involve the activation of molecular, biochemical and physiological responses (Du Jardin, 2015). For example, they often involve defence gene activation, secondary metabolism such as phenylpropanoid metabolism as well as antioxidant activity in response to biotic and abiotic stresses. Indeed, biostimulants containing non-living substances, such as humic and fulvic acids, protein hydrolysates, chitosan compounds or silicon, may exhibit positive effects on leaf chlorophyll content, plant growth, yield and fruit components.

Several studies that were not specific to organic growing systems were conducted on soil disease suppressiveness via the use of biological control agents such as *Pseudomonas* spp., *Bacillus* spp., *Streptomyces* spp. and *Trichoderma* spp. to control root diseases of horticultural crops (Jayaraj et al., 2007; Caron et al., 2002; Paulitz and Bélanger, 2001; Rankin and Paulitz, 1994; Sutton and Peng, 1993). For tomato, suppression of corky root rot was correlated with cellulolytic actinomycetes and total actinomycete populations (Workneh and van Bruggen, 1994). *Pseudomonas fluorescens* strains were shown to effectively control bacterial pathogens. The beneficial response of the plant to a growth promoter may, however, be dependent on the genotype. Abiotic factors, such as water stress and nutrient deficiency, and the indigenous microflora and microfauna also affect the efficacy of these promoters. For organic crops, however, interactions with the indigenous microflora and microfauna may influence the efficacy of these control agents. Therefore, despite the apparent benefits of some biostimulants, there is controversy about the real advantages for organic horticultural crops. For example, the use of *Bacillus subtilis* and *Pythium oligandrum* in the soil of plants infected with corky root rot (*Pyrenochaeta lycopersici*), *Verticillium alboatrum* or both had no effect on root fresh weight and yield of organic tomato (Giotis et al., 2009). For organic production of spinach, the efficacy of several commercial products containing biological control agents (*Trichoderma harzianum*, *Streptomyces griseoviridis*, *Bacillus subtilis*, *Gliocladium catenulatum*, *Bacillus pumilus*) against soil-borne damping-off pathogens (*Fusarium oxysporum* f. sp. *spinaciae*, *Pythium ultimum*, *Rhizoctonia solani*) did not provide complete protection, and some treatments exacerbated damping-off (Cummings et al., 2009). In fact, the time of administration, the choice of biocontrol agent with respect to the pathogen and the combination of different biocontrol strains are important for minimising disease pressure on the crop, as are soil properties, type of compost and organic amendments, water content and available micronutrients (Dorais and Alsanius, 2015).

On the other hand, non-living substances such as seaweed extracts (*Ascophyllum nodosum*) are widely used in organic horticulture crops for their plant growth-promoting effects and for their ameliorating effect on crop tolerance to salinity, extreme temperatures, nutrient deficiency and drought. The chemical constituents of seaweed extract include complex polysaccharide, fatty acids, vitamins, hormones (e.g. cytokinins, auxins, gibberellins) and mineral nutrients (Ca, K, P, Mg and Mn, Fe, Zn, Mo, B, Co) (Sangha et al., 2014). Due to their hormonal content, several studies showed that the application of marine plant extracts stimulates fruit growth and development, increasing the generation of photosynthates and the plant absorption of water and nutrients. Minerals, antioxidant and vitamin contents of fruits were also enhanced in plants subjected to seaweed applications. Silicon (Si) applications also alleviate abiotic and biotic stresses for a number of important crops (Liang et al., 2015; Fauteux et al., 2005). Silicon has a prophylactic role by activating plant defence mechanisms, and thus increasing the resistance of plants to pathogenic fungi (e.g. powdery mildew), and insects and mites (Liang et al., 2015; Arsenault-Labrecque et al., 2012). It was also reported that Si reduces mineral toxicity, improves nutrient imbalances, increases photosynthetic activity, influences biomass accumulation in fruit and increases tolerance to environmental stresses such as water shortage, excess salts, freezing and high temperature (Liang et al., 2015; Toresano-Sánchez et al., 2012; Liang et al., 2007; Ma et al., 2004). Many studies reported that exogenous Si supply to several horticultural plants such as tomato may result in increased growth and productivity (Liang et al., 2015; Voogt and Sonneveld, 2001). However, under organic management, although a slight growth increase (6.5%) was observed for greenhouse cucumber, no significant effect on yield was found when plants were enriched with 2-8 g L^{-1} of wollastonite ($CaSiO_3$) or K_2SiO_3, despite a Si leaf content increase of ten times compared with control leaves (Dorais and Thériault, 2018). In terms of fruit quality, it was observed that Si increased the level of sugar, vitamin C and some phenolic compounds (e.g. chlorogenic and caffeic acids), while reducing levels of NO_3 (Savvas and Ntatsi, 2015; Liu et al., 2014). However, under organic crop management, no significant effect was observed for the nutrient and health components, content of cucumber and strawberry, although higher marketable yield was found due to the reduction of powdery mildew severity (Dorais and Thériault, 2018; Ouellette et al., 2017).

10 Water management

Like conventional greenhouse horticulture (see Chapter 12), irrigation management of organic greenhouse crops should be adjusted to the crop evapotranspiration demand (De Pascale et al., 2017; Dorais et al., 2016a). However, in addition to replenishing water lost through plant transpiration and soil surface evaporation, the water requirement for organic crops should also consider the needs of the soil microbiota. Indeed, both bacterial and fungal activity rapidly decreases in drying soils (Manzoni et al., 2012), as does the mineralisation rate (Agehara and Warncke, 2005; Bloem et al., 1992). In contrast, an over-wet soil boosts anaerobic microbial activity, which has detrimental environmental effects such as gaseous losses of N by denitrification. For soil-bounded organic growing systems, irrigation management based on plant transpiration is complex and not always possible due to variations related to crop heterogeneity owing to pest and disease incidence and climate gradient within the greenhouse. Depending on the type of growing system and cultivated area, water consumption can be as high as 0.9 m^3 m^{-2} $year^{-1}$ to 2.0 m^3 m^{-2} $year^{-1}$ when considering year-round production with artificial lighting (Dorais et al., 2016a). According to the organic regulation, producers should make responsible use of water by collecting rainwater and condensed water from the greenhouse covering (i.e. latent heat from plant transpiration; up to 80% of the total evapotranspiration under Northern climate). Flushing undesirable nutrients (e.g. SO_4, Na) with non-saline water should be avoided. Consequently, irrigation scheduling, which determines the irrigation events and water supply, plays a crucial role for optimising organic greenhouse production, while also conserving water resources. As for conventional greenhouse crop, different technologies are used to supply water, ensuring spatial uniformity, high water use efficiency and product quality (Dorais et al., 2016a). When zero emission is not possible, crop effluent should then be collected, treated (e.g. slow filter, wetland, passive bioreactor, membrane filtration, ozonation, thermal or UV treatment; Dorais et al., 2016a) and reused on the farm, although it is not always possible for soil-bound growing systems.

The use of high-quality water (<0.5 mS cm^{-1} of EC; <0.2–1.0 mmol L^{-1} of Na or Cl; <1.5 mmol L^{-1} of Ca; <0.7 mmol L^{-1} of Mg; <0.7 mmol L^{-1} of SO_4; Voogt, 2009) and accurate irrigation control tools such as soil moisture sensors, lysimeters and models contribute to the optimisation of the irrigation management of greenhouse crops (De Pascale et al., 2017; Dorais et al., 2016a). Plant physiological indicators such as leaf temperature measured by infrared cameras, stem diameter and sap-flow sensors, stomatal conductance measured by leaf porometers and leaf fluorescence are interesting monitoring tools. However, these tools require proper installation and maintenance. Data interpretation for crop-water demand also require a certain level of knowledge, which limits their use in commercial practices. Selection of well-balanced sources of organic fertilisers also contributes to avoiding the build-up of soil salinity by undesirable ions (e.g. Na, SO_4, Cl) and the need to over-irrigate for maintaining an appropriate EC (Zikeli et al., 2017). Fe should be lower than 10 µmol L^{-1} to avoid any precipitation and clogging of the irrigation system, and lower than 100 µmol L^{-1} to prevent leaf burning, while micronutrients (e.g. Zn, Cu, Mn, B) should be below the toxicity range. To avoid any problem related to high pH, soil HCO_3 should be lower than 1 mmol L^{-1} (61 ppm). Citric or acetic acids can be supplied to neutralise excessive HCO_3, but will increase the soil EC and provide carbon sources for microorganisms resulting in biofilm development and then clogging of the drip irrigation system. When the water contains equivalently more (bi) carbonate (HCO_3^-) than Ca and Mg and is instead balanced with Na, it will negatively affect the soil structure (sodicity; 5% or more of the cation exchange sites of the soil are occupied by Na^+) resulting in low permeability to water and air and reduced plant growth (Bernstein, 1975). The irrigation water should also be free of plant protection chemicals, pharmaceutical and personal care products as well as plant and human pathogens (Dorais et al., 2016a).

11 Plant protection

It was observed that organically managed crops are less affected by soil-borne diseases compared with conventional production because organic management increases microorganism diversity and biological activity of the soil (Montalba et al., 2010; van Bruggen, 1995). The main mechanisms involved are often related to the decrease of the disease inoculum by competition and antibiosis, the increased defence mechanisms of organic plants and the rhizosphere colonisation by beneficial fungi and bacteria that protect the plants from diseases. However, for organic-intensive greenhouse crops such as tomato, pepper and cucumber, soil-borne diseases and pests may be a real problem due to limited crop rotation with diverse species.

Soil steaming and soil solarisation are the most widely used techniques to reduce soil-borne diseases, although both techniques negatively affect beneficial microflora and fauna. Soil anaerobic disinfection and soil biofumigation by means of *Brassica* green manure or other organic residue sources is an efficient alternative tool to control diseases along with using *Brassica* break crops such as kohlrabi and white mustard or intercropping cash crops with companion species like *Tagetes erecta* (Gamliel and van Bruggen, 2016; Giotis et al., 2009). Essential oils and individual monoterpenoids from different plant species (e.g. lavender, oregano, sage and spearmint) can help prevent mycelial growth and conidial production of several plant pathogens (e.g. *A. terreus*, *F. oxysporum*, *P. expansum* and *V. dahliae*) (Kadoglidou et al., 2011). Copper, although its common use in organic farming is controversial due to its toxic impact on non-targeted microorganisms and fauna, is an efficient tool to control anthracnose, bacterial speck, bacterial spot, early and late blight, grey leaf mould, powdery mildew and septoria leaf spot (Diver et al., 2012). While elemental S is used to control powdery mildew and mites, potassium bicarbonate is often applied to prevent anthracnose, early blight, leaf blight, leaf spot and powdery mildew. Plant extracts and biocontrol agents used as soil drench and foliar applications may also be used by some organic producers, although their efficacy for organic growing system has to be proved, considering the large microbial diversity and buffering capacity of organically managed soils that prevent the establishment and reduce the effectiveness of introduced biocontrol agents in those soils (Gamliel and van Bruggen, 2016).

Control of insects, mites and nematodes is based on frequent releases of natural predators (*Amblyseius swirskii*, *A. degenerans*, *A. californicus*, *A. andersoni*, *Amblydromalus limonicus*, *Aphidius colemani*, *Coenosia attenuata*, *Dicyphus hesperus*, *Encarsia formosa*, *Eretmocerus eremicus*, *Feltiella acarisuga*, *Neoseiulus cucumeris*, *Neoseiulus fallacis*, *Orius insidiosus*, *Phytoseiulus persimilis*, *Stethorus punctillum*, *Steinernema feltiae* etc.), the use of natural repellents (garlic, onion, zinnia, marigold, nasturtium etc.), trap crops (sweet corn, marigold, barley, ornamental peppers, chrysanthemum saskia, bean, aubergine), banker plants (*Ricinus communis*, oat, wheat, castor bean, mullen, *Brugmansia*, *Caulophyllum* etc.) which provide supplemental food for beneficial insects helping them maintain a stable population, and pheromone traps or passive dispensers. When populations reach economically damaging levels, biopesticides (e.g. pyrethrins in some cases, *Bacillus thuringiensis*, *Beauveria bassiana*, *Trichoderma* strains, nuclear polyhedrosis virus) are used according to country-permitted substances lists.

For organic soil-bound greenhouse cultivation, weed control is not as important as for organic field production as mulches are often used to limit soil water evaporation, promote soil mineralisation rate and increase light reflexion (Tarara, 2000). As high plant density (e.g. leaf area index of 3 for tomato) is used, a limited amount of light reaches the ground, limiting weed infestation. Soil solarisation, when used to control soil-borne pests, acts as a prophylactic method limiting weed infestation.

12 Quality value of organic greenhouse products

Although consumers perceive organic foods as being more nutritious, tastier and richer in health-promoting compounds than conventional products, no conclusions can be drawn from the scientific literature, except that their content in pesticide residues (quantity and number of different pesticides) is much lower, and they are usually richer in polyphenols and minerals (e.g. Cu and Zn; coming from copper use for disease control and animal-based compost or manure) and lower in nitrate (Dorais and Schwarz, 2018; Dorais and Alsanius, 2015; Barański et al., 2014; Smith-Spangler et al., 2012; Brandt et al., 2011; Lairon, 2010; Dangour et al., 2009; Benbrook et al., 2009; Rembiałkowska, 2007). Nevertheless, when plants are grown under an organic regime, their natural defence mechanisms may be more active than those of conventional plants, resulting in higher defence-related secondary metabolites, which can often be considered beneficial for human health (Orsini et al., 2016; Vallverdú-Queralt et al., 2012). Indeed, it was shown that organic tomato experiencing oxidative stress had higher vitamin C and phenolic compounds (López-Martínez et al., 2016; Oliveira et al., 2013). Differences in the nutritional value of tomatoes grown under organic and conventional practices were reported in detail (Dorais and Schwarz, 2018). Main factors influencing greenhouse crops quality are the genetic (cultivars), environmental growing conditions (light quantity and quality, temperature, CO_2) and the cultural practices such as salinity and water management, nutrient management and biostimulants (Bertin and Génard, 2018; Dorais et al., 2017a; Currey et al., 2017; Gil, 2015; Urban et al., 2014; Dorais et al., 2008). For example, the colour and phenolic content of organic greenhouse strawberries were higher when the fertilisation management was based on the use of agroecological service crops combined with compost or mature cattle manure compared with an input substitution growing system where agrochemicals were replaced by organic products (Tittarelli et al., 2016b). The effects of compost and organic amendments on fruit and vegetable quality are also well documented (Prange, 2015).

13 Environmental impact of organic greenhouse cultivation

Nowadays, the central task of agriculture is to produce sufficient food of highly nutritive quality to feed the increasing population with a minimum negative environmental impact and land use. Organic agriculture is one of the methods frequently proposed for reducing the impact of agriculture on the environment. Indeed, several meta-analyses showed the environmental benefits of organic farming over conventional agriculture in terms of increasing biodiversity, water and soil conservation, wildlife diversity, microbial population and activity, soil organic content and energy use as organic farming reduced the use of synthetic fertiliser and pesticide inputs (Clark and Tilman, 2017; Foresi et al., 2016; Reganold and Wachter, 2016; Dorais and Alsanius, 2015; Schneider et al., 2014; Tuomisto et al., 2012). However, some controversial results are reported regarding the superiority of organic farming for the C sequestration following compost and manure amendments to the soil (Kirchmann et al., 2016). It was also reported that, globally, organic systems require more land to produce the same amount of food (lower yield per cultivated area), cause more eutrophication and emit similar greenhouse gas emissions (GHGs) as conventional systems, offsetting the environmental benefits associated with this production method (Clark and Tilman, 2017; Kirchmann et al., 2016). Lack of synchronicity between release of N from organic fertilisers and demand for N by the crop may effectively result in nitrate leaching for certain cropping systems (Kirchmann et al., 2016; Tuomisto et al., 2012). For organic greenhouse cultivation, achieving an optimal fertilisation management via a better synchronisation of the nutrient availability with the plant requirements as well as reaching higher productivity are key elements of improved sustainability. Thus, increasing input-use efficiency per unit of product (e.g. nutrients, water and energy per kg of fruits and vegetables) by designing improved production systems would have environmental benefits for protected horticulture. Greenhouse production systems tend to emit almost three times more GHGs than open-field systems because of the energy required to maintain adequate greenhouse growing conditions (Clark and Tilman, 2017). Thus, the use of renewable energy shall be highly favoured, although it is not mandatory for most organic standards. Under Northern growing condition or Southern regions that request heating or cooling during a long period, the energy demand for climate control is the major environmental burden for organic as well as conventional growing systems, which represents 81–96% (Torrellas et al., 2012) or 74–99% (Dorais et al., 2014) of the total burden under European or Northern American growing conditions, respectively (see Chapter 18). According to different production systems, Table 9 presents examples of the environmental footprint of 1 kg of tomatoes.

Table 9 Environmental impact of different conventional (conv) and organic (org) growing systems according to different indicators

Production system	Ecological footprint[a] ($m^2\ kg^{-1}$)	CO_2 emission (kg kg^{-1} tomatoes)	Global warming potential (kg CO_2-eq kg^{-1} tomatoes)
Heated greenhouses			
Conv-England			9.40
Conv-Netherlands			1.77-2.47
Org-Netherlands			1.95
Conv-Netherlands/cogeneration			0.89
Org-Netherlands/cogeneration			0.89
Conv-Quebec, Canada/oil			5.79
Org-Quebec, Canada/biomass			0.81
Conv-France[b]			2.02
Conv-Europe/extra light oil	111	0.6435	0.96
Org-Europe	16.75	0.0689	0.10
Cold high tunnels			
France[c]			0.51
Spain-soil			0.12
Spain-soilless			0.09
Conv-Europe	18.26	0.0681	0.4743
Org-Europe	13.46	0.0419	0.0645

[a] The ecological footprint is a measure of how much human activity changed and charged the nature. It estimates the biologically productive area needed to produce materials and energy used by the population of a certain region (Stajnko et al., 2016).
[b] Soilless and heated tomato crops in glass or plastic multispans in the northwest (Brittany and lower Loire Valley) and in the southeast (Mediterranean area and the lower Rhone Valley) France.
[c] Soil-based tomato production under high tunnels located in south-east France.
Source: Foresi et al. (2016), Dorais et al. (2014), Torrellas et al. (2012) and Boulard et al. (2011).

14 Future trends and conclusion

A major challenge for the coming decade is the transition of farms towards sustainable production systems, ensuring food security and nutrition for all now and in the future with food that is healthy, of sufficient quality and quantity, affordable, safe and culturally acceptable (Desquilbet et al., 2018; HLPE, 2017; Johnston et al., 2014). Despite numerous studies comparing the environmental pros and cons of organic versus conventional farming were published, better-defined indicators should be developed as conventional and organic farming cannot be compared on a production-by-production basis, without considering the agroecology services in the long term as well as the impact of the composition of conventional and organic diets on human health. The conventionalisation of organic production systems, replacing conventional inputs by organic authorised inputs intended mainly for exporting products, has been questioned (Constance et al., 2015). The price pressures encouraging the industrialisation of production methods, organic monoculture and imports from countries in which labour is cheap and conditions are poor for farmers, to the detriment of local employment is an important issue related to fairness, that should be taken into account to strengthen regulations on organic standards (Desquilbet et al., 2018; Darnhofer et al., 2010).

From an organic greenhouse horticulture perspective, future research trends can be summarised as the (1) development of robust plant materials and nutritive cultivars, contributing to human health, (2) the achievement of the temporal matches between nutrient availability and nutrient-plant demand without causing soil salinity build-up and environmental burdens, (3) better understanding of the impact of the cultural practices (composts, fertilisers, biostimulants, irrigation management etc.) on the biodiversity, structure and functionality of the soil and plant microbiota to improve greenhouse crop resilience to biotic and abiotic stresses, (4) improvement of the irrigation management and water-use efficiency by using appropriate tools and decision-support systems and by water recycling on the farm, (5) development of innovative growing systems that will reduce the yield gap between organic and conventional farming, (6) development of sustainable alternatives for pest management including new biocontrol pest control agents, climate management and new technologies such as LEDs to suppress leaf pathogens (Carvalho and Castillo, 2018; Vänninen et al., 2010), (7) replacement of fossil energy use by renewable energy, while improving crop energy-use efficiency, (8) improvement of the efficiency use of inputs per unit of fruits and vegetables, (9) establishment of new sustainability assessment tools and indicators for more properly integrated resource management (composts, cover crops, rotations, amendments and fertilisers, energy, water, pesticides, nutritive-product attributes, land use etc.), (10) reduction of packaging and transport of organic greenhouse products via local and urban production and (11) improvement of social benefits by tasty, nutritive and safe products, minimising risks related to physical, chemical and biological contaminants (Alsanius et al., 2017; Dorais and Alsanius, 2015).

15 Where to look for further information

Additional sources of information on organic greenhouse horticulture can be found on the Biogreenhouse website (www.biogreenhouse.org), where experts from some 30 countries have published scientific reviews, booklets and fact sheets. The special issue of *Scientia Horticulturae* on 'Recent advances in organic horticulture technology and management' contains three reviews related to organic greenhouse vegetables (Burnett et al., 2016; Gamliel and Van Bruggen, 2016; Orsini et al., 2016), while the special issue on *Biostimulants in Horticulture* (vol. 196) provides a complete overview of compounds commonly used in organic greenhouse production. Specific information about rootstock varieties is found at http://www.vegetablegrafting.org.

The following websites provide guidelines on assessments, particularly with regard to the characterisation of crop varieties: www.eppo.int (European & Mediterranean Plant Protection Organization – EPPO); www.cpvo.europa.eu (Community Plant Variety Office – CPVO); www.upov.int (International Union for the Protection of New Varieties of Plants). Complementary information on biological pest control management can be found in the following documents: Koppert Biological Systems (2017) and Buitenhuis (2014).

The history of organic farming is well detailed in the following documents: Heckman (2006) and Bartin (2018).

Key professional organisations are the International Federation of Organic Agriculture Movements, IFOAM (https://www.ifoam.bio), the International Society of Organic Agriculture Research (http://isofar.org/isofar/) and the Commission Agroecology and Organic Horticulture of the International Society for Horticultural Science (ISHS, https://www.ishs.org/agroecology-and-organic-horticulture). Related ISHS working groups are Organic greenhouse horticulture; Organic Vegetables; Organic Temperate Fruits; and Organic Tropical and Subtropical fruits. Related conference proceedings are found at http://isofar.org/isofar/index.php/proceedings and https://www.ishs.org/acta-horticulturae.

16 References

Agehara, S. and Warncke, D. 2005. Soil moisture and temperature effects on nitrogen release from organic nitrogen sources. *Soil Sci. Soc. Am. J.* 69, 1844–55.

Ahmad, A. A., Radovich, T. J. K., Nguyen, H. V., Uyeda, J., Arakaki, A., Cadby, J., Paull, R., Sugano, J. and Teves, G. 2016. Use of organic fertilizers to enhance soil fertility, plant growth, and yield in a tropical environment. In: Larramendy, M. L. and Soloneski, S. (Eds), *Organic Fertilizers – From Basic Concepts to Applied Outcomes*. IntechOpen, London, UK, pp. 85–104. Chapter 4. doi:10.5772/62529.

Alsanius, B. W., Dorais, M. and Meijer, R. 2017. Vision of COST BioGreenhouse. *Acta Hortic*. 1164(1164), 1–8. doi:10.17660/ActaHortic.2017.1164.1.

Alsanius, B. W., Khalil, S., Tadesse, A., Rosberg, A. K., Bergstrand, K. J., Hartmann, R., Mogren, L., Alam, M., Grudén, M., Naznin, T. and Getahun, A. 2019. Sustainable systems for integrated fish and vegetable production – new perspectives on aquaponics. In: Marcelis, L. and Heuvelink, E. (Eds), *Achieving Sustainable Greenhouse Cultivation*. Burleigh Dodds Science Publishing, Cambridge, UK.

Arsenault-Labrecque, G. Menzies, J. G. and Bélanger, R. R. 2012. Effect of silicon absorption on soybean resistance to *Phakopsora pachyrhizi* in different cultivars. *Plant Disease* 96, 37–42.

Barański, M., Srednicka-Tober, D., Volakakis, N., Seal, C., Sanderson, R., Stewart, G. B., Benbrook, C., Biavati, B., Markellou, E., Giotis, C., et al. 2014. Higher antioxidant and lower cadmium concentrations and lower incidence of pesticide residues in organically grown crops: a systematic literature review and meta-analyses. *Br. J. Nutr*. 112(5), 794–811. doi:10.1017/S0007114514001366.

Barrett, C. E., Zhao, X. and McSorley, R. 2012. Grafting for root-knot nematode control and yield improvement in organic heirloom tomato production. *HortScience* 47(5), 614–20. doi:10.21273/HORTSCI.47.5.614.

Barrett, G. E., Alexandera, P. D., Robinsonb, J. S. and Bragg, N. C. 2016. Achieving environmentally sustainable growing media for soilless plant cultivation systems – a review. *Sci. Hortic*. 212, 220–34.

Bartin, G. A. 2018. *The Global History of Organic Farming*. Oxford University Press, p. 256. Available at: http://www.worldcat.org/title/global-history-of-organic-farming/oclc/1021883219.

Benbrook, C., Davis, D. R. and Andrews, P. K. 2009. Methodologic flaws in selecting studies and comparing nutrient concentrations led Dangour et al to miss the emerging forest amid the trees. *Am. J. Clin. Nutr*. 90(6), 1700–1; author reply 1701. doi:10.3945/ajcn.2009.28605].

Bernstein, L. 1975. Effects of salinity and sodicity on plant growth. *Annu. Rev. Phytopathol*. 13(1), 295–312. doi:10.1146/annurev.py.13.090175.001455.

Bertin, N. and Génard, M. 2018. Tomato quality as influenced by preharvest factors. *Sci. Hortic*. 233, 264–76. doi:10.1016/j.scienta.2018.01.056.

Bhat, N., Albaho, M., Suleiman, M., George, B. T. P. and Ali, S. I. 2013. Growing substrate composition influences growth, productivity and quality of organic vegetables. *Asian J. Agric. Sci*. 5, 62–6.

Bittman, S., Hunt, D. E. and Shaffer, M. J. 2001. NLOS (NLEAP On STELLA (R)) – A nitrogen cycling model with a graphical interface: implications for model developers and users. In: Shaffer, M. J., Ma, L. and Hansen, S. (Eds), *Modeling Carbon and Nitrogen Dynamics for Soil Management*. CRC Press Taylor and Francis Group, Boca Raton, FL, pp. 383–402.

Bloem, J., de Ruiter, P. C., Koopman, G. J., Lebbink, G. and Brussaard, L. 1992. Microbial numbers and activity in dried and rewetted arable soil under integrated and conventional management. *Soil Biol. Biochem*. 24(7), 655–65. doi:10.1016/0038-0717(92)90044-X.

Boulard, T., Raeppel, C., Brun, R., Lecompte, F., Hayer, F., Carmassi, G. and Gaillard, G. 2011. Environmental impact of greenhouse tomato production in France. *Agron. Sustain. Dev*. 31(4), 757–77. doi:10.1007/s13593-011-0031-3.

Brandt, K., Leifert, C., Sanderson, R. and Seal, C. J. 2011. Agroecosystem management and nutritional quality of plant foods: the case of organic fruits and vegetables. *Crit. Rev. Plant Sci.* 30(1–2), 177–97. doi:10.1080/07352689.2011.554417.

Buitenhuis, R. 2014. *Grower Guide: Quality Assurance of Biocontrol Products*. Vineland Research and Innovation Centre, ON, Canada. Available at: http://www.vinelandresearch.com/sites/default/files/grower_guide_pdf_final.pdf; http://www.canr.msu.edu/news/common_banker_plants_in_insect_biological_control_systems; https://www.agrireseau.net/rap.

Bulgari, R., Cocetta, G., Trivellini, A., Vernieri, P. and Ferrante, A. 2015. Biostimulants and crop responses: a review. *Biol. Agric. Hort.* 31(1), 1–17. doi:10.1080/01448765.2014.964649.

Burnett, S. E., Mattson, N. S. and Williams, K. A. 2016. Substrates and fertilizers for organic container production of herbs, vegetables, and herbaceous ornamental plants grown in greenhouses in the United States. *Sci. Hortic.* 208, 111–9. doi:10.1016/j.scienta.2016.01.001.

Campiglia, E., Mancinelli, R. and Radicetti, E. 2011. Influence of no-tillage and organic mulching on tomato (*Solanum lycopersicum* L.) production and nitrogen use in the Mediterranean environment of central Italy. *Sci. Hortic.* 130(3), 588–98. doi:10.1016/j.scienta.2011.08.012.

Canellas, L. P., Olivares, F. L., Aguiar, N. O., Jones, D. L., Nebbioso, A., Mazzei, P. and Piccolo, A. 2015. Humic and fulvic acids as biostimulants in horticulture. *Sci. Hortic.* 196, 15–27. doi:10.1016/j.scienta.2015.09.013.

Caron, J., Laverdière, L., Thibodeau, P. O. and Bélanger, R. R. 2002. Utilisation d'une souche indigène de *Trichoderma harzianum* contre cinq agents pathogènes chez le concombre et la tomate de serre au Québec. *Phytoprotection* 83(2), 73–87. doi:10.7202/706230ar.

Carvalho, S. D. and Castillo, J. A. 2018. Influence of light on plant-phyllosphere interaction. *Front. Plant Sci.* 9, 1482. doi:10.3389/fpls.2018.01482.

Castilla, N. 2013. *Greenhouse Technology and Management* (2nd edn.). CAB International, Wallingford, UK, 374pp.

Ciaccia, C., Montemurro, F., Campanelli, G., Diacono, M., Fiore, A. and Canali, S. 2015. Legume cover crop management and organic amendments application: effects on organic zucchini performance and weed competition. *Sci. Hortic.* 185, 48–58. doi:10.1016/j.scienta.2015.01.011.

Clark, A. 2007. *Managing Cover Crops Profitably* (3rd edn.). Sustainable Agriculture Research and Education (SARE) Program Handbook Series; bk. 9. University of Maryland, College Park, MD, United Book Press, Inc.

Clark, M. and Tilman, D. 2017. Comparative analysis of environmental impacts of agricultural production systems, agricultural input efficiency, and food choice. *Environ. Res. Lett.* 12(6), 064016. doi:10.1088/1748-9326/aa6cd5.

Colla, G., Rouphael, Y., Leonardi, C. and Bie, Z. 2010. Role of grafting in vegetable crops grown under saline conditions. *Sci. Hortic.* 127(2), 147–55. doi:10.1016/j.scienta.2010.08.004.

Colla, G., Rouphael, Y., Di Mattia, E., El-Nakhel, C. and Cardarelli, M. 2015. Co-inoculation of *Glomus intraradices* and *Trichoderma atroviride* acts as a biostimulant to promote growth, yield and nutrient uptake of vegetable crops. *J. Sci. Food Agric.* 95(8), 1706–15. doi:10.1002/jsfa.6875.

Constance, D. H., Choi, J. Y. and Lara, D. 2015. Engaging the organic conventionalization debate. In: Freyer, B. and Bingen, J. (Eds), *Re-Thinking Organic Food and Farming in a Changing World*. Springer, Dordrecht, pp. 161–85.

Crowder, D. W. and Reganold, J. P. 2015. Financial competitiveness of organic agriculture on a global scale. *Proc. Natl. Acad. Sci. U.S.A.* 112(24), 7611–16. doi:10.1073/pnas.1423674112.

Cummings, J. A., Miles, C. A. and du Toit, L. J. 2009. Greenhouse evaluation of seed and drench treatments for organic management of soilborne pathogens of spinach. *Plant Dis.* 93(12), 1281–92. doi:10.1094/PDIS-93-12-1281.

Currey, C. J., Kopsell, D. A., Mattson, N. S., Craver, J. K., Lopez, R. G., Erwin, J. E. and Kubota, C. 2017. Supplemental and sole-source lighting of leafy greens, herbs, and microgreens. In: Lopez, R. and Runkle, E. S. (Eds), *Light Management in Controlled Environments*. Meister Media Worldwide, Willoughby, OH, pp. 170–80.

Czaban, W., Jämtgård, S., Näsholm, T., Rasmussen, J., Nicolaisen, M. and Fomsgaard, I. S. 2016. Direct acquisition of organic N by white clover even in the presence of inorganic N. *Plant Soil* 407(1–2), 91–107. doi:10.1007/s11104-016-2896-z.

Dangour, A. D., Dodhia, S. K., Hayter, A., Allen, E., Lock, K. and Uauy, R. 2009. Nutritional quality of organic foods: a systematic review. *Am. J. Clin. Nutr.* 90(3), 680–5. doi:10.3945/ajcn.2009.28041.

Darnhofer, I., Lindenthal, T., Bartel-Kratochvil, R. and Zollitsch, W. 2010. Conventionalisation of organic farming practices: from structural criteria towards an assessment based on organic principles. A review. *Agron. Sustain. Dev.* 30(1), 67–81. doi:10.1051/agro/2009011.

De Pascale, S., Rouphael, Y., Pardossi, A., Gallardo, M. and Thompson, R. B. 2017. Recent advances in water and nutrient management of soil-grown crops in Mediterranean greenhouses. *Acta Hortic.* 1170(1170), 31–44. doi:10.17660/ActaHortic.2017.1170.3.

De Ponti, T., Rijk, B. and van Ittersum, M. K. 2012. The crop yield gap between organic and conventional agriculture. *Agr. Syst.* 108, 1–9. doi:10.1016/j.agsy.2011.12.004.

De Tender, C. A., Debode, J., Vandecasteele, B., D'Hose, T., Cremelie, P., Haegeman, A., Ruttink, T., Dawyndt, P. and Maes, M. 2016. Biological, physicochemical and plant health responses in lettuce and strawberry in soil or peat amended with biochar. *Appl. Soil Ecol.* 107, 1–12. doi:10.1016/j.apsoil.2016.05.001.

De Visser, P. H. B., Voogt, W., Heinen, M., Assinck, F. and Marcelis, L. F. M. 2006. Minimizing N losses in organic greenhouse horticulture using model-based fertiliser strategies. In: Schröder, J. J. and Neeteson, J. J. (Eds), *N Management in Agrosystems in Relation to the Water Framework Directive*. PRI-report 116. Proceedings of the 14th N Workshop, pp. 114–16.

Desquilbet, M., Maigné, E. and Monier-Dilhan, S. 2018. Organic food retailing and the conventionalisation debate. *Ecol. Econ.* 150, 194–203. doi:10.1016/j.ecolecon.2018.04.025.

Dion, P. P., Pepin, S., Thériault, M. and Dorais, M. 2017a. Nitrogen mineralization rate of five organic fertilizers commonly used for greenhouse fruit and vegetable crops. In: *Proceedings Book of the XVIII International Plant Nutrition Colloquium with Boron and Manganese Satellite Meetings*, Copenhagen, Denmark, pp. 315-316. Available at: www.ipnc2017.org.

Dion, P. P., Jämtgård, S., Thériault, M., Bertrand, A., Pepin, S. and Dorais, M. 2017b. Xylem sap organic N content is linked to soil N composition in an organic greenhouse cucumber crop. In: Carstensen, A., Laursen, K. and Schoerring, J. (Eds), *Proceedings*

Book of the XVIII International Plant Nutrition Colloquium with Boron and Manganese Satellite Meetings. University of Copenhagen, Denmark, pp. 498-9. Available at: www.ipnc2017.org.

Dion, P. P., Jämtgård, S., Bertrand, A., Pepin, S. and Dorais, M. 2018. Organic Nitrogen uptake and assimilation in *Cucumis sativus* using position-specific labeling and compound-specific isotope analysis. *Front. Plant Sci.* 9, 1596. doi:10.3389/fpls.2018.01596.

Dion, P.-P., Jeanne, T., Thériault, M., Hogue, R., Pepin, S. and Dorais, M. 2019a. Nitrogen mineralization and contrasting microbial communities of five fertilizers commonly used in greenhouse organic horticulture (*submitted*).

Dion, P.-P., Thériault, M., Hunt, D., Bittman, S., Pepin, S. and Dorais, M. 2019b. NLOS-OG: a management tool for nitrogen fertilization in organic greenhouses. *Acta Hortic*. (*submitted* GreenSys2019). https://exchange.iseesystems.com/public/pierrepauldion/nlos-og/index.html

Diver, S., Kuepper, G., Born, H., Hinman, T., Baier, A. and Riesselman, L. 2012. Organic tomato production. *ATTRA* – National Sustainable Agriculture Information Service, Rural Business – Cooperative Service, United States Department of Agriculture, p. 24. Available at: attra.ncat.org/attra-pub/summaries/summary.php?pub=33 (accessed on 19 February 2019).

Dorais, M. 2017. Organic greenhouse tomato. In: Mattoo, A. and Handa, A. K. (Eds), *Achieving Sustainable Tomato Cultivation*. Burleigh Dodds Science Publishing, Cambridge, UK, pp. 77–114. Chapter 4.

Dorais, M. 2018. Advances in greenhouse and other protected structures used for cultivation of vegetables. In: Hochmuth, G. (Ed.), *Achieving Sustainable Cultivation of Vegetables*. Burleigh Dodds Science Publishing, Cambridge, UK.

Dorais, M. and Alsanius, W. B. 2015. Advances and trends in organic fruit and vegetable farming research. *Hortic. Rev.* 43, 185–267.

Dorais, M. and Cull, A. 2017. Organic protected horticulture in the world. *Acta Hortic*. 1164(1164), 9–22. doi:10.17660/ActaHortic.2017.1164.2.

Dorais, M. and Schwarz, D. 2018. Organic tomato. In: Heuvelink, E. (Ed.), *Tomato, Crop Production Science in Horticulture Series* (2nd edn.). CAB International, Wallingford, UK, 388pp. Chapter 11.

Dorais, M. and Thériault, M. 2018. Beneficial effects of using Si for organic greenhouse cucumber. *Acta Hortic*. 1227, 443–8.

Dorais, M., Ehret, D. L. and Papadopoulos, A. P. 2008. Tomato quality and healthy-components. In: *Bioactive Phytochemicals, Connecting Farm and Health*. Phytochem. Rev. 7(2), 231–50. doi:10.1007/s11101-007-9085-x.

Dorais, M., Antón, A., Montero, J. I. and Torrellas, M. 2014. Environmental impact of soil and demarcated bed grown organic greenhouse tomatoes. *Acta Hortic*. 1041, 291–8.

Dorais, M., Alsanius, B. W., Voogt, W., Pépin, S., Tüzel, H., Tüzel, Y. and Möller, K. 2016a. Impact of water quality and irrigation management on organic greenhouse horticulture. *BioGreenhouse COST Action FA 1105*. Available at: www.biogreenhouse.org.

Dorais, M., Thériault, M., Pepin, S. and Lefort, J. 2016b. Fertilisation management for organic cucumber grown in raised demarcated beds. *Acta Hortic*. 1137(1137), 27–32. doi:10.17660/ActaHortic.2016.1137.4.

Dorais, M., Martinez, C., Diop, M., Thériault, M., Ménard, C. and Pepin, S. 2016c. Assessing the potential of biochar as a growing media component for potted plants. *Acta Hortic*. 1137, 19–26.

Dorais, M., Gagnon, F., Laurin-Lanctôt, S., Thériault, M., Ménard, C. and Pepin, S. 2017a. Short-term improvement of soil biological activity in biochar-amended organic greenhouse tomato crops. *Acta Hortic.* 1164(1164), 249–56. doi:10.17660/ActaHortic.2017.1164.32.

Dorais, M., Mitchell, A. C. and Kubota, C. 2017b. Lighting greenhouse fruiting vegetables. In: Lopez, R. and Runkle, E. (Eds), *Light Management in Controlled Environments.* Meister Media Worldwide, Willoughby, OH, pp. 159–69. Chapter 17.

Du Jardin, P. 2015. Plant biostimulants: definition, concept, main categories and regulation. *Sci. Hortic.* 196, 3–14. doi:10.1016/j.scienta.2015.09.021.

EEC. 1991. European Commission Nitrates Directive 91/676. *OJL375* 8.

European Council. 2018. Legislation. *OJL150* 61, 161.

FAO. 2017. *Good Agricultural Practices for Greenhouse Vegetable Production in the South East European Countries; Principles for Sustainable Intensification of Smallholder Farms.* Food and agriculture organization of the united nations, Rome, 449pp.

Fauteux, F., Rémus-Borel, W., Menzies, J. G. and Bélanger, R. R. 2005. Silicon and plant disease resistance against pathogenic fungi. *FEMS Microbiol. Lett.* 249(1), 1–6. doi:10.1016/j.femsle.2005.06.034.

Foresi, L., Schmutz, U., Antón, A., Vieweger, A., Bavec, M., Meier, M., Shahid, M., Peña, N., Petrasek, R., Stajnko, D., et al. 2016. Sustainability assessment tools for organic greenhouse horticulture. BioGreenhouse COST Action FA 1105. Available at: www.biogreenhouse.org.

Franklin, O., Cambui, C. A., Gruffman, L., Palmroth, S., Oren, R. and Näsholm, T. 2017. The carbon bonus of organic nitrogen enhances nitrogen use efficiency of plants. *Plant Cell Environ.* 40(1), 25–35. doi:10.1111/pce.12772.

Gagnon, F., Pepin, S. and Dorais, M. 2016. Soil biologica activity in biochar-amended organic greenhouse tomato crops – study case for six types of soil. *Canadian Organic Scientific Conference*, Longeuil, QC, Canada, 21 September 2016.

Gamliel, A. and van Bruggen, A. H. C. 2016. Maintaining soil health for crop production in organic greenhouses. *Sci. Hortic.* 208, 120–30. doi:10.1016/j.scienta.2015.12.030.

Ganeteg, U., Ahmad, I., Jämtgård, S., Aguetoni-Cambui, C., Inselsbacher, E., Svennerstam, H., Schmidt, S. and Näsholm, T. 2017. Amino acid transporter mutants of Arabidopsis provides evidence that a non-mycorrhizal plant acquires organic nitrogen from agricultural soil. *Plant Cell Environ.* 40(3), 413–23. doi:10.1111/pce.12881.

Gaskell, M., Richard, S., Mitchell, J., Koike, S. T., Fouche, C., Hartz, T., Hor-Wath, W. and Jackson, L. 2007. Soil fertility management for organic crops. University of California – Agriculture and Natural Resources. Available at: http://anrcatalog.ucanr.edu/pdf/7249.pdf.

Ge, T., Song, S., Roberts, P., Jones, D. L., Huang, D. and Iwasaki, K. 2009. Amino acids as a nitrogen source for tomato seedlings: the use of dual-labeled (^{13}C, ^{15}N) glycine to test for direct uptake by tomato seedlings. *Environ. Exp. Bot.* 66(3), 357–61. doi:10.1016/j.envexpbot.2009.05.004.

Gil, M. I. 2015. Pre- and postharvest strategies to enhance bioactive constituents of fruits and vegetables. *Acta Hortic.* 1079(1079), 95–106. doi:10.17660/ActaHortic.2015.1079.8.

Gioseffi, E., de Neergaard, A. and Schjoerring, J. K. 2012. Interactions between uptake of amino acids and inorganic nitrogen in wheat plants. *Biogeosciences* 9(4), 1509–18. doi:10.5194/bg-9-1509-2012.

Giotis, C., Markelou, E., Theodoropoulou, A., Toufexi, E., Hodson, R., Shotton, P., Shiel, R., Cooper, J. and Leifert, C. 2009. Effect of soil amendments and biological control agents (BCAs) on soil-borne root diseases caused by *Pyrenochaeta lycopersici* and *Verticillium alboatrum* in organic greenhouse tomato production systems. *Eur. J. Plant Pathol.* 123(4), 387–400. doi:10.1007/s10658-008-9376-0.

Goldberger, J. R. 2011. Conventionalization, civic engagement, and the sustainability of organic agriculture. *J. Rur. Stud.* 27(3), 288–96. doi:10.1016/j.jrurstud.2011.03.002.

Gómez-Merinoa, F. C. and Trejo-Téllez, L. I. 2015. Biostimulant activity of phosphite in horticulture. *Sci. Hortic.* 196, 82–90.

Graber, E. R., Meller Harel, Y., Kolton, M., Cytryn, E., Silber, A., Rav David, D., Tsechansky, L., Borenshtein, M. and Elad, Y. 2010. Biochar impact on development and productivity of pepper and tomato grown in fertigated soilless media. *Plant Soil* 337(1–2), 481–96. doi:10.1007/s11104-010-0544-6.

Gravel, V., Dorais, M., Dey, D. and Vandenberg, G. 2014. Suppressiveness of fish effluents against two greenhouse tomato pathogens: *Pythium ultimum* and *Fusarium oxysporum*. *Can. J. Plant Sci.* 95, 427–36.

Greer, L. and Diver, S. 2000. Organic greenhouse vegetable production. ATTRA – National Sustainable Agriculture Information Service. Rural Business – Cooperative Service, United States Department of Agriculture.

Gruffman, L., Jämtgård, S. and Näsholm, T. 2014. Plant nitrogen status and co- occurrence of organic and inorganic nitrogen sources influence root uptake by Scots pine seedlings. *Tree Physiol.* 34(2), 205–13. doi:10.1093/treephys/tpt121.

Hallmann, E. and Rembiałkowska, E. 2012. Characterisation of antioxidant compounds in sweet bell pepper (*Capsicum annuum* L.) under organic and conventional growing systems. *J. Sci. Food Agric.* 92(12), 2409–15. doi:10.1002/jsfa.5624.

Hartz, T. K., Mitchell, J. P. and Giannini, C. 2000. Nitrogen and carbon mineralization dynamics of manures and composts. *HortScience* 35, 209–12.

Hartz, T. K. and Johnstone, P. R. 2006. Nitrogen availability from high-nitrogen-containing organic fertilizers. *HortTechnology* 16, 39–42.

Heckman, J. 2006. A history of organic farming: transitions from Sir Albert Howard's War in the Soil to USDA National Organic Program. *Renew. Agric. Food Syst.* 21, 143–50.

Hershey, D. R. 1992. Sir Albert Howard and the indore process. *HortTechnology* 2, 267–9.

Hickman, G. W. 2018. *International Greenhouse Vegetable Production – Statistics* (2018 edn.), 171pp. Available at: http://cuestaroble.com/statistics.htm.

HLPE. 2017. Nutrition and food systems. A report by the High Level Panel of Experts on Food Security and Nutrition of the Committee on World Food Security. HLPE, Rome, 151pp.

IFOAM. 2014. *The IFOAM Norms for Organic Production and Processing: Version 2014.* IFOAM, Germany, 132pp. ISBN: 978F3F944372F10F5.

Jack, A. L. H., Rangarajan, A., Culman, S. W., Sooksa-Nguan, T. and Thies, J. E. 2011. Choice of organic amendments in tomato transplants has lasting effects on bacterial rhizosphere communities and crop performance in the field. *Appl. Soil Ecol.* 48(1), 94–101. doi:10.1016/j.apsoil.2011.01.003.

Janse, J. and van der Wurff, A. W. G. 2011. Suitable rootstocks for organically grown tomato and cucumber in relation to *Meloidogyne* spp. *Acta Hortic.* 915(915), 117–24. doi:10.17660/ActaHortic.2011.915.14.

Jayaraj, J., Parthasarathi, T. and Radhakrishnan, N. V. 2007. Characterization of a *Pseudomonas fluorescens* strain from tomato rhizosphere and its use for integrated management of tomato damping-off. *BioControl* 52(5), 683–702. doi:10.1007/s10526-006-9046-0.

Johnston, J. L., Fanzo, J. C. and Cogill, B. 2014. Understanding sustainable diets: a descriptive analysis of the determinants and processes that influence diets and their impact on health, food security, and environmental sustainability. *Adv. Nutr.* 5, 418–29.

Kadoglidou, K., Lagopodi, A., Karamanoli, K., Vokou, D., Bardas, G. A., Menexes, G. and Constantinidou, H.-I. A. 2011. Inhibitory and stimulatory effects of essential oils and individual monoterpenoids on growth and sporulation of four soil-borne fungal isolates of *Aspergillus terreus*, *Fusarium oxysporum*, *Penicillium expansum*, and *Verticillium dahliae*. *Eur. J. Plant Pathol.* 130, 297–309. doi:10.1007/s10658-011-9754-x.

Kalakodio, L., Bakayoko, M., Kalagodio, A., Abo, B. O., Muhoza, J. P. and Ismaila, E. M. 2018. Progress on the amendment in biochars and its effects on the soil-plant-micro-organism-biochar system. *Rev. Environ. Health* 33(3), 281–93. doi:10.1515/reveh-2018-0007.

Kasselaki, A. M., Malathrakis, N. E., Goumas, D. E., Cooper, J. M. and Leifert, C. 2008. Effect of alternative treatments on seed-borne *Didymella lycopersici* in tomato. *J. Appl. Microbiol.* 105(1), 36–41. doi:10.1111/j.1365-2672.2007.03715.x.

Kasselaki, A. M., Goumas, D., Tamm, L., Fuchs, J., Cooper, J. and Leifert, C. 2011. Effect of alternative strategies for the disinfection of tomato seed infected with bacterial canker (*Clavibacter michiganensis* subsp. *michiganensis*). *NJAS - Wagen. J. Life Sci.* 58(3–4), 145–7. doi:10.1016/j.njas.2011.07.001.

Kavitha, B., Reddy, P. V. L., Kim, B., Lee, S. S., Pandey, S. K. and Kim, K.-H. 2018. Benefits and limitations of biochar amendment in agricultural soils: A review. *J. Environ. Manag.* 227, 146–54. doi:10.1016/j.jenvman.2018.08.082.

Kirchmann, H., Kätterer, T., Bergström, L., Börjesson, G. and Bolinder, M. A. 2016. Flaws and criteria for design and evaluation of comparative organic and conventional cropping systems. *Field Crops Res.* 186, 99–106. doi:10.1016/j.fcr.2015.11.006.

Kleinhenz, M. D. 2015. Description of commercial tomato rootstocks as of February 5, 2015, USDA. Available at: http://www.vegetablegrafting.org/wp/wp-content/uploads/2015/02/usda-scri-tomato-rootstock-table-feb-15.pdf.

Koller, M., Alföldi, T., Siegrist, M. and Weibel, F. 2004. A comparison of plant and animal based fertiliser for the production of organic vegetable transplants. *Acta Hortic.* 631(631), 209–15. doi:10.17660/ActaHortic.2004.631.27.

Koller, M., Rayns, F., Cubison, S. and Schmutz, U. 2016. Guidelines for experimental practice in organic greenhouse horticulture. BioGreenhouse COST Action FA 1105. Available at: www.biogreenhouse.org.

Kolton, M., Harel, Y. M., Pasternak, Z., Graber, E. R., Elad, Y. and Cytryn, E. 2011. Impact of biochar application to soil on the root-associated bacterial community structure of fully developed greenhouse pepper plants. *Appl. Environ. Microbiol.* 77(14), 4924–30. doi:10.1128/AEM.00148-11.

Koppert Biological Systems. 2017. *Knowing and Recognizing – The Biology of Pests, Diseases and Their Natural Solutions* (3rd edn.). Koppert B.V., p. 443.

Korcak, R. F. 1992. Early roots of the organic movement: A plant nutrition perspective. *HortTechnology* 2, 263–7. doi:10.21273/HORTTECH.2.2.263.

Laber, H. 2003. N-Freisetzung aus organischen Handelsdüngern – Übersicht und eigene Versuchsergebnisse im ökologischen Gemüsebau. (Translation: N-release from organic commercial fertilisers - overview and own experimental results in organic vegetable crops). In: Rahmann, G. and Nieberg, H. (Eds), *Ressortforschung für den ökologischen Landbau 2002*. Bundesforschungsanstalt für Landwirtschaft (FAL), Special Issue 259, pp. 17–20.

Lairon, D. 2010. Nutritional quality and safety of organic food: a review. *Agron. Sustain. Dev.* 30(1), 33–41. doi:10.1051/agro/2009019.

Lammerts van Bueren, E. T., Jones, S. S., Tamm, L., Murphy, K. M., Myers, J. R., Leifert, C. and Messmer, M. M. 2011. The need to breed crop varieties suitable for organic farming, using wheat, tomato and broccoli as examples: a review. *NJAS - Wagen. J. Life Sci.* 58, 193–205.

Laurin-Lanctôt, S. 2015. Effet de l'amendement en biochar des sols biologiques pour une culture de tomates sous serre: rétention en nutriments, activité biologique et régie de fertilisation. Master thesis. Laval University, Quebec, Canada. Available at: https://corpus.ulaval.ca/jspui/bitstream/20.500.11794/25931/1/31583.pdf.

Lehmann, J., Rillig, M. C., Thies, J., Masiello, C. A., Hockaday, W. C. and Crowley, D. 2011. Biochar effects on soil biota - a review. *Soil Biol. Biochem.* 43(9), 1812–36. doi:10.1016/j.soilbio.2011.04.022.

Lévesque, V., Rochette, P., Ziadi, N., Dorais, M. and Antoun, H. 2018. Mitigation of CO_2, CH_4 and N_2O from a fertigated horticultural growing medium amended with biochars and a compost. *Appl. Soil Ecol.* 126, 129–39. doi:10.1016/j.apsoil.2018.02.021.

Lévesque, V., Jeanne, T., Dorais, M., Ziadi, N., Hogue, R. and Antoun, H. 2019. Effect of biochar amendments on yield of tomato and sweet pepper in relation to bacterial diversity in the growing medium. *Plant and Soil* (*Submitted*).

Liang, Y., Sun, W., Zhu, Y. G. and Christie, P. 2007. Mechanisms of silicon-mediated alleviation of abiotic stresses in higher plants: a review. *Environ. Pollut.* 147(2), 422–8. doi:10.1016/j.envpol.2006.06.008.

Liang, Y., Nikolic, M., Bélanger, R., Gong, H. and Song, A. 2015. *Silicon in Agriculture: From Theory to Practice*. Springer Netherlands, Dordrecht, 235pp.

Liu, M., Cai, K., Chen,Y., Luo, S., Zhang, Z. and Lin, W. 2014. Proteomic analysis of silicon-mediated resistance to *Magnaporthe oryzae* in rice (*Oryza sativa* L.). *Eur. J. Plant Pathol.* 139(3), 579–92. doi:10.1007/s10658-014-0414-9.

López-Bucio, J., Pelagio-Flores, R. and Herrera-Estrella, A. 2015. Trichoderma as biostimulant: exploiting the multilevel properties of a plant beneficial fungus. *Sci. Hortic.* 196, 109–23. doi:10.1016/j.scienta.2015.08.043.

López-Martínez, J. D., Vázquez-Díaz, D. A., Esparza-Rivera, J. R., García-Hernández, J. L., Castruita-Segura, M. A. and Preciado-Rangel, P. 2016. Yield and nutraceutical quality of tomato fruit produced with nutrient solutions prepared using organic materials. *Rev. Fitotec. Mex.* 39, 409–14.

Louws, F. J., Rivard, C. L. and Kubota, C. 2010. Grafting fruiting vegetables to manage soilborne pathogens, foliar pathogens, arthropods and weeds. *Sci. Hortic.* 127(2), 127–46. doi:10.1016/j.scienta.2010.09.023.

Ma, J. F., Mitani, N., Nagao, S., Konishi, S., Tamai, K., Iwashita, T. and Yano, M. 2004. Characterization of the silicon uptake system and molecular mapping of the silicon transporter gene in rice. *Plant Physiol.* 136(2), 3284–9. doi:10.1104/pp.104.047365.

Mäder, P., Fliessbach, A., Dubois, D., Gunst, L., Fried, P. and Niggli, U. 2002. Soil fertility and biodiversity in organic farming. *Science* 296(5573), 1694–7. doi:10.1126/science.1071148.

Maggio, A., De Pascale, S., Paradiso, R. and Barbieri, G. 2013. Quality and nutritional value of vegetables from organic and conventional farming. *Sci. Hortic.* 164, 532-9. doi:10.1016/j.scienta.2013.10.005.

Manzoni, S., Schimel, J. P. and Porporato, A. 2012. Responses of soil microbial communities to water stress: results from a meta-analysis. *Ecology* 93(4), 930–8. doi:10.1890/11-0026.1.

Möller, K. and Schultheiß, U. 2014. *Organische Handelsdüngemittel im ökologischen Landbau: Charakterisierung und Empfehlungen für die Praxis*. Kuratorium für Technik und Bauwesen in der Landwirtschaft (KTBL), Darmstadt, Germany, 392pp.

Montalba, R., Arriagada, C., Alvear, M. and Zu´ñiga, G. E. 2010. Effects of conventional and organic nitrogen fertilizers on soil microbial activity, mycorrhizal colonization, leaf antioxidant content, and Fusarium wilt in highbush blueberry (*Vaccinium corymbosum* L.). *Sci. Hortic.* 125, 775–8.

Morra, L., Pagano, L., Iovieno, P., Baldantoni, D. and Alfani, A. 2010. Soil and vegetable crop response to addition of different levels of municipal waste compost under Mediterranean greenhouse conditions. *Agron. Sustain. Dev.* 30(3), 701-9. doi:10.1051/agro/2009046.

Müller, T. and von Fragstein und Niemsdorff, P. 2006. Organic fertilizers derived from plant materials, Part I: turnover in soil at low and moderate temperatures. *J. Plant Nutr. Soil Sci.* 169(2), 255–64. doi:10.1002/jpln.200420465.

Murphy, K. M., Campbell, K. G., Lyon, S. R. and Jones, S. S. 2007. Evidence of varietal adaptation to organic farming systems. *Field Crops Res.* 102(3), 172–7. doi:10.1016/j.fcr.2007.03.011.

Nacry, P., Bouguyon, E. and Gojon, A. 2013. Nitrogen acquisition by roots: physiological and developmental mechanisms ensuring plant adaptation to a fluctuating resource. *Plant Soil* 370(1–2), 1–29. doi:10.1007/s11104-013-1645-9.

Näsholm, T., Ekblad, A., Nordin, A., Giesler, R., Hogberg, M. and Hogberg, P. 1998. Boreal forest plants take up organic nitrogen. *Nature* 392(6679), 914–6. doi:10.1038/31921.

Nemati, M. R., Simard, F., Fortin, J.-P. and Beaudoin, J. 2015. Potential use of biochar in growing media. *Vadose Zone J.* 14(6), 6. doi:10.2136/vzj2014.06.0074.

Nichols, M. A. and Savidov, N. A. 2012. Aquaponics: a nutrient and water efficient production system. *Acta Hortic.* 947, 129–32.

NOP USDA. 2018. Available at: https://www.ams.usda.gov/about-ams/programs-offices/national-organic-program.

Olivares, F. L., Aguiar, N. O., Rosa, R. C. C. and Canellas, L. P. 2015. Substrate biofortification in combination with foliar sprays of plant growth promoting bacteria and humic substances boosts production of organic tomatoes. *Sci. Hortic.* 183, 100–8. doi:10.1016/j.scienta.2014.11.012.

Oliveira, A. B., Moura, C. F. H., Gomes-Filho, E., Marco, C. A., Urban, L. and Miranda, M. R. A. 2013. The impact of organic farming on quality of tomatoes is associated to increased oxidative stress during fruit development. *PLoS One* 8(2), 1–6. doi:10.1371/journal.pone.0056354.

Orsini, F., Maggio, A., Rouphael,Y. and De Pascale, S. 2016. Physiological quality of organically grown vegetables. *Sci. Hortic.* 208, 131-9.

Ouellette, S., Goyette, M.-H., Labbe´, C., Laur, J., Gaudreau, L., Gosselin, A., Dorais, M., Deshmukh, R. and Be´langer, R. R. 2017. The benefits of silicon amendments in strawberry are dependent on fertilization regimes. *Front. Plant Sci.* 8, 949.

Ozores-Hampton, M., Roberts, P. and Stansly, P. A. 2012. Organic pepper production. In: Russo, V. M. (Ed.), *Peppers: Botany, Production and Uses*. CAB International, Wallingford, UK, pp. 165–75.

Pantanella, E., Cardarelli, M. and Colla, G. 2010. Aquaponics vs. hydroponics: potentials for sustainable aquaculture. *Aquaculture Europe*, 2010.

Partey, S. T., Quashie-Sam, S. J., Thevathasan, N. V. and Gordon, A. M. 2011. Decomposition and nutrient release patterns of the leaf biomass of the wild sunflower (*Tithonia diversifolia*): a comparative study with four leguminous agroforestry species. *Agroforest. Syst.* 81(2), 123–34. doi:10.1007/s10457-010-9360-5.

Pathak, H. and Rao, D. L. N. 1998. Carbon and nitrogen mineralization from added organic matter in saline and alkali soils. *Soil Biol. Biochem.* 30(6), 695–702. doi:10.1016/S0038-0717(97)00208-3.

Paulitz, T. C. and Bélanger, R. R. 2001. Biological control in greenhouse systems. *Annu. Rev. Phytopathol.* 39, 103–33. doi:10.1146/annurev.phyto.39.1.103.

Paull, J. 2011. The Betteshanger summer school: missing link between biodynamic agriculture and organic farming. *J. Org. Syst.* 6, 13–26.

Paungfoo-Lonhienne, C., Visser, J., Lonhienne, T. G. A. and Schmidt, S. 2012. Past, present and future of organic nutrients. *Plant Soil* 359(1–2), 1–18. doi:10.1007/s11104-012-1357-6.

Pepin, S., Dorais, M., Gruyer, N. and Ménard, C. 2008. Changes in mineral content and CO_2 release from organic greenhouse soils incubated under two different temperatures and moisture conditions. In: Neuhoff, D., Halberg, N., Alföldi, T., Lockeretz, W., Thommen, A., Rasmussen, I. A., Hermansen, J., Vaarst, M., Lueck, L., Caporali, F., et al. (Eds), *Cultivating the Future Based on Science, Vol. 1: Organic Crop Production*. International Society of Organic Agriculture Research (ISOFAR), Modena, Italy, pp. 648–51.

Pichyangkura, R. and Chadchawan, S. 2015. Biostimulant activity of chitosan in horticulture. *Sci. Hortic.*, 196, 49–65. doi:10.1016/j.scienta.2015.09.031.

Ponisio, L. C., M'Gonigle, L. K., Mace, K. C., Palomino, J., de Valpine, P. and Kremen, C. 2014. Diversification practices reduce organic to conventional yield gap. *Proc. R. Soc. B* 282(1799), 20141396. doi:10.1098/rspb.2014.1396.

Prange, R. K. 2015. Fruit and vegetable quality as affected by the use of compost and other organic amendments. *Acta Hortic.* 1076(1076), 127–36. doi:10.17660/ActaHortic.2015.1076.14.

Ranells, N. N. and Wagger, M. G. 1997. Grass-legume bicultures as winter annual cover crops. *Agron. J.* 89(4), 659–565. doi:10.2134/agronj1997.00021962008900040019x.

Rankin, L. and Paulitz, T. C. 1994. Evaluation of rhizosphere bacteria for biological control of *Pythium* root rot of greenhouse cucumbers in hydroponic culture. *Plant Dis.* 78(5), 447–51. doi:10.1094/PD-78-0447.

Raviv, M. 2010. Sustainability of organic horticulture. *Hortic. Rev.* 36, 289–333.

Raviv, M. 2017. Should we use soilless media in organic greenhouses? *Acta Hortic.* 1164(1164), 535–40. doi:10.17660/ActaHortic.2017.1164.70.

Raviv, M. and Lieth, J. H. 2008. *Soilless Culture: Theory and Practice* (1st edn.). Elsevier Science, Amsterdam, The Netherlands, 608pp.

Reddy, N. and Crohn, D. M. 2012. Compost induced soil salinity: a new prediction method and its effect on plant growth. *Compost Sci. Util.* 20(3), 133–40. doi:10.1080/1065657X.2012.10737038.

Reganold, J. P. and Wachter, J. M. 2016. Organic agriculture in the twenty-first century. *Nat. Plants* 2016, article number 15221.

Rembiałkowska, E. 2007. Quality of plant products from organic agriculture. *J. Sci. Food Agric*. 87(15), 2757–62. doi:10.1002/jsfa.3000.

Sandhu, H. S., Wratten, S. D. and Cullen, R. 2010. The role of supporting ecosystem services in conventional and organic arable farmland. *Ecol. Complex.* 7(3), 302–10.

Sangha, J. S., Kelloway, S., Critchley, A. T. and Prithiviraj, B. 2014. Seaweeds (macroalgae) and their extracts as contributors plant productivity and quality: the current status of our understanding. *Adv. Bot. Res*. 71, 189–219.

Savvas, D. and Ntatsi, G. 2015. Biostimulant activity of silicon in horticulture. *Sci. Hortic*. 196, 66–81. doi:10.1016/j.scienta.2015.09.010.

Savvas, D., Colla, G., Rouphael, Y. and Schwarz, D. 2010. Amelioration of heavy metal and nutrient stress in fruit vegetables by grafting. *Sci. Hortic*. 127(2), 156–61. doi:10.1016/j.scienta.2010.09.011.

Schmutz, U., Sumption, P. and Lennartsson, M. 2011. Economics of UK organic protected cropping. *Acta Hortic*. 915, 39–46. doi:10.17660/actahortic.2011.915.4.

Schneider, M. K., Lüscher, G., Jeanneret, P., Arndorfer, M., Ammari, Y., Bailey, D., Balázs, K., Báldi, A., Choisis, J., Dennis, P., et al. 2014. Gains to species diversity in organically farmed fields are not propagated at the farm level. *Nat. Commun*. 5(1), 4151. doi:10.1038/ncomms5151.

Schwarz, D., Colla, G., Rouphael, Y. and Venema, J. H. 2010. Grafting as a tool to improve tolerance of vegetables to abiotic stresses: thermal stress, water stress and organic pollutants. *Sci. Hortic*. 127, 162–71.

Seufert, V., Ramankutty, N. and Foley, J. A. 2012. Comparing the yields of organic and conventional agriculture. *Nature* 485(7397), 229–32. doi:10.1038/nature11069.

Shaaban, M., Van Zwieten, L., Bashir, S., Younas,A., Núñez-Delgado, A., Chhajro, M. A., Kubar, K. A., Ali, U., Rana, M. S., Mehmood, M. A., et al. 2018. A concise review of biochar application to agricultural soils to improve soil conditions and fight pollution. *J. Environ. Manag*. 228, 429–40. doi:10.1016/j.jenvman.2018.09.006.

Shekhar Sharma, H. S. S., Fleming, C., Selby, C., Rao, J. R. and Martin, T. 2014. Plant biostimulants: a review on the processing of macroalgae and use of extracts for crop management to reduce abiotic and biotic stresses. *J. Appl. Phycol*. 26(1), 465–90. doi:10.1007/s10811-013-0101-9.

Smith-Spangler, C., Brandeau, M. L., Hunter, G. E., Bavinger, J. C., Pearson, M., Eschbach, P. J., Sundaram, V., Liu, H., Schirmer, P., Stave, C., et al. 2012. Are organic foods safer or healthier than conventional alternatives?: a systematic review. *Ann. Intern. Med*. 157(5), 348–66. doi:10.7326/0003-4819-157-5-201209040-00007.

Somerville, C., Cohen, M., Pantanella, E., Stankus, A. and Lovatelli, A. 2014. Small-scale aquaponic food production integrated fish and plant farming. FAO Fisheries and aquaculture technical paper 589. FAO, Rome, Italy.

Song, S. W., Lehne, P., Le, J. G., Ge, T. D. and Huang, D. F. 2009. Yield, fruit quality and nitrogen uptake of organically and conventionally grown muskmelon with different inputs of nitrogen, phosphorus, and potassium. *J. Plant Nutr*. 33(1), 130–41. doi:10.1080/01904160903394622.

Sonneveld, C. and Voogt, W. 2009. *Plant Nutrition of Greenhouse Crops*. Springer, p. 430.

Sullivan, P. 2003. Overview of cover crops and green manures. ATTRA – National Sustainable Agriculture Information Service. Rural Business – Cooperative Service, United States Department of Agriculture.

Sutton, J. C. and Peng, G. 1993. Manipulation and vectoring of biocontrol organisms to manage foliage and fruit diseases in cropping systems. *Annu. Rev. Phytopathol.* 31(1), 473–93. doi:10.1146/annurev.py.31.090193.002353.

Tarara, J. M. 2000. Microclimate modification with plastic mulch. *HortScience* 35, 169–180.

Thies, J. E. and Rillig, M. 2009. Characteristics of biochar: biological properties. In: Lehmann, J. and Joseph, S. (Eds), *Biochar for Environmental Management: Science and Technology*. Earthscan Publications, London, UK, pp. 85–105.

Tittarelli, F., Båth, B., Ceglie, F. G., del Carmen Garcia, M., Möller, K., Reents, H. J., Védie, H. and Voogt, W. 2016a. Soil fertility management. In: *Organic Greenhouses in Europe*. BioGreenhouse COST Action FA 1105. Available at: www.biogreenhouse.org.

Tittarelli, F., Ceglie, F. G., Ciaccia, C., Mimiola, G., Amodio, M. L. and Colelli, G. 2016b. Organic strawberry in Mediterranean greenhouse: effect of different production systems on soil fertility and fruit quality. *Renew. Agric. Food Syst.* 32(6), 485–97. doi:10.1017/S1742170516000417.

Toresano-Sánchez, F., Valverde-García, A. and Camacho-Ferre, F. 2012. Effect of the application of silicon hydroxide on yield and quality of cherry tomato. *J. Plant Nutr.* 35(4), 567–90. doi:10.1080/01904167.2012.644375.

Torrellas, M., Antón, A., Ruijs, M., Victoria, N. G., Stanghellini, C. and Montero, J. I. 2012. Environmental and economic assessment of protected crops in four European scenarios. *J. Cleaner Prod.* 28, 45–55. doi:10.1016/j.jclepro.2011.11.012.

Tuomisto, H. L., Hodge, I. D., Riordan, P. and Macdonald, D. W. 2012. Does organic farming reduce environmental impacts? – A meta-analysis of European research. *J. Environ. Manag.* 112, 309–20. doi:10.1016/j.jenvman.2012.08.018.

Tuzel, Y., Gul, A., Tuncay, O., Anac, D., Madanlar, N., Yoldas, Z., Gumus, M., Tuzel, I. H. and Engindeniz, S. 2005. Organic cucumber production in the greenhouse: A case study from Turkey. *Renew. Agric. Food Syst.* 20(4), 206–13. doi:10.1079/RAF2005105.

Tuzel, Y., Oztekin, G. B. and Tan, E. 2015. Use of different growing media and nutrition in organic seedling production. *Acta Hortic.* 1107(1107), 165–75. doi:10.17660/ActaHortic.2015.1107.22.

Urbain, L. and Urbain, I. 2010. *La production sous serre. Tome 1 La gestion du climat*. 2e édition. Lavoisier, Paris, France, 336pp.

Urban, L., Jawad, A. and Nicot, P. C. 2014. Prospects for using environmental levers in greenhouses to stimulate plant defences and the quality of fruits and vegetables. *Acta Hortic.* 1041(1041), 247–52. doi:10.17660/ActaHortic.2014.1041.29.

Vaccari, F. P., Maienza, A., Miglietta, F., Baronti, S., Di Lonardo, S., Giagnoni, L., Lagomarsino, A., Pozzi, A., Pusceddu, E., Ranieri, R., et al. 2015. Biochar stimulates plant growth but not fruit yield of processing tomato in a fertile soil. *Agric. Ecosyst. Environ.* 207, 163–70. doi:10.1016/j.agee.2015.04.015.

Vallance, J., Déniel, F., Le Floch, G., Guérin-Dubrana, L., Blancard, D. and Rey, P. 2011. Pathogenic and beneficial microorganisms in soilless cultures. In: Lichtfouse, E., Hamelin, M., Navarrete, M. and Debaeke, P. (Eds), *Sustainable Agriculture Volume 2*. Springer, France, pp. 711–26.

Vallverdú-Queralt, A., Jáuregui, O., Medina-Remón, A. and Lamuela-Raventós, R. M. 2012. Evaluation of a method to characterize the phenolic profile of organic and conventional tomatoes. *J. Agric. Food Chem.* 60(13), 3373–80. doi:10.1021/jf204702f.

van Bruggen, A. H. C. 1995. Plant disease severity in high-input compared to reduced-input and organic farming systems. *Plant Dis.* 79(10), 976–84. doi:10.1094/PD-79-0976.

van der Lans, C. J. M., Meijer, R. J. M. and Blom, M. 2011. A view of organic greenhouse horticulture worldwide. *Acta Hortic.* 915(915), 15–22. doi:10.17660/ActaHortic.2011.915.1.

van der Wurff, A. W. G., Fuchs, J. G., Raviv, M. and Termorshuizen A. J. 2016. *Handbook for Composting and Compost Use in Organic Horticulture*. BioGreenhouse COST Action FA, 108pp. doi.org/10.18174/375218 1105. Available at: www.biogreenhouse.org.

Vänninen, I., Pinto, D. M., Nissinen, A. I., Johansen, N. S. and Shipp, L. 2010. In the light of new greenhouse technologies: 1. Plant-mediated effects of artificial lighting on arthropods and tritrophic interactions. *Ann. Appl. Biol.* 157(3), 393–414. doi:10.1111/j.1744-7348.2010.00438.x.

Voogt, W. 2009. Normen gietwaterkwaliteit glastuinbouw. Wageningen UR glastuinbouw. Bleiswijk, Wageningen UR glastuinbouw: 20.

Voogt, W. 2014. Soil fertility management in organic greenhouse crops: a case study on fruit vegetables. *Acta Hortic.* 1041(1041), 21–35. doi:10.17660/ActaHortic.2014.1041.1.

Voogt, W. and Sonneveld, C. 2001. Chapter 6 silicon in horticultural crops grown in soilless culture. *Stud. Plant Sci.* 8, 115–31. doi:10.1016/S0928-3420(01)80010-0.

Voogt, W., De Visser, P. H. E., van Winkel, A., Cuijpers, W. J. M. and van de Burgt, G. J. H. M. 2011. Nutrient management in organic greenhouse production: navigation between constraints. *Acta Hortic.* 915(915), 75–82. doi:10.17660/ActaHortic.2011.915.9.

Wang, Q., Klassen, W., Li, Y. and Codallo, M. 2009. Cover crops and organic mulch to improve tomato yields and soil fertility. *Agron. J.* 101(2), 345–51. doi:10.2134/agronj2008.0103.

Wichern, J., Wichern, F. and Joergensen, R. G. 2006. Impact of salinity on soil microbial communities and the decomposition of maize in acidic soils. *Geoderma* 137(1–2), 100–8. doi:10.1016/j.geoderma.2006.08.001.

Willer, H. and Lernoud, J. 2018. *The World of Organic Agriculture – Statistics and Emerging Trends 2018*. FiBL and IFOAM, Switzerland.

Workneh, F. and van Bruggen, A. H. C. 1994. Suppression of corky root of tomatoes in soils from organic farms associated with soil microbial activity and nitrogen status of soil and tomato tissue. *Phytopathology* 84(7), 688–94. doi:10.1094/Phyto-84-688.

Xu, X., Hui, D., King, A. W., Song, X., Thornton, P. E. and Zhang, L. 2015. Convergence of microbial assimilations of soil carbon, nitrogen, phosphorus, and sulfur in terrestrial ecosystems. *Sci. Rep.* 5(1), 17445. doi:10.1038/srep17445.

Yan, F., Schubert, S. and Mengel, K. 1996. Soil pH increase due to biological decarboxylation of organic anions. *Soil Biol. Biochem.* 28, 611–624.

Yang, X., Feng, L., Zhao, L., Liu, X., Hassani, D. and Huang, D. 2017. Effect of glycine nitrogen on lettuce growth under soilless culture: a metabolomics approach to identify the main changes occurred in plant primary and secondary metabolism. *J. Sci. Food Agric.* 98, 467–77. doi:10.1002/jsfa.8482.

Zikeli, S., Deil, L. and Moeller, K. 2017. The challenge of imbalanced nutrient flows in organic farming systems: a study of organic greenhouses in Southern Germany. *Agric. Ecosyst. Environ.* 244, 1–13. doi:10.1016/j.agee.2017.04.017.

Zublena, J. P., Baird, J. V. and Lilly, J. P. 1997. Nutrient content of fertilizer and organic materials. North Carolina Cooperative Extension Service publication AG-439-18.

Chapter 6

Towards sustainable plant factories with artificial lighting (PFALs): from greenhouses to vertical farms

Toyoki Kozai, Japan Plant Factory Association, Japan; Yumiko Amagai, Chiba University, Japan; and Eri Hayashi, Japan Plant Factory Association, Japan

1 Introduction

As many researchers and policymakers have pointed out, environmentally sustainable and economically viable food production systems need to be developed to solve the global as well as local food-environment-resource trilemma amid the situation of (1) increasing urban population, (2) increasing demand for nutritious, tasty, safe food, (3) decreasing agricultural population with increasing percentage of aged farmers, (4) decreasing fertile arable land area, (5) decreasing amount of available water for irrigation, (6) decreasing amount of available phosphorous and potassium fertilizers and (7) changing global and local climates.

1.1 Requirements for sustainable plant production systems

Environmentally sustainable and economically viable food production systems need to be designed and operated (1) to achieve the maximum yield with the highest quality as scheduled under the changing climate, (2) to use the minimum amount of resources (land area, electricity, water, labour, fertilizer,

http://dx.doi.org/10.19103/AS.2019.0052.06

fuel for cultivation, transportation and post-harvest processing), resulting in the minimum emission of waste including CO_2, plant residue, food loss and heat energy, (3) to achieve energy-autonomous operations through the use of electricity generated by natural energy, (4) at the minimum initial and operating costs, (5) to provide rewarding job opportunities and healthy food to a diverse range of people and (6) to achieve economically viable plant production (Fig. 1).

1.2 Types of plant production systems

Plant production systems can be categorized into three groups: (1) open fields with or without an irrigation system and/or soil cover film for mulching and tunnels, (2) plastic or glass greenhouses with or without environmental control units including light emitting diodes (LEDs) for supplemental lighting and (3) plant factories with artificial lighting (PFALs) or indoor vertical farms (Fig. 2). The PFALs include closed transplant production systems with artificial lighting. The transplants include seedlings, micropropagated plantlets, cuttings and grafted plantlets which are transported and transplanted in the open fields, greenhouses or PFALs for further growth.

Each plant production system has its advantages and disadvantages. An appropriate plant production system depends on the plant species, market prices of the produce and resources, geographical location and many other factors. Open-field agriculture has a history of over ten centuries, whereas the greenhouse industry has a history of around one century, and PFALs for commercial production have a history of only several decades. Thus, the current PFAL technologies are still in their infancy, and most of their potential advantages are not yet actualized.

Many people still consider that the production cost of plants grown in PFALs is too high to be economically viable, and that PFALs cannot be environmentally sustainable because their operation consumes too much

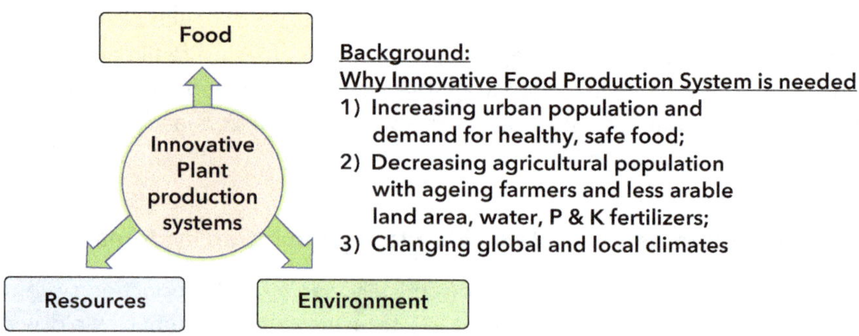

Figure 1 To solve the food-environment-resource trilemma, we need to develop innovative food production systems.

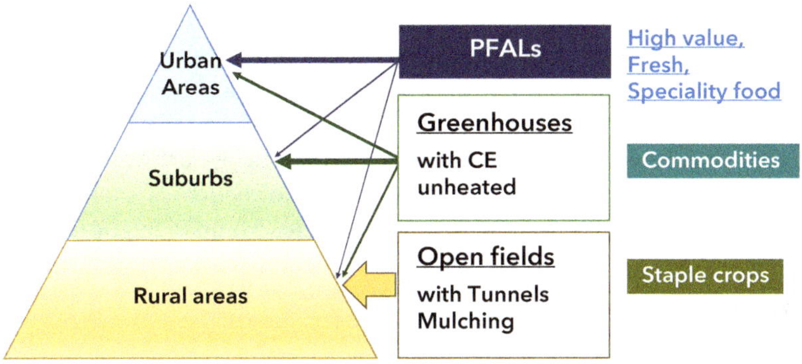

Figure 2 Types of plant production systems and their applications.

electricity which comes from fossil fuel and/or nuclear power. However, at some recently renovated PFALs, the production cost was reduced and the yield and quality were also significantly improved during 2015–17, as shown in a later section of this chapter.

1.3 Objectives of this chapter

This chapter discusses the following under the vision, mission and constraints shown in Fig. 3: (1) how and how much resources are consumed in the current PFALs and (2) how much the resource consumption and production cost can be reduced by further improving the current PFAL systems. Our tentative numerical target values to be achieved by 2023 through such improvements are (1) a 45% increase in the yield, a 5% increase in the economic value per unit food produced and a 5% increase in percentage of produce sold over total

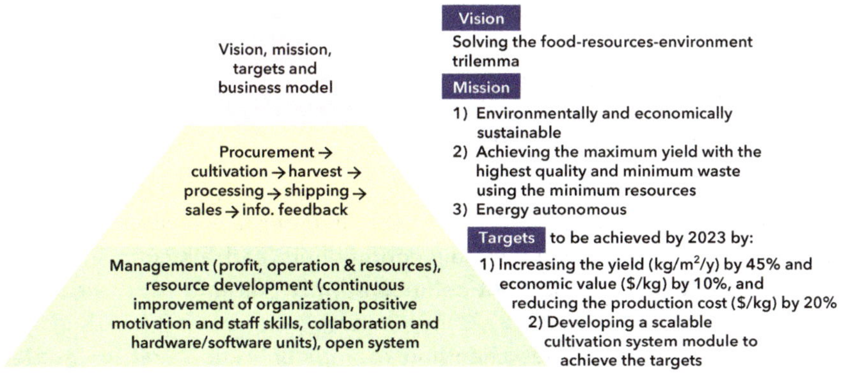

Figure 3 Factors to be considered in the design of PFALs, and the vision, mission, targets (goals) of PFAL R&D.

produce, (2) a 20% reduction in the production cost and thus (3) a two-fold (2.0 = (1.05 × 1.45 × 1.05)/(1.0 - 0.2)) increase in the current cost performance compared to that in 2018. Many people in the world are currently developing a scalable cultivation system module to actualize the target values. This chapter also discusses the concept and methodology for further improving the productivity and cost performance in the forthcoming smart PFALs by introducing the recent advanced technologies described in the next section.

1.4 Recent advanced technologies

Recent advanced industrial technologies related to PFALs include (1) white LEDs with a broad spectrum, (2) high-speed, small and inexpensive microcomputers, DNA analysers/sequencers, memory chips, wireless networks, cameras (thermal, night vision, fisheye, chlorophyll fluorescence and hyperspectral cameras), environmental sensors, ion sensors and biosensors, (3) practical software for big data mining and/or deep learning in artificial intelligence (AI), (4) global and local ICT (information and communication technology), (5) electric power generation using natural energy such as solar radiation, wind and biomass, (6) omics (genomics, metabolomics, proteomics and transcriptomics), (7) phenomics, (8) open databases, (9) self-learning via the Internet through massive open online courses (MOOCs), (10) three-dimensional (3D) printers for making parts, devices and tools to improve the PFAL and (11) 3D simulators using virtual and augmented reality.

The application of such recent advanced technologies will play an important role in improving the traceability, predictability and controllability of plant production processes at low cost, and thus significantly improve the yield and quality of produce. For example, the cost of LEDs has recently been decreasing and will continue to decrease year by year, and LEDs have been increasingly applied in plant production systems to boost yield and quality at a lower production cost (Kozai et al., 2016).

2 Characteristics of PFALs

2.1 Main components of the PFAL

The main components of the PFAL cultivation room are (1) an almost airtight and thermally well-insulated structure, (2) multi-tier racks with lighting and hydroponic cultivation units, (3) air conditioners and air circulation fans, (4) CO_2 supply unit, (5) nutrient solution supply unit and (6) environment control unit (Fig. 4; Kozai, 2013). Other units installed in most PFALs include an air shower room, clothes changing room, sterilization unit for nutrient solution (and air), movable racks for the transport of harvests, cultivation panels for hydroponics and so on. It is estimated that the cost for the

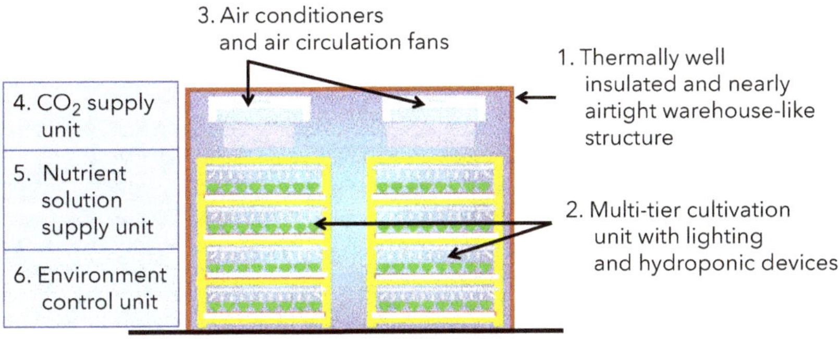

Figure 4 Six main components of the PFAL cultivation room (modified from Kozai (2013)).

cultivation room or the building itself (No. 1 in Fig. 3) accounts for about 50% of the initial investment cost; the rest (Nos. 2-6 and other units) of the facilities and equipment accounts for the other 50%. Among Nos. 2-6, the cost for No. 2 is the highest.

2.2 Potential and actualized benefits of PFALs

Since around 2017, a few PFALs in Japan have been making significant profits, and have increased their floor area to raise production capacity to meet the increasing demand. They succeeded in reducing the production costs by about 30% and improved the productivity by about 30% during 2015-17, with little use of the advanced technologies described in Section 1.4 except for LEDs. On the other hand, the number of such profitable PFALs is currently limited. In other words, most potential benefits are not actualized in most current PFALs.

To actualize the potential benefits of PFALs, the concept and methodology for their actualization must be understood before designing and operating a PFAL. At the same time, the vision, mission and goals of the PFAL design, operation and business model must be clearly established and shared with the team members. For details, refer to Kozai et al. (2015) and Kozai (2018a,b).

1 *Relatively easy measurement* of hourly and/or daily resource supply rates and production rates (Tables 1 and 2). The greatest benefit of PFALs over greenhouses and open fields is that all the resource supply rates and production rates can be measured and controlled relatively easily at a low cost. Then, the production costs can also be estimated and methods for reducing the production cost can subsequently be found online relatively easily.
2 *High resource use efficiency (RUE)* in Table 1 is defined as the amount of resource element fixed or utilized in the plants (F) divided by the

Table 1 Resource supply rates, definitions of resource use efficiencies (RUEs) and main uses of each resource element

Resource supply rate	Definition of RUE for each resource element	Main uses
Electricity (kWh)	Ratio of the chemical energy fixed in plants to the electric energy consumed by lamps and air conditioners	Lamps, air conditioners, air circulation fans, nutrient solution circulation pumps
Light energy	Ratio of the light energy fixed as chemical energy in plants to the light energy emitted by lamps	Photosynthesis and photomorphogenesis of plants
Water for irrigation	Weight ratio of the water held in plants to the water irrigated	Irrigation for hydroponic culture
Water for washing	Weight ratio of the water reused after washing to the water supplied	Cleaning of floor, cultivation panels, cultivation beds and so on
CO_2	Weight or mole ratio of the CO_2 fixed in plants by photosynthesis to the CO_2 supplied	Carbon source for carbohydrates produced by photosynthesis
Fertilizer	Mole ratio of each macronutrient element in the fertilizer absorbed by plants to the macronutrient in the fertilizer supplied	Macro- (N, P, K, Ca, Mg and S) and micronutrient elements
Seeds	Number of marketable plants over the number of seeds sown	Seeding
Other consumables	Number or weight ratio of the consumable successfully used to the consumable supplied	Substrate, cultivation panels, sanitary goods, shoes, disinfectant, disposable gloves, clothing, masks and head caps, plastic bags and so on
Land/space	Production rate per cultivation area (m^2) or cubic space (m^3)	Cultivation, maintenance, management, operation
Electric equipment	Coefficient of performance (COP) of air conditioners, power factors of pumps, motors and so on	Air conditioning, nutrient solution circulation and so on

amount of resource element supplied to the PFAL (s) (Fig. 5 and Table 1). Namely, RUE = B/A or (A − C)/A where C is the amount of resource that is released and wasted by escaping outside the PFAL (Kozai et al., 2015). The RUE for each resource element can also be estimated online.

3 *Easy online estimation of plant responses* such as net photosynthetic rate, dark respiration rate and water uptake rate based on the mass (or material) balance of CO_2 and water (Kozai, 2013). In the near future, plant transpiration rate and fresh weight increase rate, ion uptake rates by nutrient macroelements and growth rates of leaf area and roots will be estimated online. These estimations are relatively easy because of

Table 2 Production rates including waste production rates in the cultivation room

Production rate	Remarks
Marketable produce	Average fresh weight and quality, and their variations
Plant residue	Damaged leaves and stems, roots with substrate and dead roots in the cultivation beds
Wastewater	Drained nutrient solution from hydroponic units, and water used for washing and cleaning
CO_2	Respired by workers in the cultivation room
Damaged consumables	Used disposable goods
Heat energy	Generated by lamps and other electric equipment in the cultivation room to be removed by air conditioners to the outside
Microorganisms	Propagated in the cultivation room (difficult to measure)

almost airtight cultivation room, closed hydroponic unit and accurate rate measurements of resource supply and production (Table 2). Furthermore, plant traits such as canopy structure, chemical composition in leaves/plants and so on can be estimated online.

4 *Stepwise improvement* of the RUEs, productivity and economic value of the plants is possible by visualizing the flow of energy, substances and workers in the production process and other related costs/sales using the measured data.

5 *High annual productivity* per unit land area. Over 100-fold annual productivity per unit land area has already been achieved and the productivity will be improved further in the near future without using any pesticides, compared with that in open fields, mainly due to the use of multilayers, shortened cultivation period through optimal environmental control, low percentage of unused or empty cultivation

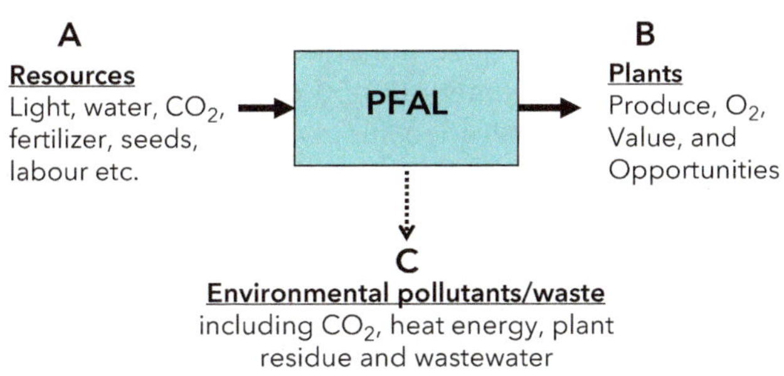

Figure 5 Resource supply rates (A) and production rates (B) in the PFAL, and definition of resource use efficiency (RUE = B/A). The RUE is estimated for each resource element supplied to the PFAL. Waste production rates (C) can also be measured for each waste element.

areas, high planting density and virtually no damage by weather and pest insects.

6 *High controllability of plant aerial and rootzone environments.*
7 *High reproducibility and predictability* of yield and quality. Because of the high controllability of the environment all year round, scheduled and/or on-demand plant production is possible regardless of the weather.
8 *High traceability* throughout the whole supply chain from seeds to table, which enables a high level of risk management.
9 *High controllability of sanitary conditions.* Because of the high controllability of environmental and sanitary conditions, pesticide-free and other contaminant-free plants are produced.
10 *High weight percentage of marketable parts* over the whole plant biomass.
11 *High-quality plants* can be produced as scheduled through the proper environmental control, cultivar selection and cultivation system.
12 *Long shelf life* due to low population density of microorganisms, which decreases the amount of vegetable garbage or loss at home and in shops.
13 *No need to wash* or cook before serving, if packed in a sealed package after harvesting in the culture room. Then, the labour cost for washing/cleaning at consumer sides can be saved.
14 *High adaptability for the location* (near or in food/meal delivery shops etc.). The PFAL can be built without any problems in a shaded area, on contaminated or infertile soil, and in vacant rooms, buildings and land in urban areas. The thermally well-insulated PFAL can also be operated in very cold, arid or hot areas with minimum cooling/heating costs. In fact, cooling is necessary to remove the heat generated by lamps in the thermally well-insulated cultivation room in cold regions where air temperature outside is -40°C.
15 *Light and safe work* under comfortable room air temperature (22–25°C) and moderate air movement (30–50 cm s^{-1}), although there are still some problems to be solved to improve the working environment.
16 *Design and environment control are simpler in the PFAL than in the greenhouse.*
17 *A small PFAL* with a floor area of 0.1–10 m^2 is a useful way to learn the principles of life science, engineering and technology at home, school or a community centre, especially when the PFAL is connected via the Internet to other small PFALs and a PFAL database is created for exchanging information and opinions (Kozai et al., 2015; Harper and Siller, 2015).

2.3 Risk management

A serious, even near-fatal, event in the culture room of a large-scale PFAL can start with a small incident. If such an event happens, it could take weeks or months to recover completely. Thus, risk management is essential to avoid small and near-fatal events including:

1. Invasion of insects carrying pathogens through a drain, air leak and on clothing/hair, seeds and other materials;
2. Propagation and growth of insects, fungi and algae in wet places;
3. Propagation of pathogenic microorganisms in culture beds containing dead roots. Cleaning of culture beds at regular intervals is necessary;
4. Fire and/or power failure caused by electric short circuit often caused by drops of nutrient solution on the power line near the culture beds;
5. Overflow of nutrient solution from the culture beds or nutrient solution tank to the floor due to clogging of a pipe or drain;
6. Shortage of nutrient solution in culture beds due to malfunction of a water pump or clogging of a pipe;
7. Physiological disorders such as tipburn due to an undesirable environment and selection of an unfavourable cultivar;
8. High room air temperature due to the malfunction of air conditioners for cooling during the photoperiod (all the lamps for plant lighting need to be turned off when the air conditioners are stopped);
9. Misunderstanding among workers due to a lack of human communication;
10. Emergency treatment and repair systems are not prepared in the event of a failure of electricity, water/nutrient solution and CO_2 delivery systems;
11. Alarm system does not work well, or the PFAL manager does not respond appropriately to the alarm. The meaning of the alarm signals must be easy to understand. The alarm system must work with a battery in case the main power line fails.

3 Resource consumption by resource elements: simulation studies

Electricity, CO_2, water and fossil fuel consumed in a PFAL producing lettuce were compared with those of a greenhouse (GH) in the Netherlands (NL), United Arab Emirates (UAE) and Sweden (SWE), using a sophisticated integrative simulation model (Graamans et al., 2018). The resource consumption in the GH with supplemental HPS (high-pressure sodium lamp) lighting during the winter season (GH + SL) is also simulated. Since the PFAL is almost airtight and thermally well insulated and the electricity is mostly used for lighting, the electricity consumption of the air conditioners is not significantly affected by the weather (although the COP of air conditioner is lower at a lower temperature outside, especially when the outside air temperature is lower than room air temperature). The simulated results by Graamans et al. (2018) are summarized below.

1 Electricity consumption per dry weight is the highest (250 kWh kg^{-1}) in PFAL, which is approximately 1.2-fold of SWE, 1.4-fold of SWE + SL, 2.2-fold of UAE and 3.5-fold of NLD (1 kg of dry weight equals to 14.3 kg of fresh weight).
2 Amount of CO_2 supply for photosynthesis per dry weight is the lowest (2.0 kg kg^{-1}) in PFAL, which is roughly 1/3rd of UAE, 1/7th of SWE + SL, 1/9th of SWE and 1/13th of NLD. CO_2 supplied is almost all fixed by plants in PFAL, while most CO_2 supplied escaped to the outside due to air infiltration through the glass walls.
3 Water consumption per dry weight is the lowest (13.3 kg kg^{-1}) in PFAL, which is 1/1.3th of UAE, 1/11th of SWE + SL, 1/14th of SWE and 1/20th of NLD. Water in SWE and NLD is used for irrigation and evaporative cooling and is lost through ventilation and infiltration. In PFAL and UAE, most transpired water vapour is condensed at the cooling pipes of air conditioners and is recycled.
4 Annual dry weight production per cultivation area is the highest (5 kg m^{-2} $year^{-1}$) in the PFAL, which is 1.2-fold of UAE, 1.7-fold of SWE + SL, 2.2-fold of NLD and 2.5-fold of SWE. The production in SWE and NLD in June are almost the same as that of PFAL, but those in December are only around 15% of PFAL.
5 Fossil fuel (natural gas) energy consumption for heating per dry weight is zero in the PFAL and UAE, while it is 240 MJ kg^{-1} in NLD, 390 MJ kg^{-1} in SWE + SL and 750 MJ kg^{-1} in SWE.

Kikuchi et al. (2018) examined the environmental and resource use properties of a PFAL in Japan and compared them with GH and open-field horticultural crop production in Japan. They found that the PFAL significantly reduced the use of phosphorous, water and land area in leaf lettuce production compared with GH and open-field production, at the expense of extra electric energy consumption. Based on their environmental and resource use analysis, they concluded that the electric energy consumption of the PFAL can be sufficiently reduced by employing emerging technologies of energy savings and electricity generation using natural energy. Then, the PFAL could become a viable or competitive plant production technology in the food, energy and water systems of the future.

4 Production cost and its components

4.1 Percentages of cost components and profit of a PFAL producing leaf lettuce in Japan

Figure 6 shows the percentages of the cost components and profit of a PFAL with LEDs producing leaf lettuce in Japan (Ijichi, 2018). The cost percentages are 25% for depreciation, 25% for labour and 21% of electricity. It should be noted that this PFAL makes a profit of 11% even though it is neither automated nor robotized for harvesting and transporting the produce in the cultivation room. The cost for LED lighting, which accounts for around 80% of the total electricity cost, is said to be about 40% lower than the electricity cost for fluorescent lamps.

4.2 Resource consumption by components in the PFAL

It is difficult to estimate the actual value of resource consumption and production cost by components in the commercial PFAL. Table 3 shows the estimated resource consumption and production cost per kg of fresh produce by components in PFALs producing leaf lettuce in Japan as of 2017 (Kozai, 2018a). The data shown in Table 3 was obtained through information exchange and discussion at the committee meeting on productivity improvements organized by the Japan Plant Factory Association and through personal interviews by the authors with each of several relatively large PFALs in Japan.

As described in Section 3, Graamans et al. (2018) showed that electricity consumption per lettuce dry weight was 250 kWh kg^{-1} in the PFAL. Assuming

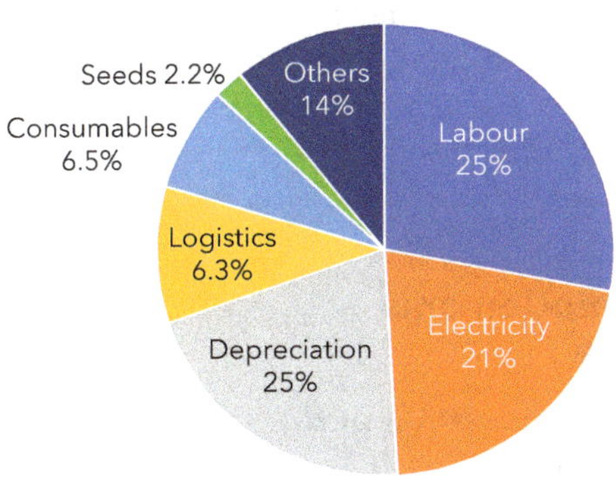

Factor	Value
Fresh weight/head	80 g
Unit price/head	0.86 USD
Cost/head	0.76 USD
Daily production	1,000 kg
% revenue/sales	9%
% sold	95%
% sold to supermarket	60%
% sold to business use	40%

Figure 6 An example of percentages of cost components and profit of a PFAL producing leaf lettuce in Japan (Ijichi, 2018). Cost for sales promotion is excluded. US$ 1 equals 105 Japanese yen. Electricity price: 15–20 yen kWh^{-1}. Labour cost: about 1200 yen h^{-1} for part-time workers, about 500 000 yen $month^{-1}$ for part-timew orkers, about 500 000 yen month−1 for regular employees.

Table 3 Estimated resource consumption and production cost per kg of fresh produce by components in representative profitable PFALs producing leaf lettuce in Japan as of 2017

Resource element	Resource consumption	Unit cost (US\$ kWh^{-1})	Cost (US\$ kg^{-1})
Electricity for lighting, air conditioning etc.	7-8 kWh kg^{-1}	0.14–0.18	1.0–1.4
Labour (personnel expenditure)	0.10–0.13 h kg^{-1}	*	1.5–1.8
Cultivation (bed) area (a major component of depreciation cost)	3-4 m^2 kg^{-1} day^{-1}	-	1.3–1.7
Others (seeds, CO_2, water, fertilizer, substrate, plastic bags etc.)	-	-	1.0–1.4
Production cost (US\$ kg^{-1})			**4.8-6.3**

Source: Redrawn from Kozai (2018a).
Cost for sales promotion is excluded in the production cost. US\$1 was equal to 105 Japanese yen as of 2017.
* Labour cost = about 1200 yen h^{-1} for part-time workers and about 500 000 yen/month for regular employees as of 2017.

dry matter of 7%, this value can be converted to electricity consumption of 17.5(=250 × 0.07) kWh kg^{-1} fresh lettuce weight, which is 2.2–2.5-fold the electricity consumption given in Table 3. In this case, the electricity cost in Japan would be 2.45–3.15 US\$ kg^{-1}. This difference in electricity consumption between Graamans et al. and Table 3 can be partly explained as follows: In Graamans et al., photosynthetic photon flux density (PPFD) of 500 μmol m^{-2} s^{-1}, butter head lettuce and 300 g of fresh weight at harvest are assumed for their simulation study, while PPFD of about 150 μmol m^{-2} s^{-1}, leaf lettuce and 100 g of fresh weigh at harvest are common in Japanese commercial PFALs. In addition, transplanting or respacing to improve the efficiency of using light energy and availability of space is not considered in the simulation. These three factors together could easily explain a factor 2–2.5 in the light use efficiency (and thus electricity use efficiency) of the Japanese PFALs and the simulated one.

5 Reducing electricity consumption and improving cost performance

5.1 Factors affecting electricity consumption and its possible percentage reduction

Table 4 shows the estimated current energy use efficiency (EUE) and expected improved EUE at each stage of the electric light to chemical energy conversion process (Nos. 1-3) and the chemical energy partitioning, affecting the electricity consumption per kg of produce. The fresh weight use ratio at Stages 4 and 5 is considered to be the chemical EUE of the produce. The estimated current and expected improved EUEs at Stages 1-3 are obtained from Kozai (2011) and Yano (2016) and the ratios at Stages 4 and 5 are obtained by inspection, oral interviews and discussion by the author at profitable commercial PFALs. It is noted that only around 4% of the electric energy is fixed in the produce sold even in the current profitable PFALs. For more details on the energy conversion process regarding the lamps in the PFAL, refer to Kozai (2011) and Yano (2016).

As shown in Table 4, by improving the EUE or the ratio at each stage by 1.05–1.20 (5–20% increase), the electricity consumption per kg of produce can be reduced by 57% (=100 × 1.00/1.74). It is considered that a higher EUE and ratio than that shown as the estimated current EUE in Table 4 can be actualized at each stage in the near future using recent advanced technologies, as discussed in a later section.

No. 1 (in Table 4): The photosynthetic radiant energy (PRE) efficiency as well as photosynthetic photon efficacy of LED lamps has improved and will continue to improve for at least the next several years. The price for LEDs per photosynthetic photons (US\$ mol^{-1}) will also keep decreasing, reducing the depreciation cost for LEDs. A wise choice of appropriate LED lamps considering many factors such as PRE efficiency, photosynthetic photon efficacy, product lifetime, price per photon, the spectrum and angular distribution of light is important.

Table 4 Current estimated energy use efficiency (EUE) and expected improved EUE by 2023 at each stage of the energy conversion process (Nos. 1-3) and the (Nos. 4 and 5) process, affecting the electricity consumption per kg of produce

No. (i)	Definition of RUE at each stage of energy conversion process	Current estimated EUE (B_i)	Expected improved EUE (A_i)	(A_i/B_i) Ratio
1	Ratio of the photosynthetic radiant energy (PRE) emitted to the electric energy consumed by LEDs (photosynthetic photon efficacy)	0.35 J J^{-1} (2.5 μmol J^{-1})	0.39 (3.2)	1.10 (1.10)
2	Ratio of the PRE absorbed by leaves to the PRE emitted by LEDs	0.70	0.77	1.14
3	Ratio of the PRE fixed as chemical energy in plants to the PRE absorbed by leaves	0.25	0.30	1.20
4	Ratio of the fresh weight of marketable leaf lettuce to the whole leaf lettuce plants at harvest (called harvest index or HI)	0.70	0.77	1.10
5	Ratio of the fresh weight of produce sold to the total produce	0.90	0.95	1.05
	EUE of marketable produce	**0.039**	**0.066**	

Improvement factor: $(A_i/B_i)^i$ (i = 1-5) = 1.10 × 1.14 × 1.20 × 1.10 × 1.05 = **1.74**.

No. 2: The ratio (or efficiency) is affected by the design and operation of the cultivation space and lighting system, leaf area index (LAI) or plant canopy structure, uniformness of PPFD distribution over and within the plant canopy, light reflectivity of inner walls and cultivation panels in the cultivation space, and lighting direction (downward, sideward and upward) (Zhang et al., 2015). Ideally, the photosynthetic photons emitted by LEDs are uniformly distributed over all the leaves to maximize the net photosynthesis of whole plants.

No. 3: The ratio is affected by the CO_2 concentration, air temperature, vapour pressure deficit (VPD), air current speed, pH (acidity/alkalinity) and EC (electric conductivity) or nutrient element composition of the nutrient solution. Basically, such environmental factors need to be controlled to maximize the net photosynthesis of the marketable parts of plants at a minimum cost. Generally, the cost of increasing the PPFD is the highest among the costs for optimizing other environmental factors.

No. 4: The fresh weight ratio of marketable produce to that of whole plants is affected by the morphology (leaf/stem/root ratios), percentage of plants with physiological disorders, percentage of plants with physically damaged leaves and so on.

No. 5: The fresh weight ratio of produce sold to the total produce is affected by shelf life, demand predictions, production management, marketing including marketing creation, cold supply chain, logistics, advertisement, wholesale/retail prices and so on.

5.2 Factors affecting electricity cost

In many countries, the electricity cost is the sum of a pay-as-you-go charge in units of US\$ kWh^{-1} and the basic charge for maximum electric current available in units of US\$ kW^{-1}. The monthly pay-as-you-go charge is proportional to the monthly electricity consumption. The basic charge is generally determined by a yearly contract between the user and the electric company, considering the maximum possible electric current required in the coming year. In this charge system, the monthly electricity cost is lower at a lower basic charge, even when the monthly electricity consumption is the same. Besides, the pay-as-you-go charge depends on the time of day (night-time discount), month of the year and seasonal fossil fuel price in some countries like Japan.

In most PFALs, the photoperiod is set at around 16 h (continuous lighting) per day with a dark period of 8 h (continuous) per day. There are some options for the lighting schedule, for example: (1) all the lamps are turned on or off at the same time, or (2) all the lamps are divided into three groups, and two groups (two-thirds of the lamps) are turned on and one group (one-third of the lamps) is turned off. In this case, the basic charge for lighting of the second option can be two-thirds that of the first option. Besides, the daily maximum cooling load and thus the basic charge for cooling is two-thirds. Furthermore, the required cooling capacity of the air conditioners can also be two-thirds, reducing the initial cost for the air conditioners.

In a case where all the lamps are turned off, the VPD of the room air reaches 0 kPa (relative humidity 100%), which may cause stem elongation and physiological disorders of the plants, and condensation on the floor and walls of the cultivation room. On the other hand, in a case where two-thirds of the lamps are always turned on, the VPD is kept at an appropriate level, because the room air is continuously cooled and dehumidified by the air conditioners. In this case, the room air temperature set point needs to be the same for both the photoperiod and dark period (even so, the air temperature under light is 1-2°C higher than that under dark). Besides, CO_2 respired by plants under dark is absorbed through photosynthesis of plants under light.

5.3 Definition of cost performance

Cost performance (CP) is simply expressed in this chapter as sales (S_T) divided by production cost (C_T) (Eq. 1). S_T is the product-sum of the unit price U_i (US\$ kg^{-1}) and the amount of produce sold P_i (kg) (Eq. 2) where n is the number of produce types. The profit is the difference between S_T and C_T.

$$CP = \frac{S_T}{C_T} \tag{1}$$

$$S_T = \sum (U_i \times P_i)(i = 1 - n) \tag{2}$$

Production cost (C_T) is expressed as the sum of the three cost components (Eq. 3).

$$C_T = C_D + C_O + C_W \tag{3}$$

where C_D is the depreciation cost for constructing the whole cultivation system and the building; C_O is the operating cost; and C_W is the cost for processing wastewater, waste nutrient solution, plant residue and so on. C_O is the sum of the costs for electricity (C_{OE}), labour (C_{OL}) and others (C_{OO}) (Eq. 4). C_{OO} includes the costs for water, fertilizer, CO_2, seeds, fertilizer, substrate, plastic bags and other consumables.

$$C_O = C_{OE} + C_{OL} + C_{OO} \tag{4}$$

5.4 Main factors affecting the CP and economic value of produce

The main factors affecting the CP are briefly described below in terms of C_{OL}, C_D for the cultivation system and the economic value of the produce, before discussing the challenges for the next generation of smart PFALs in the next section.

5.4.1 Labour cost (C_{OL})

C_{OL} is roughly expressed by the product-sum of the daily/monthly labour hours per person and the unit wage per hour/day/month for each worker. In order to reduce the labour hours of employees working in the cultivation room, it is essential to realize smooth and standardized flows for workers, plant growth, harvests, consumables and tools/machines (e.g. cultivation panels, goods delivery boxes and carts for transportation), regardless of the degree of automation/robotization. To make the flows smoother, stepwise improvements of the layout and function of tools, machines and consumables in the cultivation room is necessary. The PDCA (Plan-Do-Check-Action) cycle works well for this purpose, and also keeps the workers motivated. Automation and robotization at a large-scale PFAL can be a powerful tool for significantly reducing the C_{OL}, but at the expense of increasing the depreciation cost.

5.4.2 Depreciation cost (C_D) for the cultivation system and building

Currently, there is no globally standardized hardware or software for hydroponic cultivation systems in PFALs. This is one of the main reasons for the high initial cost and thus the high depreciation cost. Standardization of the cultivation system and/or its component parts is required. Development of a scalable hydroponic cultivation system module (CSM) (see Chapter 5) using standardized components will considerably reduce the depreciation cost. It should be noted that the current hydroponic cultivation systems used in PFALs are similar to those developed in the 1970s for greenhouses. Therefore, an innovative hydroponic cultivation system suited to PFALs is expected to be developed to reduce costs and raise yields.

To reduce the initial cost of the PFAL building, an old warehouse or vacant industrial factory is sometimes used. Even in this case, however, a significant cost for renovation is required to meet the specifications for the cultivation room, as described below:

1 The cultivation room needs to be almost airtight and thermally well insulated (see Section 2.1).
2 A relatively large power supply unit is necessary to operate the LEDs and air conditioners.
  Power consumption in the cultivation room, E_C, is roughly given by:

$$E_C = 1.3 \times E_L \times N$$

where E_L is the power consumption per LED lamp; N is the total number of LED lamps that are turned on at the same time period. The rest ($0.3(=1.3-1.0) \times E_L$) is for the electricity consumed by the air conditioners, air circulation fans, water pumps for nutrient solution and so on.

3 The floor needs to be the same type as for food processing factories (smooth surface but not slippery even when it gets wet). Drainage at the floor is necessary to remove the wastewater from washing the floor, cultivation panels and culture beds. The drainage needs to have a trap valve to prevent the entry of insects and small animals into the cultivation room. Dust must not accumulate at the corners of the floor.

5.4.3 Factors affecting the unit economic value (U) and sales (S_T)

Sales (S_T) is expressed as a product of the amount of produce (P), average economic value of the produce sold (U) and the weight ratio of the total produce to the produce sold (r). Annual produce is the product of the average annual yield (kg of produce per m^2 cultivation area or kg m^{-3} cultivation cubic space) and the total cultivation area or space.

The average annual yield is affected by many factors including:

1. Plant species (cultivar), planting density (plants m^{-2}), culture period (days) at each plant growth stage and cropping schedule (crop calendar);
2. Time courses of environmental factors;
3. Percentage of empty (or unused) cultivation space; and
4. Human or machine intervention for seeding, transplanting and cultivation.

The unit economic value (U) and sales (S_T) are affected by many factors including:

5. Quality of produce such as appearance, colour, size, flavour and chemical composition;
6. Demand/supply ratio, sales promotion, package design and supplier-customer relationship; and
7. Marketing, branding and market creation for new cooking menus, cosmetics, supplements, food and drink additives, and healthcare.

5.4.4 Expected improvement of CP

Table 5 shows how the CP could be doubled in advanced PFALs. The grounds for this estimation are: (1) C_{OE} can be reduced by 50% or more (Table 4 and Section 5.2), (2) C_{OL} could be reduced by 20% (Section 5.4), (3) C_{OO} could be reduced by 10%, and then C_O can be reduced by 22% or more, (4) C_D could be reduced by 20% (Section 5.4), (5) there is much room for improving the yield

Table 5 Expected improvement of relative cost performance in the near future

Variable	Equation	Relative reduction or increase	Itemized calculation
Relative reduction			
C_O	$C_{OE} + C_{OL} + C_{OO}$	C_{OE} (0.5), C_{OL} (0.4), C_{OO} (0.1)	0.50 × 0.20 + 0.40 × 0.24 + 0.10 × 0.27 = 0.22
C_T	$C_D + C_O + C_W$	C_D (0.2), C_O (0.22)	0.20 × 0.29 + 0.22 × 0.71 = 0.20
Relative increase			
S_T	U × P × r	U (1.05), ···, P (1.45), ···, r (1.05)	1.05 × 1.45 × 1.05 = 1.6
CP	S_T/C_T	S_T (1.6), ···, C_T (0.8 = 1.0–0.2)	1.6/0.8 = 2.0

For definitions and symbols of variables, see Eqs. (1)–(4) in Section 5.3. Percentages of C_D, C_O, C_{OE}, C_{OL} and C_{OO} over C_T are assumed to be 29%, 71% (=20 + 24 + 27), 20%, 24% and 27%, respectively, based on the data in Fig. 5. Cost for C_W is assumed to be included in C_{OO}.
CP: cost performance, C_T: production cost; S_T: product-sum of the unit price U_i (US$ kg^{-1}) and the amount of produce sold P_i (kg); n: the number of produce types; C_D: depreciation cost for constructing the whole cultivation system and the building; C_O is the operating cost (the sum of the costs for electricity (C_{OE}), labour (C_{OL}) and others (C_{OO}); CW is the cost for processing wastewater, waste nutrient solution, plant residue and so on.

and thus the amount of produce (P) by optimizing Items (1)–(4) in Section 5.4, (6) the economic value or unit price (U) and the ratio of total produce to produce sold (r) can be improved, both by 5%. This improvement of CP is expected to be realized by 2023 based on stepwise PDCA cycles mentioned in Section 5.4. Further improvement of CP by introducing recent advanced technologies is discussed in the next section.

6 Challenges for the next-generation smart PFALs

6.1 Challenges and key technologies

Challenges for the next-generation smart PFAL include developments of (1) smart LED lighting system, (2) alternating and time-dependent lighting effects of ultraviolet, blue, green, red and far-red lights on the plant growth and morphology (Jishi et al., 2016; Park et al., 2015), (3) software for searching optimal combinations of environmental factor set points, (4) scalable CSM, (5) plant phenotyping unit (Kozai, 2018a,b) including root phenotyping unit, (6) plant cohort research unit (Kozai, 2018a,b), (7) breeding of crops suited to PFALs, (8) dual (real/virtual) PFAL (Kozai et al., 2016), (9) control of each ion concentration in nutrient solution (Kozai, 2018a,b) and (10) energy-autonomous PFALs. Key technologies to develop the systems mentioned above include applications of (1) white LEDs, (2) ion sensors, (3) IoT (Internet of Things), (4) big data mining, (5) AI (deep learning), (6) modelling and simulation using mechanistic, statistic, behaviour (surrogate) and AI models, (7) robots, (8) local/global network and (9) natural energy utilization. Selected challenges and technologies are described below.

6.2 Core technology

The smart PFAL is considered to be capable of sensing and collecting all the necessary data using sensors and other means of data collection automatically. It is also capable of modifying its behaviour or functions by itself to achieve its mission and targets under the given constraints. Furthermore, it is capable of learning by itself to improve its overall performance. Such smart PFALs do not exist in 2018, but its prototype will be developed as the next-generation PFAL within a few years or so. To develop such a smart PFAL at a low cost, most of the recent advanced technologies mentioned in Section 1.4 are indispensable.

The core technology of smart PFALs is schematically shown in Fig. 7. In the figure, phenome (*P*) is a general term meaning 'plant traits' such as plant canopy architecture, physiological responses and chemical composition of the leaves (Kozai et al., 2018a). Phenome (*P*) is determined by genomic traits (*G*), plant environment over time (*E*) and management (*M*) or human and machine intervention (seed selection, grading and priming, seeding, transplanting, spacing, harvesting, trimming etc.). *E*, *M* and *P* are, respectively, time series data of multiple variables. Each factor of *E* and *M* is basically controllable as desired by the PFAL manager. On the other hand, *G* is basically not affected by *E*, *M* or *P*.

A genome (*G*) database already exists and is open to the public for major horticultural crops such as tomatoes (https://solgenomics.net/organism/Solanum_lycopersicum/genome) and will soon be open to the public for lettuce (http://lgr.genomecenter.ucdavis.edu/Links.php). Environmental factors (*E*) such as air temperature, VPD, PPFD and CO_2 concentration can be measured relatively easily and controlled continuously and automatically. Human and machine intervention or management, *M*, can be recorded (semi-)automatically. Then, if phenome (*P*) data is measured and collected automatically and non-invasively at a low cost, a data warehouse containing a dataset of *G* and time

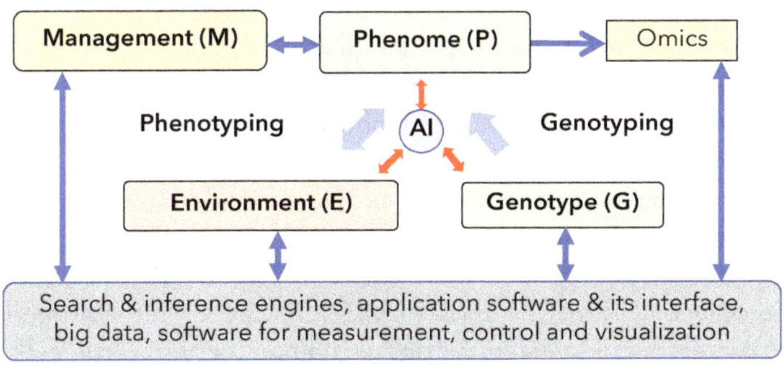

Figure 7 Phenotyping- and artificial intelligence (AI)-based environmental control and breeding for PFALs. Omics is a general term meaning metabolomics, proteomics, transcriptomics, genomics and so on.

series *E-P-M* dataset is constructed. Using deep learning technology for this dataset, *P* for *G* can be predicted as a function of *E* and *M* (Eq. 5):

$$P = \text{Function}(G, E, M) \tag{5}$$

By accumulating the *P-G-E-M* dataset over time, the prediction accuracy of *P* is improved. Using the *P-G-E-M* dataset, an inverse function of Eq. (5) can also be determined as Eq. (6). Then, *E* can be determined to achieve *P* for *G* and *M*. At the same time, activated DNA (mRNA) or genes in *G* that is expressed in *P* can be found.

$$E = \text{Function}(G, P, M) \tag{6}$$

A unique characteristic of the deep learning method is that mathematical/statistical models are not initially required. On the other hand, once Eq. (5) or (6) is determined by deep learning, the corresponding mechanical and/or statistical models can be developed relatively easily. Besides, Eq. (5) or (6) can be expanded to estimate the costs for environmental control (electricity, CO_2, water, fertilizer etc.) and labour or management under the given unit costs (Fig. 8). If this data warehouse is open to the public, democratization, transparentization and socialization of PFALs will be greatly advanced.

The AI models mentioned above are used mainly to express complicated relationships among factors with a huge dataset. For more simpler relationships, behaviour models, multivariate statistic models, mechanistic models including heat and mass balance models and plant growth models are used.

6.3 Subsystems to be developed for smart PFALs

A smart PFAL consists of a number of subsystems as shown in Table 6. Some of the subsystems are discussed in this section.

6.4 Phenotyping unit

Plant phenotyping is defined as the use of methodologies and protocols and various types of cameras and so on to non-invasively measure a plant-specific trait related to plant structure and function, ranging from the cellular level to the canopy level (Ghanem et al., 2015). Typical plant traits for phenotyping include (1) 3D plant canopy structure, (2) LAI, projected leaf area, spatial distribution of leaf angle and leaf area, plant height, colour and number of leaves, (3) traits of roots such as root mass and chemical composition, (4) spectral reflectance and absorptance of leaves, (5) chlorophyll fluorescence, stomatal conductance, content/concentration of water, nitrogen, vitamin C, chlorophyll, anthocyanins and so on, (6) rates of net photosynthesis, dark respiration transpiration and biomass increase, and (7) physiological disorders such as tipburn and intumescence (Tripodi et al., 2018).

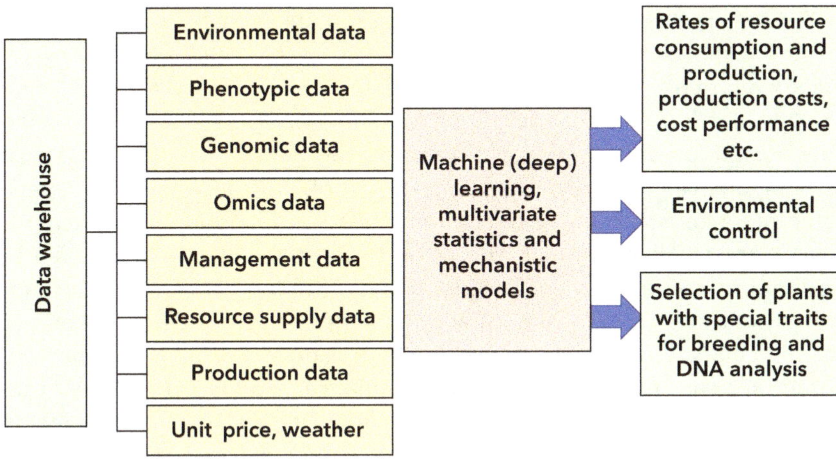

Figure 8 Big dataset obtained in the PFAL is useful for machine learning, behaviour (surrogate), multivariate statistics and mechanistic models.

Table 6 Subsystems to be developed for smart PFALs

No.	Subsystem
1	Phenotyping unit for non-invasive measurement of plant traits including those of roots
2	Data warehouse containing big data consisting of current and historical datasets of genome (*G*), phenome (*P*), environment (*E*), management (*M*), together with related data, in an integrated form
3	Cohort research unit (Kozai, 2018a,b). Time courses of plant traits (i.e. phenome) are measured non-invasively for plant individuals together with the data on environment, human and machine intervention, and resource inputs and outputs (Kozai, 2018a,b)
4	Algorithms to find the optimal set of set points for environmental factors to maximize the objective function under the given constraints
5	Smart LED lighting system to optimize the PPFD, lighting cycle and spectral distribution of LEDs depending on the objectives and constraints (Kozai et al., 2016)
6	Scalable cultivation system module (CSM) (Kozai, 2018a,b)
7	Methodology for molecular and selection breeding of cultivars suited to PFALs
8	Electric energy generation subsystem using a natural energy conversion unit for an energy-autonomous PFAL
9	Open data warehouse with local and global networks
10	Dual (real/virtual) PFAL for training, education and self-learning (Kozai, 2018a,b)
11	Interactive human-machine conversation subsystem with AI
12	Robotic and automated subsystem for plant production and handling
13	Hydroponic subsystem using organic fertilizer derived from plant residue
14	Symbiosis unit for plants with microorganisms for better plant growth
15	Food chain from seeds to plant residue and food waste for their recycling use
16	Human resource development, software for cultivation, scheduling and finance

The phenotyping unit for PFALs must be small so that it can be installed on the ceiling of the narrow cultivation space. The data collected by the phenotyping unit can be used as a powerful tool for plant trait monitoring, plant production and breeding. The variations in growth and other plant traits are attributable to either environmental, genetic or human/machine intervention. Namely, P = Function (G, E, M). By analysing the spatial distributions of P with the corresponding data on E and M, it is possible to distinguish the effects of G, E and M on P. Then, the analysed data can be used for improving the spatial distribution of E for more uniform growth and higher yield, searching for the variation in target traits for selection breeding including marker-assisted selection, and improving M. Selection of plants with target traits can be conducted in the commercial production process of the PFAL, so that the selection process is efficient with a minimum cost. Factors to be considered when evaluating the plant traits by phenotyping under a controlled environment are (1) minimization of environmental variation and increase in heritability, (2) increase in precision of critical measurements, (3) maximization of information from a minimum of replications and (4) development of hypothesis to be tested on a targeted set of traits (Cobb et al., 2013). PFALs can satisfy these factors relatively easily.

6.5 Traits of plants suited for production in PFALs

Plants cultivated in fields and greenhouses need to be resistant to environmental stress (high/low temperatures, dry/wet soil etc.) and diseases. However, such resistance is not necessary for plants cultivated in PFALs because abiotic and pathogenic environments are controlled. In the history of breeding, a large portion of selected plants were not used in practice simply because they were not resistant to such stresses. A portion of the selected plants might be used in the PFAL. On the other hand, plant traits required for field and greenhouse cultivation are different from those required for the PFAL. General characteristics of plants suited for production in PFALs are:

1 Fast growth under relatively low PPFD (100–150 μmol m^{-2} s^{-1});
2 Fast growth under relatively high CO_2 concentration (1000–2000 ppm);
3 Good growth at relatively high planting density (relatively upright leaf angle);
4 Relatively short plant height and/or short internode (dwarf type, 30–40 cm) to reduce the vertical distance between tiers (culture shelves);
5 High percentage (90% or so) of marketable, edible or saleable parts of plants,
6 High concentration of target substances (medicinal components etc.);

7 High economic value if grown under pesticide-free conditions;
8 Plant traits or economic value controllable by environmental control; and
9 High economic value per fresh weight.

PFALs with a controlled environment and a phenotyping unit will greatly accelerate the time required for breeding (Watson et al., 2018). On the other hand, a wide variety of genetic information and molecular biology techniques are necessary to create objective traits, including marker-assisted selection, quantitative trait locus analysis, genetic linkage maps and transgenic techniques (Jiao et al., 2012; Hiwasa-Tanase and Ezura, 2016).

6.6 Algorithms for the optimal set of set points for environmental factors

The optimal set point for any environmental factor needs to be determined by considering many factors: (1) set points of other environmental factors, (2) cost for environmental control, (3) current, target and predicted plant growth and other plant traits, (4) market price trends for produce and consumables, and (5) available labour for human intervention. This complex decision-making for determining the set of set points for environmental factors is usually made by a highly experienced and knowledgeable PFAL manager, considering the production schedule (Fig. 9). Thus, it is expected that an AI-based subsystem will be developed to find the optimal set of set points for environmental factors to maximize the multipurpose objective function under the given constraints to assist the PFAL manager.

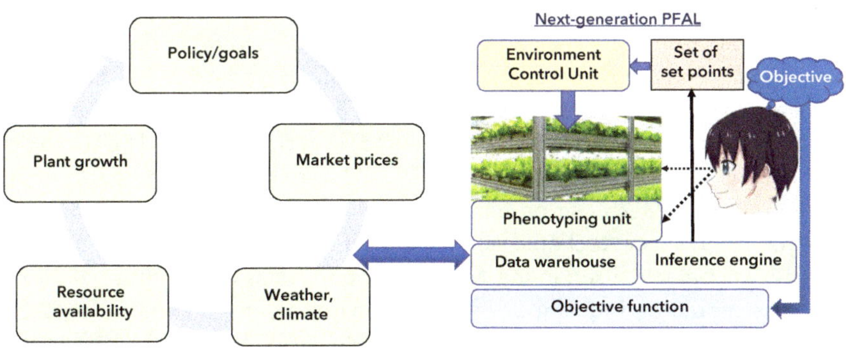

Figure 9 In the next-generation PFAL, an optimal set of environmental set points are proposed to the PFAL manager by AI-based software using a data warehouse to maximize the given multipurpose objective function.

6.7 Applications of PFALs

Various sizes of PFALs can be used in many areas for different purposes: (1) education at school, self-learning and lifetime hobbies at home and community centres, (2) production of medicinal plants for cosmetics, food and drink additives, medicine and pharmaceuticals, (3) breeding, seed (reproductive) propagation and vegetative propagation of plants, (4) the PFAL as a part of integrated biological systems consisting of units for mushroom culture, aquaculture units and biomass energy conversion for integrative bionic sustainable smart cities.

6.8 Smart LED lighting system

Light environmental factors in the PFAL include PPFD and its spatial distribution, light quality (photon percentages of ultraviolet, blue, green, red and far-red), lighting cycle (photoperiod/dark period) and lighting direction (downward, sideward and upward). Light environments affect the photosynthesis, morphology, secondary metabolite production and other physiological responses such as stomatal conductance and intumescence of plants. Thus, the light environment significantly affects the yield and quality of produce. In addition, the electricity cost accounts for around 20% of the production cost of produce. Thus, determining the optimal light environment involves a complex process; software and hardware for the smart LED lighting system need to be developed (Kozai et al., 2016).

7 Future trends and conclusion

As discussed in Subsection 2.2, PFALs have high potential to contribute to solving the global and local trilemma of food, environment and resource issues. However, much of its potential is not actualized in the current PFALs, so these PFALs are not a sustainable food production system and only a limited number of commercial PFALs are making a profit. Based on the technological and economic analyses discussed in Sections 3 to 5, it is estimated that the CP (cost performance) of current PFALs can be doubled in advanced PFALs, which will hopefully be built by around 2022. The CP of next-generation smart PFALs to be built in 2023–25 can be further improved by appropriately introducing the recent advanced technologies described in Subsection 1.4. Then, more benefits will be actualized in the next-generation smart PFALs, thus helping to solve the global and local food-resource-environment trilemma. To actualize most of the potential benefits of PFALs, interdisciplinary and trans-disciplinary collaboration with the appropriate vision and mission is essential. For this purpose, open data warehouses and worldwide active non-profit organizations consisting of academia, industry and citizens need to be established for better global communication and information sharing.

8 Acknowledgements

The authors would like to thank Cecilia Stanghellini and Luuk Graamans for their valuable comments on Section 3 and Subsection 5.3 regarding the electricity and energy consumption of greenhouse and PFAL. Section 6, of this chapter is based on a project supported by a funding from the New Energy and Industrial Technology Development Organization (NEDO). The authors sincerely thank the project members of the Japan Plant Factory Association for their technical support.

9 References

Cobb, J. N., DeClerk, G., Greenberg, A., Clark, R. and McCouth, S. (2013). Next-generation phenotyping: Requirements and strategies for enhancing our understanding of genotype-phenotype relationships and its relevance to crop improvement. *Theor. Appl. Genet.* 126, 867–87. doi:10.1007./s0012-013-2066-0.

Despommier, D. (2010). *The Vertical Farm. Feeding the World in the 21st Century*. St. Martin's Press, New York, NY, 305p.

Ghanem, M. E., Marrou, H. and Sinclair, T. R. (2015). Physiological phenotyping of plants for crop improvement. *Trends Plant Sci.* 20(3), 139–44. doi:10.1016/j.tplants.2014.11.006.

Graamans, L., Baeza, E., van den Dobbelsteen, A., Tsafaras, I. and Stanghellini, C. (2018). Plant factories versus greenhouses: Comparison of resource use efficiency. *Agric. Sys.* 160, 31–43.

Harper, C. and Siller, M. (2015). OpenAg: A globally distributed network of food computing. *IEEE Pervas. Comput.* 14(4), 24–7.

Hiwasa-Tanase, K. and Ezura, H. (2016). Molecular biology to create optimized crops: From genetic manipulation to potential application in plant factories. *Front. Plant Sci.* 7, 539. doi: 10.3389/fpls.2016.00539.

Ijichi, H. (2018). Plant factory business in Japan. *Agric. Biotechnol.* 2(6), 19–23 (in Japanese).

Jiao, Y., Zhao, H., Ren, H., Song, W., Zeng, B., Guo, J., Wang, B., Liu, Z., Chen, J., Li, W., Zhang, M., Xie, S. and Lai, J. (2012). Genome-wide genetic changes during modern breeding of maize. *Nat. Genet.* 44, 812–15. doi:10.1038.ng.2312.

Jishi, T., Kimura, K., Matsuda, R. and Fujiwara, K. (2016). Effects of temporally shifted irradiation of blue and red LED light on cos lettuce growth and morphology. *Sci. Hortic.* 198, 227–32. doi:10.1016/j.scienta.2015.12.005.

Kikuchi, Y., Kanematsu, Y., Yoshikawa, N., Okubo, T. and Takagaki, M. (2018). Environmental and resource use analysis of plant factories with energy technology options: A case study in Japan, *J. Cleaner Prod.* 186, 703–17.

Kozai, T. (2011). Improving light energy utilization efficiency for a sustainable plant factory with artificial light. *Proceedings of the Green Lighting Shanghai Forum*, pp. 375–83.

Kozai, T. (2013). Resource use efficiency of closed plant production system with artificial light: Concept, estimation and application to plant factory. *Proc. Jpn Acad. Ser. B Phys. Biol. Sci.* 89(10): 447–61.

Kozai, T. (Ed.) (2018a). *Smart Plant Factory: The Next Generation Indoor Vertical Farms*. Springer, 456pp.

Kozai, T. (2018b). Benefits, problems and challenges of plant factories with artificial lighting (PFALs): A short review. *Acta Hortic.* (GreenSys 2017) (*in press*).

Kozai, T., Niu, G. and Takagaki, M. (Eds) (2015). *Plant Factory: An Indoor Vertical Farming System for Efficient Quality Food Production*. Academic Press, Amsterdam and Boston, 405pp.

Kozai, T., Fujiwara, K. and Runkle, E. (Eds) (2016). *LED Lighting for Urban Agriculture*. Springer, Puchong, Selangor, 454pp.

Park, Y. G., Muneer, S. and Jeong, B. R. (2015). Morphogenesis, flowering, and gene expression of *Dendranthema grandiflorum* in response to shift in light quality of night interruption. *Int. J. Mol. Sci.* 16, 16497–513.

Tripodi, P., Massa, D., Venezia, A. and Cardi, T. (2018). Sensing technologies for precision phenotyping in vegetable crops: Current status and future challenges. *Agronomy* 8, 57. doi:10.3390/agronomy/8040057.

Watson, A., Ghosh, S., Williams, M. J., Cuddy, S., Simmonds, J. and Rey, M.-D. (2018). Speed breeding is a powerful tool to accelerate crop research and breeding. *Nat. Plants* 4, 23–9. doi:10.1038/s41477-017-0083-8.

Yano, A. (2016). Energy balance and energy conversion process of LEDs and LED lighting systems (Chapter 30). In: *LED Lighting for Urban Agriculture* (Kozai, T., Fujiwara, K. and Runkle, E., eds). Springer, pp. 417–27.

Zhang, G, Shen, S., Takagaki, M., Kozai, T. and Yamori, W. (2015). Supplemental upward lighting from underneath to obtain higher marketable lettuce (*Lactuca sativa*) leaf fresh weight by retarding senescence of outer leaves. *Front. Plant Sci.* 16, 1–9.

Part 2

Crop management

Chapter 7

Understanding crop responses to controlled climates in greenhouses

Chieri Kubota, The Ohio State University, USA

1 Introduction

The goal of sustainable greenhouse crop production is to maximize the crop productivity while minimizing the resource use as well as mitigating the impact to the outside environment. Therefore, understanding plant physiological and developmental responses to aerial and root-zone environmental conditions is one of the most critical components in this practice. For many years, researchers in plant sciences conducted various basic and applied research on effects of selected environmental factors affecting plant growth and development. These studies have contributed to developing the foundational knowledge base useful for advancing controlled environment agriculture, and various technologies have developed and are widely utilized as outcomes. Yet, limited study or advancement has been made of interactions of two or more factors as well as integration of the knowledge into the complex greenhouse climate control. This chapter therefore reviews the current body of knowledge regarding plant responses to environmental interactions. Taking advantage of better access to information technology as well as decreasing costs of instrumentation in data acquisition and control, I expect that this area of study will advance quickly in the next 5–10 years.

While reviewing the environmental interactions affecting greenhouse crop growth, this chapter also emphasizes the importance of understanding plant responses at the whole canopy or system level. In many horticultural

http://dx.doi.org/10.19103/AS.2019.0052.08

curricula, plant physiology has been taught at cellular or molecular levels, yet understanding of plant physiological response at whole plant or as a whole stand (canopy) is crucial for successful crop production. Systems-based understanding and the integrative approach to optimize (co-optimize) climate conditions is the only way for research outcomes to be relevant to the commercial greenhouse settings. Here systems refer to whole plant canopy (stand) and/or the whole production system including the structure, the microclimate and the plants. Researchers need to pay attention to the interactions between environmental factors and plant canopy structure as the impact of the environmental variables can be different depending on the canopy structure which varies by planting density, leaf and shoot pruning practices as well as canopy size.

Controlled environment greenhouses allow us to optimize not only a single environmental factor but also multiple factors to meet goals such as increasing productivity, produce quality and profitability. Developing better systems-based understanding of crop responses is crucial in managing greenhouse crop production sustainably.

2 Research towards integrative optimizations and control of aerial environmental factors in greenhouses

2.1 Air temperature and light intensity

Air and leaf temperatures and photosynthetically active radiation (PAR) are the most important aerial factors affecting plant growth and development. There are in fact ample amounts of research conducted to find plant temperature responses as well as light responses for greenhouse crops. Typically, plant developmental rates (e.g. leaf emergence rate) increase linearly with increasing temperature within the range of minimum temperature and the optimum temperature where the maximum rate is observed (Fig. 1). Plant dry mass accumulation is largely affected by PAR and cannot be explained by temperature alone (e.g. Ramirez-Arias et al., 2011). However, when attempting to apply our 'foundational knowledge' of temperature and light interactions with regard to plant dry mass increase and plant development, we often realize that there is very limited information available for growers to apply in their greenhouses. For example, how do we respond to a grower asking for specific temperature set points for sunny days vs. cloudy days. How cloudy should it be in order to consider reducing set point temperatures?

In the study conducted by Körner et al. (2009), plant canopy photosynthesis (whole stand photosynthesis) shifted the optimum temperature to a 1.8°C higher temperature for tomato and a 3.4°C higher temperature for cut chrysanthemum by increasing PAR from 600 to 900 μmol m^{-2} s^{-1}. However, when temperature is managed dynamically based on PAR, we must consider the effect of 24-h average temperature for plant development. For example, lowering air

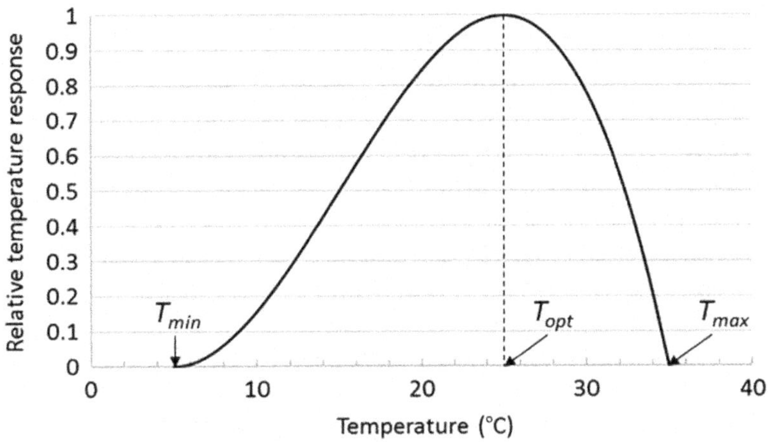

Figure 1 An example of temperature response curve showing cardinal temperatures (T_{min} and T_{max}) with the optimum temperature (T_{opt}). A model expressed by Rosso et al. (1993) was used to create the curve.

temperature by a few degrees would maximize the net photosynthetic rate during the day-time with lower PAR, but the resulting lower 24-h average temperature may slow down the leaf development and therefore reduce leaf area index (particularly during the time before the canopy closes). Also the lower 24-h average temperature may promote flower bud initiation for some species such as tomato (Ohyama et al., 2005). In such a situation, it is complicated to predict the ultimate impact of lowering day-time temperature when the PAR level is low (e.g. overcast day). Growers also apply supplemental lighting for longer hours beyond the natural day length. In such a situation, should the temperature set point be the same as day-time even though the light level is lower than natural PAR (e.g. 50–100 μmol m^{-2} s^{-1} for a typical supplemental lighting)? These results encourage us to examine model/simulation-based climate control to optimize these environmental variables together. Of interest, Kläring and Schmidt (2017) showed that cucumber crop could be grown at a lower temperature at night when compensated by higher temperatures during the day. This approach can reduce the heating set point and thereby reduce the heating fuel costs during winter as discussed later in this chapter.

Reported interactions between temperature and light seem to have inconsistent results among crop species. For tomato, Uzun (2007b) analysed the effects of combinations of 34 daily average temperatures (11–28°C) and 34 daily global radiation integrals (DGRI) on tomato flower and fruit development, and showed that, for increasing flower buds per cluster as well as average fruit mass, optimum temperatures decreased under higher DGRIs, instead of the expected increase under high DGRIs. For example, mean fruit mass was maximized at 22.5°C daily average temperature under the lowest DGRI but maximized when temperature was ~16°C at the highest DGRI occurring in the greenhouse. Similarly, Uzun (2007a) showed that optimum temperature for eggplant flower bud development decreased from 20°C to 17°C when DGRI increased from 3 to 7 MJ m^{-2} day^{-1}. However, they found that optimum temperature for the eggplant yield increased from 15–16°C to 21–22°C with increasing DGRI from 3 to 7 MJ m^{-2} day^{-1}. The range of light investigated in these studies was 2–8 MJ m^{-2} day^{-1} (equivalent to ~4 to ~16 mol m^{-2} day^{-1} daily light integral (DLI, 400–700 nm) using a conversion of 2 mol m^{-2} day^{-1} per MJ m^{-2} day^{-1}, Korczynski et al., 2002). Similar evaluations should be conducted under higher light conditions.

Thermal management by shading in greenhouse is a standard practice that growers apply during summer, in order to prevent thermal stress due to supra-optimal temperatures negatively impacting the plants. Typical shading percentages range between 10% and 50%, depending on outside climate conditions. By shading a greenhouse, solar radiation absorbed inside the greenhouse (thermal load) is reduced so that temperature inside the greenhouse can be more manageable. Excessively high temperature could induce various unfavourable crop responses, including incomplete pollination (and resulting misshapen fruit) in tomato and suppression of flower induction in many ornamental species. However, reduction of PAR by shading decreases the plant photosynthesis and thereby dry mass accumulation. Warner and Erwin (2005) showed this trade-off, namely impacts of high temperature on flower and leaf development (number of flowers, flower diameter and number of leaves) as well as effect of reduced PAR on flower development and dry mass increase for five selected herbaceous ornamental species. Increasing temperature from 20°C to 32°C caused a significant decrease in the number of flower buds of all species. They also quantified the negative response of flowering by reducing PAR, suggesting that growers consider other means (e.g. better ventilation, fans and evaporative cooling) to reduce the temperature in greenhouse.

2.2 CO_2 and light intensity

Effects of CO_2 and light intensity have also been widely studied in controlled environment crop production. Many of such studies focused on collecting valuable information for model-based climate optimization. One thing to be aware of is that instantaneous responses to CO_2, often measured as photosynthetic response using a leaf chamber approach, tend to overestimate the responses than more long-term whole plant/canopy responses. This is because plants would be acclimatized to high CO_2 concentration by reducing stomatal conductance (Drake et al., 1997; Ainsworth and Long, 2005) and would not maintain the same high photosynthesis measured as an instantaneous response to high CO_2. Kimball (1986) reviewed the effects of CO_2 enrichment reported for various greenhouse crops and showed that the efficacy of CO_2 enrichment was greater for crops at young immature stage than those at mature stages, partly due to this physiological acclimation to the high CO_2 concentration. Also biomass increase by CO_2 was greater for vegetative crops (such as lettuce) than fruiting crops (such as tomato).

While the degree of yield increase varies for different crops, CO_2 enrichment is a common practice in greenhouse. However, the practice is rather limited in commercial greenhouses as efficiency of CO_2 enrichment decreases when the greenhouse is ventilated (i.e. most CO_2 is vented out without contributing to plant photosynthesis). Also, the effect of CO_2 enrichment on photosynthesis is greater at higher PAR, at which the greenhouse may need ventilation for temperature control, limiting typical CO_2 enrichment hours to the early morning and the evening when ventilation is not required. Therefore, combination of CO_2 enrichment and 'night-time' supplemental lighting (for ~6-h light extension after the sunset) has been studied where the greatest increase in growth, yield and flower stem quality for cut roses were under the combination, compared with supplemental lighting alone or no supplemental light (Naing et al., 2016). In contrast, Mortensen and Gislerød (2012) showed that during the morning and evening hours when PAR is low, the efficacy of CO_2 enrichment is limited for cut rose production. In their study, a greater increase of photosynthesis by CO_2 enrichment (800 µmol mol^{-1}) was observed when air temperature was increased to 32°C (compared with 22°C) and PAR is high (800 µmol m^{-2} s^{-1}). Midday CO_2 enrichment (increased to 800 µmol mol^{-1} from 370 µmol mol^{-1}) increased the photosynthesis by 50% during the midday 6 h when temperature reached 32°C; however, overall diurnal CO_2 effect was only 15% due to the limited effect of CO_2 enrichment during other hours with lower temperatures. These results show that delay in ventilation might be beneficial for achieving greater efficacy of CO_2 enrichment. However, Mortensen and Gislerød (2012) also discussed potential adverse effect of high temperature on rose plant quality knowing that increase in daily average temperature reduces the shoot length of roses (Moe, 1972) and suggested 'balancing' the high temperature with high PAR in the middle of the day with lower temperature during the remaining part of the day.

If managed properly, CO_2 enrichment can be a more economical means to increase photosynthesis compared with supplemental lighting. A series of studies has been done at Cornell University (NY, USA) to optimize CO_2 enrichment and supplemental lighting. Both et al. (1998) showed that similar shoot dry mass of lettuce was achieved under 17 mol m^{-2} day^{-1} DLI at ambient CO_2 (330 µmol mol^{-1}) and under lower DLI at increased CO_2 concentration. Also, Kitaya et al. (1998) reported dry mass and morphological parameters of young lettuce plants grown under sole-source electrical lighting at four levels of PAR (100–300 µmol m^{-2} s^{-1}) and two photoperiods (16 h vs. 24 h) under two levels of CO_2 (400 μmol mol^{-1} vs. 800 μmol mol^{-1}). While their study showed that a similar plant dry mass can be obtained at lower DLI with high CO_2 and at higher DLI with low CO_2, the sensitivity of dry mass increase seems to be greater for increasing DLI, than CO_2 enrichment (400 μmol mol^{-1} vs. 800 μmol mol^{-1}). For example, using their linear regressions, increasing DLI from 10 to 20 mol m^{-2} day^{-1} increased the plant dry mass by nearly 100% under low CO_2 concentration (400 μmol mol^{-1}), but increasing CO_2 from 400 to 800 μmol mol^{-1} increased the dry mass by 43% under 10 mol m^{-2} day^{-1}. This is supporting our general physiological understanding of limited efficacy of CO_2 enrichment when light is the primary limitation of photosynthesis.

At a given target intensity, supplemental lighting costs depend on the electricity rate structure as well as lighting technologies that determine the efficiency. Based on the cost of liquid CO_2 and electricity rate as well as lettuce yield responses under varied CO_2 concentrations and DLIs, Ferentinos et al. (2000) showed that the model-based optimization of CO_2 concentration and supplemental lighting could significantly reduce the greenhouse operational costs especially during the winter when ventilation is limited and therefore CO_2 enrichment is more efficient than warm seasons. More recently, for a greenhouse without supplemental lighting, Stanghellini et al. (2012) developed an optimization algorithm for the rate of CO_2 injection to the greenhouse, considering loss of CO_2 by ventilation (driven by temperature), plant photosynthetic rate, crop value and CO_2 price. However, these approaches are feasible for mechanically ventilated greenhouses given that ventilation rate can be predicted relatively easily. When the greenhouses are naturally ventilated, another model is likely needed for estimating ventilation rate which may become problematic in commercial applications.

2.3 Canopy structure, management and light in the canopy, diffuse light

Understanding crop response at canopy level is critical in crop physiology. While single-leaf-based evaluation can bring in-depth physiological understanding, canopy level understanding is critical as it can include morphological effect in addition to physiological effect. When researchers examine factors that likely affect not only photosynthesis but also morphology (such as internode elongation or leaf expansion, leaf angle), the experiment must consider a design where planting density and production methods are representing a typical commercial setting. This is because morphological enhancement to increase whole plant or whole canopy photosynthesis by architectural improvement is depending on crops and their growing conditions. High-wire crops such as tomato reach a leaf area index (LAI) of 1.0 several weeks after transplanting, while leafy greens may not close the canopy until the later stage of crop cycle (depending on the re-spacing practices). Once the canopy closes and light interception becomes near saturation, there will be limited benefit by expanding leaf area. Theoretically, light interception will increase rapidly up to LAI = 2 and is maximized at around LAI = 4 when all leaves are oriented horizontally (Fig. 2). Enhancing leaf expansion of crops that already have large LAI may reduce the overall crop productivity by increasing shaded leaves with negative carbon balance on a daily basis. Management of light use in the canopy seems to be critical for achieving high crop productivity.

Heuvelink et al. (2005) discussed the impact of manipulating canopy LAI by altering leaf-pruning methods for tomato. Tomato as well as other high-wire crops that create several metres tall plant canopy would require periodic leaf pruning (e.g. once a week). While leaf pruning is important for disease prevention as well as sink-source management, a scientific approach to leaf pruning is limited. Growers typically follow existing guidelines such as removing lower leaves up to the ripening fruit truss or, in some cases (determined by season and cultivar), removing upper leaves. Of interest, Heuvelink et al. (2005) examined systematic alternating removal of young leaves while maintaining high LAI of 3 where the optimal light interception by the canopy is likely realized. Their simulation indicated that yield can be possibly increased by 10%. Nevertheless, optimum LAI to maximize the canopy photosynthesis is dependent on the canopy structure (i.e. extinction coefficient; Fig. 2) and incident light intensity (Saeki, 1960; Hirose, 2005). For tomato, Sarlikioti et al. (2011) conducted a theoretical simulation study and showed that changing canopy structure (e.g. leaf angle and leaf shape) could improve canopy photosynthesis by about 10%. Of interest, another theoretical simulation study (Chen et al., 2014) showed that the physiological change associated with ageing of leaves was the most

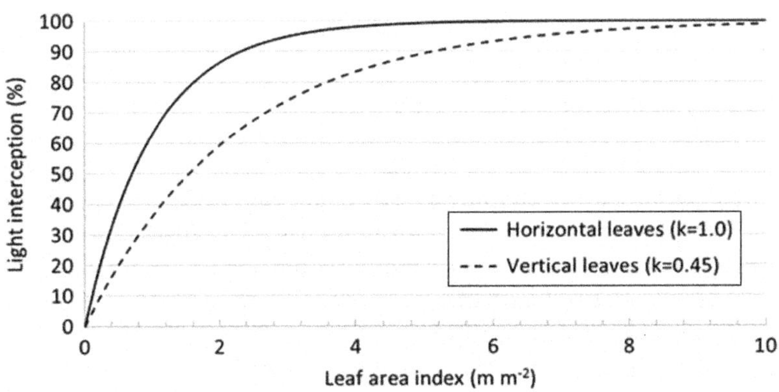

Figure 2 Light interception by a horizontal-leaved canopy and a vertical-leaved canopy. Light extinction coefficient (k) was computed at a solar elevation of 66° (k = 1.0 and 0.45 for horizontal and vertical leaves, respectively; after Jones, 2014).

prominent limitation of leaf photosynthesis at middle canopy layers, supporting the efficacy of intra-canopy light management via supplemental lighting or diffusive glazing that would direct more light into the canopy.

Installing glazing with higher haze property to enhance diffuse light in the incident light over canopy has been studied (e.g. Hemming et al., 2008; Li et al., 2014a; Li and Yang, 2015). Among these studies, Li et al. (2014a) confirmed the yield increase in tomato by glass with high per cent haze properties (45–71%) and through their various ecophysiological, morphological and biochemical analyses, they confirmed that the highest contributing factor for the tomato yield increase under diffuse light enhanced environment was more uniform PAR distribution. Diffuse light is beneficial also for ornamental plants that are typically shaded to avoid damage caused by strong radiation. Li et al. (2014b) showed that, when diffuse glass was employed, two cultivars of *Anthurium* exhibited greater plant growth as well as visual quality (compactness) under a high DLI exceeding the conventional limitation (8 mol m^{-2} day^{-1}) to avoid negative damage.

2.4 Humidity, light and air circulation for mitigating an environmental physiological disorder, tipburn

Transpiration is essential for leaf temperature control and mass flow for nutrient uptake. Understanding transpiration helps to solve many physiological disorders for greenhouse crops (e.g. tipburn, blossom end rot and fruit cracking). For example, tipburn is a disorder observed for crops including lettuce and strawberry caused by localized calcium (Ca) deficiency due to limited Ca supply to fast-growing shoot tips. As calcium uptake is driven by mass flow, management of transpiration by humidity and air circulation is known to be effective to prevent crops from exhibiting tipburn. For lettuce, tipburn incidence is known to increase when plant growth rate is greater (Cox et al., 1976; Goto and Takakura, 1992a; Sago, 2016). Therefore, tipburn incidence can be suppressed by slowing down plant growth by shading or lowering temperature; however, this is a counterproductive measure as growers always want a faster growth rate to maximize productivity. Mitigating tipburn by frequent foliar Ca spray may be effective (e.g. Corriveau et al., 2012), but consistently wet leaves may increase the risk of foliar disease. This is why environmental control is a preferred measure to manage tipburn.

Lettuce tipburn can be an issue in open-field production (e.g. Lafta et al., 2017) as well as under controlled environment (e.g. Both et al., 1997; Cox and McKee, 1976; Frantz et al., 2004; Goto and Takakura, 1992a,b; Sago, 2016). Intensive studies for managing tipburn by environmental control were performed through efforts by scientists and engineers for developing technologies to grow food crops in space such as NASA's Controlled

Ecological Life Support System (CELSS) Program (Wheeler, 2004). Collier and Tibbitts (1984) showed that lower relative humidity (e.g. 51%) during the day or higher relative humidity (>95%) at night increased Ca concentration of inner leaves and thereby mitigated lettuce tipburn. However, plant growth was also compromised at the low relative humidity due to water stress and so they considered that the tipburn under low humidity during the day was attributed to the lower growth rate (rather than enhanced transpiration rate under the low relative humidity). The lower incidence of tipburn under a near-saturated relative humidity at night was attributed to increased turgor pressure, which promoted Ca transport to inner leaves. These results were the first report indicating the distinct difference in regulatory effects on tipburn during the night and during the day (Collier and Tibbitts, 1984). The key role of transpiration has been understood for tipburn as it is a localized Ca deficiency due to the obstructed transpiration rate by leaf enclosure, resulting in less than ten times lower Ca concentration in enclosed leaves compared with outer exposed leaves (Collier and Huntington, 1983; Barta and Tibbitts, 1986). In order to examine the direct effect of transpiration rate, Barta and Tibbitts (1986) examined a modification of microclimate around the shoot tip by enclosing young lettuce leaves with aluminized polyethylene sheaths. This unique approach demonstrated that the artificial enclosure reduced tissue Ca concentration and induced tipburn even at this young stage.

Soon after these milestone physiological studies were done, two independent engineering groups at University of Tokyo (Goto/Takakura group) and Cornell University (Both/Albright group) contributed to develop a practical technology of tipburn control by applying vertical air circulation under controlled environment growth chambers and greenhouses. Specifically, Goto and Takakura (1992a) did the critical proof-of-concept study and discovered that targeted air circulation around the shoot tip of lettuce effectively reduced the tipburn incidence of lettuce plants grown under high light. In their follow-up study (Goto and Takakura, 1992b) they showed that this air circulation around shoot tip of lettuce dramatically increased the Ca concentration of inner leaves thereby preventing tipburn. Baseline understanding is that air circulation reduced the humid boundary layer around the shoot tip and enhanced the transpiration rate and Ca accumulation of leaves around the shoot tip. Following these findings, Both et al. (1997) examined a vertical airflow fan in greenhouse to blow air downward over the lettuce canopy and reported that tipburn was successfully mitigated. Today, vertical (downward) airflow fans are well adopted in commercial lettuce greenhouses.

As tipburn is induced by imbalance of Ca supply and demand, enhancement of Ca supply by vertical airflow can mitigate tipburn only when the supply meets or exceeds the demand of Ca driven by growing conditions such as DLI (Fig. 3). The upper DLI limitation should be different under different

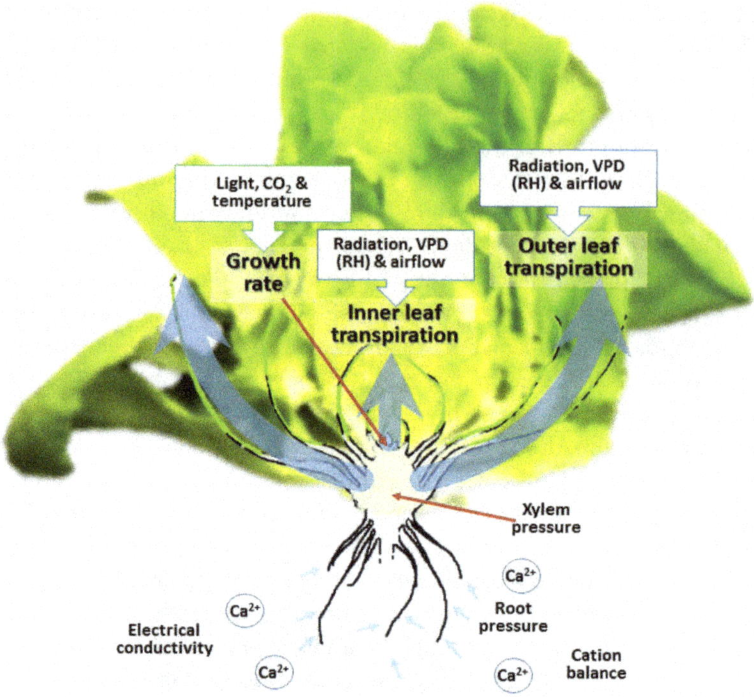

Figure 3 Diagram of a causal factors of tipburn. Ca supply is affected by root-zone conditions affecting mass flow (e.g. root pressure and cation balance) and by aerial conditions affecting transpiration rate (e.g. radiation, vapour pressure deficit (VPD) and airflow). The Ca demand is affected by factors affecting growth rate (e.g. light, CO_2 and temperature). Tipburn occurs when Ca supply to the inner leaves does not meet the Ca demand.

environmental conditions as well as cultivar-specific sensitivities to tipburn. Among limited information, Both et al. (1997) showed that an upper threshold of DLI existed between 16 and 20 mol m^{-2} day^{-1} in the greenhouse conditions examined in their study. They also reported that, in order to prevent tipburn, a minimum amount of transpiration rate of 400 mL g^{-1} biomass accumulation was likely needed. This information of the minimum transpiration per biomass is crucial as it suggests the ideal balance of demand (i.e. growth) and supply (i.e. transpiration) of Ca in lettuce. More recently, Sago (2016) reported the highly correlated relationships between incidence of tipburn and lettuce growth rate or light intensity (DLI). As Ca concentrations of inner leaves (enclosed leaves) were unaffected by light intensity, he also described that the high incidence of tipburn under high DLI resulted from an imbalance between the Ca demand and supply for the inner leaves.

While it is practical to set a target DLI for lowering the risk of tipburn, DLI limit should be dependent on daily transpiration and resulting Ca availability in the shoot tip. As a consequence, many growers still do not have greenhouse condition or cultivar-specific information nor any reliable means to predict the risk. For example, a U.S. Midwest commercial greenhouse experienced tipburn when they started using supplemental lighting in late fall season. Others in southern California observed that when overcast weather continues they typically see tipburn. Preventing tipburn while maximizing production is a common challenge among greenhouse growers. For a better understanding of the tipburn-inducing conditions, we need to take a more quantitative approach of measuring key variables affecting tipburn, including stomatal (day-time) and non-stomatal (night-time) transpiration rate, growth rate and Ca uptake rate. The mitigation measures may not need to be applied everyday and this point of view is discussed in the case study in Section 3.

2.5 Timing and environmental control, integrative control

The term 'integrative control' refers to the consideration of multiple variables in the control logic simultaneously, and was first proposed when computerization was introduced to control climate in greenhouse in the 1970s (e.g. Takakura et al., 1971; Udink ten Cate et al., 1978). During this period of rapid development of computer-based climate control, outstanding visions and concepts such as 'speaking plant' climate control, 'simulation model-based dynamic climate control', 'adaptive control' and 'integrative control to optimize plant response and minimize resource use' were developed. While the technological infrastructure developed rapidly, implementations of technologies based on plant physiological responses, especially environmental interactions, seem to lag behind.

Greenhouse climate control based on temperature integration has been examined by many research groups (e.g. Cockshull et al., 1982; Langhans et al., 1981; Hurd and Graves, 1984; Kläring and Schmidt, 2017). For example, Kläring and Schmidt (2017) examined various diurnal temperature settings for cucumber plants using growth chambers. Their findings suggest that short-term temperature increase during the day could mitigate the negative impact of a low night temperature. They examined a constant 16/16°C day/night temperature setting, a conventional 19/13°C temperature setting and finally a combination of a midday high temperature of 25°C for 3 h with a low night temperature of 11°C with linear temperature increase/decrease between 11°C and 25°C for the rest of day-time. All temperature settings were made to achieve the same 24-h average temperature. While fruit yield was lower in the constant temperature setting than the settings with positive day–night temperature difference, likely due the thermo-periodicity of plants, similar overall growth and yield between the two diurnal temperature settings suggest possible temperature control regimes to reduce night-time heating costs or day-time cooling costs. Dieleman et al. (2005) integrated average temperature-based diurnal temperature setting and heat-demand-based CO_2 management, showing that by both simulation and greenhouse experiment, a significant reduction of energy use (6–7%) was achieved while no difference or small increase in yield was suggested. In both simulation and the experiment, more CO_2 was injected in the greenhouse in the morning than in the afternoon when more loss of CO_2 by ventilation was expected.

The concept of temperature integration can be further extended to a longer period (several days). For example, one or two days of lower-than-optimum average temperature may be able to be compensated for when followed by warmer temperature for multiple days as long as temperatures are within the linear temperature range of plant growth. The baseline concept is that plant growth is affected by dynamically changing climate conditions, but the cumulative response (e.g. yield) can be expressed by cumulative variables (such as cumulative radiation or cumulative temperature). De Koning (1990) examined temperature integration times (6, 12 or 24 days) and except for the 24-day cycle, mature tomato plants exhibited similar growth and fruit biomass. Körner and Challa (2003) developed a more sophisticated control regime of longer temperature integration (over 6 days) combined with nesting shorter (24-h) temperature integrations and showed a potential energy saving (up to 9%) and a potential increase in crop photosynthesis (by allowing longer time for CO_2 enrichment) by simulating greenhouse climate, energy use and plant photosynthesis using mathematical models. The same research group later validated the efficacy of this regime in chrysanthemum production greenhouse and showed that heating energy costs were significantly reduced and the plant growth was enhanced (Körner and Challa, 2004). As the extended temperature integration is a logical and attractive way to reduce energy costs, more validation studies are needed for this concept to be more widely accepted among growers. Short cycle crops such as leafy greens and transplants may be less tolerant to fluctuating temperature or light in an extended integration time, while longer cycle crops (e.g. tomato) may be more manageable. Species-specific evaluations are needed in this area of study.

In fact, effects of tentative exposure of reduced PAR and high or low temperature have been studied. For example, Boldt (2018) showed that temporary short-term reduction of photosynthetically active radiation and temperature (1–7 days $week^{-1}$) had minimum impact on plant growth and development of five selected floricultural species, yet significantly reduced energy costs. Cold-sensitive species (e.g. *Angelonia*) could tolerate one-day-per-week suboptimal conditions of light and temperature, while cold-tolerant species (e.g. pansy) did not show measurable impact for fewer than 4 days exposure per week.

3 Case study: intermittent night-time humidity in integration control can manage tipburn

In an earlier section of this chapter, mitigating tipburn by aerial environmental control is discussed, based on abundant knowledge for lettuce tipburn. In strawberry, tipburn is also a problematic environmental disorder but no practical measures were known for greenhouse production in the past. Tipburn on strawberry occurs on edges of leaves as well as the calyx (small leaves around berry). Calyx tipburn is particularly problematic as it reduces the market value of fruits. Earlier studies showed that tipburn of strawberry leaves occurred due to inadequate Ca transportation (Mason and Guttridge, 1974) and that high relative humidity during the day (i.e. 95% vs. 65%) and higher salinity in root zone increased tipburn (Mason and Guttridge, 1975). Unlike lettuce, however, strawberry tipburn incidence seems to be associated more with plant water relation (transpiration and mass flow) (Ca supply) rather than plant growth rate (Ca demand). This may be due to the relatively slow growth of strawberry compared with lettuce (e.g. Kubota et al., 2016).

We have described our early effort (2009–16) to mitigate tipburn incidence that we encountered in a greenhouse located in a semi-arid region by controlling night-time relative humidity (Kroggel and Kubota, 2017). The idea of night-time humidity control was developed as an approach to manage tipburn, based on the study reported by Bradfield and Guttridge (1979), showing that plants grown under high night-time humidity (>95% relative humidity or <0.1 kPa vapour pressure deficit, VPD) exhibited the least incidence of tipburn. In their study, guttation, a sign of high turgor pressure, was also observed under the high night-time humidity but not under low night-time humidity (~60% relative humidity or ~0.6 kPa VPD). As a proof of concept, during the 2009/2010 season, we confirmed the reduction of tipburn by covering the entire plant canopy with a polyethylene film every night to reach near-saturation humidity. This approach was implemented as movable curtains over the growing troughs and more recently as an under-trough misting system to mitigate tipburn (Kroggel and Kubota, 2017). However, when near-saturation humidity is attempted, there are always concerns of plant disease outbreak. Therefore, we targeted the duration of. high humidity (>95%) to 3 h, a duration apparently sufficient to induce guttation. While this approach worked well to manage tipburn, we looked further into a possible 'integrative control' of night-time humidity, based on the climate conditions over an extended integration period. Namely, achieving high Ca supply conditions can compensate for low Ca supply conditions within a pre-determined integration time (days). Based on our experiences, we developed a set of humidity control regimes in 2016/2017 season, where we allowed a maximum of two consecutive nights of low relative humidity to be followed by two nights in which plants were fully covered to restore the needed

Ca supply by achieving full turgor conditions. This strategy worked well in the semi-arid climate and kept tipburn under control (Kroggel and Kubota, 2017).

Later (2017/2018 crop season), this night-time humidity control was slightly modified to accommodate a new greenhouse condition using heating and supplemental lighting. The experimental glasshouse was located in Columbus (Ohio, USA), where night-time heating is essential due to subfreezing outdoor temperatures. Without the under-trough misting system, 'Albion' strawberry plants exhibited a high incidence of tipburn in November (73–88% plants showing tipburn; McKean and Kubota, *unpublished data*) as soon as intensive heating began to maintain night-time temperature at 10°C. The modified strategy we applied for the rest of the production period (to April 2018) was to achieve 3 h of high humidity (>95%) for three nights with under-trough misting, followed by two dry nights (without misting). As there was supplemental lighting used in the greenhouse, the window of time that allows this night-time humidification was limited to the true dark hours (8 h a day). Early in the season (November to December), 3 h of high humidity was achievable during the early part of the dark hours. However, it took longer to achieve the target humidity level when outside temperatures were lower and condensation on cold glazing removed more vapour from internal air during misting. The strategy worked for most of the crop season except during a few weeks of deep freezing in January. Over the years, we have constantly adjusted the onset of misting as well as duration based on the outside weather condition impact on internal greenhouse condition. More advanced computational approaches (i.e. artificial intelligence) using climate forecasts and greenhouse thermal and water mass balances should be used to manage this control strategy.

Achieving nearly saturated humidity levels inside a plant production system is typically unwelcomed by growers for the possible risk of fungal disease outbreak. For strawberry, such a disease could be grey mould or *Botrytis* spp. Plant pathologists investigated effects of temperature, relative humidity, wind and duration of leaf or fruit wetness (condensation) on *Botrytis* growth and infection (e.g. Ciliberti et al., 2015). Although our under-trough misting system avoids directly wetting plants and duration of high night-time humidity is no more than 3 h, additional measures such as air circulation to dry the canopy immediately after the high humidity treatment may keep the fungal outbreak risk low, although the incidence of *Botrytis* was low in our greenhouses over the past 10 years in both greenhouses in Arizona and Ohio. A similar attempt of tipburn management using under-bench misting was reported for lettuce (Vanhassel et al., 2015). While the all-night high humidity treatment they applied was shown to increase the risk of downy mildew (*Bremia* spp.) incidence, it better mitigated tipburn than the 4-h night-time humidification. In my experience with lettuce tipburn inside an indoor production facility using electric lighting, night-time high humidity treatment needs to be longer than 6 h, likely due to greater demand of Ca attributed to greater growth rate of lettuce than strawberry and/or difference in root-to-shoot relations between these crops.

Future studies necessary for implementing this knowledge to commercial growers include developing tools to predict demand and supply of Ca in the growing tip. A remote-sensing approach as well as crop modelling can be used for quantifying the plant growth (Ca demand) and potential inner leaf transpiration and root pressure can be used for quantifying the Ca supply. With a better understanding of the physiology of plants and pathogens under various conditions, night-time climate control can be accomplished using more computational approaches to mitigate both physiological disorder and disease incidence.

4 Conclusion and future trends

There is no doubt that greenhouse climate control technology will become more sophisticated by integrating plant sciences, engineering, mathematics and computer sciences. However, what is currently lacking is a good understanding of interactions between key climate conditions; large amounts of data using various species and cultivars under different climate conditions are also needed. In this chapter, only a few examples of two or more factor interactions were presented, while there are many other combinations of factors, not only of aerial factors but also of root-zone factors. Single-factor experiments are simple and easier to design, but we should be encouraged to work on more complex questions and should not be afraid to evaluate a set of factors considering a target response (phenotype). An effective methodology to conduct such experiments with multiple factors needs to be developed also. Collaborations with computer scientists as well as statisticians will be crucial to achieve such goals.

5 Where to look for further information

The following conferences are useful to understand the current research status for controlled environment crop production:

- Greensys – ISHS International Symposium on Advanced Technologies and Management for Innovative Greenhouses. This symposium held every other year includes research updates in both greenhouse engineering and greenhouse crop production physiology. https://www.greensys2019.org/
- NCERA-101 – The North American Committee on Controlled Environment Technology and Use. Members meet once a year to discuss various topics of controlled environment technology. https://www.controlledenvironments.org/

The following is a project including integrative control of lighting and other environmental parameters (such as shading or CO_2):

- Greenhouse Lighting and Systems Engineering (GLASE). https://glase.org

6 References

Ainsworth, E. A. and Long, S. P. 2005. What have we learned from 15 years of free-air CO_2 enrichment (FACE)? A meta-analytic review of the responses of photosynthesis, canopy properties and plant production to rising CO_2. *New Phytol.* 165(2), 351–71. doi:10.1111/j.1469-8137.2004.01224.x.

Barta, D. J. and Tibbitts, T. W. 1986. Effects of artificial enclosure of young lettuce leaves on tipburn incidence and leaf calcium concentration. *J. Amer. Soc. Hort. Sci.* 111, 413–6.

Boldt, J. K. 2018. Short-term reductions in irradiance and temperature minimally affect growth and development of five floriculture species. *HortScience* 53(1), 33–7. doi:10.21273/HORTSCI10289-17.

Both, A. J., Albright, L. D., Langhans, R. W., Reiser, R. A. and Vinzant, B. G. 1997. Hydroponic lettuce production influenced by integrated supplemental light levels in a controlled environment agriculture facility: experimental results. *Acta Hortic.* 418(418), 45–52. doi:10.17660/ActaHortic.1997.418.5.

Both, A. J., Albright, L. D. and Langhans, R. W. 1998. Coordinated management of daily PAR integral and carbon dioxide for hydroponic lettuce production. *Acta Hortic.* 456(456), 45–52. doi:10.17660/ActaHortic.1998.456.3.

Bradfield, E. G. and Guttridge, C. G. 1979. The dependence of calcium transport and leaf tipburn in strawberry on relative humidity and nutrient solution concentration. *Ann. Bot.* 43(3), 363–72. doi:10.1093/oxfordjournals.aob.a085644.

Chen, T. W., Henke, M., de Visser, P. H. B., Buck-Sorlin, G., Wiechers, D., Kahlen, K. and Stützel, H. 2014. What is the most prominent factor limiting photosynthesis in different layers of a greenhouse cucumber canopy? *Ann. Bot.* 114(4), 677–88. doi:10.1093/aob/mcu100.

Ciliberti, N., Fermaud, M., Roudet, J. and Rossi, V. 2015. Environmental conditions affect *Botrytis cinerea* infection of mature grape berries more than the strain or transposon genotype. *Phytopathology* 105(8), 1090–6. doi:10.1094/PHYTO-10-14-0264-R.

Cockshull, K. E., Hand, D. W. and Langton, F. A. 1982. The effects of day and night temperature on flower initiation and development in chrysanthemum. *Acta Hortic.* 125(125), 101–10. doi:10.17660/ActaHortic.1982.125.12.

Collier, G. F. and Huntington, V. C. 1983. The relationship between leaf growth, calcium accumulation and distribution, and tipburn development in field-grown butterhead lettuce. *Sci. Hortic.* 21(2), 123–8. doi:10.1016/0304-4238(83)90157-7.

Collier, G. F. and Tibbitts, T. W. 1984. Effects of relative humidity and root temperature on calcium concentration and tipburn development in lettuce. *J. Amer. Soc. Hort. Sci.* 109, 128–31.

Corriveau, J., Gaudreau, L., Caron, J., Jenni, S. and Gosselin, A. 2012. Testing irrigation, day/night foliar spraying, foliar calcium and growth inhibitor possible as cultural practices to reduce tipburn in lettuce. *Can. J. Plant Sci.* 92(5), 889–99. doi:10.4141/cjps2011-242.

Cox, E. F. and McKee, J. M. T. 1976. A comparison of tipburn susceptibility in lettuce under field and green house conditions. *J. Hort. Sci.* 51(1), 117-22. doi:10.1080/00221589.1976.11514671.

Cox, E. F., McKee, J. M. T. and Dearman, A. S. 1976. The effect of growth rate on tipburn occurrence in lettuce. *J. Hort. Sci.* 51(3), 297-309. doi:10.1080/00221589.1976.11514693.

De Koning, A. N. M. 1990. Long-term temperature integration of tomato. Growth and development under alternating temperature regimes. *Sci. Hortic.* 45(1-2), 117-27. doi:10.1016/0304-4238(90)90074-O.

Dieleman, J. A., Meinen, E., Marcelis, L. F. M., de Zwart, H. F. and van Henten, E. J. 2005. Optimization of CO_2 and temperature in terms of crop growth and energy use. *Acta Hortic.* 691, 149-54.

Drake, B. G., Gonzàlez-Meler, M. A. and Long, S. P. 1997. More efficient plants: A consequence of rising atmospheric CO_2? *Annu. Rev. Plant. Physiol. Plant. Mol. Biol.* 48(1), 609-39. doi:10.1146/annurev.arplant.48.1.609.

Ferentinos, K. P., Albright, L. D. and Ramani, D. V. 2000. Optimal light integral and carbon dioxide concentration combinations for lettuce in ventilated greenhouses. *J. Agric. Eng. Res.* 77(3), 309-15. doi:10.1006/jaer.2000.0593.

Frantz, J. M., Ritchie, G., Cometti, N. N., Robinson, J. and Bugbee, B. 2004. Exploring the limits of crop productivity: beyond the limits of tipburn in lettuce. *J. Amer. Soc. Hort. Sci.* 129(3), 331-8. doi:10.21273/JASHS.129.3.0331.

Goto, E. and Takakura, T. 1992a. Prevention of lettuce tipburn by supplying air to inner leaves. *Trans. Amer. Soc. Agr. Eng.* 35, 641-5.

Goto, E. and Takakura, T. 1992b. Promotion of Ca accumulation in inner leaves by air supply for prevention of lettuce tipburn. *Trans. Amer. Soc. Agr. Eng.* 35, 641-5.

Hemming, S., Dueck, T. A., Janse, J. and Van Noort, F. 2008. The effect of diffuse light on crops. *Acta Hortic.* 801(801), 1293-300. doi:10.17660/ActaHortic.2008.801.158.

Heuvelink, E., Bakker, M. J., Elings, A., Kaarsemaker, R. C. and Marcelis, L. F. M. 2005. Effect of leaf area on tomato yield. *Acta Hortic.* 691(691), 43-50. doi:10.17660/ActaHortic.2005.691.2.

Hirose, T. 2005. Development of the Monsi-Saeki theory on canopy structure and function. *Ann. Bot.* 95(3), 483-94. doi:10.1093/aob/mci047.

Hurd, R. G. and Graves, C. J. 1984. The influence of different temperature patterns having the same integral on the earliness and yield of tomatoes. *Acta Hortic.* 148(148), 547-54. doi:10.17660/ActaHortic.1984.148.69.

Jones, H. G. 2014. *Plants and Microclimate: A Quantitative Approach to Environmental Plant Physiology* (3rd edn.). Cambridge University Press, Cambridge, UK, 407pp.

Kimball, B. A. 1986. Influence of elevated CO_2 on crop yield. In: Enoch, H. Z. and Kimball, B. A. (Eds), *Carbon Dioxide Enrichment of Greenhouse Crops. II. Physiology, Yield, and Economics*. CRC Press, Boca Raton, FL, pp. 105-15.

Kitaya, Y., Niu, G., Kozai, T. and Ohashi, M. 1998. Photosynthetic photon flux, photoperiod, and CO_2 concentration affect growth and morphology of lettuce plug transplants. *HortScience* 33(6), 988-91. doi:10.21273/HORTSCI.33.6.988.

Kläring, H.-P. and Schmidt, A. 2017. Diurnal temperature variations significantly affect cucumber fruit growth. *HortScience* 52(1), 60-4. doi:10.21273/HORTSCI11215-16.

Korczynski, P. C., Logan, J. and Faust, J. E. 2002. Mapping monthly distribution of daily light integrals across the contiguous United States. *HortTechnology* 12, 12-6. doi:10.21273/HORTTECH.12.1.12.

Körner, O. and Challa, H. 2003. Design for an improved temperature integration concept in greenhouse cultivation. *Comput. Electron. Agric.* 39(1), 39–59. doi:10.1016/S0168-1699(03)00006-1.

Körner, O. and Challa, H. 2004. Temperature integration and process-based humidity control in chrysanthemum. *Comput. Electron. Agric.* 43(1), 1–21. doi:10.1016/j.compag.2003.08.003.

Körner, O., Heuvelink, E. and Niu, Q. 2009. Quantification of temperature, CO_2, and light effects on crop photosynthesis as a basis for model-based greenhouse climate control. *J. Hortic. Sci. Biotechnol.* 84(2), 233–9. doi:10.1080/14620316.2009.11512510.

Kroggel, M. and Kubota, C. 2017. Controlled environment strategies for tipburn management in greenhouse strawberry production. *Acta Hortic.* 1156(1156), 529–36. doi:10.17660/ActaHortic.2017.1156.78.

Kubota, C., Kroggel, M., Both, A. J., Burr, J. F. and Whalen, M. 2016. Does supplemental lighting make sense for my crop? – empirical evaluations. *Acta Hortic.* 1134(1134), 403–12. doi:10.17660/ActaHortic.2016.1134.52.

Lafta, A., Turini, T., Sandoya, G. V. and Mou, B. 2017. Field evaluation of green and red leaf lettuce genotypes in the Imperial, San Joaquin, and Salinas Valleys of California for heat tolerance and extension of the growing seasons. *HortScience* 52(1), 40–8. doi:10.21273/HORTSCI10835-16.

Langhans, R. W., Wolfe, M. and Albright, L. D. 1981. Use of average night temperatures for plant growth for potential energy savings. *Acta Hortic.* 115(115), 31–8. doi:10.17660/ActaHortic.1981.115.3.

Li, T. and Yang, Q. 2015. Advantages of diffuse light for horticultural production and perspectives for further research. *Front. Plant Sci.* 6, 704. doi:10.3389/fpls.2015.00704.

Li, T., Heuvelink, E., Dueck, T. A., Janse, J., Gort, G. and Marcelis, L. F. M. 2014a. Enhancement of crop photosynthesis by diffuse light: quantifying the contribution factors. *Ann. Bot.* 114(1), 145–56. doi:10.1093/aob/mcu071.

Li, T., Heuvelink, E., van Noort, F., Kromdijk, J. and Marcelis, L. F. M. 2014b. Responses of two *Anthurium* cultivars to high daily integrals of diffuse light. *Sci. Hortic.* 179, 306–13. doi:10.1016/j.scienta.2014.09.039.

Mason, G. F. and Guttridge, C. G. 1974. The role of calcium, boron and some divalent ions in leaf tipburn of strawberry. *Sci. Hortic.* 2(3), 299–308. doi:10.1016/0304-4238(74)90039-9.

Mason, G. F. and Guttridge, C. G. 1975. The influence of relative humidity and nutrition on leaf tipburn on strawberry. *Sci. Hortic.* 3(4), 339–49. doi:10.1016/0304-4238(75)90048-5.

Moe, R. 1972. Effect of day length, light intensity, and temperature on growth and flowering in roses. *J. Amer. Soc. Hort. Sci.* 97, 796–800.

Mortensen, L. M. and Gislerød, H. R. 2012. The effect of high CO_2 concentration on diurnal photosynthesis at high daytime temperatures in small stands of cut roses. *Europ. J. Hort. Sci.* 77(S), 163–9.

Naing, A. H., Jeon, S. M., Park, J. S. and Kim, C. K. 2016. Combined effects of supplementary light and CO_2 on rose growth and the production of good quality cut flowers. *Can. J. Plant Sci.* 96(3), 503–10. doi:10.1139/cjps-2015-0304.

Ohyama, K., Omura, Y. and Kozai, T. 2005. Effects of air temperature regimes on physiological disorders and floral development of tomato seedlings grown under continuous light. *HortScience* 40(5), 1304–6. doi:10.21273/HORTSCI.40.5.1304.

Ramirez-Arias, A., Pineda-Pineda, J., Rodriguez-Diaz, F. and Berenguel-Soria, M. 2011. Empirical models for tomato crop using temperature and photosynthetically active radiation. *Acta Hortic.* 893(893), 779–84. doi:10.17660/ActaHortic.2011.893.84.

Rosso, L., Lobry, J. R. and Flandrois, J. P. 1993. An unexpected correlation between cardinal temperatures of microbial growth highlighted by a new model. *J. Theor. Biol.* 162(4), 447–63. doi:10.1006/jtbi.1993.1099.

Sago, Y. 2016. Effects of light intensity and growth rate on tipburn development and leaf calcium concentration in butterhead lettuce. *HortScience* 51(9), 1087–91. doi:10.21273/HORTSCI10668-16.

Saeki, T. 1960. Interrelationships between leaf amount, light distribution and total photosynthesis in a plant community. *Bot. Mag. (Tokyo)* 73(860), 55–63. doi:10.15281/jplantres1887.73.55.

Sarlikioti, V., de Visser, P. H. B., Buck-Sorlin, G. H. and Marcelis, L. F. M. 2011. How plant architecture affects light absorption and photosynthesis in tomato: towards an ideotype for plant architecture using a functional-structural plant model. *Ann. Bot.* 108(6), 1065–73. doi:10.1093/aob/mcr221.

Stanghellini, C., Bontsema, J., de Koning, A. and Baeza, E. J. 2012. An algorithm for optimal fertilization with pure carbon dioxide in greenhouses. *Acta Hortic.* 952(952), 119–24. doi:10.17660/ActaHortic.2012.952.13.

Takakura, T., Jordan, K. A. and Boyd, L. L. 1971. Dynamic simulation of plant growth and environment in the greenhouse. *Trans. Amer. Soc. Agr. Eng.* 14, 964–71.

Udink ten Cate, A. J., Bot, G. P. A. and van Dixhoorn, J. J. 1978. Computer control of greenhouse climates. *Acta Hortic.* 87(87), 265–72. doi:10.17660/ActaHortic.1978.87.28.

Uzun, S. 2007a. Effect of light and temperature on the phenology and maturation of the fruit of eggplant (*Solanum melongena*) grown in greenhouses. *N. Z. J. Crop Hort. Sci.* 35(1), 51–9. doi:10.1080/01140670709510167.

Uzun, S. 2007b. The effect of temperature and mean cumulative daily light intensity on fruiting behavior of greenhouse-grown tomato. *J. Amer. Soc. Hort. Sci.* 132(4), 459–66. doi:10.21273/JASHS.132.4.459.

Vanhassel, P., Bleyaert, P., Van Lommel, J., Vandevelde, I., Crappe, S., Van Hese, N., Hanssens, J., Steppe, K. and Van Labeke, M. -C. 2015. Rise of nightly air humidity as a measure for tipburn prevention in hydroponic cultivation of butterhead lettuce. *Acta Hortic.* 1107, 195–202. doi:10.17660/ActaHortic.2015.1107.26.

Warner, R. M. and Erwin, J. E. 2005. Prolonged high temperature exposure and daily light integral impact growth and flowering of five herbaceous ornamental species. *J. Amer. Soc. Hort. Sci.* 130(3), 319–25. doi:10.21273/JASHS.130.3.319.

Wheeler, R. M. 2004. Horticulture for Mars. *Acta Hortic.* 642, 201–15. doi:10.17660/ActaHortic.2004.642.22.

Chapter 8

Developments in growing substrates for soilless cultivation

Youbin Zheng, University of Guelph, Canada

1 Introduction

With the increasing world population and rapidly decreasing arable land base, soilless crop production has been increasingly playing an important role in providing plant materials for modern society as food, medicine, ornamentals or for environment quality improvement. Good arable land is not essential for soilless production; soilless production can be set up anywhere, even in places with no topsoil. Soilless production systems are more resource efficient than most open field soil production systems. For example, Barbosa et al. (2015) demonstrated that, in comparison to a conventional soil-based system, a hydroponic system (e.g. nutrient film technology) uses 13 times less water for lettuce production.

In as early as the eighteenth century, people started to realize that plants can grow in a nutrient solution without soil. Plants can grow well as long as the rootzone has adequate nutrients, water, and oxygen; is within the right temperature range; and is without biotic (e.g. pathogens) or abiotic (e.g. unwanted chemicals) limitations (Zheng, 2019). Rootzone, as defined by Zheng (2019), is the environment where roots reside. Soil and soilless growing substrates are not essential for plant growth and health; however, for certain plant production systems they are necessary or preferred for use. For example, for potted ornamental plants, substrates have to be used in pots to enable the

http://dx.doi.org/10.19103/AS.2019.0052.11

plants to be transported and sold in the marketplace. For high water demand vegetable (e.g. cucumber) production, hydroponic or aeroponic systems can be used; however, with growing substrates, there is less risk of crop damage or failure in the case of a power outage. When substrates are used in greenhouse and other controlled-environment production systems, they are mainly used for physical support, as a reservoir for water and nutrients, for providing air space for oxygen to diffuse into the rootzone and for harmful gases to move out freely, and also for providing habitat for beneficial microorganisms. It is less ideal to grow plants in soil compared to in soilless growing substrates in large-scale greenhouses, since natural soil properties tend to vary spatially, even within a short distance, which makes horticultural management such as automated fertigation difficult.

In the majority of modern advanced greenhouse operations, soilless growing substrates are used. To produce plants in a profitable way and in an environmentally friendly manner, substrate selection is an important process. Currently, some of the most commonly used substrate constituents are not, or are perceived as not, environmentally friendly or sustainable. Recent research efforts have been devoted to finding alternative materials to, either entirely or partially, replace those unsustainable materials. Research has also been focused on developing methods to evaluate the quality and suitability of the materials used in forming growing substrates.

This chapter discusses current developments in soilless growing substrates for greenhouse cultivation, including (1) materials used in forming growing substrates; (2) methods and technologies for assessing growing substrates; (3) using beneficial microorganisms to create a more resilient rootzone; and (4) an integrated rootzone management (IRM) approach for creating and maintaining a healthy rootzone.

2 Materials used in forming growing substrates

Currently, in advanced greenhouse production operations, the most commonly used growing substrates are peat or coir, either mixed with or without perlite, and rockwool (Savvas, 2009; Zheng et al., 2009b; Schmilewski, 2017). According to Schmilewski (2017), peat accounts for more than 75% of the total amount of constituents used for the production of growing substrates for the professional and hobby markets in 16 EU countries, followed by 11% organic constituents other than peat (excluding composts), 8% composts, and 6% mineral constituents. The drivers of material selection for making growing substrates include whether the substrate can be easily managed to provide a rootzone with adequate and balanced levels of water, nutrients, and air (Zheng, 2019) and material availability, as well as economic and environmental concerns (Barrett et al., 2016). Materials such as peat, coir, and rockwool have a proven record for their excellence in providing a healthy plant rootzone and can also be managed by suppliers to provide consistent and homogenous products. However, these materials are perceived as either not environmentally friendly or not sustainable. For example, coir materials are mostly produced in Southeast Asia (e.g. Sri Lanka), which results in a large carbon footprint when shipped from the production sites to Europe and North America. Both rockwool and perlite need high energy inputs during the manufacturing process and also need to be disposed of or reused, which can be an environmental issue and can also be costly. Although certain regions are repurposing used rockwool for brick making or other applications, most regions still have to transport used rockwool to landfills.

To overcome some of the disadvantages for commonly used substrate materials, research has been exploring alternatives to either partially or entirely replace the aforementioned materials (Barrett et al., 2016). All of the recent research efforts can be roughly divided into the following categories: (1) composting municipal organic waste materials to be used in growing substrates, (2) composting locally available agricultural or natural materials to be used in growing substrates, (3) processing locally available materials to partially or entirely replace the currently used materials, and (4) developing synthetic biodegradable materials or recyclable materials. The following sections discuss the recent developments made in each category.

2.1 Composting municipal organic waste materials

Urbanization has been resulting in increasing amounts of household organic waste materials, such as food and garden wastes. Large composting facilities have been established in many cities around the world to process these waste materials. Research has been looking into how to compost (Tognetti et al., 2007; Zheng et al., 2009a) and how much compost to incorporate into peat- (Herrera et al., 2009) or coir- (Ribeiro et al., 2013) based substrates, or has studied mixing compost with other non-peat or coir materials (Lu, 2008).

The advantages of including municipal compost in growing substrates can include: (1) getting rid of municipal waste in an environmentally friendly way, (2) the potential to creating substrates with plant pathogen-suppression properties (Van der Gaag et al., 2007; Raviv, 2016), and (3) short distance transportation. However, there are also some challenges: (1) high salt, especially Cl and Na, content from food waste and (2) potential contaminants, such as heavy metals (Moral et al., 2013), 3) variability in compost quality, including physical, chemical, and biological properties, from season to season and batch to batch. For example, in the autumn, compost made from municipal waste contains more tree leaves and other garden plant residues than in other seasons. These variations can be challenging when a homogenous and consistent-quality product is required for growing substrates in large-scale greenhouse production. Large-scale greenhouse production normally requires the growing substrate to be as homogenous as possible, both temporally and spatially, for easy rootzone management.

2.2 Composting locally available agricultural or natural organic materials

Different regions have different specialized and concentrated agricultural activities. For example, there are wine and olive industries in Mediterranean regions and in Texas, the United States that produce large volumes of cattle manure. Researchers in different regions have been conducting various trials to find ways to utilize locally available agricultural and natural organic materials to produce growing substrates for greenhouse production.

Luo et al. (2015) incorporated locally produced pig litter compost to replace peat in a peat, perlite, and vermiculate mix. Their results showed that pig litter compost can replace peat up to 40%; however, higher than 40% is not as good as without compost for cabbage seedling production. Similarly, Al Naddaf et al. (2011) demonstrated that composted pig manure either alone or mixed with perlite can be used to produce crops such as cucumber under Mediterranean greenhouse conditions, as long as the growing bag height is appropriate, and the substrate has appropriate physical properties. Gizas et al. (2012) demonstrated that a growing substrate of composted *Posidonia oceanica* alone or a mix with non-composted *Posidonia* at 50:50 can produce a good yield for lettuce crops. Carmona et al. (2012) demonstrated that 100% compost from the mixtures of de-alcoholized local grapevine marc and grape stalks in a 1:1 ratio (v:v) can be used to grow geranium, petunia, carnation, and gerbera crops that are as good as the control in peat- or coir-based substrates. Ceglie et al. (2011) found that up to 70% of green waste compost and 45% of olive pomace compost can be used to replace peat for producing organic tomato seedlings.

Again, most of the studies have tested how much compost to incorporate into peat- or coir-based substrates or evaluated mixing compost with other non-peat or coir materials (Ceglie et al., 2015). A review by Moral et al. (2013) indicated that many studies showed that an array of composted materials can replace 30% peat from peat-based growing substrates, with comparable crop performances.

Some advantages of using composted local agricultural and natural organic materials can include: (1) added macro- and micro-nutrients, (2) beneficial microorganisms in the compost providing pathogen-suppression properties (Raviv, 2016; Tian and Zheng, 2013a,b), and (3) helping other agricultural sectors to get rid of their wastes. Alternatively, using composted materials comes with certain disadvantages such as (1) possible highly soluble salt content (Moral et al., 2013), (2) limited amount available, and (3) temporal and spatial variations in quality, depending on the available sources.

2.3 Processing locally available materials

Processing locally available materials includes both organic and mineral materials. For example, bark, wood fiber (Carlile et al., 2015), rice hulls (Bartz et al., 2017), as organic materials; sand, tuff, pumice, expanded clay, zeolite as mineral materials (for details see Bar-Tal et al., 2019). In this category, wood fiber and biochar are probably two of the most actively studied materials in recent years.

2.3.1 Wood fiber

Wood fiber has been tested as a material to replace peat, coir, and even perlite in growing substrates for soilless plant production since the 1980s (Frangi et al., 2008; Schmilewski, 2017). Even though not many peer-reviewed research papers have been published, the use of wood fiber has been increasing in popularity in recent years. This increased interest is due to some of its desirable properties as a growing substrate component, being regarded as a renewable resource if a region has enough wood, and reducing the need for long distance transportation of substrate constituents. The qualities (e.g. physical and chemical properties) of wood fibers differ depending on many factors, such as the type and age of the wood, as well as the process used for making the wood fibers. Generally speaking, wood fibers are less hydrophobic and water moves along the fibers more easily, compared to peat. In addition, wood fibers have high saturated hydraulic conductivity and air space (De Boodt and Verdonck, 1972), with large shrinkage (Gruda and Schnitzler, 2004a), and are chemically more inert compared to some other materials (AHDB, 2018; Domeño et al., 2009). For example, Domeño et al. (2009) showed that when growing plants in a wood fiber substrate, the pH can be maintained within the acceptable range throughout the 7-week test period. Their data also demonstrated that both coir and wood fiber have the capacity to retain NO_3, SO_4, Ca, and K. Since coir contains K, their trial showed that K was released from the coir, but not from the wood fiber substrate. Gruda and Schnitzler (2004b) demonstrated that a wood fiber substrate can produce tomato transplants with a comparable quality to those grown in white peat when wood fibers were not heavily compacted.

2.3.2 Biochar

Biochar is a black, fine-grained, and porous substance produced by thermal decomposition of organic matter such as wood chips, in partial or complete absence of oxygen, through a process called pyrolysis. Biochar use in growing substrates has been increasing in the literature. It has been described as a material which has many beneficial properties, such as high nutrient retention capacity (Altland and Locke, 2012), high cation exchange capacity (Cheng et al., 2008), and a low bulk density (Downie et al., 2009). However, its physical and chemical properties are highly variable depending on the feedstock, the actual pyrolysis process, etc. For example, the pH of biochar can range from 4.2 to 13.0 (Cha et al., 2016). In addition, few biochar manufacturers are able to produce and supply large amounts of biochar with consistent quality. Due in part to these limitations, as well as others, it seems that there is limited potential for biochar to be widely adapted and used as a component of growing substrates in the horticulture industry, especially in large-scale soilless greenhouse production systems.

2.4 Developing synthetic biodegradable materials or recyclable materials

Several groups have worked on developing synthetic materials, such as urethane and polymers, to use as growing substrates. For example, Zheng et al. (2009b) reported that a recyclable polymer-based growing substrate and biodegradable urethane-based foam were developed and tested in greenhouses for plant production. Their results demonstrated that both substrates are able to produce the same or higher yields of vegetable fruits (e.g., cucumber and sweet pepper) than rockwool. Khoshgoftarmanesh et al. (2012) mixed acid-washed shredded waste-tire rubber with vermiculite or zeolite and perlite at a 25:25:50 ratio and found that tomato plant growth and yield are better than in a peat and perlite mix, when grown in pots. Even scientifically proven that these types of materials can be used in greenhouse soilless production, currently the applications are still limited in area and scale. Different crop, substrate, and growing environment combinations need different rootzone management strategies (e.g. fertigation) to be successful (Huber et al., 2009; Zheng, 2019). Therefore, one of the reasons that these products are not widely used in the greenhouse industry is most likely due to the lack of financial support to develop specific rootzone management strategies tailored specifically for these types of substrates.

3 Methods and technologies for assessing growing substrates

There is no ideal growing substrate for every situation (e.g. plant species/cultivar and irrigation systems). To select growing substrates for different production systems, people do measure physical, chemical, and biological properties to serve as a guide. This section discusses some of the most commonly used methods and technologies for evaluating growing substrates.

3.1 Physical property measurement

Commonly measured physical properties include total porosity, container capacity, air porosity, bulk density, and, less often, particle size distribution. Recently, researchers have started to put emphasis on measuring water and gas dynamics in the growing substrate, which includes measuring material wettability, hydraulic conductivity, and air diffusivity (Caron et al., 2014; Gizas et al., 2012). For example, Gizas et al. (2012) demonstrated that the actual water availability to plants grown on soilless substrate depends more on the water flux toward roots than on the actual content of easily available water. Water flux toward roots is measured as substrate hydraulic conductivity.

Quite often total porosity, container capacity, air porosity, and bulk density are measured at the same time using one device (Fonteno and Hardin, 1995). However, currently there is no standard on methods or devices for measuring the aforementioned physical properties (Zheng, 2019). Most of the laboratories in North America are using the NCSU Porometer developed by W.C. Fonteno's lab at the North Carolina State University (Fonteno and Hardin, 1995); and most of the European countries are using the European Standard CEN TC 223 WG4 N151 (CEN, 2006). These two groups are using two different sizes (e.g., diameter and height) of devices and different procedures. For a given substrate, using different procedures and devices can provide varying results for parameters such as water holding capacity and air porosity (Caron et al., 2015). This measurement variability makes it difficult to compare results measured using different methods. It would be advisable for the international soilless substrate community to choose a single measurement standard for the world to follow in order to facilitate comparison and worldwide application of research results.

3.2 Chemical property measurement

Unlike producing plants using conventional soil, in soilless production, nutrient supply from the growing substrate is of low importance. Growing substrates primarily serve as reservoirs for nutrients, rather than a source of nutrients. However, the chemical properties of growing substrates can affect the form, concentration, and availability of the supplied plant nutrients through fertigation/fertilization. During greenhouse plant production, it is essential to ensure plant roots have access to the proper ratio of different nutrient elements at appropriate concentrations. It would be ideal to be able to measure these nutrient concentrations in situ in the plant rootzone; however, currently there is a limited selection of technologies that can do so and they are not commonly used. Presently, the most common practices used to measure rootzone nutrient concentrations are various methods that extract nutrient solutions from the substrate or rootzone, then either immediately analyze its pH and electronic conductivity (EC) before sending the extracted solution to a laboratory for individual nutrient element content analysis, or sending the extracted solution directly to a laboratory to have pH, EC, and all of the nutrient elements analyzed.

There are different extraction methods used to obtain nutrient solutions for analysis. These methods have different advantages and disadvantages and can also result in different values. Therefore, it is important to understand the differences among these extraction methods in order to interpret the results most appropriately. Again, the predominant methods used in North America are different from the predominant methods used in Europe and in other regions. In North America, there are three commonly used extraction methods for onsite measuring of substrate pH and EC: the 1:2 water:substrate (v:v) suspension test (1:2), the Pour-through (PT) method, and the saturated media extract (SME) method (Torres et al., 2010). The last two methods are often used to extract nutrient solutions for sending to a laboratory for nutrient content analysis, while the 1:2 test is not preferable for this analysis. The PT method (Wright, 1986) leaves the plant and substrate intact and involves pouring enough distilled or deionized water onto the surface of the growing substrate, 1–2 h after normal irrigation or fertigation, to displace about 50 ml of leachate for analysis. The SME method involves putting a certain amount of growing substrate into a container, saturating the substrate with distilled/deionized water until glistening, and allowing the solution to equilibrate for 30 min before vacuum filtering to collect the solution for analysis (Warncke, 1998). The PT method is faster and easier than SME, with minimal differences in pH measured by these two methods. However, there are linear relationships between nutrient concentrations measured using SME and PT, with SME results higher than those of PT (Cavins et al., 2004). Because of the differences among the analysis results for these three methods, considering the method used is important when determining whether the plant rootzone has adequate nutrient levels and when making recommendations.

In Europe and some other regions, the 1:1.5, 1:3, and 1:5 volume extract methods are commonly used for water soluble nutrient content analysis in growing substrates (for more detail see Blok et al., 2019). The 1:1.5 volume extract method is intended to be used for peaty (or similar) substrates and the 1:3 volume extract method is intended to be used for non-peaty substrates. The results from both methods are comparable. The 1:5 volume extract method is designed for all granular growing substrates. The procedures of these three methods are more complicated than the three methods used in North America, and all the results, especially considering nutrient concentrations, can be different. Again, it is important to keep the specifically used extraction method in mind when making recommendations for fertilization.

3.3 Biological tests

When developing or using a new material for growing substrates, it is important to make sure the newly formed substrate does not immobilize nitrogen to a large extent, is stable without degrading too fast, is not phytotoxic, and contains no weed seeds. Biological tests are needed to assess these aspects of a newly developed substrate. Detailed descriptions of currently used methodologies can be found in a recent book chapter by Blok et al. (2019).

4 Use of beneficial microorganisms to create a resilient rootzone

For sustainable soilless plant production, reusing leachate/runoff is an increasingly adopted practice. The potential risk accompanying this practice is that water- and soil-borne pathogens can spread from infected plants to other healthy plants. There are many effective water treatment technologies (e.g. chlorination, ozonation, hydrogen peroxide, copper ionization, pasteurization, sand filtration, and UV) currently being used in the greenhouse industry; however, almost all of the existing technologies can only kill pathogens in solution and are not effective in solid growing substrates, especially substrates containing organic materials such as peat (Zheng, 2018). In order to keep plants healthy in a nutrient-solution-reusing system, a suggested approach is inoculation of the plant rootzone (e.g. growing substrate) with beneficial microorganisms or incorporating materials which contain pathogen-suppression properties (e.g. compost) when mixing growing substrates. Research has shown that by inoculating beneficial microorganisms into the growing substrate, certain plant pathogens in soilless greenhouse production systems can be controlled (Postma, 2010; Tian and Zheng, 2013a,b). Research has also shown that microorganism population size and diversity increase over time in soilless production systems, in both the nutrient solution and the growing substrate, and some of these microorganisms can suppress plant pathogens (Postma, 2010). For example, by analyzing a nutrient solution, which had been reused for more than 20 years in a Canadian ornamental commercial greenhouse, Tian and Zheng (2013a) found that it contained large and diverse microbial populations, with the predominant microorganisms being bacteria and filamentous fungi. This study also demonstrated that this reused nutrient solution can significantly suppress *Fusarium foetens* and *Rhizoctonia solani*. The effectiveness of using beneficial microorganisms in soilless production has also been reported by many other researchers (Gravel et al., 2009, 2015; Vallance et al., 2011). Use of beneficial microorganisms to suppress plant pathogens in soilless production system has also been discussed in detail in a book chapter recently authored by Alsanius and Wohanka (2019). More research is needed to investigate which beneficial microorganisms can be used for controlling individual pathogens on different plant species (Raviv, 2016) and which beneficial microorganisms function beneficially in combination. Another area requiring research is whether or not and how beneficial microorganisms should be used in conjunction with water disinfection technologies. For example, H_2O_2 is effective in killing many plant pathogens; however, research is still needed to determine how to apply H_2O_2 in soilless production system so it is safe for beneficial microorganisms as well as plants (Eicher-Sodo et al., 2019).

5 Integrated rootzone management (IRM)

Practitioners often ask "What is the best growing substrate for my operation?" or "How often and how much should I water my plants?" To correctly answer these types of questions, all the related growing environment factors (e.g. microclimate, fertigation system) must be considered and an integrated approach must be taken. Otherwise, even with the 'best' growing substrate, a crop can still have disorders. For example, over-watering can cause rootzone oxygen deficiency, which can lead to diseases such as pythium root rot proliferation, while over-fertilization can cause salt stress to plants. To achieve a healthy rootzone, an integrated approach is required. The need to acknowledge this integrated approach is the reason why Zheng (2019) coined the term 'Integrated Rootzone Management' (IRM). The IRM considers all of the interconnected elements of the rootzone (e.g. physical, chemical and biological) and how changing any one element can affect the others.

In soilless plant production, an IRM approach requires growers to select the correct container (including size, shape, and color), growing substrate, fertigation strategies, etc. based on the plant species/cultivar, as well as the climate and microclimate conditions. Since there are numerous rootzone elements (e.g. oxygen, water, and nutrients) and management strategies (e.g. irrigation/fertigation) available, and all the rootzone elements are interconnected and changing constantly, it is difficult for a grower to determine the best combinations of all factors for different production systems in real time. To create an optimal rootzone, horticulturists can collaborate with computer scientists to take advantage of the rapidly developing artificial intelligent (AI) technologies. To take the IRM approach, regardless of using AI or not, reliable in situ sensors are essential. The measurement capabilities of sensors can include, but are not limited to, evaluating temperature, water, gases (e.g. oxygen), and nutrients (ideally which can measure individual ion concentrations). Currently, there are in situ (can be inserted to growing substrates for continuous sensing) sensors available for measuring some of these parameters such as temperature, pH, EC, and water status (both water content and water potential); however, more research and development is needed to advance technologies for in situ monitoring of other parameters such as nutrient composition and concentration as well as oxygen levels (Zheng, 2019).

6 Future trends and conclusion

Peat- and coir-based organic substrates and rockwool as a mineral substrate are the most commonly used substrates in modern advanced greenhouse production systems. Efforts have been made to develop and test more environmentally friendly materials to replace or reduce use of the aforementioned materials. The main challenge is to find materials which are available in large amounts with consistent quality. Growing substrate qualities are often assessed by measuring their physical, chemical, and biological properties. Standardized methods for measuring growing substrate qualities need to be developed, clearly communicated, and consistently followed in order to benefit both growers and researchers worldwide. Rootzone elements (e.g. water and oxygen) are interconnected; therefore, an IRM approach is necessary to achieve optimal rootzone conditions. Future research should focus on developing in situ rootzone sensors (e.g. oxygen sensor) and using AI technologies to manage rootzone in order to more sustainably produce plants in greenhouses.

7 Where to look for further information

7.1 Books

1 Handreck, K. and Black, N. 2010. *Growing Media for Ornamental Plants and Turf* (4th edn.). UNSW Press.
2 Raviv, M., Lieth, H. and Bar-Tal A. (Eds). 2019. *Soilless Culture: Theory and Practice* (2nd edn.). Elsevier.
3 Reed, D.W. 1996. *Water, Media, and Nutrition for Greenhouse Crops*. Ball Publishing, Batavia, IL.

7.2 Articles

1 Barrett, G. E., Alexander, P. D., Robinson, J. S. and Bragg, N.C. 2016. Achieving environmentally sustainable growing media for soilless plant cultivation systems - A review. *Sci. Hortic.* 212, 220–34.
2 Caron, J., Pepin, S. and Periard, Y. 2014. Physics of growing media in a green future. *Acta Hortic.* 1034, 309–17.

3 Raviv, M. 2016. Compost as a tool to suppress plant diseases: established and putative mechanisms. *Acta Hortic.* 1146, 11-18.
4 Zheng, Y. 2019. Integrated rootzone management for successful soilless culture. *Acta Hortic.* (in press).

7.3 Websites and conferences

Within the International Society for Horticultural Science (ISHS; https://www.ishs.org/), there is a commission called Plant Substrate and Soilless Culture which has four working groups: (1) Growing media, (2) Composting for horticultural applications, (3) Substrate analysis, and (4) Hydroponics and aquaponics. This commission organizes some international symposia such as: (1) International Symposium on Growing Media, Soilless Cultivation, and Compost Utilization in Horticulture (occurring on average once every two years), (2) International Symposium on Soilless Culture, and (3) International Symposium on Soil and Substrate Disinfestation. There are also other symposia regularly organized by the ISHS such as GreenSys. All of these symposia discuss the science and technologies related to soilless growing substrates. These symposia normally result in proceedings published through Acta Horticulturae (https://www.actahort.org/), which is a great source for current advances in growing substrate research and development.

7.4 Research laboratories

1 North Carolina State University, Horticultural Substrate Lab: http://www.substrateslab.com/
2 Dr. Jean Caron's lab at the University of Laval: http://www.vrrc.ulaval.ca/fileadmin/ulaval_ca/Images/recherche/bd/chercheur/fiche/12762.html
3 Dr. Jean-Charles Michel's lab: http://ephor.agrocampus-ouest.fr/infoglueDeliverLive/
4 Dr. James Altland's lab: https://www.ars.usda.gov/people-locations/person/?person-id=41447
5 Chris Blok at Wageningen University and Research: https://www.vcard.wur.nl/Views/Profile/View.aspx?id=2517

8 References

AHDB. 2018. All things fiber: an American perspective. *Growing Media Review*. Available at: https://projects.ncsu.edu/project/woodsubstrates/documents/press/all-things-fiber.pdf (accessed on 13 April 2019).

Al Naddaf, Q., Livieratos, I., Stamatakis, A., Tsirogiannis, I., Gizas, G. and Savvas, D. 2011. Hydraulic characteristics of composted pig manure, perlite, and mixtures of them, and their impact on cucumber grown on bags. *Sci. Hortic.* 129, 135–41.

Alsanius, B. W. and Wohanka, W. 2019. Root zone microbiology of soilless cropping systems. In: Raviv, M., Lieth, H. and Bar-Tal, A. (Eds), *Soilless Culture: Theory and Practice* (2nd edn.). Elsevier, pp. 149–94.

Altland, J. E. and Locke, J. C. 2012. Biochar affects nacronutrient leaching from a soilless substrate. *HortScience* 47(8), 1136–40. doi:10.21273/HORTSCI.47.8.1136.

Barbosa, G. L., Gadelha, F. D. A., Kublik, N., Proctor, A., Reichelm, L., Weissinger, E., Wohlleb, G. M. and Halden, R. U. 2015. Comparison of land, water, and energy requirements of lettuce grown using hydroponic vs. conventional agricultural methods. *Int. J. Environ. Res. Publ. Health* 12(6), 6879-91. doi:10.3390/ijerph120606879.

Barrett, G. E., Alexander, P. D., Robinson, J. S. and Bragg, N. C. 2016. Achieving environmentally sustainable growing media for soilless plant cultivation systems – a review. *Sci. Hortic.* 212, 220–34. doi:10.1016/j.scienta.2016.09.030.

Bar-Tal, A., Saha, U. K., Raviv, M. and Tuller, M. 2019. Inorganic and synthetic organic components of soilless culture and potting mixtures. In: Raviv, M., Lieth, H. and Bar-Tal, A. (Eds), *Soilless Culture: Theory and Practice* (2nd edn.). Elsevier, pp. 259–301.

Bartz, W. C., Pill, W. G. and Evans, T. A. 2017. Yield of greenhouse-grown tomato in substrates containing coir and parboiled rice or burnt rice hulls. *J. Hort. Sci. Biotech.* 92(3), 231-9. doi:10.1080/14620316.2016.1271289.

Blok, C., De Kreij, C., Baas, R. and Wever, G. 2019. Analytical methods used with soilless substrates. In: Raviv, M., Lieth, H. and Bar-Tal, A. (Eds), *Soilless Culture: Theory and Practice* (2nd edn.). Elsevier, pp. 509–64.

Carlile, W. R., Cattivello, C. and Zaccheo, P. 2015. Organic growing media: constituents and properties. *Vadose Zone J.* 14(6), doi:10.2136/vzj2014.09.0125.

Carmona, E., Moreno, M. T., Avilés, M. and Ordovás, J. 2012. Composting of wine industry wastes and their use as a substrate for growing soilless ornamental plants. *Span. J. Agr. Res.* 10(2), 482–91. doi:10.5424/sjar/2012102-320-11.

Caron, J., Pepin, S. and Periard, Y. 2014. Physics of growing media in a green future. *Acta Hortic.* 1034(1034), 309–17. doi:10.17660/ActaHortic.2014.1034.38.

Caron, J., Price, J. S. and Rochefort, L. 2015. Physical properties of organic soil: adapting mineral soil concepts to horticultural growing media and histosol characterization. *Vadose Zone J.* 14(6). doi:10.2136/vzj2014.10.0146.

Cavins, T. J., Whipker, B. E. and Fonteno, W. C. 2004. Establishment of calibration curves for comparing pour-through and saturated media extract nutrient values. *HortScience* 39(7), 1635-9. doi:10.21273/HORTSCI.39.7.1635.

Ceglie, F. G., Elshafie, H., Verrastro, V. and Tittarelli, F. 2011. Evaluation of olive pomace and green waste composts as peat substitutes for organic tomato seedling production. *Compost Sci. Util.* 19(4), 293–300. doi:10.1080/1065657X.2011.10737011.

Ceglie, F. G., Bustamante, M. A., Amara, M. B. and Tittarelli, F. 2015. The challenge of peat substitution in organic seedling production: optimization of growing media formulation through mixture design and response surface analysis. *PLoS ONE* 10(6), e0128600. doi:10.1371/journal.pone.0128600.

CEN (European Committee for Standardization). 2006. Soil improvers and growing media – Determination of physical properties – Dry bulk density, air volume, water volume, shrinkage value and total pore space, CEN TC 223 WG4 N 151.

Cha, J. S., Park, S. H., Jung, S., Ryu, C., Jeon, J., Shin, M. and Park, Y. 2016. Production and utilization of biochar: a review. *J. Ind. Eng. Chem.* 40, 1–15. doi:10.1016/j.jiec.2016.06.002.

Cheng, C. H., Lehmann, J. and Engelhard, M. H. 2008. Natural oxidation of black carbon in soils: changes in molecular form and surface charge along a climosequence. *Geochim. Cosmochim. Acta* 72(6), 1598–610. doi:10.1016/j.gca.2008.01.010.

De Boodt, M. and Verdonck, O. 1972. The physical properties of the substrates in horticulture. *Acta Hortic.* 26(26), 37–44. doi:10.17660/ActaHortic.1972.26.5.

Domeño, I., Irigoyen, N. and Muro, J. 2009. Evolution of organic matter and drainages in wood fibre and coconut fibre substrates. *Sci. Hortic.* 122(2), 269–74. doi:10.1016/j.scienta.2009.05.006.

Downie, A., Crosky, A. and Munroe, P. 2009. Physical properties of biochar. In: Lehmann, J. and Joseph, S. (Eds), *Biochar for Environmental Management Science and Technology*. Earthscan, London, pp. 13–32.

Eicher-Sodo, M., Gordon, R. and Zheng, Y. 2019. Characterizing the phytotoxic effects of hydrogen peroxide on common microgreen species and lettuce cultivars. *HortTechnology* 29(3), 283–9, doi:10.21273/HORTTECH04255-18.

Fonteno, W. C., Hardin, C. T. and Brewter, J. P. 1995. Procedures for determining physical properties of horticultural substrates using NCSU porometer. Horticultural Substrates Laboratory, North Carolina State University.

Frangi, P., Amoroso, G., Ferrini, F. and Fini, A. 2008. Growth of ornamental shrubs in wood fibre-based growing media. *Acta Hortic.* 801(801), 1571–6. doi:10.17660/ActaHortic.2008.801.194.

Gizas, G., Tsirogiannis, I., Bakea, M., Mantzos, N. and Savvas, D. 2012. Impact of hydraulic characteristics of raw or composted *Posidonia* residues, coir, and their mixtures with pumice on root aeration, water availability, and yield in a lettuce crop. *HortScience* 47(7), 896–901. doi:10.21273/HORTSCI.47.7.896.

Gravel, V., Ménard, C. and Dorais, M. 2009. Pythium root rot and growth responses of organically grown geranium plants to beneficial microorganisms. *HortScience* 44, 1622–7.

Gravel, V., Dorais, M., Dey, D. and Vandenberg, G. 2015. Fish effluents promote root growth and suppress fungal diseases in tomato transplants. *Can. J. Plant Sci.* 95(2), 427–36. doi:10.4141/cjps-2014-315.

Gruda, N. and Schnitzler, W. H. 2004a. Suitability of wood fiber substrate for production of vegetable transplants I. Physical properties of wood fiber substrates. *Sci. Hortic.* 100, 309–22.

Gruda, N. and Schnitzler, W. H. 2004b. Suitability of wood fiber substrates for production of vegetable transplants. II. The effect of wood fiber substrates and their volume weights on the growth of tomato transplants. *Sci. Hortic.* 100, 333–40.

Herrera, F., Castillo, J. E., López-Bellido, R. J. and López Bellido, L. L. 2009. Replacement of a peat-lite medium with municipal solid waste compost for growing melon (*Cucumis melo* L.) transplant seedlings. *Compost Sci. Util.* 17(1), 31–9. doi:10.1080/1065657X.2009.10702397.

Huber, J., Zheng, Y. and Dixon, M. 2009. Hydroponic cucumber production using a new recyclable growth substrate and irrigation system. *Acta Hortic.* 843(843), 303–10. doi:10.17660/ActaHortic.2009.843.40.

Khoshgoftarmanesh, A. H., Eshghizadeh, H. R. and Chaney, R. L. 2012. Using acid-washed shredded waste-tire rubber in soilless media for tomato production. *J. Plant Nutr.* 35(9), 1341–8. doi:10.1080/01904167.2012.684126.

Lu, W. 2008. Utilization of municipal solid waste compost in horticulture. PhD Thesis. Auburn University, Auburn, AL.

Luo, J., Fan, R. Q., Wang, T., Gao, Y., Liu, L. Z., Yan, S. H. and Zhang, Z. H. 2015. Evaluation of spent pig litter compost as a peat substitute in soilless growth media. *Biol. Agric. Hortic.* 31(4), 219–29. doi:10.1080/01448765.2015.1005669.

Moral, R., Pareded, C., Perez-Murcia, M. D., Perez-Espinoza, A. and Bustamante, M. A. 2013. Challenges of composting for growing media purposes in Spain and the Mediterranean area. *Acta Hortic.* 1013(1013), 25–39. doi:10.17660/ActaHortic.2013.1013.1.

Postma, J. 2010. The status of biological control of plant diseases in soilless cultivation. In: Gisi, U., Chet, I. and Gullino, M. (Eds), *Recent Developments in Management of Plant Diseases. Plant Pathology in the 21st Century (Contributions to the 9th International Congress)* (vol. 1). Springer, Dordrecht, pp. 133–46.

Raviv, M. 2016. Compost as a tool to suppress plant diseases: established and putative mechanisms. *Acta Hortic.* 1146(1146), 11–24. doi:10.17660/ActaHortic.2016.1146.2.

Ribeiro, H. M., Freire, C., Cabral, F., Vasconcelos, E. and Brito, L. M. 2013. Production of lettuce seedlings in coconut coir amended with compost and vermicompost. *Acta Hortic.* 1013(1013), 417–22. doi:10.17660/ActaHortic.2013.1013.52.

Savvas, D. 2009. Modern developments in the use of inorganic media for greenhouse vegetable and flower production. *Acta. Hortic.* 819(819), 73–86. doi:10.17660/ActaHortic.2009.819.7.

Schmilewski, G. 2017. Growing media constituents used in the EU in 2013. *Acta Hortic.* 1168(1168), 85–92. doi:10.17660/ActaHortic.2017.1168.12.

Tian, X. and Zheng, Y. 2013a. Compost teas and reused greenhouse nutrient solution suppress several plant pathogens *in vitro*. *HortScience* 48(4), 510–2. doi:10.21273/HORTSCI.48.4.510.

Tian, X. and Zheng, Y. 2013b. Evaluation of biological control agents for Fusarium wilt in *Hiemalis begonia*. *Can. J. Plant Pathol.* 35(3), 363–70. doi:10.1080/07060661.2013.812580.

Tognetti, C., Mazzarino, M. J. and Laos, F. 2007. Improving the quality of municipal organic waste compost. *Bioresour. Technol.* 98(5), 1067–76. doi:10.1016/j.biortech.2006.04.025.

Torres, A. P., Mickelbart, M. V. and Lopez, R. G. 2010. Leachate volume effects on pH and electrical conductivity measurements in containers obtained using the pour-through method. *HortTechnology* 20(3), 608–11. doi:10.21273/HORTTECH.20.3.608.

Vallance, J., Déniel, F., Floch, G., Guérin-Dubrana, L., Blancard, D. and Rey, P. 2011. Pathogenic and beneficial microorganisms in soilless cultures. *Agron. Sust. Develop.* 31(1), 191–203. doi:10.1051/agro/2010018.

Van der Gaag, D. J., van Noort, F. R., Stapel-Cuijpers, L. H. M., de Kreij, C., Termorshuizen, A. J., van Rijn, E., Zmora-Nahum, S. and Chen, Y. 2007. The use of green waste compost in peat-based potting mixtures: fertilization and suppressiveness against soilborne diseases. *Sci. Hortic.* 114(4), 289–97. doi:10.1016/j.scienta.2007.06.018.

Warncke, D. 1998. Greenhouse root media. In: *Recommended Chemical Soil Test Procedures for the North Central Region*. Mo. Agr. Expt. Sta. SB1001, pp. 61–4.

Wright, R. D. 1986. The pour through nutrient extraction procedure. *HortScience* 21, 227–9.

Zheng, Y. 2018. Current nutrient management practices and technologies used in North American greenhouse and nursery industries. *Acta Hortic.* 1227(1227), 435–42. doi:10.17660/ActaHortic.2018.1227.54.

Zheng, Y. 2019. Integrated rootzone management for successful soilless culture. *Acta Hortic.* (in press).

Zheng, Y., Dombrowsky, M.. and Dixon, M. 2009a. The use of compost or composting process to suppress *Pythium aphanitdermatum* in used rockwool. *Acta Hortic.* 843, 327–32.

Zheng, Y., Huber, J., Zhang, P. and Dixon, M. 2009b. Searching for recyclable or biodegradable growing media. *Acta Hortic.* 819, 435–42.

Chapter 9

Advances in irrigation management in greenhouse cultivation

Stefania De Pascale, University of Naples Federico II, Italy; Luca Incrocci, University of Pisa, Italy; Daniele Massa, Council for Agricultural Research and Economics, Italy; Youssef Rouphael, University of Naples Federico II, Italy; and Alberto Pardossi, University of Pisa, Italy

1 Introduction

Over the next few years, the world's horticultural systems will face an important balancing act between two needs - increasing horticultural production to sustain a rapidly increasing global population, projected to reach 10 billion by 2050, and reducing horticulture's impact on the environment by improving resource use efficiency in terms of water and nutrients (De Pascale et al., 2011; Rouphael et al., 2015). In the case of horticultural crops such as vegetables and ornamentals, maximal productivity is attained under controlled environments and indoor conditions where production may expand vertically and microclimate factors such as air and root zone temperature, light conditions (quality and intensity), carbon dioxide enrichment as well as vapour pressure deficit can be controlled (Rouphael et al., 2018). Besides the advantages of protected cultivation in terms of increasing off-season product and buffering against external climatic conditions, maximizing productivity by unit of water use is an urgent need among scientists, extension specialists and horticultural growers to face the shortage of water as a result of climate change (De Pascale et al., 2018). Therefore, in the years to come, greenhouse growers will have to be prepared to use less water, and in some cases that of lesser quality

http://dx.doi.org/10.19103/AS.2019.0052.12

(i.e. saline water), meaning they will have to improve irrigation methods and techniques as well as management practices in order to maximize water use efficiency (WUE). In crop production systems, WUE expressed as kg m^{-3} is usually defined as '*the relationship between crop produced and the amount of water involved in crop production, expressed as crop production per unit volume of water*' (Molden, 2003; Ali and Talukder, 2008). The volume of water applied to the crop to obtain a unit of crop yield includes crop evapotranspiration (referring to the water loss by both soil evaporation and crop transpiration), leaching fraction and water present in the crop tissues. Increasing WUE under greenhouse conditions can be achieved by adopting efficient water delivery systems and irrigation methods such as microirrigation and subirrigation as well as adopting semi-closed- and closed-loop systems (De Pascale et al., 2017). Additional improvements in WUE can be attained through innovative and accurate tools and strategies able to optimize irrigation scheduling in terms of water amount and timing (De Pascale et al., 2018). Among these innovative tools and techniques, the use of soil moisture and crop water status sensors, as well as the calculation of crop water requirement and water balance through simplified models, are gaining interest among researchers and greenhouse growers (De Pascale et al., 2018). In addition to the above-mentioned technological strategies to boost WUE, several studies have revealed that adopting good agricultural practices, such as grafting and the use of natural substances and compounds as well as beneficial microorganisms, have also been reported to reduce the amount of water applied to the cropping system, enhancing WUE (Colla and Rouphael, 2015; Kumar et al., 2017).

This chapter provides an updated overview on the recent advances in water/irrigation management in controlled environments of horticultural crops including both vegetables and ornamentals. A technical design of a typical greenhouse irrigation system will be presented first. Tools and techniques (water balance and crop evapotranspiration) as well as the use of high-tech moisture sensors for irrigation scheduling will be covered. In the context of enhancing the WUE of greenhouse crops, innovative management practices (biostimulants and grafting) will also be discussed. Finally, the chapter will conclude with prospects and research breakthroughs that have to be considered in order to improve water management and WUE in greenhouse crops.

2 Irrigation systems

2.1 Introduction and reference characteristics

The irrigation system has as its main task delivering water to plants. This operation can be accomplished using different techniques and technologies that can vary largely depending on the cultivation system (soilless or soil growing conditions), production type (plant growth or propagation), environmental conditions (climate, soil and water physico-chemical characteristics etc.). The growing environment, *in toto*, is therefore of high relevance to identify the most efficient irrigation system. For instance, irrigation through water infiltration in the root zone is highly inefficient in soil cultivation and much more valuable in closed-loop soilless culture (e.g. subirrigation). On the other hand, some irrigation techniques are required for their particular characteristics, such as, for example, mist or fog systems for their effects on microclimate as well as liquid culture and aeroponics where water represents the growing medium in which nutrients are added following fertigation principles. Only irrigation systems for water nutrition and the utilization of smart and high-efficiency management methods (i.e. allowing high irrigation/WUE) will be discussed in this section. Detailed descriptions of basic components, manufacturing design, installation schemes and material characteristics have been meticulously summarized in previous works covering the main issues and potentialities of different cropping systems, structure types and growing areas of greenhouse production (Dasberg and Or, 1999; Raviv and Lieth, 2008; Baudoin et al., 2013; Thompson et al., 2018). The main intention of this section is therefore to provide an overview of those insights and novelties that in the last few years have made it possible to achieve more and more sustainable use of water and through the adoption of smart irrigation technologies in greenhouse systems.

In a strict sense, the irrigation system consists of an apparatus that allows the distribution of water, or rather, of the nutritive solution (in fertigated crops) to each single culture unit. From the perspective of achieving highly sustainable irrigation management, however, the same apparatus should be thought of and managed as part of a more complex system composed of three main sectors, that is (1) water collection for both water initial use and reuse, (2) water storage system and (3) water delivery system. An exhaustive overview of different water storage methods has been given by Thompson et al. (2018). Water collection and storage is strategic in greenhouse systems to support constant and high-quality production, especially during the hot season when water availability is limited (Fig. 1). This operation is usually accomplished by collecting water from the environment, that is by pumping municipal tap water, surface water and groundwater, and/or collecting rainfall.

However, a more circular and sustainable greenhouse management entails the collection and reuse of drainage water and transpired water vapour after condensation (Fig. 2). Both techniques allow water saving with high WUE (Thompson et al., 2018). Nevertheless, the management of recycled water can be difficult, especially of that drained from the substrate after irrigation, which may show increased (with respect to the input values) (1) nutrient

Figure 1 Examples of water silos (left) and lined water storage (right) for the collection of rain and/or drainage water.

and non-nutrient concentration, which in turn causes increasing electrical conductivity, (2) accumulation of plant protection products and (3) presence of allelopathic compounds (Baudoin et al., 2013; Thompson et al., 2018). A major concern regarding the reuse of water is the accumulation of ballast ions. Different strategies can be adopted to control growing systems that inevitably face problems due to poor quality (e.g. saline) water (Massa et al., 2010). Concerns about the reuse of condensed water has also been the subject of research in greenhouse crops due to the possible accumulation of metal elements, such as zinc, from which greenhouse structures are made (Dannehl et al., 2014).

Apart from its operational management, the reuse of water entails the adoption of additional equipment and technologies for water collection and treatment (biological and physical) with additional costs for installation and maintenance (Thompson et al., 2018). Depending on the quality and characteristics of available water, however, some of the above-mentioned apparatus are already present at farm level as basic components of the irrigation system. In this case, the technological level of the implemented equipment mostly depends on (1) the initial quality of water sources available

Figure 2 Collection of water drained from the cultivation system after irrigation (left) and water transpired by plants as vapour and successively condensed on greenhouse roof (right).

for irrigation purposes, (2) the irrigation delivery system and last but not least, (3) crop requirements and sensitivity to stress possibly caused or increased by the occurrence of inadequate water chemo-physical characteristics.

Parts that are common to all water delivery systems for irrigation are:

1 pumping station to pump water into the irrigation circuit;
2 computer station to control automatic or manual valve opening and closure;
3 water treatment station for particle removal, chemical element removal and disinfection; and
4 network of valves, tubes and emitters (Fig. 3).

The main task of filtration apparatus is to eliminate suspended solids that could damage irrigation systems and components or cause them to clog. Many methods are available (Thompson et al., 2018), however the most widespread are those based on the application of hydrocyclones, discs, sand bed (slow filtration) system and sieve filters. In general, filtration operations remove particles within the range of 0.05–1 mm while microfiltration methods are required in the range of 0.1–10 μm. Microfiltration is a filtration process that implies the use of microporous membranes but is different from nanofiltration and reverse osmosis because it does not require active pressurization and is not suitable for the removal of contaminants. To correct the chemical characteristic of water, more sophisticated techniques and technologies are required, which include reverse osmosis, nanofiltration, ion precipitation and deionization (Thompson et al., 2018).

The main task of the irrigation system is to deliver water to plants to prevent drought conditions. Such a task is accomplished by restoring field/substrate capacity through the supply of water volume evapotranspirated in the time between two irrigations. However, for the correct crop management, additional (extra) water volumes could be required to induce ad hoc leaching

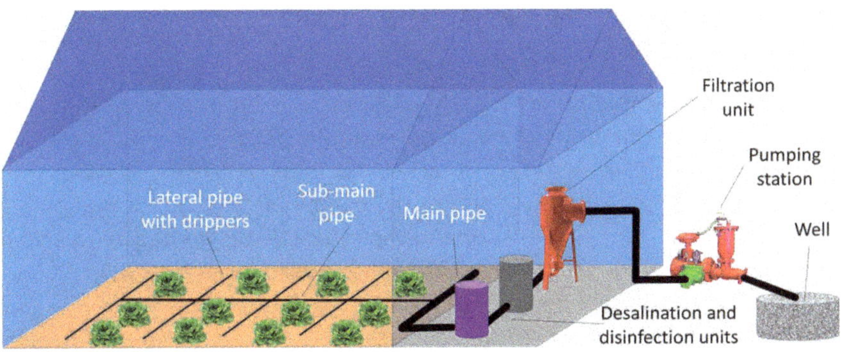

Figure 3 Schematic representation of a typical drip irrigation system for a greenhouse.

phenomena during irrigation. Depending on the cropping system, a network of valves, tubes and emitters must be realized to connect water sources to the growing environment, thereby ensuring optimal distribution of water to each growing unit. Sustainable water use in greenhouse systems mostly depends on how irrigation systems are designed in order to perform at the highest level as a function of crop needs, final production targets, soil/substrate characteristics and climatic conditions. Efficient use of water for irrigation systems is mainly concerned with the level of technology and irrigation engineering. It can be generally evaluated in terms of irrigation efficiency (IE) calculated as the ratio between 'water beneficially used' and 'total water applied' (van Halsema and Vincent, 2012). Typical uses for different irrigation systems have been reported in the literature as well as more specific indices for water distribution and conveyance efficiency (Hsiao et al., 2007; van Halsema and Vincent, 2012).

2.2 Microirrigation

The highest levels of efficiency and distribution precision in irrigation can be achieved in greenhouse systems, through the application of microirrigation (Evans and Sadler, 2008; Levidow et al., 2014). This technique can be adopted in a very broad variety of cultivation systems, from soil to soilless conditions, and different crop types from ornamental species to edible plants. Microirrigation implies the adoption of irrigation systems characterized by low flow rate and low pressure (i.e. 0.5–4 bar) in which water is delivered within a very limited area surrounding the plant thereby maximizing the interception of water by roots while limiting evaporation and percolation phenomena. The main advantages of microirrigation are

1. high IE due to the low risk of water runoff and evaporation;
2. indirect control of weeds;
3. lower risk of the occurrence of pathogens;
4. improved irrigation scheduling and water supply that can be accomplished over the day including during the warmest hours;
5. reduced soil compaction and erosion; and
6. fertigation.

Moreover, this technique is the only one that can be applied to carry out particular irrigation strategies such as, for example, deficit irrigation (Evans and Sadler, 2008) or other agronomic practices to improve WUE and other crop performances (Dasberg and Or, 1999). The combination of drip irrigation and plastic mulching is, for example, one of the most reliable strategies to ensure efficient nutrient, water and weed management (Vázquez et al., 2005, 2006).

The very high theoretical uniformity performed by microirrigation systems (up to 90%) is sometimes not achieved on farm level due to poor maintenance and insufficient system design (Hsiao et al., 2007). Therefore,

a sub-immersion of the pots (roughly 5 cm). The pots are then arranged on benches periodically flooded for a period sufficient to restore optimal moisture conditions in the root zone (5–20 min). Immersion depth and time of exposure to irrigation strongly depends on the physical characteristics of the growing medium: water depth and irrigation duration have an inverse relationship with unsaturated hydraulic conductivity of substrates. Benches are usually made of plastic or metallic material, with a fluted base to facilitate water outflow after irrigation. The system is quite economical and requires moderate maintenance interventions; moreover, using mobile benches it is possible to exploit up to 80–90% of the greenhouse surface for cultivation. This technique can be applied in the cultivation of containerized crops (Cardarelli et al., 2010) and for plant propagation in nurseries (Santamaria et al., 2007). A cheaper option for flooded systems is the adoption of flooded floor. In this case, the irrigation solution is dispensed directly onto the floor of the greenhouse appositely built to the scope. Water ebb and flow usually occur through a central channel. The system allows for high mechanization of the cultivation process through the adoption of machineries that distribute pots in the area according to a desired plant density. The main drawback of the flooded systems consists of potentially high air humidity at canopy level, especially when the duration of flooding is high.

The trough system entails the arrangement of pots or slabs on sloping channels (0.5–1%) where the irrigation solution flows discontinuously for a time sufficient to restore in the root zone water lost by plant transpiration; the excess volume is collected in the drainage tank and reused in a closed loop. Plastic and metallic channels can be used in trough subirrigation; it is however fundamental that they do not bend under the weight of the pots to avoid waste stagnation. Space efficiency is slightly lower than other flooded systems (70–80%). The application of trough subirrigation was found suitable for tomato crops grown under moderate salinity allowing higher nutrient and water saving compared with drip irrigation (Incrocci et al., 2006; Montesano et al., 2010).

In the case of mat systems, a plastic film isolates the base of the cultivation area (e.g. plastic or metallic benches, concrete floor or even well-levelled ground) on which a fibreglass mat is positioned. The irrigation solution is supplied from one side and flows on the mat, which will soak, thereby distributing the irrigation solution to the pots. However, manufactured mats are available that integrate driplines with the fibres to improve water distribution. In the mat system, the holes of the pots must be on the bottom of the pot to allow contact between the potting medium and the mat. Furthermore, uneven levelling can bring about low irrigation uniformity due to the different moisture tension possibly occurring in the mat due to different height levels.

3 Irrigation management strategies

Smart and efficient irrigation faces contemporaneous challenges related to climate change, water demand for human activities other than agricultural production, and consumer and market requirements. This implies sustainable management strategies in a dynamic and flexible way as a function of variables that may also vastly differ from each other, such as human tendencies and behaviours to actual (measurable) needs and/or limitations in the use of water. In fact, if improving efficiency of water use is imperative to the responsible and sustainable use of global resources, the resulting social benefits can also be exploited as a marketing strategy, as well as obtaining added-value products. The adoption of adequate irrigation strategies is therefore a priority to achieve high economic and environmental sustainability of greenhouse cropping systems. New approaches arise and old ones can be adapted or revised to modulate water management, thereby facing limitations while exploiting new opportunities in the irrigation of greenhouse crops. Different irrigation strategies, herein reported, are available (1) to face limited water availability; (2) to prevent salt accumulation in the root zone; (3) to handle different water sources (e.g. municipal reclaimed water); and (4) to reuse water in the same cropping system or recycle it for the cultivation of successive crops in cascade-production cropping systems.

3.1 Strategies to optimize free-drain systems

3.1.1 Deficit irrigation (including RDI and PRD)

Deficit irrigation (DI) is a broad term used to identify those irrigation techniques that aim to reduce crop water supply in time and/or in space. The working principle of DI lies in the main assumption that plants exposed to drought are physiologically stimulated to improve plant-soil interaction mechanisms for enhancing water relations. Plants that are undergoing drought stress promptly reduce stomatal activity and trigger metabolic processes aimed at improving the uptake of water (Golldack et al., 2014). Based on the above assumptions, it has been demonstrated that it is possible to induce temporary water stress during specific phases of the cultivation cycle and/or in specific portions of the root zone in such a way to trigger plant reactions without causing permanent (irreversible) water stress, which would otherwise result in reduced produce yield and quality (Jones, 2004; Costa et al., 2007).

It should be highlighted that DI was primarily developed for tree crops in which the botanical characteristics of plants allow the safe management of this technique since ligneous organs, typically present in those species, generally make plants less sensitive to water stress compared with herbaceous plants. Moreover, the cultivation phases of tree crops include plant phenological stages from vegetative to reproductive phases during which water supply can be regulated *ad hoc* taking into account production targets. In practice, water stress is induced in these crops mainly during the vegetative phase. Such conditions and characteristics are not always present in the cultivation of vegetable and ornamental species such as, for example, leafy vegetables in which vegetative organs represent the harvested produce.

From an operational point of view, DI is more viable in soil-grown crops since applications in containerized production systems are highly risky due to the very limited volume of the root zone and relatively limited water buffer capacity available per plant. DI can be effectively managed by irrigation systems that allow high precision in water delivery and distribution, such as that of drip irrigation systems. The adoption of sensing techniques and technologies for monitoring crop water stress is strategic to carry out DI safely and fruitfully. This task can be accomplished in horticultural systems by the use of remote and proximity touch sensors for monitoring canopy or root zone water status (Incrocci et al., 2017; Tripodi et al., 2018).

The term DI is classically associated with a technique that implies irrigation volumes lower than actual crop evapotranspiration. However, for the sustainability of intensive protected cultivation, such a generalization would not be compatible with farmer economic targets due to the potential reduction in yield and quality depending on the extent of water limitation and crop-specific characteristics (Pulupol et al., 1996; Dorji et al., 2005). DI has two main principles of application, which consist of the regulated DI (RDI) and partial root zone drying (PRD). Table 1 reports a critical overview of different DI techniques tested for the irrigation of various greenhouse crops; in many experimental works, the application of DI results in enhanced WUE but lower yield. In summary, DI can be adopted in greenhouse cropping systems, although careful management and meticulous crop monitoring must be taken into account for successful application.

The physiological justification for the adoption of RDI lies in the fact that plant response to water stress varies during the cultivation cycle. Therefore, irrigation can either be regulated to cause plant reactions through a moderate water stress, or to reduce water supply when the plant is less sensitive; the two strategies or their combination would lead to enhanced crop WUE. The main difficulty consists in the identification of water stress thresholds as a function of the crop-specific response over the cultivation period. The main risk is that the narrow limits between light and severe stress could be suddenly surpassed if the greenhouse climate is not well controlled due to the high correlation between crop evapotranspiration and temperature or other correlated climatic variables (Jones, 2004).

In the PRD approach, water is applied to obtain the same effects of RDI in terms of plant response and water saving but alternately to different sides of the root zone. This expedient allows safer and more cost-effective plant watering since potential errors in the technique are counteracted by the well-watered

Table 1 Overview of works assessing the effects of different deficit irrigation (DI), regulated deficit irrigation (RDI) and partial root zone drying (PRD) strategies on produce yield and quality of typical greenhouse vegetable crops. The table reports a non-exhaustive compendium of experiments published in the period 2000–18

Crop	Strategy	Growing conditions/Treatments	Main results	Reference
Cucumber	PRD	Glasshouse, container experiment, perlite PRD 65% full irrigation in open and closed system	Comparable produce yield and quality in open system, 12% lower yield in closed system	Dasgan et al. (2012)
Green bean	RDI	Plastic greenhouse, soil experiment, sandy-loam soil covered with 0.3-m layer of loamy soil, 0.02-m layer of manure, 0.1-m mulch layer of coarse sand	Comparable yield in the autumn–winter season, 31% lower yield in the spring season	González et al. (2009)
Hot pepper	RDI	Plastic greenhouse, soil experiment, sandy-loam soil, drip and furrow irrigation RDI 33% and 66% full irrigation in three different growth stage = six treatments + control	Comparable yield in the drip irrigation with exception when applied during 'late fruit bearing and harvesting' stage, comparable yield in the drip irrigation only if applied during 'flowering and fruit setting' stage	Yang et al. (2017)
	PRD	Greenhouse, sand-loam-peat substrate PRD 70% full irrigation	Comparable yield and quality, significantly higher WUE	Sharma et al. (2015)
	RDI PRD	Greenhouse, soil experiment, clay-loam soil PRD × RDI 50% full irrigation in four different growth stages = eight treatments + control	Comparable yield only when PRD × RDI not imposed during 'bloom and fruit setting stage' or 'vigorous fruit-bearing stage'	Guang-Cheng et al. (2010)
	DI PRD	Glasshouse, soil experiment, clay-loam soil DI 75% and 50% commercial irrigation PRD 50% applied to one side or alternately	Comparable yield only for DI 75% significantly reduced otherwise, 21% higher WUE	Shao et al. (2008)
	PRD DI	Glasshouse, container experiment, bark-pumice-peat substrate PRD 50% commercial irrigation DI 50% commercial irrigation	PRD: 19% yield reduction, higher TSS, lower fruit number DI: 35% yield reduction, higher TSS, lower fruit number	Dorji et al. (2005)

Tomato	PRD	Multi-tunnel greenhouse, container experiment, sandy-silty substrate PRD 70% full irrigation PRD 50% full irrigation	PRD 70%: comparable yield, 27% higher water use efficiency PRD 50%: comparable yield, 65% higher water use efficiency	Affi et al. (2014)
	PRD	Glasshouse, container experiment PRD 70% and 50% full irrigation in open and closed system	Comparable yield and quality	Dasgan et al. (2009)
	DI PRD	Glasshouse and plastic tunnel, soil experiment DI and PRD 50% full irrigation	22% lower yield on average	Sirigu et al. (2007)
	PRD DI	Plastic greenhouse, soil experiment, clay soil PRD 50% ETc vs. full irrigated DI 50% ETc vs. full irrigated	PDR: 20% yield reduction, 50% water saving, reduced gaseous exchanges DI: 30% yield reduction, 38% water saving, reduced gaseous exchanges	Topcu et al. (2007)
	DI PRD	Plastic greenhouse, soil experiment, clay soil DI and PRD 50% full irrigation	Comparable yield	Kaman et al. (2006)
	DI PRD	Plastic greenhouses, soil experiment, clay soil DI and PRD 50% full irrigation	No significant difference for the statistics in yield, although 23% decrease was observed on average	Kirda et al. (2004)
Watermelon	RDI	Plastic greenhouse, soil experiment, sandy-loam soil RDI upper and lower limitations and sole upper limitation	Comparable yield in the winter-spring and autumn season, 11% lower yield in the winter season, comparable or better fruit quality	Yang et al. (2017)
	RDI	Plastic greenhouse, soil experiment, sandy-loam soil covered with 0.3-m layer of loamy soil, 0.02-m layer of manure, 0.1-m mulch layer of coarse sand	Comparable marketable yield	González et al. (2009)

part of the root zone; therefore, also less precision in monitoring crop water status is possible (Jones, 2004; Costa et al., 2007). By contrast, PRD requires a more complex irrigation net due to the duplication of sub-main pipes to allow switching irrigation from one dripline to another during cultivation (Thompson et al., 2018).

3.1.2 Leaching fraction control

Plant water uptake is driven by the difference in salt concentration between the water contained in the plant root and in the soil. If the concentration of the salts in the soil water is too high, the plant water uptake is reduced or stopped, leading to plant dehydration, yield reduction or, even, death of the plant. Another important issue of soil salinity is also the ion toxicity (mainly chloride) that could cause leaf necrosis.

The salt tolerance is well-studied in literature (Maas and Hoffman, 1977; Shannon and Grieve, 1999; Cassaniti et al., 2012) and crops show different sensibility to salt stress. Maas and Hoffman introduced in 1977 a general linear model to describe the general salt tolerance of a crop in terms of reduction of relative yield (Y) versus the electro-conductivity of soil saturated extract (SSE):

$$Y = \text{Min}\left[100;\ 100 - s\left(EC_{SSE} - EC_{SSE\text{-}TR}\right)\right] \qquad (1)$$

therefore, salt tolerance can be adequately measured on the basis of only two parameters: (1) the threshold ($EC_{SSE\text{-}TR}$), that is the maximum salinity without yield reduction and (2) the slope (s), that is the percentage yield decrease per unit increase in salinity above the threshold (Fig. 6). According to this

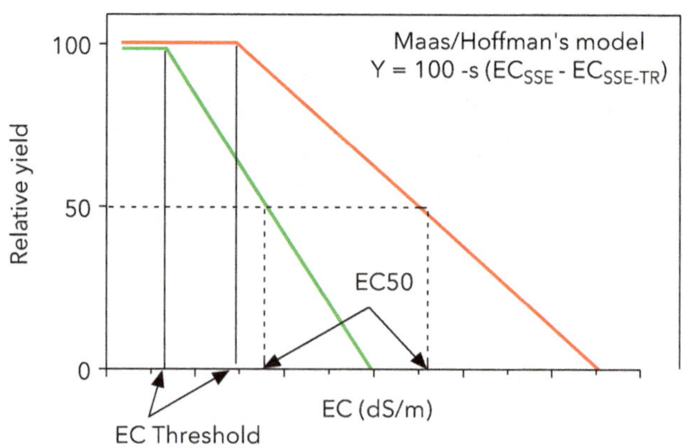

Figure 6 Schematic representation of the linear Maas and Hoffman's model. The green and red lines represent the yield-EC_{SSE} relationship of a moderately sensitive crop and a moderately salt-resistant crop, respectively.

model, the plants can be classified as sensitive, moderately salt sensitive or moderately salt tolerant: a classification of the main greenhouse crops is reported in Table 2.

The effect of salinity on soil-grown crops has been largely studied and expressed on the basis of the EC of SSE: in contrast, less information regarding the effect of substrate salinity on crop yield and quantity is present in the literature, and this information is put in relation to the EC of the substrate solution or the recirculating nutrient solution (EC_{SS}) instead of the SSE. Nevertheless, data obtained in soil salinity trials could also be used for an indicative estimate of crop salinity resistance into soilless systems using the relationship between the EC of the solution in the recirculating nutrient solution or in the substrate (SS) and the SSE, as proposed by Sonneveld (2000):

$$EC_{SS} = 1.6EC_{SSE} - 0.18 \tag{2}$$

Crop salt tolerance may depend on many environmental and genotypic factors, and sometimes the soilless culture could mitigate the negative effects of the salinity on the crop, as reported by Marcelis and Van Hooijdonk (1999) for radish.

Irrigating soil with water with a medium-high salt content (sodium and chloride - that are not essential ions for plant life - and calcium and magnesium) is one of the main causes of soil salinization. Control of salt build-up in the root zone is possible by leaching the salt accumulated in the soil/substrate: the correct calculation of the leaching fraction (the ratio between the drained and supplied water) can reduce crop salinity, thus minimizing the amount of water necessary.

The optimal leaching fraction (LF) can be calculated using the formula (Eq. (3)) proposed by Ayers and Westcot (1985):

$$LF = \frac{EC_I}{(5EC_{SSE\text{-}TR} - EC_I)} \tag{3}$$

where EC_I and $EC_{SSE\text{-}TR}$ are respectively the electrical conductivity of the irrigation water and the EC maximum value tolerated by the crop for the SSE.

3.2 Strategies to optimize water reuse systems

3.2.1 Closed- and semi-closed-loop systems

Closed-loop systems are growing systems where the water drained from the root zone can be collected and reused for irrigation of the same crop. Sometimes, when the water used for irrigation contains more salts than those absorbed by the crop, these will accumulate in the recirculating nutrient solution (or in the substrate) with the consequence that it is necessary to discharge (leaching) and

Table 2 Salt tolerance of the main vegetable and cut flower species grown in soil under greenhouse, according to the Maas-Hoffman model

	EC_{SSE} (mS cm^{-1})		EC_{SS} (mS cm^{-1})		
	EC_{TR}	Slope	EC_{TR}	Slope	References
Salt-sensitive crop					
Bean	1.9	19.0	2.9	11.9	Maas and Hoffman (1977)
Carrot	1.0	14.0	1.4	8.7	Maas (1986) - and instances below
Chrysanthemum	0.9	8.7	1.3	5.4	Barbieri and De Pascale (1992)
Eggplant	1.1	6.9	1.6	4.3	Maas 1986
Fennel	1.2	14.0	1.7	8.8	Graifenberg et al. (1996)
Lettuce	1.3	13.0	1.9	8.1	Shannon and Grieve (1999)
Onion	1.4	17.0	2.1	10.6	Shannon and Grieve (1999)
Pepper	1.5	14.0	2.2	8.7	Maas 1986
Radish	1.2	13.0	1.7	7.9	Shannon and Grieve (1999)
Rose	1.5	9.7	2.2	6.1	Barbieri and De Pascale (1992)
Strawberry	1.0	33.0	1.4	20.6	Maas and Hoffman (1977)
Moderately salt-sensitive crop					
Broccoli	2.8	9.2	4.3	5.8	Shannon and Grieve (1999)
Cabbage	1.8	9.7	2.7	6.1	Shannon and Grieve (1999)
Carnation	2.5	3.9	3.8	2.4	Barbieri and De Pascale (1992)
Celery	1.8	6.2	2.7	3.9	Maas 1986
Cucumber	2.5	13.0	3.8	8.1	Maas and Hoffman (1977)
Melon	2.2	7.3	3.3	4.6	Maas and Hoffman (1977)
Spinach	2.0	7.6	3.0	4.7	Maas and Hoffman (1977)
Tomato	2.5	9.9	3.8	6.2	Maas and Hoffman (1977)
Moderately salt-tolerant vegetables					
Artichoke	4.9	10.7	7.7	6.7	Graifenberg et al. (1993)
Asparagus	4.1	2.0	6.4	1.25	Maas (1986)
Cherry tomato	6.0	5.8	9.4	3.6	Gough and Hobson (1990)
Garden orach	6.4	4	10.0	2.5	Shannon and Grieve (1999)
Purslane	6.3	9.6	9.9	6.0	Shannon and Grieve (1999)
Rocket salad	3.5	9.0	5.4	5.6	Shannon and Grieve (1999)
Swiss chard	7.0	9.1	11.0	5.7	Shannon and Grieve (1999)
Table beet	4.0	9.0	6.2	5.6	Shannon and Grieve (1999)
Turnip	3.3	4.8	5.1	3.0	Shannon and Grieve (1999)
Zucchini, squash	5.1	11.6	8.0	7.2	Graifenberg et al. (1996)

Indicative data about the possible crop salt tolerance in soilless culture was also calculated applying the Sonneveld formula ($EC_{(SS)} = 1.6\ EC_{(SSE)}-0.18$).
Abbreviations: EC_{SSE} = EC (dS m^{-1}) of the soil saturated extract; EC_{SS} = EC (dS m^{-1}) of nutrient solution in the substrate; EC_{TR} = maximum EC tolerate by crop without growth reduction; slope = % of yield reduction per each dS m^{-1} increase in salinity above the EC_{TR}.

replace it with fresh nutrient solution. In this case, the system is defined as a semi-closed system.

The adoption of a closed-loop system may significantly increase water and nutrient use efficiency, and simplifies the irrigation scheduling, since potential over-irrigation events are buffered by the collection of excessive drainage. Nevertheless, four main drawbacks limit the large diffusion of closed-loop systems (Ehret et al., 2001; Bar-Yosef, 2008; Van Os, 2010): (1) the potential risks of the diffusion of root-borne diseases and the possible accumulation of plant and microbial metabolites with phytotoxic effects; (2) the difficulty in the nutrient replenishment of recirculating nutrient solution, due to the accumulation of non-essential and/or potentially harmful ions (sodium, chloride, sulphate, microelements and, in the case of hard water, calcium and magnesium as well); (3) the lack of technicians able to manage closed-loop systems; and (4) the high investment costs due to the equipment necessary to collect, disinfect and replenish nutrient elements in the recirculating nutrient solution.

In commercial soilless greenhouses, the disinfection of recirculating nutrient solution is strongly recommended to minimize risks of development of soil-borne diseases. According to Van Os (2010), the methods to disinfect the recirculating nutrient solution could be classified as physical (heat treatment, UV radiation, membrane filtration), physical-biological (slow sand filtration) and chemical (addition of ozone, hydrogen peroxide, sodium hypochlorite, chlorine dioxide, copper silver ionization and active carbon adsorption).

As a matter of fact, only four disinfection methods are commonly used: heat treatment, UV radiation, slow sand filtration and ozone addition.

Heat treatment consists of heating the drain water up to 95°C for a minimal exposure of 10 s: the method requires a cheap energy source. UV irradiation is based on the application of UV-C rays to a thin layer of nutrient solution: for a complete elimination of all possible pathogens, an irradiation dose of 250 mJ cm^{-2} is necessary. Slow sand filtration is a low-cost method to eliminate soil-borne pathogens, but *Fusarium* spp., viruses and nematodes are only partly (90–99.9%) removed by this method: pathogens are partially blocked by the sand in the upper layers of the filter, and are destroyed by a biologically active film that covers the sand grains. To ensure the filter works properly, the size of the sand particles (0.15–0.35 mm; D10 < 0.4 mm) and the flow rate (from 100 L m^{-2} h^{-1} to a maximum of 300 L m^{-2} h^{-1}) are fundamental parameters. Among the chemical methods, sodium hypochlorite and hydrogen peroxide are also cheap solutions, but their performance is sometimes insufficient. The addition of ozone at the dose of 10 g m^{-3} h^{-1} is a promising method to disinfect recirculating nutrient solution, since it presents no residue at the end of the treatment. Nevertheless, ozone treatment is not often used due to side effects including harmfulness to workers (irritation of mucous membranes), the inability to process large quantities of water at the same time, the investment cost (need of an ozone generator) and the partial oxidation of iron chelate.

3.2.2 Accumulation of phytotoxic substances

Another problem related to the adoption of closed soilless cultivation is the possible accumulation of phytotoxic substances from root secretion and organic substrate decomposition (Lee et al., 2006) during the long-term recycling of nutrient solution.

A lot of these compounds are simple water-soluble organic acids, straight chain alcohols, fatty acids, complex quinones, simple phenols, alkaloids, cyanogenic glycosides, phenolic acids such as benzoic acid, 4-hydroxy-benzoic acid, cinnamic acid, ferulic acid, salicylic acid, gallic acid, tannic acid, acetic acid, palmitic acid, stearic acid and so on (Lee et al., 2006). The accumulation of phytotoxic substances is one of the key factors causing continuous cropping obstacles of field and horticultural crops. Today, the removal of phytotoxic substances can be carried out using activated carbon adsorption and nano-TiO_2 photocatalysis (Qiu et al., 2013).

3.2.3 Cascade crops

The concept of the cascade cropping system is to use freshwater to irrigate the most sensitive crops (lettuce or strawberry, donor crops) and to employ the exhausted drainage (mainly for the high content in NaCl) collected from this culture for irrigating more tolerant crops such as cherry tomato, asparagus and zucchini (named user crops). Usually, greenhouse farms specialize in the cultivation of a few similar crops. It is therefore important that the crops used in the cascade belong to the same crop category (e.g. for cut flower farms a possible cascade crop could be roses or gerbera as donor crops and *Lisianthus* or Japanese limonium as user crops).

4 Irrigation scheduling

The term 'irrigation scheduling' identifies a systematic procedure that calculates the expected future water requirement over relatively short periods of time to meet all crop needs and avoid under- or over-application of water (Evans, 2008). Under-irrigation could result in a moderate plant wilt stress with a reduction in crop produce quality and yield, while over-irrigation could promote the development of plant diseases, an increase in energy cost for pumping as well as environmental pollution due to the large amount of water leached that also contains nutrients (Pardossi and Incrocci, 2011). For example, Thompson et al. (2007a) identified the poor management of drip irrigation in greenhouse crops as one of the main causes of nitrate leaching in the Almeria district (southeastern Spain), one of the most concentrated greenhouse areas in the world. Generally, the main cause of over-irrigation is the regulation of irrigation based on the grower's experience rather than on the real water crop needs (Incrocci et al., 2014). Since nutrient use and water are closely related, better water management generally results in improved nutrient use efficiency. Moreover, precise scheduling is crucial for the RDI, which aims to induce a mild water stress to the plant to save water and improve crop yield and produce quality (FAO, 2000; Jones, 2004).

As a matter of fact, irrigation scheduling must answer two basic questions: (1) how much water must be applied for each irrigation event (optimal irrigation volume) and (2) the timing for this application (irrigation turn or irrigation frequency). Generally, the optimal irrigation volume is quite easy to determine, while the right time of application of this volume remains the main weak point and problematic issue to solve in the farm in order to obtain a real improvement in water use in horticulture.

4.1 Determination of irrigation volume

The optimal irrigation volume (I) is divided in a net irrigation volume (I_{NET}, depending mainly on the water-holding capacity of the soil or substrate) and a second amount of water necessary to face possible (1) salinity build-up in the root zone due to the use of saline water, (2) uneven crop uniformity (i.e. differences in transpiration among plants) and (3) uneven water distribution of the irrigation system (i.e. difference in discharge rate of emitters, IE and uniformity etc.). Generally, this extra irrigation water volume is expressed as a safety percentage of the net irrigation (safety irrigation coefficient, Ks).

$$I = I_{NET} \cdot K_S \tag{4}$$

In other words, I_{NET} represents the amount of water depleted from the root zone between two subsequent irrigation events and corresponds to the water evapotranspirated by the crop during this time.

Soil and substrate moisture can be measured as volumetric water content (expressed as m^3 of water per m^3 of soil or substrate volume), or water (matric) potential (expressed as kPa or hPa). As the soil or substrate becomes drier, water potential, which is assumed to be close to zero at saturation, decreases and the value is more negative.

In soil, the amount of water available to the crop (AW) is defined as the difference between the moisture content at field capacity (water potential: -33 kPa) and at the wilting point (-1500 kPa). Water content at field capacity represents the water that the soil can hold against gravity, which is the water that remains in the soil after (over)irrigation and drainage. In general, horticultural crops are sensitive to water stress; consequently, to obtain the best productive performance the irrigation should be operated when only a fraction F (ranging from 0.30 to 0.50, depending on the crop) of AW is depleted. Another important parameter for soil irrigation scheduling is water permeability (or infiltration rate, expressed as mm rain that can be absorbed by the soil in an hour), which determines the rate of water application, especially in the case of overhead irrigation. The available water in a soil is strictly correlated with its texture, organic matter content and salt content. Some values of hydrological parameters are reported in Table 3 for different soils; a better estimation of an AW of a soil could be obtained using various online simulators (i.e. http://www.dynsystem.com/netstorm/soilwater.html) and software (i.e. SPAW software developed by the USDA-Hydrology and Remote Sensing Laboratory) able to simulate soil water tension, water conductivity and water-holding capability based on the soil texture, with adjustments to account for gravel content, compactness, salinity and organic matter.

In soilless culture, where substrates are used instead of soil, the calculation of AW is possible through the knowledge of the moisture retention curve (MRC), a relationship between the volumetric water content (θ) and the water matric potential (ψ_m). Generally, the MRC is determined in a laboratory by a suction table system (de Boodt and Verdonck, 1972; Kipp et al., 2001).

Conventionally, AW is defined as the difference between water content at -1 and -10 kPa matric potential, while easily available water (EAW) corresponds to the difference between water content at -1 and -5 kPa (de Boodt and Verdonck, 1972). Table 4 reports the main physical characteristics of some widely employed substrates in horticulture.

The range of matric potential values that define AW in a substrate is very different from those used in the soil (−1/−10 kPa compared to −33/−1500 kPa): this is due to the larger dimension of the particles of the substrate with respect to those of the soil, which results in a decrease of several orders of magnitude in the hydraulic conductance over a narrow range of suction values. In other words, when the ψm is more negative than −10/−15 kPa, the residual water

Table 3 Representative soil volumetric water content and water potential at saturation, field capacity, wilting point and infiltration rate for different soils

	Sand	Loam	Clay
Bulk density (kg L^{-1})	1.4-1.6	1.2-1.4	1.1-1.2
Soil water content at field capacity (L L^{-1})	0.10-0.18	0.25-0.35	0.35-0.45
Soil water content at wilting point (L L^{-1})	0.03-0.09	0.12-0.16	0.18-0.22
Available water (L L^{-1})	0.07-0.09	0.13-0.19	0.17-0.23
Infiltration rate (mm h^{-1})	> 40	20-40	3-15

Table 4 Physical properties of some widely used substrates in soilless culture

Substrate	Peat	Perlite	Pumice	Peat-perlite (1:1)	Peat-pumice (1:1)	Rockwool
Bulk density (kg L^{-1})	0.06–0.10	0.15–0.17	0.65–0.95	0.11–0.13	0.40–0.50	0.15–0.20
Porosity (% vol.)	92	75	68	78	77	93
Air capacity (% vol.)	38	70	29	32	20	15
AW[a] (% vol.)	33	9	4	28	18	78
EAW[b] (% vol.)	21	8	3	22	13	77

[a] Available water.
[b] Easily available water.
Source: Adapted from Reed (1996) and Kipp et al. (2001).

present in the substrate is unable to move across the substrate, and thus to reach the root system. These low values of water matric potential are consistent with the effects on water retention due to the gravimetric potential, related to the height of the water above the reference point (normally the bottom of the container). The force of gravity translates to approximately 1 kPa per 10 cm; so, at the end of an irrigation event, the layer of substrate at the bottom of the container will be completely saturated by water, while the substrate layer at the height of 10 cm will have a water retention equal to −1 kPa. Thus, the shape of the container is very important in determining the water-holding capacity of the substrate-container system. In this case, the term 'field capacity' is substituted by 'container capacity' (CC; water hold at the matric potential of -1 kPa). In general, for optimal plant growth, the water content in the container must not fall below 60% of the total container water-holding capacity (AWcc).

The knowledge of the AW of the soil and the crop root depth (RD, expressed in m) or the AW_{CC} of the container-substrate system and the container density (CD, expressed as number of containers per m^2) permits to calculate the optimal net irrigation (expressed as mm or L m^{-2}) for the soil (Eq. (5)) or the soilless (Eq. (6)) culture:

$$I_{NET} = AW \cdot RD \cdot F \quad (5)$$

$$I_{NET} = AW_{CC} \cdot CD \cdot F \quad (6)$$

where F is the percentage of the available water that could be depleted in the root zone between two subsequent irrigation events. For most greenhouse crops F ranges between 0.25 and 0.35 and 0.1 and 0.6 in soil and substrate, respectively.

The safety irrigation coefficient ranges from 1.15 (i.e. uniform crop and water distribution; use of low salinity water and salt-tolerant crops) to 2.0, which results in an LF (defined as the ratio between drainage water and applied irrigation water) from 13% until 50%. The safety irrigation coefficient could be optimized by an analytic calculation of the optimal LF according to the electrical conductivity of irrigation water (EC_I) and crop tolerance to salinity, as discussed in Section 3.2.1.

4.2 Irrigation turn or irrigation frequency

The irrigation frequency (I_F) is the number of irrigation events in a given period (generally one day or one week), while the irrigation interval (I_I) is the time (in days or hours) between two irrigation events.

In both cases, a precise hourly or daily knowledge of the crop evapotranspiration (ET_C, expressed as mm or L m^{-2}) is required:

$$I_F\left(\text{number of watering per day}\right)=\frac{E_{TC}}{I_{NET}} \quad (7)$$

$$I_I\left(\text{hours or days}\right)=\frac{I_{NET}}{ET_C} \quad (8)$$

Crop evapotranspiration is affected by many factors, either environmental (e.g. air temperature, radiation, humidity, wind speed) or plant related (e.g. growth phase, leaf area). Any method used for an accurate estimate of plant water requirements must take these environmental and plant factors into account. During the last few decades, many approaches have been proposed and tested in order to obtain robust and effective irrigation scheduling. These methods could be divided into four main groups according to how the ETc was estimated (see Table 5): (1) timer-based; (2) sensor and gravimetric-based (direct measurement of ET_C); (3) climate-based; and (4) 'speaking plant'. The first method is quite empiric and requires extensive experience of the grower. The methods belonging to the second group are based on a direct measurement of the ET_C and are generally more expensive, but easier to manage at the farm level. The third group involves methods based on an indirect measure of ET_C using models that are able to describe the ET_C through the measurements of various plant and climate parameters. Finally, in the last group the ET_C is determined by the measurement of transpiration or plant water status.

The *timer-based method* is a very cheap and easy approach to automatize irrigation, especially in soilless culture, where many irrigation events are necessary each day: it consists of the use of timers for many irrigation sectors, able to turn on and off with a pre-set time and irrigation length. The system is based on the use of highly porous media that are able to easily discharge the surplus of water supplied with irrigation. This represents the most used scheduling system, but is not very efficient in detecting the variation in ET_C during the day, resulting in high drainage percentage (sometimes above 40–50%) not justified by the water quality. The irrigations

Table 5 Values for the coefficients A and B of Baille equation (Eq. (12)) reported in the literature for some greenhouse crops

Crop	Growing conditions	LAI	A	B	Reference
Begonia	Pot plants	2.7	0.20	0.026	Baille et al. (1994)
Cyclamen		2.9	0.32	0.019	
Hibiscus		2.4	0.37	0.037	
Impatiens		5.1	0.67	0.013	
Pelargonium		5.7	0.61	0.017	
Poinsettia		2.0	0.12	0.017	
Schefflera		4.4	0.60	0.014	
Gardenia		4.5	0.46	0.019	
Gardenia		6.6	0.53	0.013	
Cucumber	Mediterranean regions (Almeria); autumn and spring; perlite pot substrate	0.5-2.6	0.26	0.034	Medrano et al. (2005)
			0.42	0.042	
			0.24	0.032	
			0.24	0.055	
Geranium	Mediterranean regions (Spain)	2.5	0.56	0.018	Montero et al. (2001)
Zucchini	Mediterranean regions (Italy); autumn and spring; pumice culture	0.5-5.5	0.63	0.009	Rouphael and Colla (2004)
Gerbera	Mediterranean regions (Almeria, Spain); autumn and spring; semi-closed rockwool culture	1.0-2.2	0.55	0.019	Carmassi et al. (2013)
Rose	Mediterranean regions (Greece); perlite pot culture	2.5-3.5	0.236	0.026	Kittas et al. (1999)
Tomato	Mediterranean regions (Spain); autumn and spring; perlite culture	2.5	0.580	0.025	Medrano, *pers. comm.*

must be distributed over 24 h, taking into account that 70%, 23% and 7% of daily ET_C is evapotranspirated respectively during the sunny hours of the day (10.00 am-6.00 pm), in the early morning and late afternoon, and, also, during the night.

4.3 Direct determination of crop evapotranspiration

An easy system to schedule irrigation is the direct and automatic measurement of the soil/substrate moisture variation in the root zone of some representative areas (sentinel plants) of the crop. The methods used are based on the measurement of the variation in weight (gravimetric method) of the root zone, the volumetric water content (dielectric sensors) or the water matric potential (tensiometers).

4.3.1 Weighing system

Irrigation scheduling may be controlled by one or more devices that measure continuously the weight of the whole substrate-plant system. This system is based on the assumption that in a short timeframe, changes in weight due to plant growth are negligible with respect to evapotranspiration losses (Fig. 7). After the end of each irrigation event, the system records the actual weight of the whole plant-substrate-container system; the user must set the maximum value of weight loss (the net irrigation volume) after which a new irrigation event will start. In the literature different examples of plant weighing systems are reported for tomato (Takaichi et al., 1996), gerbera (Baas and Slootweg, 2004), cucumber and tomato (CropAssist®, Helmer et al., 2005), Boston and coral lettuce (Chen et al., 2016). The system is very good for small potted plants, while some problems are encountered when it is applied to species with large or tall canopies: in all these cases, the oscillation of the plant canopy can result in an unstable and erroneous measurement of the plant weight. In the last few years, this system has been re-evaluated by the growers thanks to the huge reduction in the cost of load cells, and the development of specific software able to process automatically the raw data collected from the weighing.

Figure 7 Weighing gutter for automated monitoring of crop evapotranspiration and irrigation control in tomato substrate culture. The suspended trough with some tomato plants is connected to a load cell: the difference in weight between two consecutive irrigations represents the crop evapotranspiration in this interval. Photo: A. De Koning, Hortimax, Pijnacker, NL.

4.3.2 Tensiometers and dielectric sensors

Soil moisture sensors (SMSs) can be divided into two main groups: (1) sensors able to measure the matric potential of the soil/substrate (i.e. water-filled tensiometers or porous matric sensors) and (2) sensors able to measure the volumetric water content (dielectric sensors) (Pardossi et al., 2009).

A conventional tensiometer is a tube filled with water with a porous cup that maintains equilibrium between the pressure inside the cup and the suction potential in the surrounding soil: the measure of the matric potential is obtained by measuring the depression at the upper part of the water column using a pressure transducer. The hydraulic tensiometer could measure a pressure range from 0 to a theoretical value of −80 kPa (saturation vapour pressure), below which the cavitation phenomenon occurs (Nolz et al., 2013). Tensiometers are good tools for regulating irrigation in the wet band of the soil, as in the case of vegetable crops or in soilless culture (here, the wilting point is −10 kPa). Despite their sound performance and high accuracy (appreciated by scientists), tensiometers are not much used by growers due to the necessity of regular maintenance and fragileness. The matric potential limit of hydraulic tensiometers was overcome by the use of electrical resistance sensors (the most famous and economic is the Watermark® by Irrometer company, Inc.), able to measure from −10 to −20 kPa (Centeno et al., 2010), and porous matric sensors able to measure from −50 to −300 kPa (Whalley et al., 2007). These sensors have the drawback of difficult calibration, and appear to respond slowly in rapidly drying soil (Thompson et al., 2007b). The main advantages of sensors measuring the matric potential is the ability to choose a reproducible set point value of matric potential for irrigation management.

The second group of SMSs are able to estimate the volumetric water content by measuring the dielectric permittivity of the material surrounding the sensor, which is subjected to an electromagnetic field. For this reason, they are also known as 'dielectric' sensors.

There are different types of dielectric sensors, which use different types of output signal for estimating the volumetric water content (θ), and are based on time domain reflectometry (TDR), frequency domain (FD) (reflectometry and capacitance), time domain transmission (TDT), amplitude domain reflectometry or phase transmission methods. All these sensors differ in terms of use and maintenance, calibration requirements, accuracy and price.

Expensive and complex SMSs, such as TDR instruments, are available for soil and plant scientists, while low-cost and practical devices are needed for irrigation control of commercial crops. New types of SMSs that measure soil dielectric properties have opened up interesting possibilities (Pardossi et al., 2009). In particular, new dielectric SMSs based on FDR or FRC are cheaper and

Figure 8 Two examples of dielectric sensors for measuring substrate water content: SM 200™ (left) and WET™ (right), both commercialized by Delta-T device Ltd (UK). The latter sensor also measures substrate salinity (bulk EC) and temperature.

need much less maintenance and user expertise than traditional water-filled tensiometers (Fig. 8).

The main drawback of these sensors is that their calibration changes as a function of substrate and soil characteristics, and requires knowledge of the MRC for the choice of the right water capacity threshold to start the irrigation event. In any case, the grower can overcome this problem by an empirical, although quite effective, *in situ* calibration of the sensor: after placing the dielectric sensor in the soil, the grower observes the soil/substrate volumetric water content measured by the dielectric sensor during each irrigation event, in order to choose the right set point soil/substrate moisture value.

The use of dielectric sensors for fertigation management also makes possible the automatic modulation of fertilizer addition to the recirculating nutrient solution as well as the mix of different water quality sources.

Recently, an automated fertigation device was designed and successfully tested to modulate both irrigation frequency and EC of fertigation water, based on the simultaneous measurement of θ and pore water EC (EC_{PW}) of the growing medium by means of the WET sensor (Incrocci et al., 2010). Specific algorithms were implemented in the control software with the intention to activate irrigation when a pre-set θ threshold was reached and to modulate irrigation dose and/or nutrient solution EC (also by mixing different sources of water such as recirculated water, groundwater, rainwater etc.), to avoid salt accumulation in the substrate and minimize water drainage.

4.4 Modelling crop evapotranspiration in greenhouse

Crop evapotranspiration can be predicted by climatic (i.e. global radiation, air humidity and temperature, wind) and crop parameters (mainly leaf area index, LAI, and crop resistance) using different approaches as for example proposed by the FAO-56 methodology with the Penman-Monteith (PM) equation. This is commonly called the 'two-step approach', since the crop evapotranspiration (ET_C) is calculated through a previous estimation of the potential evapotranspiration (ET_0) and a successive adjustment obtained by multiplying the estimated value by a crop-specific coefficient (Kc):

$$ET_C = ET_0 \times K_C \tag{9}$$

Bacci et al. (2011) reported an exhaustive discussion on how to calculate ET_0 using the PM equation. A further useful equation for ET_0 calculation of crops with high LAI and plant density is the CIMIS equation (CIMIS, 2009), which runs on an hourly basis and therefore is more accurate than the PM equation under greenhouse or nursery conditions (Bacci et al., 2011).

In the past, the application of the FAO-ET model was strongly limited by scarce information on both daily ET_0 and crop coefficient. Today, the knowledge of the daily or hourly ET_0 is not a problem, since many commercial companies offer affordable weather stations, while the choice of the right crop coefficient still remains the main obstacle that limits the widespread use of this model.

In fact, crop coefficients depend on LAI but also on leaf stomatal resistance. This could be modelled as a function of some environmental condition, mainly irradiance and vapour pressure deficit. For example, Rouphael and Colla (2005) reported that in a pot-grown zucchini greenhouse cultivation, the stomatal resistance could be correlated to global radiation by an exponential model.

Consequently, many authors (e.g. Shuttleworth, 2007) have suggested that a better approach to ET_C evaluation would consist of directly calibrating the PM equation on the basis of crop evapotranspiration. It is called the 'one-step approach' in which the canopy surface resistance (r_s) of the crop would play the same role as the crop coefficient Kc present in the two-step approach.

4.4.1 PM model and simplified model: a case study of *Gerbera* evapotranspiration

Carmassi et al. (2013) compared the PM model, to estimate the evapotranspiration of a greenhouse gerbera crop under typical Mediterranean climate conditions, with the one-step PM model and other two simplified methods.

The PM equation is reported below:

$$\lambda \cdot ET_C = \frac{\Delta \cdot 0.981 I \cdot \left(1 - \exp^{-k \cdot LAI}\right)}{\Delta + \gamma \cdot \left(1 + r_c / r_a\right)^{*}} + \frac{\left(\frac{\rho \cdot c_p}{r_a}\right) \cdot \left(e_a^{*} - e_a\right)}{\Delta + \gamma \cdot \left(1 + r_c / r_a\right)} \quad (10)$$

where crop ET_C (kg m^{-2} s^{-1}) is based on ground area, λ (J kg^{-1}) is the latent heat of vapourization of water; I (W m^{-2}) is the global radiation over the crop; k (dimensionless, determined as 0.60 ± 0.02 with $n = 40$) is the light interception coefficient; LAI is the leaf area index; ρ (kg m^{-3}) is air density; c_p (J kg^{-1} K^{-1}) is the specific heat of air at constant pressure; e_a (kPa) is the air vapour pressure; e_a^{*} (kPa) is the saturated air vapour pressure; Δ (kPa °C^{-1}) is the slope of the linear relation of saturation vapour pressure versus temperature, γ (kPa °C^{-1}) is the psychrometric constant; and r_c and r_a (s m^{-1}) are, respectively, the mean canopy resistance and the aerodynamic resistance to vapour transfer.

Considering that in an unheated greenhouse I_n (net radiation) matches I during the light period (Baille et al., 1994), in Eq. (10), global radiation (I) was used instead of I_n. In another experiment, conducted in 2006 with gerbera grown in the same glasshouse (unpublished results), a close linear relationship was found between I_n and I ($I_n = 0.981\ I$; $R^2 = 0.902$; $n = 487$) in the range between 25 and 545 W m^{-2}. Leaf area index was estimated as a function of growing degree days (GDD) assuming a base temperature of 8°C, using the following experimental relationship (validated in the 1–2.4 LAI value range):

$$LAI = (1.043 - 2.448) \cdot \exp^{-0.0066 \cdot GDD} + 2.448 \quad (11)$$

In this experiment, no clear relationship between r_a and climatic variables was found, therefore the Ra was assumed as a constant value throughout the seasonal period. The PM calibrated equation was able to predict the *Gerbera* ET_C: the slope of the regression equation between predicted and measured data was nearly 1, the intercept was negligible and the R^2 was close to 0.90.

4.4.2 The simplified models for the prediction of crop evapotranspiration

The FAO PM equation is currently considered a standard reference (Stanghellini, 1987; Allen et al., 1998; Walter et al., 2002). However, its application is not straightforward as it requires the knowledge of several variables and parameters. Therefore, several authors proposed simplified equations for predicting ET as a function of LAI, intercepted radiation (MJ m^{-2} h^{-1}) and vapour pressure deficit (kPa).

Transpiration rate was also modelled using the regression equation proposed by Baille et al. (1994):

$$ET_C = A \cdot \frac{\left(1-\exp^{-k \cdot LAI}\right)}{\lambda} + B \cdot LAI \cdot VPD \tag{12}$$

Carmassi et al. (2013) also calibrated the above equation using the multiple regression method for the measured ET_C collected on cut flower gerbera against the variables $\left(1-\exp^{-k \cdot LAI}\right)$ (MJ m^{-2} h^{-1}) and (LAI·VPD). They obtained the values 0.546 and 0.019 respectively for coefficients A (dimensionless) and B (kg m^{-2} h^{-1} kPa^{-1}): LAI and k values were the same as used in the previous P-M calibration.

The low value of B coefficient in Eq. (12) suggests that in the Mediterranean area, especially in unheated greenhouses, the ET_C is driven mainly from the incident global solar radiation (radiative component), and thus the aerodynamic term of the Eq. (12) (LAI·VPD) could be excluded without substantial errors in the predicted values of ET_C. Upon this evidence, Carmassi et al. (2013) proposed and validated a simplified Baille's model on gerbera, where the A value could be calculated applying a linear regression (forced by the origin) of ET_C against the light effectively intercepted by the crop:

$$ET_C = A \cdot \frac{I \cdot \left(1-\exp^{-k \cdot LAI}\right)}{\lambda \cdot} \tag{13}$$

The equation predicted satisfactorily the ET_C during the day, while it produced some errors during the night: nevertheless, the formula is very easy to manage and requires only a radiometer to collect data.

Similar results were found on a semi-closed soilless tomato culture (Carmassi et al., 2007; Massa et al., 2011), using a similar equation to predict ET_C on a daily basis:

$$ET_C = A \cdot \left(1-e^{-k \cdot LAI}\right) \cdot \frac{I}{\lambda} + B \tag{14}$$

where A (0.946, dimensionless) and B (0.188 L m^{-2}) are empirical constants, k is the canopy light extinction coefficient (0.69, dimensionless; Carmassi et al., 2007) and λ (2.45 MJ kg^{-1}) is the latent heat of water vaporization.

In this case LAI was assumed to obey a sigmoid function of thermal time expressed as GDD:

$$\mathrm{LAI} = a_1 + \frac{(a_2 - a_1)}{1 + e^{\left(\frac{a_3 - \mathrm{GDD}}{a_4}\right)}} \quad (15)$$

where a_1 (−0.335), a_2 (4.803), a_3 (755.3) and a_4 (134.7) are regression coefficients.

GDD was computed since sowing from *T* using 8°C as basal temperature. Eq. (15) is valid for *GDD* ranging from 400 (approximately the value at transplanting) to 1600 and LAI up to 4.8.

The use of these simplified methods instead of PM equation aims to simplify the use as well as the auto-calibration for crop evapotranspiration. Many authors have confirmed the reliability of the Baille equation, as summarized in Table 5. In practice, in many commercial greenhouses, growers supply an irrigation volume when a cumulated global solar radiation threshold is reached, and then adjust this threshold according to the information obtained from the LF measurements.

4.4.3 Methods based on plant water status

An alternative approach to irrigation management is the monitoring of plant water status, known also as 'plant stress sensing' (Jones, 2004; Fernandez and Cuevas, 2010; Ruger et al., 2010; Cahn and Johnson, 2017). According to Jones (2004), the methods based on this approach can be divided into two categories: (1) methods measuring the plant water status either as visible wilting leaf thickness, fruit or stem diameter, or by means of pressure chamber, leaf pressure probe (ZIM pressure probe), psychrometer; and (2) methods measuring some physiological responses such as stomatal conductance (using porometer, thermal sensing camera or sap flow sensor) or growth rate.

The pressure chamber is the main system used for determining the water status of the plant. The method is based on the pressure necessary to counterbalance the xylem pressure of a leaf or leafy twig. The measurement is very simple, but is time consuming and destructive, so continuous recordings cannot be carried out. A new solution for the continuous measurement of turgor pressure of plant leaves over long periods of time was proposed by Zimmermann et al. (2013), with a non-invasive, magnetic leaf patch clamp pressure probe (also known as ZIM-probe, commercialized by YARA GmbH).

The leaf thickness approach consists of the measurement of leaf turgidity. The system was tested and can be used as an alarm for potential water deficits (e.g. Leaf-Sen Irrigation Systems, Givat Hayim Ichud, Israel).

The sap flow system was developed mainly for trees. Basically, a flow sap sensor consists of a central needle that produces a heat pulse, and two additional symmetrically placed needles that act as temperature sensors. The flow of the sap is estimated by the temperature ratio of the two sensing needles, following the release of a heat pulse by the heater needle (e.g. Dynamax, Houston, TX).

In addition, the stem diameter variation has interesting applications mainly for woody plants, but is not suitable for most greenhouse crops.

By measuring the ratio of heat transported to two symmetrically placed temperature sensors, the magnitude and direction of the water flux can be calculated.

The use of infrared thermal cameras for measuring temperature differences between the air and the crop canopy is used to detect the early symptoms of a DI. At present the use of these cameras to schedule irrigation does not appear appropriate for greenhouse crops, since when the crop canopy temperature starts to increase, the water stress phenomenon has already begun. On the other hand, these devices are very interesting as a preliminary alarm to prevent the effects of irrigation failure.

Until now, most methods based on plant water status are mainly used for research purposes, and their implementation in commercial farms is limited due to the expensive investment costs and/or by necessity of qualified technicians for their use.

Nowadays, the most important irrigation companies have integrated different kinds of plant water status sensors in a control system for the management of both irrigation and climate in greenhouse production systems (e.g. Hortimax, Pijnacker, the Netherlands; PhyTech, Yad Mordechai, Israel), together with other more traditional sensors such as tensiometers or dielectric sensors. In the future, thanks to a reduction in costs and developments in electronics, these kinds of sensors could be used extensively in commercial greenhouses.

5 Coupling crop management practices with IE

5.1 Vegetable grafting

In vegetable crops, grafting was essentially rediscovered in the past 20 years and expanded on a large scale as a sustainable alternative to soil sterilization by means of methyl bromide (Kumar et al., 2017). In spite of the use of grafting as agro-technology for reducing disease damage, recent research reported the potential of this crop management practice to alleviate several abiotic factors (Colla et al., 2011; Rouphael et al., 2012) including water stress (Kumar et al., 2017). This is especially important for greenhouse vegetable cropping systems characterized by a shallow-rooted system and their high crop water requirements (Rouphael et al., 2008). Therefore, understanding how grafted plants can minimize water use without significant reduction in crop productivity (increasing WUE) is a basic requirement for the continued success of this agronomic practice. In a recent greenhouse experiment, Liu et al. (2016) demonstrated that luffa rootstock (*Luffa cylindrica* Roem. cv. Xiangfei No. 236) exhibited higher crop growth parameters, in particular leaf area and dry biomass, leading to a higher instantaneous WUE (defined as the ratio of the CO_2 assimilation rate and transpiration rate) when grafted with its own scion or with cucumber (*Cucumis sativus* L. cv. Jinyan No. 4). The beneficial effect of grafting on several fruit vegetables such as tomato, mini-watermelon and pepper onto vigorous rootstocks was also reported in a set of studies conducted mainly under greenhouse conditions. For instance, Ibrahim et al. (2014) and Al-Harbi et al. (2016) reported an increase in yield and yield WUE in tomato (*Solanum lycopersicum* L.) cv. Faridah grafted onto the interspecific tomato hybrid 'Unifort' (*Solanum lycopersicum* L.; *Solanum pimpinellifolium* L.), grown under both full (100% of crop evapotranspiration) and DI regimes (80%, 60% and 40% of crop evapotranspiration). Similarly, in watermelon Rouphael et al. (2008) observed a significant increase in marketable yield and yield WUE (by 60% and 10%, respectively) in pumpkin rootstock (*Cucurbita maxima* Duch. × *Cucurbita moschata* Duch.; cv. PS1313) grafted mini-watermelon [*Citrullus lanatus* (Thunb.) Matsum. and Nakai; cv. Ingrid] than in the non-grafted control. Moreover López-Marín and co-workers (2017) assessed the morpho-physiological responses of greenhouse sweet pepper (*Capsicum annuum* L.) cv. Herminio non-grafted or grafted onto three hybrid rootstocks, Atlante, Creonte and Terrano, under two well-watered (100% of crop evapotranspiration) and water stress conditions (50% of crop evapotranspiration). Pepper yield was higher in the three grafting combinations than in non-grafted plants, regardless of the irrigation regime used (constitutive response). However, among the three hybrid rootstocks tested, Creonte was the most effective in improving yield and yield WUE of grafted plants.

The effectiveness of vegetable grafting using vigorous rootstocks to increase yield and yield WUE of fruit vegetables has been associated with several improved traits of grafted vegetables such as (1) more vigorous root system architecture in terms of density, area and length; (2) improved water and nutrient uptake and assimilation; (3) enhanced photosynthetic capacity and water relations; (4) more resilient oxidative defence system; (5) modulation of hormonal balance; and (6) long-distance movement of mRNAs, small RNAs and proteins (Albacete et al., 2015; Kumar et al., 2017).

5.2 Plant biostimulants

The use of plant biostimulants has also been proposed as a sustainable and meaningful approach to boost yield and WUE of greenhouse vegetables and ornamentals (Colla and Rouphael, 2015). The term 'Plant biostimulants' has been adopted by many scientists to denote '*substances and/or microorganisms applied to plants with the intention to enhance nutrition efficiency, abiotic stress tolerance and/or crop quality traits, regardless of its nutrients content*' (du Jardin, 2015). Plant biostimulants include bioactive substances (humic acids, protein hydrolysates and seaweed extracts) and microorganisms (mycorrhizal fungi and plant growth promoting rhizobacteria of strains belonging to the genera *Azospirillum*, *Azotobacter* and *Rhizobium* spp.). In the last few years, plant biostimulants have been claimed to activate, under a limited irrigation regime, a series of molecular and physiological mechanisms such as the enhancement of leaf-water relations, an increase in relative water content and pigments biosynthesis as well as reducing stomatal limitation, which in turn led to higher water uptakes (Colla et al., 2013, 2014, 2015a,b, 2017; Xu and Leskovar, 2015; Abd El-Mageed et al., 2017).

In a recent research study, Abd El-Mageed et al. (2017) reported that the foliar application of leaf extract of *Moringa oleifera* (3%) was effective in increasing WUE in zucchini squash grown under mild and moderate water stress conditions (60% and 80% of crop evapotranspiration, respectively). Similarly, foliar spraying or root drench application of brown seaweed (*Ascophyllum nodosum*) extracts also improved the yield WUE of important leafy greens such as spinach (Xu and Leskovar, 2015). A putative mode of action involved in the stimulation of crop performance and increasing WUE under water stress conditions might be the capability of maintaining higher chlorophyll synthesis and fluorescence, improvement of photosynthetic activity and leaf water relations as well as a reduction in stomatal limitations. Saving water and increasing WUE could also be attained by microbial biostimulants inoculation in particular with arbuscular mycorrhiza fungi. For instance, the WUE was higher in inoculated (*Funneliformis mosseae* BEG25, *Funneliformis geosporus* BEG11 or a 50:50 mixed inoculation treatment of both species) strawberry plants than in non-inoculated plants under water stress conditions (Boyer et al., 2015).

Similarly, inoculation with a mixed inoculum of *Glomus intraradices* and *Glomus mosseae* has been reported to improve yield and WUE by 20% in watermelon under water stress regimes compared to non-inoculated plants (Omirou et al., 2013). The increase in root growth, the improvement in root system morphology (root diameter, total root length and diameter) and the development of external mycelium may have enhanced nutrient availability, uptake and translocation of the inoculated plants (Rouphael et al., 2015).

6 Future trends and conclusion

The constant pressure on vegetable growers to produce food for an increasing world population while minimizing excessive use of water and consequently fertilizers is a major challenge for the scientific community, which is expected to develop sustainable approaches and tools that increase WUE. Future studies should focus on the optimized use of water resources through the application of strategies for collection, storage and reuse. For instance, a fundamental issue for increasing the WUE of greenhouse crops will be the application of accurate irrigation scheduling, especially in soil-grown crops and in soilless open systems. In actual fact, the main problem concerning irrigation scheduling still remains the correct calculation of crop evapotranspiration: different approaches such as the direct measurements of the moisture in the root zone, the crop water balance or the monitoring of plant water status are available, each having both advantages and disadvantages. In the near future, the development of low-cost technologies for measuring crop water status could be the most important challenge in irrigation management. Another important research field will be the development of decision support systems that could help the greenhouse growers to manage soil salinity and irrigation scheduling.

Nevertheless, the main constraints to significant improvements in WUE in greenhouse crops relate to the high cost of these modern technologies and lack of knowledge of their use by the end users. As a matter of fact, in most worldwide countries, growers still decide when to irrigate their crop depending on personal experience, or at fixed intervals. In this sense, the improvement of IE seems related to a reduction in the overall cost of these technologies and to the policies adopted for their dissemination and transfer to professional growers. Finally, potential cultural practices such as vegetable grafting and the use of natural plant biostimulants can also appreciably affect the WUE of greenhouse crops. However, the knowledge about the potential benefits derived from applications of these strategies is far from being clear and widespread.

In the coming few years, future research should focus on innovative, simple and sustainable tools to control soil/plant water status and also to shed light on the genotype × environment × management practice interactions to define and select the best combination(s) that are able to enhance the water use efficiency in vegetable cropping systems. The overall impact of this chapter could be of interest for all horticultural actors including researchers, extension specialists and growers dealing with the vegetable sectors by identifying recent scientific studies regarding tools and techniques to determine the amount and timing of irrigation and of water, and thus increasing water use efficiency.

7 References

Abd El-Mageed T., Semida W. M. and Rady M. M. (2017). Moringa leaf extract as biostimulant improves water use efficiency, physico-biochemical attributes of squash plants under deficit irrigation. *Agricultural Water Management* 193: 46–54.

Affi N., El-Fadl A., El-Otmani M., Benismail M. C. and El Mastor A. (2014). Improvement of tomato water productivity by the use of a partial rootzone drying irrigation strategy. *Acta Horticulturae* 1038: 359–66.

Al-Harbi A., Hejazi A. and Al-Omran A. (2016). Responses of grafted tomato (*Solanum lycopersicon* L.) to abiotic stresses in Saudi Arabia. *Saudi Journal Biology Science (online, in press)*. https://doi:10.1016/j.sjbs.2016.01.005.

Albacete A., Martinez-Andujar C., Martinez-Pérez A., Thompson A. J., Dodd I. C. and Pérez-Alfocea F. (2015). Unravelling rootstock × scion interactions to improve food security. *Journal of Experimental Botany* 66: 2211–26.

Ali M. H. and Talukder M. S. U. (2008). Increasing water productivity in crop production – A synthesis. *Agricultural Water Management* 95: 1201–13.

Allen R. G., Pereira L. S., Raes D. and Smith M. (1998). Crop evapotranspiration: Guidelines for computing crop water requirements. Irrigation Drainage Paper 56. Food and Agriculture Organization of the United Nations (FAO), Rome.

Amin M. S. M. and Ekhmaj A. I. M. (2006). DIPAC-drip irrigation water distribution pattern calculator. *7th International Micro-Irrigation Congress.* Kuala Lumpur (Malaysia).

Ayers R. S. and Westcot D. W. (1985). Water quality for agriculture. FAO Irrigation and Drainage Paper No. 29 (Rev. 1). FAO, Rome, Italy, 176pp.

Baas R. and Slootweg C. (2004). On-line acquisition of plant related and environmental parameters (plant monitoring) in gerbera: Determining plant responses. *Acta Horticulturae* 654: 139–46.

Bacci L., Battista P., Cardarelli M., Carmassi G., Rouphael Y., Incrocci L., Malorgio F., Pardossi A., Rapi B. and Colla G. (2011). Modelling evapotranspiration of container crops for irrigation scheduling. In: Gerosa G. (Ed.), *Evapotranspiration – From Measurements to Agricultural and Environmental Applications*. InTech. ISBN: 978-953-307-512-9.

Baiamonte G., Provenzano G. and Rallo G. (2015). Analytical approach determining the optimal length of paired drip laterals in uniformly sloped fields. *Journal of Irrigation and Drainage Engineering* 141: 04014042.

Baille M., Baille A. and Laury J. C. (1994). A simplified model for predicting evapotranspiration rate of nine ornamental species vs climate factors and leaf area. *Scientia Horticulturae* 59: 217–32.

Bar-Yosef B. (2008). Fertigation management and crops response to solution recycling in semi-closed greenhouses. In: Raviv M. J. and Lieth H. (Eds), *Soilless Culture: Theory and Practise*. Elsevier, Amsterdam, The Netherlands, pp. 343–424.

Barbieri G. and De Pascale S. (1992). Salinità dell'acqua di irrigazione e colture ortofloricole. *Colture Protette* 2: 75–81.

Baudoin W., Nono-Womdim R., Lutaladio N. and Hodder A. (2013). Good Agricultural Practices for greenhouse vegetable crops: Principles for Mediterranean climate areas. FAO Food and Agriculture Organization of the United Nations, Roma, Italy.

Boyer L. R., Brain P., Xu X. M. and Jeffries P. (2015). Inoculation of drought-stressed strawberry with a mixed inoculum of two arbuscular mycorrhizal fungi: Effects on population dynamics of fungal species in roots and consequential plant tolerance to water deficiency. *Mycorrhiza* 25: 215–27.

Cahn M. D. and Johnson L. F. (2017). New approaches to irrigation scheduling of vegetables. *Horticulturae* 3(2): 28.

Cardarelli M., Rouphael Y., Rea E., Salerno A. and Colla G. (2010). Nutrient solution concentration and growing season affect growth and quality of potted petunia in a recirculating subirrigation and drip-irrigation system. *Journal of Food, Agriculture and Environment* 8: 312–20.

Carmassi G., Incrocci L., Maggini R., Malorgio F., Tognoni F. and Pardossi A. (2007v An aggregated model for water requirements of greenhouse tomato grown in closed rockwool culture with saline water. *Agricultural Water Management* 88: 73–82.

Carmassi G., Bacci L., Bronzini M., Incrocci L, Maggini R., Bellocchi G., Massa D. and Pardossi A. (2013). Modelling transpiration of greenhouse gerbera (*Gerbera jamesonii* H. Bolus) grown in substrate with saline water in a Mediterranean climate. *Scientia Horticulturae* 156: 9–18.

Cassaniti C., Romano D. and Flowers T. J. (2012). The response of ornamental plants to saline irrigation water. In: Garcia-Garizabal, I. (Ed.), *Irrigation Water Management, Pollution and Alternative Strategies*. InTech, Rijeka, pp. 131–58

Centeno A., Baeza P. and Lissarrague J. R. (2010). Relationship between soil and plant water status in wine grapes under various water deficit regimes. *HortTechnology* 20, 585–93

Chen W. T., Yeh Y. H. F., Liu T. Y. and Lin T. T. (2016). An automated and continuous plant weight measurement system for plant factory. *Frontiers in Plant Science* 7: 392.

CIMIS. (2009). *PM equation*. Available from: http://wwwcimis.water.ca.gov/cimis/infoEtoPmEquation.jsp.

Colla G. and Rouphael Y. (2015). Biostimulants in horticulture. *Scientia Horticulturae* 196: 1–2. https://doi.org/10.1016/j.scienta.2015.10.044.

Colla, G., Rouphael, Y., Mirabelli, C., and Cardarelli, M. (2011). Nitrogen-use efficiency traits of mini-watermelon in response to grafting and nitrogen-fertilization doses. *Journal of Plant Nutrition and Soil Science* 174: 933–94. doi:10.1002/jpln.201000325.

Colla G., Svecova E., Cardarelli M., Rouphael Y., Reynaud H., Canaguier R., and Planques B. (2013). Effectiveness of a plant-derived protein hydrolysate to improve crop performances under different growing conditions. *Acta Horticulturae* 1009: 175–9.

Colla G., Rouphael Y., Canaguier R., Svecova E. and Cardarelli M. (2014). Biostimulant action of a plant-derived protein hydrolysate produced through enzymatic hydrolysis. *Frontiers in Plant Science* 5, 1–6.

Colla G., Rouphael Y., Di Mattia E., El-Nakhel C. and Cardarelli M. (2015a). Co-inoculation of Glomus intraradices and Trichoderma atroviride acts as a biostimulant to promote growth, yield and nutrient uptake of vegetable crops. *Journal of the Science of Food and Agriculture* 95: 1706–15.

Colla G., Nardi S., Cardarelli M., Ertani A., Lucini L., Canaguier R. and Rouphael Y. (2015b). Protein hydrolysates as biostimulants in horticulture. *Scientia Horticulturae* 96: 28–38.

Colla G., Hoagland L., Ruzzi M. Cardarelli M., Bonini P., Canaguier R. and Rouphael, Y. (2017). Biostimulant action of protein hydrolysates: Unraveling their effects on plant physiology and microbiome. *Frontiers in Plant Science* 8: 2202.

Costa J. M., Ortuño M. F. and Chaves M. M. (2007). Deficit irrigation as a strategy to save water: Physiology and potential application to horticulture. *Journal of Integrative Plant Biology* 49: 1421–34.

Dannehl D., Suhl J., Huyskens-Keil S., Ulrichs C. and Schmidt U. (2014). Effects of a special solar collector greenhouse on water balance, fruit quantity and fruit quality of tomatoes. *Agricultural Water Management* 134: 14–23.

Dasberg S. and Or D. (1999). *Drip Irrigation*. Springer-Verlag Berlin Heidelberg.

Dasgan H. Y., Kusvuran S. and Kirda C. (2009). Effects of short duration partial rootzone drying (PRD) on soilless grown tomato crop. *Journal of Food, Agriculture and Environment* 7: 83–91.

Dasgan H. Y., Kusvuran S. and Kirda C. (2012). Use of short duration partial root drying (PRD) in soilless grown cucumber by 35% deficit irrigation. *Acta Horticulturae* 927: 163–70.

De Boodt M. and Verdonck O. (1972). The physical properties of the substrate in horticulture. *Acta Horticulturae* 26:37–44.

De Pascale S., Dalla Costa L., Vallone S., Barbieri G., and Maggio A. (2011). Increasing water use efficiency in vegetable crop production: From plant to irrigation systems efficiency. *HortTechnology* 21: 301–7.

De Pascale S., Rouphael Y., Pardossi A., Gallardo M. and Thompson R. B. (2017). Recent advances in water and nutrient management of soil-grown crops in Mediterranean greenhouses. *Acta Horticulturae* 1170: 31–44.

De Pascale S., Rouphael Y., Gallardo M. and Thompson R. B. (2018). Water and fertilization management of vegetables: State of art and future challenges. *European Journal of Horticultural Science (in press).*

Dorji K., Behboudian M. H. and Zegbe-Domínguez J. A. (2005). Water relations, growth, yield, and fruit quality of hot pepper under deficit irrigation and partial rootzone drying. *Scientia Horticulturae* 104: 137–49.

Du Jardin P. (2015). Plant biostimulants: Definition, concept, main categories and regulation. *Scientia Horticulturae* 196: 3–14.

Ehret D., Alsanius B., Wohanka W., Menzies J. and Utkhede R. (2001). Disinfestation of recirculating nutrient solutions in greenhouse horticulture. *Agronomie* 21: 323–39.

Evans R. G. and Sadler E. J. (2008). Methods and technologies to improve efficiency of water use. *Water Resources Research* 44(7).

Fernandez J. E. and Cuevas M. V. (2010). Irrigation scheduling from stem diameter variations: A review. *Agriculture, Forest and Meteorology* 150:135–51.

Food and Agriculture Organization of the United Nations. 2000. Deficit irrigation practices. 23 December 2010. http://www.fao.org/docrep/004/y3655e/y3655e00.htm

Golldack D., Li C., Mohan H. and Probst N. (2014). Tolerance to drought and salt stress in plants: Unraveling the signaling networks. *Frontiers in Plant Science* 5: 1–10.

González A. M., Bonachela S. and Fernández M. D. (2009). Regulated deficit irrigation in green bean and watermelon greenhouse crops. *Scientia Horticulturae* 122: 527–31.

Gough C. and Hobson G. E. (1990). A comparison of the productivity, quality, shelf-life characteristics and consumer reaction to the crop from cherry tomato plants grown at different levels of salinity. *Journal of Horticultural Science* 65: 431–9.

Graifenberg A., Lipucci di Paola M., Giustiniani L. and Temperini O. (1993). Yield and growth of globe artichoke under saline-sodic conditions. *HortScience* 28: 791–3.

Graifenberg A., Botrini L., Giustiniani L. and Lipucci Di Paola M. (1996). Yield, growth and element content of zucchini squash grown under saline-sodic conditions. *Journal of Horticultural Science* 71: 305–11.

Guang-Cheng S., Na L., Zhan-Yu Z., Shuang-En Y. and Chang-ren C. (2010). Growth, yield and water use efficiency response of greenhouse-grown hot pepper under Time-Space deficit irrigation. *Scientia Horticulturae* 126: 172–9.

Helmer T., Ehret D. L. and Bittman S. (2005). CropAssist, an automated system for direct measurement of greenhouse tomato growth and water use. *Computers and Electronics in Agriculture* 48: 198–215.

Hsiao T. C., Steduto P. and Fereres E. (2007). A systematic and quantitative approach to improve water use efficiency in agriculture. *Irrigation Science* 25: 209–31.

Ibrahim A., Wahb-Allah M., Abdel-Razzak H. and Alsadon A. (2014). Growth, yield, quality and water use efficiency of grafted tomato plants grown in greenhouse under different irrigation levels. *Life Science Journal* 11: 118–26.

Incrocci L. and Riccò E. (2004). Impianti per l'irrigazione e per la fertirrigazione. In: Pardossi A., Incrocci L. and Marzialetti P. (Eds), *Uso razionale delle risorse nel florovivaismo: l'acqua*. Quaderno ARSIA 5/2004, 107–24 (in Italian).

Incrocci L., Malorgio F., Della Bartola A. and Pardossi A. (2006). The influence of drip irrigation or subirrigation on tomato grown in closed-loop substrate culture with saline water. *Scientia Horticulturae* 107: 365–72.

Incrocci L., Marzialetti P., Incrocci G., Balendonck J., Spagnol S. and Pardossi A. (2010). Application of WET sensor for management of reclaimed wastewater irrigation in container-grown ornamentals (*Prunus laurocerasus* L). In: Paltineau I. C. and Muñoz J. V. (Eds). Transactions of the '*Third International Symposium on Soil Water Measurement Using Capacitance, Impedance and TDT*'. Murcia, Spain, 7–9 April 2010. Paper 2.8, p. 12.

Incrocci L., Marzialetti P., Incrocci G., Di Vita A., Balendonck J., Bibbiani C., Spagnol S. and Pardossi A. (2014). Substrate water status and evapotranspiration irrigation scheduling in heterogenous container nursery crops. *Agricultural Water Management* 131: 30–40.

Incrocci L., Massa D. and Pardossi A. (2017). New trends in the fertigation management of irrigated vegetable crops. *Horticulturae* 3: 37.

Jones H. G. (2004). Irrigation scheduling: Advantages and pitfalls of plant-based methods. *Journal of Experimental Botany* 55: 2427–36.

Kaman H., Kirda C., Cetin M. and Topcu S. (2006). Salt accumulation in the root zones of tomato and cotton irrigated with partial root-drying technique. *Irrigation and Drainage* 55: 533–44.

Kipp J. A., Wever G. and De Kreij C. (2001). *International Substrate Manual*. Elsevier, The Netherlands.

Kirda C., Cetin M., Dasgan Y., Topcu S., Kaman H., Ekici B., Derici M. R. and Ozguven A. I. (2004). Yield response of greenhouse grown tomato to partial root drying and conventional deficit irrigation. *Agricultural Water Management* 69: 191–201.

Kittas C., Katsoulas N. and Baille A. (1999). Transpiration and canopy resistance of greenhouse soilless roses: Measurements and modelling. *Acta Horticulturae* 507: 61–8.

Kumar P., Rouphael Y., Cardarelli M. and Colla G. (2017). Vegetable grafting as a tool to improve drought resistance and water use efficiency. *Frontiers in Plant Science* 8: 1130.

Lee J. G., Lee B. Y. and Lee H. J. (2006). Accumulation of phytotoxic organic acids in reused nutrient solution during hydroponic cultivation of lettuce (*Lactuca sativa* L.). *Scientia Horticulturae* 110: 119–28.

Levidow L., Zaccaria D., Maia R., Vivas E., Todorovic M. and Scardigno A. (2014). Improving water-efficient irrigation: Prospects and difficulties of innovative practices. *Agricultural Water Management* 146: 84–94.

Liu S., Li H., Lv X., Ahammed G. L., Xia X., Zhou J, Shi K., Asami T., Yu J. and Zhou Y. (2016). Grafting cucumber onto luffa improves drought tolerance by increasing ABA biosynthesis and sensitivity. *Scientific Reports* 6: 1.

López-Marin J., Gálvez A., del Amor F. M., Albacete A., Fernandez J. A., Egea-Gilabert C. and Pérez-Alfocea F. (2017). Selecting vegetative/generative/dwarfing rootstocks for improving fruit yield and quality in water stressed sweet peppers. *Scientia Horticulturae* 214, 9–17.

Maas E. V. (1986). Salt tolerance of plants. *Applied Agricultural Research* 1: 12-26.

Maas E. V. and Hoffman G. J. (1977). Crop salt tolerance: Current assessment. *Journal of Irrigation and Drainage Division* 103: 115-34.

Marcelis L. and Van Hooijdonk, J. (1999). Effect of salinity on growth, water use and nutrient use in radish (*Raphanus sativus* L.). *Plant and Soil* 215, 57.

Massa D., Incrocci L., Maggini R., Carmassi G., Campiotti C. A. and Pardossi A. (2010). Strategies to decrease water drainage and nitrate emission from soilless cultures of greenhouse tomato. *Agricultural Water Management* 97: 971-80.

Massa D., Incrocci L., Maggini R., Bibbiani C., Carmassi G., Malorgio F. and Pardossi A. (2011). Simulation of crop water and mineral relations in greenhouse soilless culture. *Environmental Modelling and Software* 26: 711-22.

Medrano E., Lorenzo P., Sanchez-Guerrero M. C. and Montero J. I. (2005). Evaluation and modelling of greenhouse cucumber-crop transpiration under high and low radiation conditions. *Scientia Horticulturae* 105:163-75.

Molden D. 2003. A water-productivity framework for understanding and action. In: Kijne J. W., Barker R. and Molden D. (Eds), *Water Productivity in Agriculture: Limits and Opportunities for Improvement*. International Water Management Institute, Colombo, Sri Lanka, pp. 1-18.

Montero J. J., Antón A., Muñoz P. and Lorenzo P. (2001). Transpiration from geranium grown under high temperatures and low humidities in greenhouses. *Agricultural and Forest Meteorology* 107: 323-32

Montesano F., Parente A. and Santamaria P. (2010). Closed cycle subirrigation with low concentration nutrient solution can be used for soilless tomato production in saline conditions. *Scientia Horticulturae* 124: 338-44.

Nolz R., Kammerer G. and Cepuder P. (2013). Calibrating soil water potential sensors integrated into a wireless monitoring network. *Agricultural Water Management* 116: 12-20.

Omirou M., Ioannides I. M. and Ehaliotis C. (2013). Mycorrhizal inoculation affects arbuscular mycorrhizal diversity in watermelon roots, but leads to improved colonization and plant response under water stress only. *Applied Soil Ecology* 63: 112-19.

Pardossi A. and Incrocci L. (2011). Traditional and new approaches to irrigation scheduling in vegetable crops. *HortTechnology* 21: 309-13.

Pardossi A., Incrocci L., Incrocci G., Malorgio F., Battista P., Bacci L., Rapi B., Marzialetti P., Hemming J. and Balendonck J. (2009). Root zone sensors for irrigation management in intensive agriculture. *Sensors (Basel Switzerland)* 9: 2809-35.

Pulupol L. U., Behboudian M. H. and Fisher K. J. (1996). Growth, yield, and postharvest attributes of glasshouse tomatoes produced under deficit irrigation. *HortScience* 31: 926-9.

Qiu Z., Yang Q. and Liu W. (2013). Photocatalytic degradation of phytotoxic substances in waste nutrient solution by various immobilized levels of nano-TiO_2. *Water, Air, & Soil Pollution* 224: 1.

Raviv M. and Lieth J. H. (2008). *Soilless Culture: Theory and Practice*. Elsevier, Amsterdam, The Netherlands.

Reed D. W. (1996). *A Grower's Guide to Water, Media, and Nutrition for Greenhouse Crops*. Ball Publishing, Batavia, IL.

Rouphael Y. and Colla G. (2004). Modelling the transpiration of a greenhouse zucchini crop grown under a Mediterranean climate using the Penman-Monteith equation and its simplified version. *Australian Journal of Agricultural Research* 55: 931–7.

Rouphael Y. and Colla G. (2005). Growth, yield, fruit quality and nutrient uptake of hydroponically cultivated zucchini squash as affected by irrigation systems and growing seasons. *Scientia Horticulturae* 105: 177–95.

Rouphael Y, Cardarelli M., Rea E., Battistelli A. and Colla G. (2006). Comparison of the subirrigation and drip-irrigation systems for greenhouse zucchini squash production using saline and non-saline nutrient solutions. *Agricultural Water Management* 82: 99–117.

Rouphael Y., Cardarelli M., Rea E. and Colla G. (2008a). The influence of irrigation system and nutrient solution concentration on potted geranium production under various conditions of radiation and temperature. *Scientia Horticulturae* 118: 328–37.

Rouphael Y., Cardarelli M., Colla G. and Rea E. (2008b). Yield, mineral composition, water relations, and water use efficiency of grafted mini-watermelon plants under deficit irrigation. *HortScience* 43: 730–6.

Rouphael Y., Cardarelli M., Rea E. and Colla G. (2012). Improving melon and cucumber photosynthetic activity, mineral composition, and growth performance under salinity stress by grafting onto Cucurbita hybrid rootstocks. *Photosynthetica* 50, 180–8.

Rouphael Y., Franken P., Schneider C., Schwarz D., Giovanetti M., Agnolucci M., De Pascale S., Bonini P. and Colla G. (2015). Arbuscular mycorrhizal fungi act as biostimulants in horticultural crops. *Scientia Horticulturae* 196: 91–108.

Rouphael Y., Kyriacou M. C. and Colla G. (2018). Vegetable grafting: A toolbox for securing yield Stability under multiple stress conditions. *Frontiers in Plant Science* 8: 2255.

Ruger S., Netzer Y., Westhoff M. D., Zimmermann D., Reuss R., Ovadiya S., Gessner P. and Zimmermann G. (2010). Remote monitoring of leaf turgor pressure of grapevines subjected to different irrigation treatments using the leaf patch clamp pressure probe. *Australian Journal of Grape and Wine Research* 16: 405–12.

Santamaria P., Ayala O., Buttaro D., Parente A. and Kirkayak B. (2007). N-rate and N-source for producing artichoke cv. concerto transplants in float bed under greenhouse conditions. *Acta Horticulturae* 730: 315–19.

Schwartzman M. and Zur B. (1986). Emitter spacing and geometry of wetted soil volume. *Journal of Irrigation and Drainage Engineering* 112: 242–53.

Shannon M. C. and Grieve C. M. (1999). Tolerance of vegetable crops to salinity. *Scientia Horticulturae* 78: 5–38.

Shao G.-C., Zhang Z.-Y., Liu N., Yu S.-E. and Xing W.-G. (2008). Comparative effects of deficit irrigation (DI) and partial rootzone drying (PRD) on soil water distribution, water use, growth and yield in greenhouse grown hot pepper. *Scientia Horticulturae* 119: 11–16.

Sharma H, Shukla M. K., Bosland P. W. and Steiner R. L. (2015). Physiological responses of greenhouse-grown drip-irrigated chile pepper under partial root zone drying. *HortScience* 50: 1224–9.

Shuttleworth W. J. (2007). Putting the 'vap' into evaporation. *Hydrology and Earth System Sciences* 11: 210–44.

Sirigu A., Mameli M. G., Chessa F and Meloni S. (2007). Effect of partial root zone drying on growth, yield and fruit quality in greenhouse tomato cultivation. *Acta Horticulturae* 747: 219–26.

Sonneveld C. (2000). Effect of salinity on substrate grown vegetables and ornamentals in greenhouse horticulture. PhD Thesis, Agricultural University, Wageningen, The Netherlands, 151p.

Stanghellini C. (1987). Transpiration of greenhouse crops: An aid to climate management. PhD thesis, Wageningen Agricultural University, Wageningen, The Netherlands, p. 150.

Takaichi M, Shimaji H. and Higashide T. (1996). Monitoring of the change in fresh weight of plants grown in water culture. *Acta Horticulturae* 440: 413–18.

Thompson R. B., Martınez-Gaitan C., Gallardo M., Giménez C. and Fernández M. D. (2007a). Identification of irrigation and N management practices that contribute to nitrate leaching loss from an intensive vegetable production system by use of a comprehensive survey. *Agricultural Water Management* 89: 261–74.

Thompson R. B., Gallardo M, Valdez L. C., Fernandez M. D. (2007b). Using plant water status to define threshold values for irrigation management of vegetable crops using soil moisture sensors. *Agricultural Water Management* 88: 147–58.

Thompson R. B., Delcour I., Berckmoes E. and Stavridou E. (2018). *The Fertigation Bible*. Available at: http://www.fertinnowa.com/the-fertigation-bible/.

Topcu S., Kirda C., Dasgan Y., Kaman H., Cetin M., Yazici A. and Bacon M. A. (2007). Yield response and N-fertiliser recovery of tomato grown under deficit irrigation. *European Journal of Agronomy* 26: 64–70.

Tripodi P., Massa D., Venezia A. and Cardi T. (2018). Sensing technologies for precision phenotyping in vegetable crops: Current status and future challenges. *Agronomy* 8.

van Halsema G. E. and Vincent L. (2012). Efficiency and productivity terms for water management: A matter of contextual relativism versus general absolutism. *Agricultural Water Management* 108: 9–15.

Van Os E. A. (2010). Comparison of some chemical and non-chemical treatments to disinfect a recirculating nutrient solution. *Acta Horticulturae* 843: 229–34.

Vázquez N., Pardo A., Suso M. L. and Quemada M. (2005). A methodology for measuring drainage and nitrate leaching in unevenly irrigated vegetable crops. *Plant and Soil* 269: 297–308.

Vázquez N., Pardo A., Suso M. L. and Quemada M. (2006). Drainage and nitrate leaching under processing tomato growth with drip irrigation and plastic mulching. *Agriculture Ecosystems & Environment* 112: 313–23.

Walter I. A., Allen R. G., Elliott R, Itenfisu D., Brown P., Jensen M. E., Mecham B., Howell T. A., Snyder R. L., Eching S., Spofford T., Hattendorf M., Martin D., Cuenca R. H. and Wright J. L. (2002). *The ASCE standardized reference evapotranspiration equation*. Report Task Com. on Standardized Reference Evapotranspiration 9 July 2002, EWRI-Am. Soc. Civil. Engr., Reston, VA, USA.

Whalley W. R., Clark L. J., Take W. A., Bird N. R. A., Leech P. K., Cope R. E. and Watts C. (2007). A porous-matrix sensor to measure the matric potential of soil water in the field. *European Journal of Soil Science* 58, 18–25.

Xu C., and Leskovar D. (2015). Effects of A. nodosum seaweed extracts on spinach growth, physiology and nutrition valued under drought stress. *Scientia Horticulturae* 183, 39–47.

Yan D., Bai Z., Rowan M., Gu L., Shumei R. and Yang P. (2009). Biofilm structure and its influence on clogging in drip irrigation emitters distributing reclaimed wastewater. *Journal of Environmental Sciences* 21: 834–41.

Yang H., Du T., Qiu R., Chen J., Wang F., Li Y., Wang C., Gao L. and Kang S. (2017). Improved water use efficiency and fruit quality of greenhouse crops under regulated deficit irrigation in northwest China. *Agricultural Water Management* 179: 193–204.

Zimmermann U., Bitter R., Marchiori P. E. R., Rüger S., Ehrenberger W, Sukhorukov V. L. and Ribeiro R. V. (2013). A non-invasive plant-based probe for continuous monitoring of water stress in real time: A new tool for irrigation scheduling and deeper insight into drought and salinity stress physiology. *Theoretical and Experimental Plant Physiology* 25: 2–11.

Chapter 10

Advances in nutrient management in greenhouse cultivation

Neil S. Mattson, Cornell University, USA; and Christopher J. Currey, Iowa State University, USA

1 Introduction

It has long been known that plant productivity may be limited by availability of mineral elements (Liebig, 1862; Mitscherlich, 1909). The goal of greenhouse nutrient management is to efficiently deliver the essential mineral elements to plants in a way that optimizes crop health and quality and enhances postharvest life. Beyond carbon, oxygen, and hydrogen (supplied by the air and water), there are generally considered to be at least 14 mineral elements which must be supplied to plant roots: nitrogen (N), phosphorus (P), potassium (K), calcium (Ca), magnesium (Mg), sulfur (S), chlorine (Cl), iron (Fe), boron (B), manganese (Mn), zinc (Zn), copper (Cu), molybdenum (Mo), and nickel

http://dx.doi.org/10.19103/AS.2019.0052.13

(Ni). Collectively, these mineral elements must be supplied to the plant from the water source, fertilizer, or the substrate. In addition to these 14 essential elements, cobalt (Co) appears to be involved in N-fixing microorganisms associated with legumes (Jayakumar, 2008) and as we will discuss later in this chapter there are additional beneficial elements. Because greenhouse crops are intensively managed, they necessitate high water and nutrient inputs, for example year-round intensive tomato production consumes 1185 and 284 $kg \cdot ha^{-1}$ of N and P, respectively (Sonneveld and Voogt, 2009). A recent survey of greenhouse container ornamental producers in Maryland found they used on average 198 and 53 $kg \cdot ha^{-1}$ annually of N and P, respectively (Majsztrik et al., 2018). Careful management is required to limit environmental nutrient runoff (especially N and P) and to comply with local regulatory requirements. In the study by Majsztrik et al. (2018), a wide variation in fertilizer usage even within crop type suggested that adoption of best management practices would help further reduce nutrient inputs. This chapter addresses advances in nutrient management of soilless greenhouse crops including commonly used practices in commercial greenhouse production and with particular emphasis on new research trends and bottlenecks in research and commercialization.

2 Controlled-release fertilizers

Controlled- (CRF) or slow-release fertilizers (SRF) are fertilizers packaged as small prills, incorporated into growing substrate prior to planting or placed on the substrate surface after planting. Both CRFs and SRF are fundamentally similar: they mete out nutrients over a defined period of time. However, prill coatings, fertilizer analyses, and estimated duration of release vary widely among products.

The most common CRFs are polymer-coated fertilizers (Nelson, 2012). Once polymer-coated CRF prills are hydrated, the fertilizer solution inside the prill is released through the porous membrane, with the rate of release being positively related to substrate temperature, whereby water temperatures enhance the nutrient-release rate. While most commercial CRFs use polymer coatings, some prills are coated with sulfur; as microorganisms oxidize the prill coating, nutrients are hydrated upon exposure to water and released. These SRF have a less-predictable release than polymer-coated CRFs.

The nutrient-release characteristics of CRF impact fertilizer utility and nutrient use. CRF are formulated with a range of target fertilizer release time, ranging from as short as 14 to 21 days up to 270+ days. Utilizing different release rates allows specific needs for different crops and cropping systems. For example, a 60-day release may be suitable for herbaceous annual plants in small containers, while a 180-day release is more useful for semi-woody plants with a long production time or life cycle (Pilon, 2006). When using CRFs with longer release periods, greater fertilizer amounts are usually necessary to provide sufficient nutrients to plants earlier in production when less total cumulative fertilizer is minimal, especially when compared to an equivalent amount of fertilizer of the same analysis with a shorter release period (Andriu, 2013a).

Across fertilizer manufacturers, there can be variation in release patterns even for products with similar durations of nutrient release (Adams et al., 2013), so variation among comparable products from commercial manufacturers should be considered when using CRFs. In addition to the overall or entire period of nutrient release, CRFs may provide specially designed release patterns, such as those formulations which delay release until later in the release period, or those which release more fertilizer in the initial phases (Nelson, 2012).

In addition to the analysis and release period, the physical size of prills vary, with regular or standard sized prills, as well as smaller 'micro' prills (Khadduri, 2014). These small prills are useful for incorporating CRF into substrate to fill small containers such as small containers, cell packs, or plug trays to ensure more even distribution of nutrients across containers to improve crop uniformity, and they may be used successfully to produce rooted cuttings when incorporated into substrate prior to propagation (Currey and Lopez, 2014).

3 Water-soluble fertilizers

The most common method of providing mineral nutrients to greenhouse crops is through water-soluble fertilizers (WSF) provided when crops are irrigated. WSF are comprised of one or more salts and are dissolved to a specific concentration. Single-salt fertilizers such as calcium nitrate (NO_3) providing a limited number of elements may be used. However, most commonly, WSF are formulated using multiple salts to produce formulations that are complete, including all or most primary and secondary macronutrients and micronutrients, and are balanced, with respect to ratios (by mass) of specific nutrients (i.e., N:K, Ca:Mg, macro:micronutrients, etc.). Fertilizers may be mixed on-site by using an array of different salts and compounds. While this approach requires more labor and skill than other methods, it allows for complete flexibility in creating desired nutrient solutions which may be essential for conducting research. More commonly, premixed commercial formulations are selected to suit crop needs. A comprehensive list of commonly available commercial fertilizer formulations may be found in Scoggins and Williams (2018) detailing the composition of a wide assortment of commercial formulation. These fertilizers vary in the proportion of N in different forms (NO_3^- and NH_4^+), specific nutrient ratios, and macro- and micronutrient composition.

The form of N in fertilizers is frequently used to help manage root-zone pH (Johnson et al., 2013). Those formulations with more NH_4^+ -nitrogen have the potential to decrease root-zone pH. These acidic fertilizers maintain a low pH for crop-specific requirements or for suppressing the pH-raising effects from irrigation water with high alkalinity. Alternatively, fertilizers with greater proportions of NO_3^- nitrogen can increase substrate pH, and these high-NO_3^- fertilizers are used when irrigation water alkalinity is low (<40 mg·L^{-1}) or crops have a higher root-zone pH requirement.

The balance or ratio (by mass) of different nutrients in fertilizers is another important consideration. For example, N and K are commonly kept in a 1:1 ratio for general production of container-grown floriculture crops, while a 2:1 ratio may be used for foliage plants where vegetative growth is desirable or a 1:2 ratio may be used for cyclamen due to their higher K requirements (Nelson, 2012). The Ca:Mg ratio is another important ratio to consider, and targeting 3–5 ppm Ca for every ppm Mg can avoid unwanted antagonisms in uptake between these divalent cations. Traditionally, the N:P ratio is from approximately 2:1 (i.e., 20-8.7-16.6) to 4:1 (i.e., 15-2.2-12.5). However, with a growing realization that P rather than nitrogen form has a larger influence on stem elongation (Nelson et al., 2012) has led this ratio to be reexamined. Restricting P can be an effective height control method for container-grown herbaceous plants (Justice, 2014; Henry et al., 2017, 2018). As a result, fertilizers are being formulated to have P concentrations resulting in N:P concentrations much lower than traditionally used, from 1:0.05 to 1:0.20 (Peters, 2018). Micronutrient concentrations in these commercial formulations have increased overtime (Scoggins and Williams, 2018). Initially this occurred along with the adoption of soilless growing substrates comprised of peat and perlite, micronutrient concentrations in common formulations such as 15-2.2-12.5 and 20-4.4-16.6 approximately doubled, to provide the additional micronutrients no longer provided by mineral field soils. Recently, there are fertilizers being formulated with even greater micronutrient concentrations, as the trend to provide lower fertilizer concentrations to suppress growth and runoff may result in inadequate micronutrient provision to crops (Currey, 2019). In addition to fertilizers with higher concentrations of all micronutrients, recent fertilizer formulations are featuring higher Fe concentrations for the increasing popularity of taxa with greater Fe requirements. Another step toward better-managing nutrition for containerized crops growing in soilless mixes is the development of fertilizer formulations that incorporate phosphorus acidifying chemistry to manage root-zone pH for water sources high in alkalinity (Mattson, 2017).

While WSF can be mixed to the final desired concentration and applied directly to crops, they are commonly mixed at a strong concentration and, through the use of a fertilizer injector, diluted into irrigation water upon delivery to plants. The use of fertilizer injectors saves both time and space when providing mineral nutrients for crops. For many ornamental crops, stock fertilizer concentrations may be mixed into a single concentrated stock solution. However, fertilizer stock solutions for hydroponically grown food crops and cut flowers are commonly separated into two stock tanks–one for calcium nitrate and another for phosphates and sulfates–to avoid unwanted nutrient precipitation in stock tanks (Resh, 2013). A third stock tank may be used for pH adjustment. Additionally, new products such as supplemental potassium silicate, which is used to enhance plant quality and resiliency to biotic infestations (Kamenidou et al., 2008), must also be mixed in separately from other stock fertilizers to avoid Si precipitation.

4 Foliar fertilizers

Greenhouse fertilizers are sometimes applied as a foliar spray, typically, not as the primary fertilizer source but to complement substrate-applied fertilizer when plant or environmental conditions limit sufficient nutrient supply. Traditionally foliar fertilization delivers phloem-immobile nutrients to prevent or correct specific disorders. Examples of foliar fertilization include $CaCl_2$ sprays to prevent bract edge necrosis of poinsettias (Whipker, 2014) and Fe chelate sprays to correct high pH-induced Fe deficiency in calibrachoa and other Fe-inefficient plants (Fisher et al., 2003). In the above applications, foliar fertilization is a short-term solution and must be reapplied as new growth occurs. Foliar fertilization enhances fruit quality when root uptake and supply through xylem is insufficient to meet needs of rapidly developing fruit. Once-weekly foliar/fruit application of either KCl_2 chloride or a glycine amino acid-complexed K enhances netted muskmelon (*Cucumis melo* L.) fruit firmness, sugar content, and ascorbic acid (Lester et al., 2006). Foliar fertilization may also play a role in precision fertilization, reducing the quantity of substrate-applied fertilizer, which is more subject to environmental losses (Fernández and Eichert, 2009). Foliar-fertilized tomato growing in sawdust substrate provided with a blended macro- and micronutrient fertilizer every 2 weeks, allowed 75% of control fertigation to be used while maintaining yield and fruit nutritional quality (Maboko and Du Plooy, 2017). However, foliar fertilization has limited effects on mini-cucumber growing in sawdust, but does reduce fruit yellowing (Maboko et al., 2017). New fertilizer materials offer promise for more effective delivery/absorption of foliar fertilizers such as amino acid- or organic-chelate fertilizers (Souri and Aslani, 2018) or use of nano-fertilizers. Foliar-supplied nano-calcium (average particle size 50 nm) alleviates fruit cracking of pomegranate (*Punica granatum*) as effectively as traditional $CaCl_2$, but at five times lower Ca concentrations for nano-calcium (Davarpanah et al., 2018). While foliar fertilization is beneficial in many cases, there is a large variability in response and phytotoxicity can occur. More research is needed to understand which conditions (substrate, plant, and environment) are conducive to positive benefits (Fernández and Eichert, 2009).

5 Organic nutrient management of container-grown crops

Certified organic production of greenhouse crops requires use of fertilizer inputs compliant with national/regional organic standards. In the European Union, organic greenhouse production must take place in soil-based systems (Voogt, 2013), whereas soilless substrate production in containers are allowed in North America (Burnett et al., 2016). Organic fertilizer materials may be derived from animal or plant matter (often derived from waste streams) or from naturally occurring minerals (Treadwell et al., 2007). When animal-based inputs are used for food crops, applications must also meet food safety requirements. For example, the US National Organic Program requires raw manure be applied 90 days prior to harvest for crops that don't come in contact with soil and 120 days prior for crops that come in contact with soil (US FDA, 2018). Traditionally, single-source materials (such as alfalfa [*Medicago sativa*] meal, blood meal, feather meal, or compost) were used. Over time, blended organic fertilizers, in which several components are combined together to provide a more-balanced nutrient form or release period, have become more common (Möller and Schultheiß, 2015). Organic fertilizers are available in solid form and incorporated into a substrate prior to transplanting or top-dressed afterwards, as well as in liquid form as water-soluble or -dispersible products to be delivered through the irrigation system. In an analysis of 77 commercial organic fertilizers in Germany, single-source fertilizers showed high variability in composition of nutrients and heavy metals (Möller and Schultheiß, 2015), while pesticides and persistent organic pollutant concentrations were low.

Matching nutrient supply to plant demand is difficult, as release from organic fertilizers is less predictable than conventional materials. Some fraction of nutrients in organic fertilizers may be readily available, while other nutrients are bound in complex organic molecules which require microbial activity to mineralize. Relatively rapid initial N mineralization was found using a laboratory incubation method comparing fish powder, blood meal, feather meal, and seabird guano (Hartz and Johnstone, 2006). After 2 weeks at 20°C, 52–60% of nitrogen was mineralized, and thereafter mineralization rates slowed down; after 8 weeks of incubation, 30–40% of N was still remaining. Nitrogen form also changes over time. In a study with two organic fertilizer treatments (blood meal plus alfalfa meal and dried poultry manure), 1 week after planting both materials had high NH_4^+ content (100–130 mg·L^{-1}) but low NO_3^- content (<10 mg·L^{-1}); however, NH_4^+decreased rapidly over the next 4 weeks while NO_3^- increased (Bergstrand et al., 2018). In the study, the blood and alfalfa meal fertilizer was found to cause poor germination and slow growth of basil, likely due to high NH_4^+.

For short-term crops (such as herbs, flowers, or vegetable transplants), substrate-incorporated organic fertilizer prior to transplanting may be sufficient. For, longer-term crops (such as flowering potted plants, cut flowers, or fruiting vegetables) substrate-incorporated fertilizer is often not sufficient due to bottlenecks with how much one can initially add due to high salinity/ electrical conductivity (EC) if all the fertilizer were added at once or losses due to leaching and denitrification (Voogt, 2013). In this case, additional fertility is often resupplied during the crop cycle using top-dressing or fertigation. For example, for container-grown tomatoes, long-term tomato fruit yield of an organic fertilization treatment could match a conventional fertilizer treatment, when pre-plant amendment with a fertile mushroom compost as well as continuous organic fertigation regime are used (Zhai et al., 2009). In soil-based organic systems in the European Union, pre-plant amendment with animal manures and composts form the base fertilizer and these are complemented with commercial organic fertilizers, both pre-plant and as post-plant top-dressing or liquid application (Zikeli et al., 2017). It has been proposed that no more than 50% of nutrients should be provided after planting and a maximum of 25% of the nutrients from liquid form (Zikeli et al., 2017).

There are several limitations to organic nutrient management. Correction of specific nutrient deficiencies is difficult in organic production systems. For example, limestone or gypsum is allowed to correct for Ca deficiency but these materials are slowly available in substrates (Voogt, 2013). Another common issue is Fe or Mn deficiency. While $FeSO_4$ or $MnSO_4$ iron or manganese sulfate applications are allowed (if a deficiency is documented) often the primary cause of deficiency is high substrate pH. As many organic fertilizers have a neutral or basic pH, high substrate pH is a common problem and corrective options are limited (such as elemental sulfur or naturally fermented citric acid in the United States). In a study of German organic soil-based greenhouse vegetable operations, common issues included nutrient imbalances, especially a K deficit and P and Na accumulation, and increasing soil pH (Zikeli et al., 2017). Clearly, there is need for a greater toolkit of materials such as to correct for individual nutrient deficiencies as well as to decrease soil pH. Monitoring plant nutrient status in organic production systems includes similar tools to conventional production including periodic substrate and tissue analysis. One exception is that EC does not closely track fertility in organic systems because nutrients bound in uncharged organic forms do not impact EC (Möller and Schultheiß, 2015). The EC should still be tracked as an indicator or nutrient availability (Scoggins and Williams, 2018) and to indicate the need for leaching due to high salts. We expect use of organic fertilizers in greenhouse crops to increase over time to both meet market demand for organic produce but also to mitigate environmental issues with conventional fertilizers including energy intensive production (nitrogen) and declining availability of mined materials (ex: phosphorus). Organic cultivation of greenhouse crops is further elaborated elsewhere in this book.

6 Open vs closed irrigation systems

Irrigation systems greatly impact nutrient management practices. In open irrigation systems, excess irrigation water leaches through the substrate and is allowed to enter the environment. Examples of open irrigation include overhead hand watering, sprinklers, irrigation booms, or drip irrigation where water is not captured. In closed irrigation systems, water is captured and recirculated, and nutrients are not allowed to enter the environment. Examples of closed irrigation include subirrigation (flood floor, ebb-and-flow benches, troughs, and capillary mats), drip irrigation where leachate is captured, nutrient-film technique (NFT), and deep-water culture (DWC). Closed irrigation systems are viewed as environmentally friendly due to increased efficiency of water and fertilizer use and are in some regions being required due to environmental regulations (Katsoulas and Voogt, 2013). However, it should be noted that a well-designed open drip irrigation system, such as one relying on microirrigation strategies based on substrate volumetric water content soil moisture sensors, could nearly eliminate nutrient leaching (Alem et al., 2015).

In open irrigation systems, excess irrigation water can be used to leach soluble salts from a substrate, thereby allowing poorer-quality irrigation water or higher fertilizer concentrations to be used. In closed irrigation systems, more careful attention to water quality and nutrient management is required as salts can build up either in the substrate (as in subirrigation) or in the recirculating nutrient solution (as in closed drip irrigation, NFT, or DWC) (Rouphael and Colla, 2009; Stanghellini et al., 2005). In closed irrigation, supraoptimal fertilizer concentration is more likely to reduce plant performance vs. open irrigation due to salt accumulation (Ferrarezi et al., 2015; Liu et al., 2012). Because nutrients are not lost to the environment in subirrigation, fertilizer concentration can often be reduced by half without effecting yield and quality (Rouphael et al., 2008; Zheng et al., 2004), though this varies greatly by crop and environment (Ferrarezi et al., 2015) and the reference situation (currently used irrigation strategy and nutrient concentration).

In closed irrigation systems, nutrient solution concentration is typically monitored and adjusted by EC. However, nutrient imbalances can occur over time, therefore the elemental composition of the nutrition solution must be periodically monitored so that refinements to the nutrient solution can be made over time (Incrocci et al., 2017). Monitoring has traditionally used commercial analytical laboratories but this can be cost and time prohibitive. Alternative approaches include development of decision support systems (DSS) which simulate plant growth and nutrient uptake (Incrocci et al., 2017) or use of real-time measurement and adjustments using ion-selective electrodes (ISE) (Jung et al., 2015). An ISE approach with integrated controller was successful in control of NO_3^-, K, and Ca in an NFT lettuce system (Jung et al., 2015). ISE adoption by commercial industry has been limited due to cost, sensor life span, and frequent need for calibration requiring further research progress (Knight and Lefsrud, 2017).

In closed irrigation, Na and Cl can accumulate over time especially when poor-quality irrigation water is used resulting in crop yield losses. This accumulation necessitates periodic leaching of a fraction of the nutrient solution according to crop-specific EC thresholds (Stanghellini et al., 2005). Research into novel management techniques reducing the need to periodically discharge nutrient solutions due to accumulation of unwanted elements are required to fully realize the benefits of closed irrigation. Several crop management techniques can be used to mitigate salt accumulation and stress, such as decreasing vapor pressure deficit or CO_2 enrichment to limit transpiration (Katsoulas and Voogt, 2013). Real-time monitoring of sodium chloride or DSS tools to model their accumulation in a nutrient solution could result in more precise discharge (Massa et al., 2008; Varlagas et al., 2010). Longer-term, nutrient discharge may be reduced through novel approaches to selectively remove sodium and Cl as by ion-selective membranes (Qian et al., 2018) or use of halophytic plants in a constructed wetland (Rozema et al., 2014).

7 Nutrient management of crops in soilless substrates

As compared with soil, the use of soilless substrates allows for more careful control of water availability, pH, and nutrient concentration with the trade-off of reduced buffering capacity and increased management required due to a more limited root volume (Silber and Bar-Tal, 2019). As previously described, a range of nutrient supply strategies are employed, and for conventional producers this is most commonly through use of one or more concentrated WSF stock solutions diluted by fertilizer injectors. Soilless substrates vary greatly in their chemical properties which impacts WSF management. Further information on substrate properties are presented by Y. Zheng in Chapter 17 of this book. For relatively inert, low cation-exchange capacity (CEC) components such as stonewool and perlite, each irrigation cycle must supply the necessary nutrients. In substrates with high nutrient-retaining components, such as peat, coir, or bark/wood, more flexibility is available in nutrient delivery such as pre-plant incorporation of a starter nutrient charge, use of SRF, or WSF that does not need to be strictly applied at each irrigation cycle (Sonneveld and Voogt, 2009). In these substrates, the starter nutrient charge (e.g., base dressing) traditionally may contain dolomitic limestone (to increase pH of acidic substrates and contribute Ca and Mg), gypsum ($CaSO_4 \cdot 2H_2O$, to contribute Ca and S when no pH increase is desired), Epsom salts ($MgSO_4$, to contribute Mg and S when no pH change is required), triple superphosphate, and a micronutrient package (Scoggins and Williams, 2018). With the realization that P and N are readily leached, a SRF is sometimes added to the starter nutrient charge. The starter nutrient charge is prepared at lower nutrient concentrations for propagated plants and may be increased for longer-term higher fertility requiring crops (Sonneveld and Voogt, 2009). The use of wood substrates may require the use of a higher starter nutrient charge than peat-based substrates. For example, for marigold (*Tagetes patula* L.), a lower fertility requiring plant, a starter nutrient charge at 2.37 kg·m^{-3} with 7N–1.3P–8.3K, was sufficient for plants in either peat or WholeTree substrate made from loblolly pine (*Pinus taeda* L.); while for petunia (*Petunia* × *hybrida* Vilm.) a 50% greater starter nutrient charge was required for the WholeTree substrate than a peat substrate (Fain et al., 2008).

In commercial production of containerized ornamental plants (landscape annuals and perennials and indoor plants), often dozens (or even hundreds!) of species and cultivars are grown in the same operation. While nutrient demand can vary greatly by species/cultivar, it is often not practical to fertilize each crop individually. Most commonly, crops are grouped together according to fertilizer and pH requirements. For example, Straver et al. (1999) note a framework where crops are first placed in 11 groups based on specific nutrient requirements and nutrient ratios; second, plants are divided into three groups based on their salt sensitivity; and third, they are divided into five groups based on pH. Such a grouping could result in 165 categories, though it is noted that in practice many combinations are not required. A similar but more streamlined approach was presented by Scoggins and Williams (2018), whereby container ornamentals are divided into three pH groups and three groups based on fertilizer concentration. In this strategy, macronutrient ratios remain the same, though crops with substantially different macronutrient ratio requirements can be fertilized separately. In commercial practice, in these diverse operations there may be only a few different fertilizer regimes (often based on pH requirements) with reliance on CRF to amend the substrate of higher fertilizer requiring plants.

In fertilizer management of fruiting crops (cucumbers, peppers, and tomatoes), it is common to adjust the nutrient solution over time in synch with plant developmental stages. This approach is used to direct the plant toward reproductive growth as well as to adjust for changing nutritional needs of the plant as it develops. For example, Kroggel and Kubota (2018) suggest three different fertilizer formulations for tomato that alters the concentration and ratio of some nutrients based on developmental stage. The first stage (from germination until the second truss has open flowers) has a lower nitrogen concentration to 'steer' the plant away from vegetative growth. As the plant reaches fruiting stages, several adjustments are made which also has the net effect of increasing EC; N is increased to meet needs of a more mature plant, Ca is increased to prevent blossom end rot, and K increased to promote fruit quality (Kroggel and Kubota, 2018). A similar strategy, but with nutrient formulations for six different stages of tomato development, is presented by

Sonneveld and Voogt (2009). In practice, the fertilizer solution is also adjusted based on whether irrigation is open or closed. The EC nutrient solution target for closed irrigation is lower than the EC target in open irrigation with a standard 30% leaching fraction. As discussed in a later section, substrate and tissue monitoring is important to ensure a fertilizer program is meeting plant needs under its current environmental conditions.

8 Nutrient management of hydroponic crops in water culture

Water-culture hydroponic systems such as NFT and DWC or deep-flow technique (DFT) systems both use a continuously circulating nutrient solution as a primary substrate as well as for providing mineral nutrients and water to plants (Resh, 2013). For NFT systems, plant roots are continuously bathed in a small volume of nutrient solution which is delivered to the head of a channel placed on a 1–3% slope to promote solution flow. The maximum growing channel length for NFT systems is 15 m, as channels beyond this length may result in a gradient in mineral nutrient concentrations in nutrient solutions due to plant uptake (Resh, 2013). In DWC and DRF systems, seedlings are transplanted into rafts that float on nutrient solution in ponds or pools 10- to 30-cm deep. For DWC systems, nutrient solution composition is more homogenous across the growing area, which can be attributed to the larger volume of nutrient solution available per plant in DWC and DFT systems compared to those grown in NFT systems.

Nutrient solutions in both NFT and DWC systems are recirculated throughout production. After being captured and before being delivered to the crop again, nutrient solutions are adjusted with an acid or alkali to decrease or increase pH, while clear water is added to maintain the reservoir volume and stock fertilizer solutions are added to increase EC. Managing recirculating nutrient solutions varies between commercial production and laboratory research. Bugbee (2004) states that nutrient solutions require only periodic adjustment and that dumping solution may not be necessary for nutrient solutions used in hydroponic research. In commercial water-culture systems, leafy greens and herbs with staggered planting dates to provide continually available product are planted in the same production systems with a common nutrient solution. While periodic adjustments may suffice for crops with the same production schedule, continual adjustment is likely better-suited for systems with several, staggered plantings.

The effect of EC on crop growth in recirculating water-culture systems is conflicting. Ranges for target EC and pH are published for many lettuce, leafy green, and culinary herb species (Morgan, 2005, 2012); however, the importance of nutrient solution EC in recirculating culture is not clear. Walters and Currey (2018) and Currey et al. (2019) report nutrient solution EC between

0.5 and 4.0 $dS{\cdot}m^{-1}$, when consistently maintained at their setpoint, had no effect on fresh mass of culinary herbs grown in hydroponic systems when EC was adjusted daily to maintain target values. While others report increasing EC from 1.2 to 3.6 $dS{\cdot}m^{-1}$ promotes culinary herb growth (Udagawa, 2004), the EC was infrequently maintained and likely restricted growth compared to well-controlled systems restricted growth. The frequency of solution adjustment likely affects acceptable EC ranges, whereby lower-than-normal ranges can result in sufficient growth with frequent adjustments to maintain EC, whereas higher EC ranges may be required when solutions are infrequently to provide adequate nutrients to plants. Additionally, higher ECs may be recommended to maintain a critical concentration of a specific nutrient such as Ca or Mg, but this can lead to imbalances in nutrient ratios and deficiencies may result from antagonisms. As a result, adjusting nutrient solution formulations would be more efficient than simply increasing the total nutrient concentration.

One of the primary challenges in recirculating hydroponic nutrient solutions is maintaining appropriate concentrations of individual elements. As previously mentioned, solutions are recirculated and amended with one or two concentrated fertilizer solutions to maintain EC and acid and/or alkali to adjust pH in automated systems (Rius-Ruiz et al., 2014). The fertilizers added to increase solution EC usually contain several elements in fixed ratios. However, plant nutrient uptake does not occur in the same ratio that nutrients are replenished in, regardless of how well-designed or -matched a fertilizer is for a specific crop. Over time, as nutrients are used and replenished with stock fertilizer(s), imbalances in nutrient concentrations can develop. The development of ion-specific sensor for important elements such as NO3- -nitrogen, K, Ca, and Mg allows for precise maintenance and adjustment of these nutrient concentrations (Domingues, 2012; Kim et al., 2013).

9 The role of beneficial microbes and biostimulants in crop nutrient management

Beyond mineral elements, other materials such as microbes and biostimulants may be added to the root-zone/plant to enhance plant nutrient supply and growth. Varying definitions of biostimulants have been proposed and there is not currently a legal or regulatory framework for their registration and use in Europe or North America (Du Jardin, 2015). One commonly used definition is 'Plant biostimulants contain substance(s) and/or micro-organisms whose function when applied to plants or the rhizosphere is to stimulate natural processes to enhance/benefit nutrient uptake, nutrient efficiency, tolerance to abiotic stress, and crop quality' (EBIC, 2018). Several categories of biostimulants were identified by Du Jardin (2015) including 'humic and fulvic acids, protein hydrolysates and other N-containing compounds, seaweed extracts and botanicals, chitosan and other biopolymers, inorganic compounds, beneficial fungi, and beneficial bacteria'. While some research is available on the use of biostimulants for specific crops and environments, there are many more products on the market than what has been tested by peer-reviewed research. There are some efforts to develop a registry of materials (European Union) as well as databases of commercially available products (Jones and Chapin, 2016). Here we discuss some ways in which biostimulants have been used in relationship with nutrient management in greenhouses.

Humic substances (including both humic and fulvic acid) are the result of microbial decomposition and chemical degradation of organic matter in soils (Calvo et al., 2014). The incorporation of humic acid into soilless media has been found, in some cases, to enhance plant growth, root elongation/lateral roots, and plant nutrient content (N, P, Fe, etc.) (Calvo et al., 2014). Protein hydrolysates are prepared from plant or animal by-products and contain a mixture of peptides and amino acids (Calvo et al., 2014). Although these materials contain N, when used as a biostimulant, the assumption is their benefit to the plant is more than just from the nutrients they contain, for example, by enhancing C and N metabolism, influencing plant defense responses, or chelating micronutrients (Calvo et al., 2014). Extracts prepared from seaweed and vascular plants (botanicals) can enhance plant growth and stress tolerance. Complex modes of action have been associated with these extracts. Relevant to nutrient management, seaweed extracts have been reported to alter rhizosphere microbial community including mycorrhizae, promote root growth, and increase endogenous cytokinin concentration which can impact nutrient partitioning and mobilization (Khan et al., 2009). Substrate drenches of a commercial seaweed extract enhanced flower number and postharvest life of petunia and tomato extracts (Li and Mattson, 2015). Less information is available in the literature regarding botanical extracts as biostimulants. An extract of garlic applied as foliar spray or through fertigation enhanced root and shoot growth and antioxidant activity of eggplant and pepper seedlings (Hayat et al., 2018).

Beneficial elements, which are not commonly considered essential for all plants, are considered as biostimulants when they promote plant growth or enhance stress tolerance (Calvo et al., 2014; Du Jardin, 2015). Beneficial elements with biostimulant properties include aluminum (Al), cobalt (Co), sodium (Na), selenium (Se), and silicon (Si) (Du Jardin, 2015). Silicon is an especially intriguing beneficial element with a wide range of reported responses in greenhouse crops including nutrient imbalances such as mitigation of high sodium chloride (Zhu et al., 2004), copper toxicity (Frantz et al., 2011), and iron deficiency (Pavlovic et al., 2013), and mitigation of a wide range of other abiotic and biotic stresses (Etesami and Jeong, 2018). Selenium and iodine are examples of essential elements for animals including humans which may be added to the nutrient solution as part of a biofortification strategy (Voogt et al., 2010). In some cases, the addition of selenium to the nutrient solution has decreased plant oxidative stress and enhanced biomass of greenhouse crops (Haghichi et al., 2016; Ramos et al., 2010).

In the context of biostimulants, beneficial fungi and bacteria do not refer to materials used for control of diseases and arthropods but to microbes that facilitate nutrient transfer to the plant such as mycorrhizae or enhance plant growth such as plant growth-promoting bacteria (PGPB). For example, inoculation with two strains of the bacterium *Rhodopseudomonas* sp. led to nearly doubling in fruit size of greenhouse tomatoes growing in containers with sandy-loam soil (Lee et al., 2008). In many cases, more research is necessary to understand the environment factors in which beneficial responses are observed (rhizosphere and aerial conditions including abiotic stresses, substrate, interactions with microbial community). For example, the beneficial effects of a PGPB, *Pseudomonas stuzeri*, and humic acids, were affected by environment (growth chamber, greenhouse, or field), saline conditions, and use of salt-sensitive or tolerant pepper cultivars. A comprehensive review of arbuscular mycorrhizae in vegetable crops is done by Baum et al. (2015). While there is more literature on microbial biostimulant response in field soil-grown plants, there is a growing literature on microbe-based biostimulants in plants grown in soilless substrates. Two mycorrhizae (one laboratory produced and one commercially available product) enhanced the leaf dry mass of container-grown lettuce by up to 40% and in some cases enhanced nutritional compounds (phenolics, carotenoids, anthocyanins, and ascorbate) (Baslam et al., 2011). A commercially available arbuscular mycorrhizae enhanced the net photosynthetic rate, leaf elemental content (P, N, Fe, Cu, and Zn), and biomass of bush morning glory (Ipomoea carnea spp. fistulosa) growing in soilless media with an organic fertilizer source (Amaya-Carpio, 2009). Regarding P status, mycorrhizae inoculation led to increased activity of root acid phosphatase (ACP) and alkaline phosphatase (AP) activity.

10 Nutrient management to improve postharvest shelf life

Commercial greenhouse crop producers aim for products with not only a high-quality retail appearance, but also good postharvest performance for end users and consumers. However, the wide variety of crops produced in greenhouses and sold implies there are several definitions of good postharvest performance. For food crops, the shelf life or duration that products are considered acceptably consumable is the primary consideration. The duration which crops will maintain attractive appearance free of pathogens and senescence is important for flowering potted plants and cut flowers, while the maintenance of appropriate size (especially avoiding excessive stem elongation) is also important for potted plants. For container-grown annuals and perennial, the postharvest quality is determined by the transplant success and subsequent growth of these landscape plants.

For flowering potted plants, it can be a common practice to reduce the N or total fertilizer concentration for the last few weeks of production (Nell, 1993). This can diminish unwanted vegetative growth for consumers who value potted plants for the aesthetic balance between plant and container size. Additional fertilizer for the consumer phase may enhance vegetative growth, but not flowering (Chiari et al., 1999). Postharvest nutrient management for containerized herbaceous bedding plants can be more challenging compared to flowering potted plants. Following the greenhouse production phase, crops are transported to retail locations until they are purchased. After purchase, plants are transplanted into landscapes or into containers by end users. While nutrients may be restricted at the end of the production phase to acclimate plants for the retail and landscape environment (Armitage, 1993), the diminished nutrient reservoirs may negatively impact landscape performance, as consumers generally under- or do not fertilize in the landscape. The use of CRF with release periods longer than the time required to produce a marketable, containerized crop can improve the posttransplant performance of landscape plants (Andiru, 2013b). Similarly, increasing WSF concentration during greenhouse production of container crops also improves containerized bedding plant landscape establishment and growth (Mattson et al., 2009). There has been research on how to utilize CRF in both the production and consumer phases for bedding plants, including CRFs with a delayed nutrient-release coating designed to enhance landscape performance following production, with successful results (de Oliveria et al., 2016).

11 Strategies to reduce fertilizer leaching

Increasing awareness about the fate of nutrients in watersheds and the environment, combined with reducing the greenhouse crop production costs, is increasing interest in strategies to reduce nutrient leaching from greenhouse and controlled environments. There are several approaches that may be taken to reduce fertilizer leaching while growing plants in greenhouses and other controlled environments. The first is to synchronize nutrient provision with plant demands (El-Jaoual and Cox, 1998). For instance, early in a plant life cycle, when plants are seedlings, nutrient demand is less than compared to a more mature plant and lower fertilizer concentrations should be used to avoid excess nutrient application. Similarly, if plant growth is slower under cool air temperatures and/or low light intensity fertilizer concentrations may be reduced due to diminished uptake and use relative to water uptake. Another approach to reduce fertilizer leaching is to modify leaching fractions and/or fertilizer concentrations (McAvoy, 1994). When leaching fractions are reduced, the EC in the root zone will likely increase due to diminished nutrient loss, so more careful monitoring of substrate EC is needed and lower fertilizer concentrations may be supplied (Haver and Schuch, 1996). Lower fertilizer concentrations used (especially when using closed irrigation or reduced leaching fractions) diminish the amount of fertilizer leaching required to manage excess salt. When lower fertilizer concentrations are used, careful monitoring of crops with greater nutrient requirements is required to ensure adequate nutrition is provided.

In addition to the frequency and volume of irrigation, the system used to irrigate crops, planting densities, and type of fertilizer can interact to affect nutrient leaching. If WSF is being applied, the greatest nutrient losses occur in overhead irrigation from water that doesn't even make it to a container. A static system such as overhead sprinklers on a fixed pattern will waste the most of the irrigation solution and, therefore, waste the most nutrients. Some booms can be programmed to only irrigate those sections of the greenhouse-containing plants. Regardless of a fixed or mobile overhead irrigation system, the greater the crops planting density, the more efficient use of applied WSF. Nutrient efficiency improves with increasingly precise application of nutrients to the root zone. Pressure-compensated emitters or drippers (Burnett and van Iersel, 2018) deliver nutrient solutions to the surface of the substrate, eliminating non-target irrigation. The most efficient method of applying WSF is through closed irrigation such as subirrigation.

Another important practice for limiting nutrient leaching is to account for all nutrients in irrigation water and/or substrate. For instance, high-alkalinity water with moderate hardness will provide Ca and Mg to plants, therefore Ca and Mg supplied from fertilizers may be reduced. Similarly, acidified water, most commonly treated with sulfuric, phosphoric, or nitric acids, will contribute S, P, or N, respectively, to water; again, fertilizer additions for these nutrients may be reduced. Routing equipment maintenance is essential for greenhouses and all their associated equipment, and fertilizer injectors are no exception. Periodic, regular injector calibration is essential for not only knowing adequate mineral nutrients are being provided to plants, but also to avoid excessive applications. Finally, nutrients can be provided based on crop needs, not by calendar or habit. Periodic sampling of growing substrate and plant tissue will inform how fertilizer programs can and should be modified based on the current nutrient availability in growing substrate and concentrations of elements in shoot tissues.

12 Monitoring and managing nutrition

Not only are nutrient monitoring programs essential to prevent disorders but, should they develop, it can be essential in determining what causes them. A routine water, fertilizer, substrate, and tissue testing program is paramount in developing a successful mineral nutrition program for greenhouse crops, and these activities can occur both on- and off-site.

Simpler, in-house approaches to monitoring and testing provide the necessary data to help maintain adequate fertility throughout production, primarily EC and pH of irrigation water and fertilizer solutions and growing substrate. The EC and pH of un-amended irrigation water as well as solutions amended with acid and/or fertilizer is easily measured using a handheld or benchtop combination pH/EC meter (Merhaut et al., 2018). Several different procedures may be used to measure the pH and EC of growing substrate, including the saturated media extract (SME), 1:2, and Pour Thru methods for most container sizes and the plug extract method for young plants in small containers; the procedures for these different approaches are outlined in detail in Nelson (2012) and Cavins and Williams (2018). The least-destructive and quickest way to measure substrate pH and EC in a wide variety of container sizes is the Pour Thru test (Torres et al., 2010), whereby the substrate solution in the root zone is displaced by clear (de-ionized or reverse-osmosis) water. However, care should be taken when comparing the results of Pour Thru tests to published recommendations, since many recommendations have historically come from SME tests which generally have lower EC values compared to results from Pour Thru extractions. There are some simple meters for measuring petiole sap nutrient concentrations for specific nutrients, including N and K (Rosen et al., 1996). Technological innovations seek to promote early detection of disorders allowing for more precise fertilization. For example, a mobile-phone nitrogen recommendation tool was tested for cultivars of greenhouse ornamentals (Swearengin et al., 2018). The tool makes recommendations based on user inputs of SPAD or atLEAF relative chlorophyll content measurements. The tool was successfully used to delay fertilizer applications without compromising plant quality for two of the tested cultivars, but did not respond quickly enough or had too high action threshold values for two other cultivars (Swearengin et al., 2018).

For a more comprehensive analysis of mineral nutrients in fertilizer solutions, substrates, and plant tissues, samples should be submitted to commercial laboratories for analysis. Water or nutrient solution samples to be submitted for commercial analyses should be collected from the hose-end after a few minutes of running water to ensure the sample is either free of contaminants or is amended with acid and/or fertilizer as per crop requirements (Merhaut et al., 2018). For commercial substrate tests, specific procedures will be outlined by the company that will process samples (Cavins and Williams, 2018). For testing plant tissues, different protocols may be followed depending on which tissue(s) and/or nutrient(s) are being quantified (Peters, 2018). Generally, the newest growth is sampled when performing routine tissue tests to confirm sufficient tissue concentrations. However, sampling procedures are different for deficient plants and vary with the mobility of the deficient nutrient; mobile nutrient deficiency symptoms will appear on older growth, whereas immobile nutrient deficiency symptoms occur on new growth. When diagnosing symptoms, submit separate samples for both affected plants as well as unaffected plants if available for comparison.

If deficiencies are detected, either in substrate or tissue, corrective actions should be taken immediately to prevent or diminish any deficiency or toxicity symptoms which may reduce marketability. A comprehensive guide to remedying substrate and tissue problems is presented by Whipker and Gibson (2018). The degree of intervention is directly proportional to the degree of the deficiency, toxicity, or imbalance. Less-involved corrective or preventative procedures can be more successful and have more predictable results compared to larger interventions.

13 Future trends

In the future, nutrient management will continue to move toward more precise applications to limit environmental nutrient loading and to enhance plant yield. Precisely meeting plant nutritional needs will require an integrated approach relying on the adoption of several technologies. Greater ability to sense the nutrient solution composition, root-zone environment, or plant status in real time will allow for more agile decision-making and correcting of nutrient disorders before yield and quality are impaired. DSS will act to supplement knowledge available from sensors, for example predicting plant nutrient uptake based on environmental conditions, but also allow for decision-making in real time. A comprehensive list of DSSs for fertigation management of vegetable crops was presented by Incrocci et al. (2017). It was noted that while many DSS exist as scholarly exercises, relatively few have seen commercial adoption (Incrocci et al., 2017). Precision irrigation systems such as pressure-compensated drip emitters regulated by wireless sensor networks using substrate moisture, pH, and EC sensors can also deliver nutrients precisely (Lea-Cox et al., 2013). New plant and cultural management techniques, such as grafting, can also improve salinity tolerance, thereby reducing the need for water discharge in closed irrigation systems (Di Gioia, 2013).

Finally, the development of new types of fertilizers and biostimulants/microbial applications will facilitate greater use of renewable resources and more efficient nutrient absorption by the plant. New types of fertilizers are continually being developed to use by- and coproducts of other industries. For example, bioplastics and polymers are being explored as alternatives to petroleum for producing horticultural containers, and some of these materials provide mineral nutrients during production. Specifically, containers molded from biopolymer composites including by-products from soy (*Glycine max*) reduce the amount of fertilizers required to produce healthy crops (McCabe et al., 2016a); and when formulated into a pellet and incorporated into growing substrate, the same soy-amended biopolymers are effective fertilizers (McCabe et al., 2016b). Additionally, algae harvested from wastewater treatment systems applied as a paste to the surface of growing substrate or formed into pellets and incorporated into substrate can effectively fertilize container-grown greenhouse crops (Gimondo et al., 2019). It is clear that innovation in nutrient management will rise to the challenge of meeting increasing global demand for greenhouse crops in a sustainable manner.

14 Where to look for further information

14.1 Further reading

- A comprehensive resource for greenhouse practitioners and educators is *Water, Root Media, and Nutrient Management for Greenhouse Crops*. University of California, Agriculture and Natural Resources, Publication 3551 by Merhaut, D. J., Williams, K. A. and Mangiafico, S. S.
- *Plant Nutrition of Greenhouse Crops* by Sonneveld, C. and Voogt, W. is a wide-reaching book on greenhouse nutrient management.
- 'Fert, Dirt, & Squirt' is a website with comprehensive nutritional monitoring information for a wide range of floriculture crops, www.fertdirtandsquirt.com
- A comprehensive guide to nutritional analysis and interpretation of agronomic and horticultural crops is the *Plant Analysis Handbook IV* by Bryson, G. M. and Mills, H. M.
- An excellent manual including nutrient solution recipes for a wide range of greenhouse crops is *Nutrient Solutions for Greenhouse Crops*, available at: https://www.eurofins.com/agro/news/new-manual-nutrient-solutions-for-greenhouse-crops-available/

14.2 Key journals/conferences

- *Communications in Soil Science and Plant Analysis*
- *Plant and Soil*
- *Journal of Plant Nutrition*
- GreenSys is a well-attended conference held every other year encompassing a wide range of greenhouse technology, innovation, and management

14.3 Major international research projects

- SCRI-MINDS: Managing Irrigation and Nutrients via Distributed Sensing was a multi-year, multi-institutional research project that developed advanced wireless sensor networks for precision water and nutrient management, http://www.smart-farms.org/
- BioGreenhouse a European Union project which ran from 2012 to 2016 which focused on organic greenhouse horticulture https://www.agricology.co.uk/resources/biogreenhouse-publications
- Horti-blueC is a collaborative project on upcycling agricultural and fishery waste streams to develop sustainable growing media, and chemical fertilizer and pesticide replacement, https://www.interreg2seas.eu/en/Horti-blueC

15 References

Adams, C., Frantz, J. and Bugbee, B. 2013. Macro- and micronutrient-release characteristics of three polymer-coated fertilizers: theory and measurement. *Journal of Plant Nutrition and Soil Science* 176, 76–88.

Alem, P., Thomas, P. A. and van Iersel, M. W. 2015. Substrate water content and fertilizer rate affect growth and flowering of potted petunia. *HortScience* 50(4), 582–9.

Amaya-Carpio, L., Davies, F. T., Fox, T. and He, C. 2009. Arbuscular mycorrhizal fungi and organic fertilizer influence photosynthesis, root phosphatase activity, nutrition, and growth of *Ipomoea carnea* ssp. *fistulosa*. *Photosynthetica* 47(1), 1–10.

Andiru, G. A., Pasian, C. C., Frantz, J. M. and Jourdan, P. 2013a. Longevity of controlled-release fertilizer influences the growth of bedding impatiens. *HortTechnology* 23, 157–64.

Andiru, G. A., Pasian, C., Frantz, J. M. and Jones, M. L. 2013b. Greenhouse production of Impatiens wallerana using a controlled-release fertiliser produces quality finished plants with enhanced garden performance. *Journal of Horticultural Science and Biotechnology* 88, 216–22.

Armitage, A. M. 1993. *Bedding Plants: Prolonging Shelf Performance: Postproduction Care and Handling*. Ball Publishing, Batavia, IL, USA.

Baslam, M., Garmendia, I. and Goicoechea, N. 2011. Arbuscular mycorrhizal fungi (AMF) improved growth and nutritional quality of greenhouse-grown lettuce. *Journal of Agricultural and Food Chemistry* 59(10), 5504–15. doi:10.1021/jf200501c.

Baum, C., El-Tohamy, W. and Gruda, N. 2015. Increasing the productivity and product quality of vegetable crops using arbuscular mycorrhizal fungi: a review. *Scientia Horticulturae* 187, 131–41.

Bergstrand, K. J., Löfkvist, K. and Asp, H. 2018. Dynamics of nitrogen availability in pot grown crops with organic fertilization. *Biological Agriculture and Horticulture* 35, 143–50.

Bugbee, B. 2004. Nutrient management in recirculating hydroponic culture. *Acta Horticulturae* 648, 99–112.

Burnett, S. E. and van Iersel, M. W. 2018. Water, irrigation systems, and their control. In: Merhaut, D. J., Williams, K. A. and Mangiafico, S. S. (Eds), *Water, Root Media, and*

Nutrient Management for Greenhouse Crops. University of California Agriculture and Natural Resources, Davis, CA, pp. 11–45.

Burnett, S. E., Mattson, N. S. and Williams, K. A. 2016. Substrates and fertilizers for organic container production of herbs, vegetables, and herbaceous ornamental plants grown in greenhouses in the United States. *Scientia Horticulturae* 208, 111–9. doi:10.1016/j.scienta.2016.01.001.

Calvo, P., Nelson, L. and Kloepper, J. W. 2014. Agricultural uses of plant biostimulants. *Plant and Soil* 383(1–2), 3–41.

Cavins, T. and Williams, K. A. 2018. Root media: testing and interpretation. In: Merhaut, D. J., Williams, K. A. and Mangiafico, S. S. (Eds), *Water, Root Media, and Nutrient Management for Greenhouse Crops*. University of California Agriculture and Natural Resources, Davis, CA, pp. 119–35.

Chiari, A., Elliott, G. C. and Bridgen, M. P. 1999. Resin-coated fertilizers affect postproduction growth, but not flowering, of potted *Alstroemeria*. *HortScience* 34(4), 657–9. doi:10.21273/HORTSCI.34.4.657.

Currey, C. J. 2019. Micronutrient considerations. *Greenhouse Management* 39(3), 24–5.

Currey, C. J. and Lopez, R. G. 2014. Controlled-release fertilizer during propagation affected growth and tissue nutrient concentrations of rooted cuttings of annual bedding plants. *HortScience* 49, 152–9.

Currey, C. J., Walters, K. J. and Flax, N. J. 2019. Nutrient solution strength does not interact with the daily light integral to affect hydroponic cilantro, dill, and parsley growth and tissue mineral nutrient concentrations. *Agronomy* 9(7), 389. doi:10.3390/agronomy9070389.

Davarpanah, S., Tehranifar, A., Abadía, J., Val, J., Davarynejad, G., Aran, M. and Khorassani, R. 2018. Foliar calcium fertilization reduces fruit cracking in pomegranate (*Punica granatum* cv. Ardestani). *Scientia Horticulturae* 230, 86–91.

de Oliveria, S. F., Fisher, P. R., Huang, J. and Mello, S. da C. 2016. Strategies to provide fertilizer for both production and consumer phases of petunia. *HortTechnology* 26, 164–75.

Di Gioia, F., Signore, A., Serio, F. and Santamaria, P. 2013. Grafting improves tomato salinity tolerance through sodium partitioning within the shoot. *HortScience* 48(7), 855–62.

Domingues, D. S., Takahashi, H. W., Camara, C. A. and Nixdorf, S. L. 2012. Automated system developed to control pH and concentration of nutrient solution evaluated in hydroponic lettuce production. *Computers and Electronics in Agriculture* 84, 53–61.

Du Jardin, P. 2015. Plant biostimulants: definition, concept, main categories and regulation. *Scientia Horticulturae* 196, 3–14.

El-Jaoual, T. and Cox, D. 1998. Effects of plant age on nitrogen uptake and distribution by greenhouse plants. *Journal of Plant Nutrition* 21(5), 1055–66.

Etesami, H. and Jeong, B. R. 2018. Silicon (Si): review and future prospects on the action mechanisms in alleviating biotic and abiotic stresses in plants. *Ecotoxicology and Environmental Safety* 147, 881–96. doi:10.1016/j.ecoenv.2017.09.063.

European Biostimulants Industry Council. 2018. 5 December 2018. Available at: http://www.biostimulants.eu/.

Fain, G. B., Gilliam, C. H., Sibley, J. L., Boyer, C. R. and Witcher, A. L. 2008. Wholetree substrate and fertilizer rate in production of greenhouse-grown petunia (*Petunia × hybrida* Vilm.) and marigold (*Tagetes patula* L.). *HortScience* 43(3), 700–5.

Fernández, V. and Eichert, T. 2009. Uptake of hydrophilic solutes through plant leaves: current state of knowledge and perspectives of foliar fertilization. *Critical Reviews in Plant Sciences* 28(1-2), 36-68.

Ferrarezi, R. S., Weaver, G. M., Van Iersel, M. W. and Testezlaf, R. 2015. Subirrigation: historical overview, challenges, and future prospects. *HortTechnology* 25(3), 262-76.

Fisher, P. R., Wik, R. M., Smith, B. R., Pasian, C. C., Kmetz-González, M. and Argo, W. R. 2003. Correcting iron deficiency in Calibrachoa grown in a container medium at high pH. *HortTechnology* 13(2), 308-13.

Frantz, J. M., Khandekar, S. and Leisner, S. 2011. Silicon differentially influences copper toxicity response in silicon-accumulator and non-accumulator species. *Journal of the American Society for Horticultural Science* 136(5), 329-38.

Gimondo, J. A., Currey, C. J., Jarboe, D. H., Gross, M. and Graves, W. R. 2019. Wastewater-grown algae pellets and paste as fertilizers for containerized crops. *HortScience* 54(3), 528-36.

Haghighi, M., Sheibanirad, A. and Pessarakli, M. 2016. Effects of selenium as a beneficial element on growth and photosynthetic attributes of greenhouse cucumber. *Journal of Plant Nutrition* 39(10), 1493-8.

Hartz, T. K. and Johnstone, P. R. 2006. Nitrogen availability from high-nitrogen-containing organic fertilizers. *HortTechnology* 16(1), 39-42. doi:10.21273/HORTTECH.16.1.0039.

Haver, D. L. and Schuch, U. K. 1996. Production and postproduction performance of two new guinea impatiens cultivars grown with controlled-release fertilizer and no leaching. *Journal of the American Society for Horticultural Science* 121(5), 820-5.

Hayat, S., Ahmad, H., Ali, M., Hayat, K., Khan, M. and Cheng, Z. 2018. Aqueous garlic extract as a plant biostimulant enhances physiology, improves crop quality and metabolite abundance, and primes the defense responses of receiver plants. *Applied Sciences* 8(9), 1505; doi:10.3390/app8091505.

Henry, J. B., McCall, I., Jackson, B. and Whipker, B. E. 2017. Growth response of herbaceous ornamentals to phosphorous fertilization. *HortScience* 52, 1362-7.

Henry, J. B., McCall, I. and Whipker, B. E. 2018. Phosphorous restriction as an alternative to plant growth retardants in Angelonia and New Guinea impatiens. *HortTechnology* 28, 136-42.

Incrocci, L., Massa, D. and Pardossi, A. 2017. New trends in the fertigation management of irrigated vegetable crops. *Horticulturae* 3(2), 37.

Jayakumar, K., Vijayarengan, P., Changxing, Z., Gomathinayagam, M. and Jaleel, C. A. 2008. Soil applied cobalt alters the nodulation, leg-haemoglobin content and antioxidant status of *Glycine max* (L.) Merr. *Colloids and Surfaces B: Biointerfaces* 67(2), 272-5. doi:10.1016/j.colsurfb.2008.08.012.

Johnson, C. N., Fisher, P. R., Huang, J., Yeager, T. H., Obreza, T. A., Vetanovetz, R. P., Argo, W. R. and Bishko, A. J. 2013. Effect of fertilizer potential acidity and nitrogen form on the pH response in a peat-based substrate with three floricultural species. *Scientia Horticulturae* 162, 135-43.

Jones, M. L. and Chapin, L. 2016. A new resource for greenhouse growers to help identify microbe-containing products advertised to enhance growth of greenhouse floriculture crops. 5 December 2018. Available at: https://u.osu.edu/greenhouse/2016/11/10/a-new-resource-for-greenhouse-growers-to-help-identify-microbe-containing-products-advertised-to-enhance-growth-of-greenhouse-floriculture-crops/.

Jung, D. H., Kim, H. J., Choi, G. L., Ahn, T. I., Son, J. E. and Sudduth, K. A. 2015. Automated lettuce nutrient solution management using an array of ion-selective electrodes. *Transactions of the ASABE* 58(5), 1309–19.

Justice, A. and Faust, J. E. 2014. Phosphorous-restriction as a potential technique to control Impatiens stem elongation. *Acta Horticulturae* 1104, 9–14.

Kamenidou, S., Cavins, T. J. and Marek, S. 2008. Silicon supplements affect horticultural traits of greenhouse-produced ornamental sunflowers. *HortScience* 43(1), 236–9.

Katsoulas, N. and Voogt, W. 2013. Recent trends in salinity control for soilless growing systems management. *International Symposium on Growing Media and Soilless Cultivation 1034*, 17 June 2013, pp. 433–42.

Khadduri, N. 2014. Benefits of small-prill controlled-release fertilizer in container production. *National Proceedings: Forest and Conservation Nursery Associations–2013*, pp. 54–6.

Khan, W., Rayirath, U. P., Subramanian, S., Jithesh, M. N., Rayorath, P., Hodges, D. M., Critchley, A. T., Craigie, J. S., Norrie, J. and Prithiviraj, B. 2009. Seaweed extracts as biostimulants of plant growth and development. *Journal of Plant Growth Regulation* 28(4), 386–99.

Kim, H. J., Kim, W. K., Roh, M. Y., Kang, C. I., Park, J. M. and Sudduth, K. A. 2013. Automated sensing of hydroponic macronutrients using a computer-controlled system with an array of ion-selective electrodes. *Computers and Electronics in Agriculture* 93, 46–54.

Knight, R. S. and Lefsrud, M. 2017. Automated nutrient sensing and recycling. *2017 ASABE Annual International Meeting*. American Society of Agricultural and Biological Engineers, p. 1.

Kroggel, M. and Kubota, C. 2018. Hydroponic nutrient solution for optimized greenhouse tomato production. Ohio State University Extension. Factsheet HYG-1437. Available at: https://ohioline.osu.edu/factsheet/hyg-1437.

Lea-Cox, J. D., Bauerle, W. L., van Iersel, M. W., Kantor, G. F., Bauerle, T. L., Lichtenberg, E., King, D. M. and Crawford, L. 2013. Advancing wireless sensor networks for irrigation management of ornamental crops: an overview. *HortTechnology* 23(6), 717–24.

Lee, K. H., Koh, R. H. and Song, H. G. 2008. Enhancement of growth and yield of tomato by *Rhodopseudomonas* sp. under greenhouse conditions. *The Journal of Microbiology* 46(6), 641–6. doi:10.1007/s12275-008-0159-2.

Lester, G. E., Jifon, J. L. and Makus, D. J. 2006. Supplemental foliar potassium applications with or without a surfactant can enhance netted muskmelon quality. *HortScience* 41(3), 741–4.

Li, Y. and Mattson, N. S. 2015. Effects of seaweed extract application rate and method on post-production life of petunia and tomato transplants. *HortTechnology* 25(4), 505–10.

Liebig, J. V. 1862. *Die Chemie in ihrer Anwendung auf Agricultur und Physiologie* (7e Aufl. Vol. 2). F. Vieweg und Sohn, Braunschweig.

Liu, J., Leatherwood, W. R. and Mattson, N. S. 2012. Irrigation method and fertilizer concentration differentially alter growth of vegetable transplants. *HortTechnology* 22(1), 56–63.

Maboko, M. M. and Du Plooy, C. P. 2017. Response of hydroponically grown cherry and fresh market tomatoes to reduced nutrient concentration and foliar fertilizer application under shadenet conditions. *HortScience* 52(4), 572–8.

Maboko, M. M., Du Plooy, C. P. and Chiloane, S. 2017. Yield and mineral content of hydroponically grown mini-cucumber (*Cucumis sativus* L.) as affected by reduced nutrient concentration and foliar fertilizer application. *HortScience* 52(12), 1728–33.

Majsztrik, J. C., Ristvey, A. G., Ross, D. S. and Lea-Cox, J. D. 2018. Comparative water and nutrient application rates among ornamental operations in Maryland. *HortScience* 53(9), 1364–71.

Massa, D., Mattson, N. S. and Lieth, H. 2008. An empirical model to simulate sodium absorption in roses growing in a hydroponic system. *Scientia Horticulturae* 118(3), 228–35.

Mattson, N. S., 2017. Fertilizer 2.0. *GrowerTalks* 80(10), 50–3.

Mattson, N. S., Leatherwood, W. R. and Peters, C. 2009. Planting the seeds for consumer success: are your irrigation and fertility practices backfiring on the end consumer and causing garden failure? *GrowerTalks* 73(1), 56–9.

McAvoy, R. J. 1994. Nitrate nitrogen movement through the soil profile beneath a containerized greenhouse crop irrigated with two leaching fractions and two wetting agent levels. *Journal of the American Society for Horticultural Science* 119(3), 446–51.

McCabe, K. G., Schrader, J. A., Currey, C. J., Grewell, D. and Graves, W. R. 2016a. Soy-composite biocontainers allow for reduced fertilizer inputs during container-crop production. *HortScience* 51(7), 927–34.

McCabe, K. G., Currey, C. J., Schrader, J. A., Grewell, D., Behrens, J. and Graves, W. R. 2016b. Pelletized soy-based bioplastic fertilizers for container-crop production. *HortScience* 51(11), 1417–26.

Merhaut, D. J., Williams, K. A. and Mangiafico, S. S. (Eds). 2018. Water, root media, and nutrient management for greenhouse crops. Publication 3551. University of California Agriculture and Natural Resources Press, Oakland, CA. ISBN-13: 978-1-60107-983-1.

Mitscherlich, E. A. 1909. Des Gesetz des Minimums und das Gesetz des abnehmended Bodenertrages. *Landwirsch Jahrb.* 3, 537–52.

Möller, K. and Schultheiß, U. 2015. Chemical characterization of commercial organic fertilizers. *Archives of Agronomy and Soil Science* 61(7), 989–1012.

Morgan, L. M. 2005. *Fresh Culinary Herb Production*. Suntec, Tokomaru, New Zealand.

Morgan, L. M. 2012. *Hydroponic Salad Crop Production*. Suntec, Tokomaru, New Zealand.

Nell, T. A. 1993. *Flowering Potted Plants: Prolonging Shelf Performance: Postproduction Care and Handling*. Ball Publishing.

Nelson, P. V. 2012. *Greenhouse Operation and Management* (7th edn.). Pearson, Upper Saddle River, NJ.

Nelson, P. V., Songg, C.-Y., Huang, J., Niedziela Jr., C. E. and Swallow, W. H. 2012. Relative effects of fertilizer nitrogen form and phosphate level on control of bedding plant seedling growth. *HortScience* 47, 249–53.

Pavlovic, J., Samardzic, J., Maksimović, V., Timotijevic, G., Stevic, N., Laursen, K. H., Hansen, T. H., Husted, S., Schjoerring, J. K., Liang, Y. and Nikolic, M. 2013. Silicon alleviates iron deficiency in cucumber by promoting mobilization of iron in the root apoplast. *New Phytologist* 198(4), 1096–107. doi:10.1111/nph.12213.

Peters, C. 2018. Tissue analysis: testing and interpretation. In: Merhaut, D. J., Williams, K. A. and Mangiafico, S. S. (Eds), *Water, Root Media, and Nutrient Management for Greenhouse Crops*. University of California Agriculture and Natural Resources, Davis, CA, pp. 195–206.

Pilon, P. 2006. *Perennial Solutions: a Grower's Guide to Perennial Production*. Ball Publishing, Batavia, IL.

Qian, Z., Miedema, H., de Smet, L. C. P. M. and Sudhölter, E. J. R. 2018. Modelling the selective removal of sodium ions from greenhouse irrigation water using membrane technology. *Chemical Engineering Research and Design* 134, 154–61.

Ramos, S. J., Faquin, V., Guilherme, L. R. G., Castro, E. M., Ávila, F. W., Carvalho, G. S., Bastos, C. E. A. and Oliveira, C. 2010. Selenium biofortification and antioxidant activity in lettuce plants fed with selenate and selenite. *Plant, Soil and Environment* 56(12), 584–8. doi:10.17221/113/2010-PSE.

Resh, H. M. 2013. *Hydroponic Food Production: a Definitive Guidebook for the Advanced Home Gardener and the Commercial Hydroponic Grower*. CRC Press, Boca Raton, FL.

Rius-Ruiz, F. X., Andrade, F. J., Riu, J. and Rius, F. X. 2014. Computer-operated analytical platform for the determination of nutrients in hydroponic systems. *Food Chemistry* 147, 92–7. doi:10.1016/j.foodchem.2013.09.114.

Rosen, C. J., Errebhi, M. and Wang, W. 1996. Testing petiole sap for nitrate and potassium: a comparison of several analytical procedures. *HortScience* 31(7), 1173–6. doi:10.21273/HORTSCI.31.7.1173.

Rouphael, Y. and Colla, G. 2009. The influence of drip irrigation or subirrigation on zucchini squash grown in closed-loop substrate culture with high and low nutrient solution concentrations. *HortScience* 44(2), 306–11.

Rouphael, Y., Cardarelli, M., Rea, E. and Colla, G. 2008. The influence of irrigation system and nutrient solution concentration on potted geranium production under various conditions of radiation and temperature. *Scientia Horticulturae* 118(4), 328–37.

Rozema, E. R., Gordon, R. J. and Zheng, Y. 2014. Plant species for the removal of Na^+ and Cl^- from greenhouse nutrient solution. *HortScience* 49(8), 1071–5.

Scoggins, H. and Williams, K. 2018. Formulating fertilizer programs. In: Merhaut, D. J., Williams, K. A. and Mangiafico, S. S. (Eds), *Water, Root Media, and Nutrient Management for Greenhouse Crops*. University of California Agriculture and Natural Resources, Davis, CA, pp. 139–64.

Silber, A. and Bar-Tal, A. 2019. Nutrition of substrate-grown plants. In: Raviv, M., Lieth, J. H. and Bar-Tal, A. (Eds), *Soilless Culture*. Elsevier, pp. 197–257.

Sonneveld, C. and Voogt, W. 2009. *Plant Nutrition of Greenhouse Crops*. Springer Verlag, Dordrecht, The Netherlands, p. 431.

Souri, M. K. and Aslani, M. 2018. Beneficial effects of foliar application of organic chelate fertilizers on French bean production under field conditions in a calcarous soil. *Advances in Horticultural Science* 32(2), 265–72.

Stanghellini, C., Kempkes, F. L. K., Pardossi, A. and Incrocci, L. 2005. Closed water loop in greenhouses: effect of water quality and value of produce. *Acta Horticulturae* 691, 233–42.

Straver, N., De Kreij, C. and Verberkt, H. 1999. *Bemestingsadviesbasis potplanten*. Proefstation voor Bloemisterij en Glasgroenten, Naaldwijk, The Netherlands, 55p.

Swearengin, L., Dunn, B. L., Singh, H. and Goad, C. 2018. Evaluation of a mobile phone plant nitrogen recommendation application in the greenhouse. *Journal of Plant Nutrition* 41, 2615–25.

Torres, A. P., Mickelbart, M. V. and Lopez, R. G. 2010. Leachate volume effects on pH and electrical conductivity measurements in containers obtained using the pour-through method. *HortTechnology* 20(3), 608–11.

Treadwell, D. D., Hochmuth, G. J., Hochmuth, R. C., Simonne, E. H., Davis, L. L., Laughlin, W. L., Li, Y., Olczyk, T., Sprenkel, R. K. and Osborne, L. S. 2007. Nutrient management in organic greenhouse herb production: where are we now? *HortTechnology* 17(4), 461–6.

Udagawa, Y. 2004. Some responses of dill (*Anethum graveolens*) and thyme (*Thymus vulgaris*), grown in hydroponic, to the concentration of nutrient solution. *Acta Horticulturae* 396, 203–10.

U. S. Food and Drug Administration. 2018. Food safety modernization act final rule on produce safety. Available at: https://www.fda.gov/food/food-safety-modernization-act-fsma/fsma-final-rule-produce-safety.

Varlagas, H., Savvas, D., Mouzakis, G., Liotsos, C., Karapanos, I. and Sigrimis, N. 2010. Modelling uptake of Na^+ and Cl^- by tomato in closed-cycle cultivation systems as influenced by irrigation water salinity. *Agricultural Water Management* 97(9), 1242–50.

Voogt, W. 2013. Soil fertility management in organic greenhouse crops; a case study on fruit vegetables. *II International Symposium on Organic Greenhouse Horticulture 1041*, pp. 21–35.

Voogt, W., Holwerda, H. T. and Khodabaks, R. 2010. Biofortification of lettuce (*Lactuca sativa* L.) with iodine: the effect of iodine form and concentration in the nutrient solution on growth, development and iodine uptake of lettuce grown in water culture. *Journal of the Science of Food and Agriculture* 90(5), 906–13. doi:10.1002/jsfa.3902.

Walters, K. J. and Currey, C. J. 2018. Effects of nutrient solution concentration and daily light integral on growth and nutrient concentration of several basil species in hydroponic production. *HortScience* 53, 1319–25.

Whipker, B. E. 2014. Poinsettia fertilization: calcium deficiency. *e-GRO Alert* 3(58), 1-5. Available at: www.e-gro.org/pdf/358.pdf.

Whipker, B. and Gibson, J. 2018. Corrective procedures for root media and tissue. In: Merhaut, D. J., Williams, K. A. and Mangiafico, S. S. (Eds), *Water, Root Media, and Nutrient Management for Greenhouse Crops*. University of California Agriculture and Natural Resources, Davis, CA, pp. 225–48.

Zhai, Z., Ehret, D. L., Forge, T., Helmer, T., Lin, W., Dorais, M. and Papadopoulos, A. P. 2009. Organic fertilizers for greenhouse tomatoes: productivity and substrate microbiology. *HortScience* 44(3), 800–9.

Zheng, Y., Graham, T., Richard, S. and Dixon, M. 2004. Potted gerbera production in a subirrigation system using low-concentration nutrient solutions. *HortScience* 39(6), 1283–6.

Zhu, Z., Wei, G., Li, J., Qian, Q. and Yu, J. 2004. Silicon alleviates salt stress and increases antioxidant enzymes activity in leaves of salt-stressed cucumber (*Cucumis sativus* L.). *Plant Science* 167(3), 527–33.

Zikeli, S., Deil, L. and Möller, K. 2017. The challenge of imbalanced nutrient flows in organic farming systems: a study of organic greenhouses in Southern Germany. *Agriculture, Ecosystems and Environment* 244, 1–13.

Chapter 11

Advances in pest and disease management in greenhouse cultivation

Gerben J. Messelink and H. Marjolein Kruidhof, Wageningen University and Research, The Netherlands

1 Introduction

Better protection of crops against the influx of pests is often seen as one of the benefits of greenhouse horticulture compared to outdoor crop cultivation. In reality, however, greenhouse crops in all parts of the world are under pressure of infestation by a wide range of pests and diseases (Heinz et al., 2004; Eilenberg et al., 2001; van Lenteren and Woets, 1988; Heimpel and Mills, 2017). Even in greenhouses with all kinds of physical barriers and strict hygiene measures, several pests and pathogens will somehow eventually enter. A perhaps more important benefit of modern greenhouses is that arthropod biological control agents (BCAs, see Box 1 for definitions) that are introduced as part of an integrated pest management (IPM) approach have a much better chance of establishing compared to outdoor cropping conditions. In the protected environment of the greenhouse, their tendency

http://dx.doi.org/10.19103/AS.2019.0052.17

Box 1 Definitions of BCAs

BCAs can be subdivided into different categories, based on their mode of action, diet breadth or field of application. Widely used alternative terms for BCAs are 'natural enemies' in the field of entomology and 'antagonists' in the field of plant pathology.

The most-used terms for different categories of BCAs are defined below:

- *Predators* are animals that that attack and feed on other animals (the prey). The prey is killed and usually mostly or entirely eaten.
  - *Specialist predators* are defined here as predators that for their development, survival and/or reproduction are restricted to the consumption of arthropod pests belonging to a single family.
  - *Generalist predators* are defined here as predators that can sustain their development, survival and reproduction with a suite of arthropod pest species belonging to different families. They usually also feed on nectar and pollen.
  - *Omnivorous predators* are officially defined as consumers of more than one trophic level. True omnivores feed on both plant and prey. Here we use omnivores for predators that actively feed on living plant tissue and prey, which are also defined as zoophytophagous predators (Coll and Guershon, 2002).
- *Parasitoids* are holometabolous insects that have a unique suite of life history characteristics, including (a) free-living adults, (b) egg deposition in or near a host organism and (c) consumption of the host organism by the larvae, culminating in the death of the host organism, whereby at most a single host insect is consumed by an individual parasitoid larva.
- *Entomopathogens* are pathogens of insects (and sometimes mites) – including fungi, bacteria, nematodes, viruses and protozoans – that reproduce within their host by causing the breakdown of its tissues and incorporating the breakdown products into additional pathogen biomass. Due to the release of toxins, they cause the death of the infected insect within a relatively short time span.
- *Antagonists* are microorganisms used in the control of plant-pathogenic bacteria, fungi, yeasts and viruses. They can act through one or more of the following mechanisms:
  1. direct consumption (e.g. actinomycete parasitism of fungal spores);
  2. competition between fungal or bacterial pathogens and nonpathogenic fungi or bacteria for colonization sites and/or nutrients;
  3. antibiosis or the production of antibiotic compounds primarily by bacteria; and
  4. induction of systemic (acquired) resistance (ISR/SAR) in crop plants.

to disperse away from the release site is lower, and in climate-controlled greenhouses they are better protected against adverse climatic conditions. Many important arthropod pests are extremely polyphagous and infest almost every crop, such as some species of thrips, aphids, spider mites, whiteflies and mealybugs. Others, such as the tomato russet mite *Aculops lycopersici* (Massee) and the tomato borer, *Tuta absoluta* (Meyrick), are more host-plant specific, but their impact is nonetheless significant as tomato is worldwide the most produced greenhouse-grown crop. Also, plant diseases can cause severe losses to greenhouse crops. They are caused by pathogens such as fungi, bacteria and viruses and can infect both the aboveground and belowground plant parts. Common aboveground fungal diseases are *Botrytis* grey mould, powdery mildew, downy mildew and rusts. Soilborne pathogens, such as *Pythium* sp., *Phytophthora* sp., *Fusarium* sp., *Rhizoctonia solani* Kühn, *Verticillium* sp., *Agrobacterium* sp., as well as soilborne plant-pathogenic nematodes, can cause severe damage especially in soil-grown crops. Finally, several viral diseases, either transmitted mechanically or through arthropod (or fungal) vectors, can harm greenhouse-grown crops. The list of pests and diseases that can occur in greenhouse crops is extremely long and it would be impossible to discuss them all in detail in this chapter. We will discuss the main pests and diseases that occur in tomato in more detail later on in the case study. For a detailed description of all pests and diseases in the most important greenhouse crops, we recommend to read the recent book of Gullino et al. (2019). In the remainder of the chapter, we will describe the myriad of practices that can contribute to managing pests and diseases, and how to integrate these practices in an ecologically based IPM approach (see Box 2 for definitions) to ensure sustainable crop protection. We start with giving a historical perspective on the development of biological pest control and ecologically based IPM, its underlying rationale and the main challenges that need to be tackled.

2 History of ecologically based IPM

Biological pest control in greenhouse horticulture was first commercially implemented in the 1920s, with the English discovery and subsequent production of the parasitoid *Encarsia formosa* for the control of the greenhouse whitefly in tomato (Speyer, 1927). However, the interest in biological control (BC) quickly dropped following the rise in pesticide availability after the Second World War, and the production of *E. formosa* ceased completely in 1956 (Hussey and Scopes, 1985). The use of pesticides has thereafter dominated the practices for pest and disease control for several decades, and in many parts of the world this is still the case. However, the indiscriminate use of pesticides promotes the development of pesticide resistance, which makes chemical control ineffective (Denholm and Rowland, 1992). In addition, the removal of natural BC through the adverse effects of broad-spectrum pesticides on natural enemies also contributes to pest outbreaks. In the late 1950s, the concerns about these adverse effects of pesticides led to the introduction of 'integrated control' defined as 'applied pest control which combines and integrates biological and chemical control' (Stern et al., 1959). In greenhouse horticulture, the development of spider mite (*Tetranychus urticae* Koch) resistance against several pesticides (Bravenboer, 1959), and the subsequent discovery of the predatory mite *Phytoseiulus persimilis* as an effective spider mite control agent (Bravenboer and Dosse, 1962) revived the interest in biological pest control. Severe whitefly outbreaks in the early 1970s led to the rediscovery of *E. formosa*, and improved methods for rearing and shipping this parasitoid led to an effective and widely accepted whitefly BC programme (Van Driesche and Heinz, 2004). In the 1980s, the finding that many phytoseiid predatory mites could be mass-produced on a system of bran and astigmatid prey mites was a major breakthrough, because this enabled the mass production of these predatory mites in a cheap and massive way (Ramakers and van Lieburg, 1982). Predatory mites mass-reared in this way nowadays play a major role in the control of thrips, spiders mites and whiteflies. Since the 1990s, the number of researchers studying biological pest control in protected cultures, and the variety of greenhouse crops and countries in which BC methods are used, has steadily increased. In addition to specialist natural enemies, generalist predators also became more popular because of their ability to control multiple pest species and to establish populations in the absence of the pest. Besides generalist predatory mites, generalist predatory bugs belonging to the families Anthocoridae and Miridae are now extensively used in the BC of a wide variety of pest species (Fig. 1). This development has gone hand in hand with the development and application methods of alternative food sources, allowing for the introduction and population development of these generalist predators in the crop before the pest arrives ('standing army approach') (Messelink et al., 2014; Janssen and Sabelis, 2015). By the time this book chapter was written, an estimated number of 350 arthropod and nematode BCAs were commercially available worldwide (van Lenteren et al., 2018). Moreover, more than 200 strains of microbial BCAs (fungi, bacteria and viruses) were registered for pest and disease management, although this is probably a large underestimation of the real number of strains used in biocontrol (van Lenteren et al., 2018). In general, control strategies have evolved from 'prophylactic calendar spraying' to the restriction in the use of pesticides based on the monitoring of pest and disease damage and action thresholds ('threshold-based pesticide application') towards (ecologically

Figure 1 Different types of natural enemies that are used for biological control in greenhouses: (a) the specialist aphid parasitoid, *Aphidius ervi*; (b) emerging larvae of the aphid lacewing predator, *Chrysopa pallens*; (c) the specialist predatory mite, *Phytoseiulus persimilis*, and its prey, the two-spotted spider mite, *Tetranychus urticae*; (d) the generalist predatory mite, *Amblyseius swirskii*, preying upon whitefly eggs; (e) the generalist predatory bug, *Orius majusculus*, preying upon the aphid, *Myzus persicae*; and (f) the omnivorous predator, *Macrolophus pygmaeus*.

based) IPM that incorporates multiple non-chemical methods and allows for the use of compatible pesticides only as a last resort (see Box 2 for definitions). The definition of IPM provided by the European Union (EU) directive explicitly emphasizes the use of natural pest control mechanisms (Box 2), and can thus be regarded as 'ecologically based IPM'. Apart from the development of pesticide resistance and adverse effects on natural enemies, several factors have contributed to this transition. An essential factor is the adverse impact of pesticides on the environment, food safety and the health of workers who are in direct contact with them (Anon., 2015; Casals-Casas and Desvergne, 2011). Customers, retailers and governments in many countries are nowadays concerned about the use of chemical control, resulting in the request for pesticide residue-free products and a reduction in the list of legally allowed pesticides. Moreover, some scandals with illegal pesticides have forced growers to shift their pest management tactics into a more environmentally friendly direction (Pilkington et al., 2010). Improvements in crop production methods, such as the introduction of bumblebees (*Bombus terrestris* L.) as pollinators of major horticultural crops (e.g. tomato, eggplant, strawberry and melon), also promoted the use of BC due to their incompatibility with many pesticides. In addition, policy has stimulated the use of non-chemical control measures; for example, the EU has been advocating this since 2009 in its Sustainable Use of Pesticides Directive (EC, 2009).

Box 2 Terms and definitions of IPM components

BC is defined here as the use of living organisms and viruses, hereafter called BCAs, to suppress populations of pests, weeds and disease-causing organisms. Different forms of BC can be distinguished, including *'natural BC'*, *'classical BC'*, *'augmentative BC'* and *'conservation BC'* (van Lenteren and Woets, 1988; Eilenberg et al., 2001; Heimpel and Mills, 2017).

- *Natural BC* takes place without any human intervention.
- *Importation (= classical) BC* aims for long-term control of exotic pests in a certain area through the introduction of specialized BCAs from the exotic pests' native geographic range.
- *Augmentative BC* involves the mass-rearing and release of BCAs against pests with the aim of relatively short-term suppression. Augmentative BC can be further subdivided into *'inundation BC'* and *'inoculation BC'*.
    - *Inundative BC* is an augmentative BC practice in which no BCA reproduction is expected.
    - *Inoculative BC*, often used seasonally, is an augmentative BC practice in which the reproduction of a few BCA generations is expected.
- *Conservation BC* includes tactics that will improve existing levels of BC, including habitat manipulation and the selection of pesticides that do not harm BCAs.

Cultural control is defined here as 'changes in crop production methods that affect pests directly or indirectly through an effect on their BCAs and/or an effect on (induced) crop resistance' (Kruidhof and Elmer, 2019).

Chemical control concerns the use of synthetic or natural pesticides for pest and disease control. They can be subdivided into insecticides (against pests), acaricides (against mites), fungicides (against fungal diseases) and nematicides (against nematodes). Pesticides can be applied prophylactically or in a curative manner.

- *Prophylactic calendar spraying* concerns the application of pesticides on a regular basis independent of the presence or densities of pests and pathogens.
- *Threshold-based application* is defined here as a restriction in the use of pesticides based on the monitoring of pest damage and action thresholds.

IPM incorporates multiple non-chemical methods and allows for the use of pesticides only as a last resort. It has been defined by the EU as 'the careful consideration of all available plant protection methods and subsequent integration of appropriate measures that discourage the development of populations of harmful organisms and keep the use of plant protection products and other forms of intervention to levels that are economically and ecologically justified and reduce or minimise risks to human health and the environment. IPM emphasises the growth of a healthy crop with the least possible disruption to agro-ecosystems and encourages natural pest control mechanisms'. (European Union Framework Directive on the Sustainable Use of Pesticides (Directive 2009/128/EC))

3 Main challenges in ecologically based IPM

The transition towards ecologically based IPM has been most successful in fruit vegetables, such as tomatoes, cucumbers and peppers, as these crops have sufficiently long crop-cycles for populations of natural enemies of arthropod pests to build-up and can sustain relatively high levels of leaf damage without economic consequences. Moreover, they often have a (short) crop-free period during which routine sanitation measures can be carried out. In more recent years, BC is being developed for ornamental crops as well, especially for cut flowers that have lower damage thresholds than fruit vegetables, but can still sustain some pest damage to non-harvested parts. Developing ecologically based IPM for potted ornamentals and leafy vegetables poses a stronger challenge, as (to date) retailers accept virtually no pest damage on these crops. In addition, many ornamental crops and leafy vegetables have relatively short cropping cycles, thus limiting the possibility for populations of introduced BCAs to build-up. However, also in these crops, advances have been made. For example, the replacement of broad-spectrum pesticides against sciarid flies in potted flower crops by entomopathogenic nematodes (Georgis et al., 2006) and microbial BCAs has allowed for the introduction of natural enemies against other pests in these crops. Another case in which pests should be kept under tight control is when they are able to transmit viruses to the crop. Greenhouse construction and technology level also influence the success of BC (Lindquist and Short, 2004). Year-round pressure from pests that are active outside of the greenhouse in subtropical and tropical areas, in combination with greenhouse structures that offer little protection against the influx of pests, pose a relatively strong challenge to biological pest control. Moreover, extreme temperature events in these non-climate-controlled greenhouses can harm BCAs. On the contrary, greenhouses in temperate regions containing proper means of heating and ventilation, as well as computer-controlled environmental monitoring and control systems, offer good opportunities for keeping climatic conditions that promote the establishment and efficacy of natural enemies. Finally, ecologically based IPM is often complex due to the occurrence of multiple pests in a single crop. This implies that not only the alignment of multiple control strategies against a single pest organism is required, but also the alignment of these control strategies with the various control strategies against other pests. Fine-tuning of all these control strategies into an overall crop protection strategy is often referred to as a 'systems approach'. It requires extensive knowledge of all aspects of crop protection to develop ecologically based IPM system approaches for specific cropping systems, and thus a knowledge-intensive infrastructure that connects growers with crop protection consultants, extension workers and (applied) scientists is essential. Growers in countries and regions that lack such a knowledge-intensive infrastructure will face severe difficulties in implementing such approaches. In the next section, we will further illustrate the components and rationale of ecologically based IPM system approaches.

4 Components and rationale of ecologically based IPM

The IPM approach that is advocated by the EU centres around eight general principles: (1) prevention measures, (2) monitoring, (3) decision-making, (4) non-chemical methods, (5) pesticide selection, (6) reduced pesticide use, (7) anti-resistance strategies and (8) evaluation (see European Union Framework Directive on the Sustainable Use of Pesticides (Directive 2009/128/EC)). Barzman et al. (2015) have further elaborated these eight IPM principles, focussing primarily on arable cropping. Here, we tailor the IPM principles specifically to greenhouse horticulture. Just like Barzman et al. (2015), we focus primarily on prevention measures, as they form the basis of ecologically based IPM. These prevention measures include all the crop protection measures that can be applied before the pest or pathogen has established in the crop and that enhance the overall cropping system resilience. They encompass a myriad measure that falls into the category of cultural control and into certain forms of BC (see Box 2) and are opposed to so-called curative measures, that constitute intervention measures that will only take effect after the pest has established. While in the integrated control of arthropod pests in greenhouse horticulture, augmentative BC has historically clearly prevailed over cultural control methods, the opposite is true for the integrated control of crop diseases. In disease control, the so-called disease triangle, that is the interactions between the abiotic environment, the crop and the pathogen, has traditionally formed the basis. Nowadays, the focus of integrated management of both plant pathogens and pests has expanded, and both encompass not just three but all four major natural components, namely (a) the crop, (b) the BCAS, (c) the crop biotic and abiotic environment and (d) the pest or pathogen itself, as well as all the interactions between them. In addition to expanding the disease triangle to also encompass pests and BCAs, resulting in the so-called pest and disease tetrahedron (Fig. 2), we broadened the scope of the crop environment from its mere abiotic components that directly relate to crop growth to also include the 'ecological structure' (i.e. the availability of (alternative) food, shelter and oviposition sites for BCAs, pests and pathogens not offered by the crop itself).

According to the holistic scope of ecologically based IPM system approaches, all these four natural components should be tailored such that

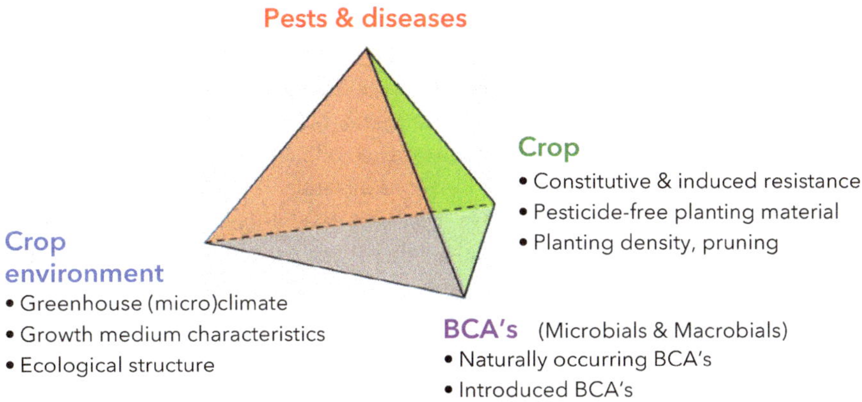

Figure 2 The pest and disease tetrahedron showing the direct and indirect effects of the crop, BCAs and the crop environment on pest and diseases.

– together – they constitute a firm cropping system resilience, and thus provide a strong first line of defence against the establishment of any incoming pests and diseases. With regard to the crop, it is for example important to choose a crop cultivar with a high level of constitutive and/or inducible resistance against the most noxious pests and pathogens, as well as to obtain pesticide-free planting material in order to allow for the early establishment of BCAs. Moreover, crop pruning and training are important to maintain a balanced microclimate that is not conducive to plant disease development. With regard to BCAs, it is vital to introduce generalist BCAs from the start of the cropping cycle (inoculative BC, see Box 1) and provide them with a functional ecological structure, that is the provision of alternative food, shelter as well as suitable oviposition sites, in case the crop itself fails to provide these in sufficient amounts or quality. Providing a functional ecological structure is also important for the establishment of naturally occurring BCAs (conservation BC). Moreover, it is important to select a crop cultivar whose resistance traits do not inhibit the functioning of BCAs of key pests, and ideally to select a crop cultivar that contains traits that enhance the functioning of BCAs (so-called indirect defence traits) (Cortesero et al., 2000). With regard to the abiotic crop environment, it is important to establish a greenhouse (micro)climate (with regard to light, temperature and relative humidity (RH)) that provides a good balance between healthy crop growth and the development and activity of BCAs in relation to the development of diseases and pests. Moreover, the choice of the growth medium is essential with respect to the risk posed by soilborne pathogens, and fertilization and irrigation practices will co-determine the propensity of pest and disease development. Tailoring the four natural components of the cropping system to enhance the cropping system's resilience, together with sanitation practices that aim to eliminate any harmful organisms present in and around the greenhouse or on the seed or planting material, forms the first general IPM principle. It is important to bear in mind that indiscriminate use of sanitation practices, independent of the pests, pathogens and BCAs already present in and around the greenhouse, may debilitate the level of resilience in the greenhouse environment. It is therefore essential to find the right balance between conserving the cropping system's resilience and implementing sanitation measures. While in practice, this is still difficult to attain, and the removal of pest and pathogen sources generally prevails as the safer option over the conservation of the cropping system's resilience, the advancement of high-throughput monitoring techniques for the detection of pest, pathogen and BCA populations on planting material, as well as in and around the greenhouse, may aid better decision-making in the future.

Monitoring, which constitutes the second general IPM principle, provides the basis for decision-making on the need for intervention with curative control measures, and ideally also for the tweaking of the applied prevention measures during the cropping cycle and the need for sanitation practices at the start of the cropping cycle. While monitoring has traditionally focussed on the incidence of harmful organisms, and is still described in this way by the EU Framework Directive, we would like to emphasize that it is essential to broaden the scope of monitoring to also include all the important factors that determine the resilience of the cropping system, that is the incidence of BCAs, the (micro)climatic conditions, the plants nutritional state and, ideally, the level of (induced) crop resistance. Only in this way it will be possible to spot any aberrations in the cropping system's resilience at an early enough stage for restoration of this resilience being a feasible option.

Decision-making, which constitutes the third general IPM principle, is inextricably linked with monitoring, and involves deciding whether, when, how and where to apply plant protection measures. It has traditionally been linked to decision-making on intervention with curative control measures, and has centred around economic pest threshold values in specific regions, crops and climatic conditions. While threshold values remain important for deciding when to apply full-field (potentially) disruptive pesticide applications, their importance for ecologically based IPM that is centred around the enhancement of the cropping system's resilience is limited. In ecologically based IPM, early warning signs that the balance of the cropping system is disturbed are usually sufficient reason for intervention with non-disruptive measures, starting with the tweaking of the prevention measures that have already been taken at the start of the cropping cycle and if needed, supplemented with inundative augmentative BC (see Box 2). Moreover, quick intervention will often allow for the implementation of curative control methods to be applied to hot spots only, thus reducing the cost of intervention as well as limiting unwanted side-effects in the case of chemical control. The fact that four of the eight general principles of IPM are related to curative control methods (principles four to seven), of which principles five to seven are related to the use of pesticides, suggests a strong emphasis on curative control within an IPM approach. However, we believe that reduced pesticide use (principle six) and anti-resistance strategies (principle seven) automatically follow when ecologically based IPM is practised well. While the EU Framework Directive has incorporated 'evaluation' as the eighth general IPM principle, and stresses the use of records on the use of pesticides and monitoring data of harmful organisms for the evaluation of the success of the applied plant protection measures, we believe that this description of evaluation does not do justice to the holistic approach of ecologically based IPM. We instead advocate the role of evaluation to be broadened, and instead of basing the success of the plant protection measures taken on the use of pesticides only, to take into account the effect of preventative measures on the cropping systems resilience (including the establishment of BCAs and the crop health status) as well. Evaluation then evolves from an activity carried out at the end of the cropping system to an activity that is tightly linked with monitoring and decision-making during the cropping season.

5 Cultural control practices

Cultural control practices (see Box 1) can be categorized as preventative control measures that can take effect before the pest or pathogen has established in the crop. They encompass (a) practices that are usually applied for general agricultural purposes, but can have a direct or indirect and positive or negative effect on pest incidence (e.g. greenhouse climate control, fertilization, irrigation, crop density, training and pruning), (b) practices that are used solely or mainly for pest and disease control (e.g. sanitation, insect screens) and (c) practices that are used for both agricultural as well as pest control purposes (e.g. the choice of crop cultivar and growing medium, grafting, crop rotation and composting) (after Katan, 1996).

5.1 Sanitation

Enhancing the cropping system's resilience can only suppress pests and diseases when pest and disease incidence is sufficiently low at the beginning of the crop production cycle. The practice of sanitation is therefore an integral component of any IPM strategy. Sanitation practices that aim to eliminate harmful organisms present in and around the greenhouse can be subdivided into (a) sanitation of greenhouse structures, (b) disinfestation of the soil or substrate, (c) disinfestation of used irrigation water, (d) disinfestation of seed and planting material, (e) removal of non-crop plants and crop residues, (f) using barriers to exclude pests and pathogens from greenhouses and (g) implementing measures to prevent the spread of pathogens and pests by equipment or people. With regard to disinfestation of soil and/or substrate prior to planting a new crop, several practices exist. Because the use of chemical soil fumigants has become very restricted in many parts of the world, many growers have shifted to soilless cultures, thus allowing for an easy replacement of used substrates. Others have resorted to alternative measures for soil disinfestation, triggering a renewed interest in non-chemical alternatives including steam disinfestation, solarization and anaerobic (or biological) soil disinfestation. Steam disinfestation is mainly applied in high-tech greenhouses in Northern regions, while solarization is usually practised in warmer climates in between cropping cycles during the summer season. Crop rotation, which constitutes another well-known measure that can break cycles of crop-specific pests and diseases, is not commonly practised by conventional greenhouse growers. Both solarization and steam disinfestation practices are based on reaching relatively high soil temperatures for a long-enough time period to kill or weaken most important plant-pathogenic fungi, nematodes and weeds (Bollen, 1969; Runia, 1983; D'Addabbo et al., 2010). With solarization, solar radiation is trapped under a plastic mulch on moist soil during a period of approximately 4–6 weeks of high ambient temperature. Combining solarization with the addition of an easily degradable carbon source (i.e. anaerobic/ biological soil disinfestation) can enhance its effectiveness, as well as reduce the solarization period and/or soil temperature that is needed to reach effective control of harmful organisms (Butler et al., 2012; Morra et al., 2018). The danger of creating a 'biological vacuum' with the application of soil disinfestation methods was already recognized early on (Baker, 1962). Fortunately, many species of antagonistic microbes, such as *Trichoderma* sp. and *Bacillus* sp., were found to be more heat-tolerant in comparison with pathogenic fungi (Stapleton and Devay, 1984; Katan, 1987; Bollen, 1969). However, excessive soil heating (i.e. temperatures exceeding 70°C, and particularly approaching 100°C) becomes detrimental to most soil biota, including most antagonistic microorganisms. Therefore, an important aspect to take care about when applying heat-based non-chemical soil disinfestation methods is to avoid temperatures exceeding 70°C as much as possible, for example by taking measures that ensure a more even distribution of temperature in the soil profile (Katan, 2000).

Clean seed and planting material is paramount to beginning a new crop, and the trade in seeds is subjected to strict phytosanitary regulations. Moreover, seeds can be relatively easily disinfected from many common pathogens. Because growers are under increasing pressure to produce chemical-free products for the market, they are in turn increasing their demands on greenhouse nurseries for the production of chemical-free planting material that allows for the quick establishment of beneficial predators and microbial antagonists. However, sanitation of planting material without negative effects for BCAs poses an important challenge. Biopesticides that are compatible with the use of natural enemies, such as mineral oils, insecticidal soaps and entomopathogenic fungi and nematodes, may be applied to reduce pest numbers prior to planting (Buitenhuis et al., 2016). Some plants propagated by bulbs, corms and rhizomes may respond well to hot water treatment. However, close attention to temperature and timing is essential to guarantee eradication of the pathogen and/or insect pest without damage to the plant (Magie, 1971). This is also true for Controlled Atmosphere Temperature Treatment, that is the exposure of plant material to a combination of relatively high temperature, high CO_2 and low O_2 levels (van Kruistum et al., 2014). However, sterility of the planting material is not the sole criterion for a good start of the cropping cycle. Transfer of natural enemies from the propagation site to the cultivation site increases the resilience of the system and should, in our opinion, allow for the presence of some (non-quarantine) pest individuals that can be controlled relatively easily.

Moreover, the removal of weeds, crop residues and algae that serve as reservoirs for pests and pathogens should be carried out as a first line of defence (e.g. Northfield et al., 2008). However, situations exist where maintaining certain plants in and around the greenhouse could provide a benefit by supporting beneficial predators. In some Mediterranean regions, an important part of pest control is carried out by naturally occurring predators. In these cases, the growth and conservation of plants that can harbour natural enemies in and around the greenhouse is an important component of a conservation biocontrol strategy (Perdikis et al., 2011). Pruning of plants that contain pests can also remove developing natural enemies. When the parasitoid *Encarsia formosa* is introduced in tomato, and black parasitized whitefly nymphs can be detected on the older leaves, it is sometimes advised to keep pruned leaves on the floor for at least 2 weeks to allow for the emergence of new parasitoids. However, this practice may lead to build-up of pathogens that profusely sporulate on decaying plant residues, such as *Botrytis*.

5.2 Exclusion of insect pests from the greenhouse

Insect screens are the most commonly used means to exclude insect pests from the greenhouse during the cropping cycle. Mesh size determines the type of insects that can be excluded (Bethke and Paine, 1991). However, smaller mesh sizes of the insect screen interfere negatively with the greenhouse microclimate due to the lowered ventilation capacity (Teitel, 2007). It is therefore generally not advantageous to prevent invasions of small-bodied insects such as thrips, and to a lesser extent whiteflies and aphids, by fixed use of insect screens. Alternatively, the positioning of vents and ventilation schemes can be designed based on information from studies that monitored when and where pests enter greenhouses and growing tunnels. An Israeli study, for example, revealed that both thrips and sweet potato whiteflies enter greenhouses almost exclusively during the morning. Moreover, both these pest species were observed to fly upwind (with thrips flying only at low wind speed), causing a greater risk of pest invasion at the leeward side of the greenhouse (Ben-Yakir et al., 2008). A completely alternative way to reduce the immigration of pests into greenhouses is to create a UV-deficient greenhouse environment. While most traditional greenhouse cover materials poorly transmit UV-B (280–315 nm), they transmit a large part of UV-A (315–400 nm). However, several polyethylene materials are nowadays on the market that block the whole UV spectrum below 360–400 nm. Multiple studies have shown a strongly reduced immigration of thrips, whiteflies, aphids and leafminers into tunnel greenhouses covered by these types of UV blocking polyethylene materials, as well as a reduced incidence of insect vector-transmitted crop viruses and a suppression of several foliar diseases (Raviv and Antignus, 2004; Antignus, 2000). Studies have shown that these pests have photoreceptors with maximum sensitivity in the UV-A spectrum (Matteson et al., 1992; Mellor et al., 1997; Kirchner et al., 2005). Whether it is beneficial to block the UV from glass greenhouses used in Northern conditions needs further investigation. On the one hand, blocking UV from glass greenhouses can create a similar or potentially larger contrast in UV spectrum compared to the above studies, as the UV-A transmission through glass is higher than through many UV-transmitting polyethylene covers (Hemming, 2004). On the other hand, the impact on pest immigration in these more closed glass greenhouses may be lower compared to the semi-open tunnel greenhouses into which pests can easily enter. Moreover, blocking the entire UV spectrum from low-iron glass greenhouses, which in contrast to traditional glass greenhouses does allow for partial transmission of the UV-B spectrum (Hemming, 2004), can also cancel beneficial effects of UV-B on plant-induced resistance. Finally, a UV-deficient environment can have negative side-effects on pollinators (Blacquiere et al., 2006; Dyer and Chittka, 2004).

5.3 Choice of crop cultivar

Crop cultivars can differ in their degree of constitutive and inducible resistance to pathogens and pests. However, to date, complete resistance against pathogens exists only for some crop-pathogen combinations, including, for example, tomato cultivars that are resistant against tomato yellow leaf curl virus and cucumber cultivars that are resistant against powdery and downy mildew. Often, resistance that is constitutively expressed is costly for the plant, and can thus have a trade-off with growth and productivity. More frequently, however, cultivars offer partial resistance towards pathogens or pests. Moreover, crop production methods, microbial and chemical elicitors of plant defences, as well as greenhouse climate can have a profound effect on induced resistance, that is the level to which crop resistance is expressed. Crop (induced) resistance traits do not only affect pests and/or diseases, but can also directly and indirectly affect natural enemies and microbial antagonists in both positive and negative ways (e.g. Riddick and Simmons, 2014; Berendsen et al., 2018; Cortesero et al., 2000). Crop resistance promotes BC whenever the negative effect on the growth rate of the pest or pathogen population exceeds the negative effect on the abundance of natural enemies and their attack rate. In many cases, so-called bottom-up control through partial plant resistance is expected to supplement 'top down control' by BCAs. In other cases, however, plant resistance traits are incompatible with certain BCAs. The presence of glandular trichomes on tomato, for example, severely inhibits the movement of most predatory mite species, but not that of the tomato russet mite *Aculops lycopersici* (van Houten et al., 2013). This pest mite has thus acquired an enemy-free space on tomato, and can currently only be controlled chemically. Moreover, the BC system in tomato is shaped around the use of mirid predatory bugs that have specifically adapted to foraging on hairy plants. When selecting a crop cultivar, it will be impossible to choose a cultivar that has resistance against all important pests and diseases. While choosing a crop cultivar is most often inspired by commercial factors, it will be most beneficial for the implementation of an ecologically based IPM systems approach to select a cultivar with enhanced constitutive and/or inducible defences against pathogens and/or pests that are most difficult to control biologically or by other non-chemical means. With regard to soilborne diseases, grafting of leaf vegetables using resistant rootstocks forms a common practice that does not interfere with commercial objectives (Louws et al., 2010).

5.4 Greenhouse (micro)climate

Traditionally, greenhouse climate has been generally managed with the sole objective of optimizing crop productivity. However, with decreasing options for chemical pest and disease control, the value of integrating knowledge on the interaction between greenhouse climate and pest and disease incidence is increasingly acknowledged. The light, temperature and humidity conditions that pests, pathogens and BCAs experience largely depend on the microhabitats where these organisms live. Depending on the species or developmental stage, the organism can reside inside or outside of the plant tissue, on the abaxial, or adaxial leaf surface, higher or lower in the crop canopy, hidden away inside flowers, bulbs, leaf axils, in the root zone or elsewhere in the greenhouse.

Hence, climate data collected by sensors above the canopy cannot be readily related to pest, disease and/or natural enemy population dynamics.

Temperature is a key factor driving the population dynamics of pests, pathogens and their BCAs. Each arthropod species has its own 'thermal window', that is the temperature range under which it can grow and reproduce (Dixon et al., 2009). Near the upper and lower limits of this thermal window, oxygen supply and anaerobic metabolism restrict the arthropod's performance. Temperature-driven models are an essential tool in helping to predict population dynamics of pests and their BCAs, and can assist in the selection of BCAs that are most suitable for the thermal conditions in the greenhouse. Numerous studies have focussed on the relationship between temperature and arthropod developmental rates, and both linear and non-linear models have been developed to describe these relationships (Campbell et al., 1974; Kontodimas et al., 2004). As temperature does not only affect arthropod developmental rate, but also affects all other arthropod life history traits, an often-used measure to characterize the impact of temperature on arthropod performance is the intrinsic rate of increase (r_m). This measure integrates all the different life history parameters including developmental rate, immature survival, adult fecundity and longevity and sex ratio, and is therefore more reliable than any single life history parameter, as these are often differentially affected by the microclimate. However, with respect to biological pest control, the r_m does not capture the effect of the temperature on the predator's attack rate. Only for a few natural enemy–prey interactions, the effect of temperature on natural enemies' functional response (i.e. the relationship between the number of prey available and the number of prey killed) has been assessed (Jalali et al., 2010; Skirvin and Fenlon, 2003). Moreover, only a few studies have assessed the flight capability of parasitoids or predators of greenhouse pests over a range of temperatures (e.g. Jerbi-Elayed et al., 2015; Zhang and Shipp, 1998). Especially for specialist natural enemies operating at low pest densities, long-range search rate is expected to have a large influence on the efficacy of biocontrol. As temperatures near the borders of the thermal window usually constrain activity more than growth and development, leaving searching rate and prey handling time out of the equation may overestimate the efficacy of natural enemies when greenhouse temperatures are close to these threshold values. This is especially important in warmer regions, where temperatures often exceed the upper temperature limits, and in greenhouses of the Northern regions where energy-saving practices result in the reduction of temperatures during winter time. Lowering the temperature set-points in the greenhouse may have adverse effects on pest control, as the activity of natural enemies may be restrained before the development of the pest species comes to a halt. The identification of natural enemies that perform well under cool climatic conditions will become important for improving the sustainability of greenhouse-grown crops through the reduction of both energy expenditure and pesticide use. Furthermore, it is important to note that in the context of energy savings, often not just the greenhouse temperature set-points are lowered, but temperatures are also allowed to fluctuate more with ambient temperatures, resulting in a larger difference in day and night temperatures and/or differences in mean temperatures over the days. The impact of fluctuating temperatures on greenhouse pests and their natural enemies has thus far not received much attention (but see Bayu et al., 2017; Ullah and Lim, 2015; Vangansbeke et al., 2013, 2015a,b). The impact of fluctuations in night- and daytime temperatures on the searching efficacy of parasitoids and predators may strongly depend on the predator's and parasitoid's diel activity. Temperature fluctuations with low night temperatures should affect the searching efficacy of day-foraging parasitoids and predators in a positive way, and affect the searching efficacy of night/twilight-active predators, such as *Aphidoletes aphidimyza*, in a negative way.

Free moisture and RH play key roles in the incidence of plant diseases, as they affect the infection and sporulation processes, and in some cases also colonization of fungi, oomycetes and bacteria (see e.g. Huber and Gillespie, 1992: and references therein). Especially the point when water will condense to form a film of water over the leaf, flower or fruit surface is a crucial factor determining infection rates by many important plant pathogens (Huber and Gillespie, 1992; Jewett and Jarvis, 2001). Pathogens differ in the duration of surface wetness they require for completing the infection process. Information on the pathogen-specific surface wetness periods required to complete infection has been summarized in Huber and Gillespie (1992) and Magarey et al. (2005). Preventing wetness of leaf, flower and fruit surfaces, or at least limiting the duration of the surface wetness period or interrupting it, can thus provide a powerful tool for disease management of many important pathogens. Crop surface wetness results primarily from water that has been intercepted by the canopy during overhead irrigation or fogging, dew that forms on the leaf, flower or fruit surfaces as a result of water vapour condensation or, in some cases, as a result of guttation (Rowlandson et al., 2015). Jewett and Jarvis (2001) have listed several studies that addressed the epidemics of plant pathogens in relation to surface wetness duration and ambient air temperature, including the epidemiology of *Botrytis cinerea* (grey mould) in cucumber, tomato, roses and gerbera. Körner and Holst (2005) developed and tested a mathematical model for *Botrytis cinerea* that avoided disease by keeping the RH low with heating and ventilation. They calculated the energy costs of this practice and calculated the time for leaf surfaces to be dry based on the energy balance of the leaf surface. Climate control was then optimized for maximum energy saving that provides grey mould prevention. As more research on pest damage/risk analysis advances, algorithms in these models can be refined to provide growers with more informed decision-making tools. Free moisture and RH can also affect the performance of arthropod species such as spider mites (Duso et al., 2004), but the impact in general is much less determining than for fungal pathogens.

For plants, light is a primary resource that is essential for growth and development. However, other organisms also, including pests, pathogens and their natural enemies, use light to steer important developmental and behavioural processes. They all possess photoreceptors that capture specific parts of the light spectrum. Light perception in arthropods, fungi and bacteria have been reviewed by, respectively, Johansen et al. (2011), Schumacher (2017) and Kraiselburd et al. (2017). Changing light conditions can be used to manipulate pests and diseases either directly or indirectly (through light-induced plant defences). Light can affect plant defences in multiple ways. To start with, the mechanisms that plants have evolved to protect themselves from harmful wavelengths and peak light intensities, that is through the production of secondary metabolites, such as phenols and flavonoids, as well as through thickened cell walls, can also enhance defences against certain pathogens and pests (Ballare, 2014; Le Gall et al., 2015). Moreover, every aspect of light (i.e. its intensity, wavelength distribution, direction and photoperiod) can give a plant specific information about its surroundings. A high red:far-red ratio, for example, is a signal that there is competition for light. This will induce a shade-avoidance response that manifests itself by a strong investment in aboveground growth and (rapid) shoot elongation. This usually goes at the expense of the formation of plant defensive metabolites and structures, thus rendering the plant more susceptible to both pests and diseases (Ballare, 2009, 2014). Moreover, light can affect the density and composition of glandular trichomes (Nihoul, 1993; Escobar-Bravo et al., 2018). Nihoul (1993) found tomato plants grown in the light to have larger type VI glandular trichome heads, resulting in a higher number of entrapped spider mites (*Tetranychus urticae*) and predatory mites (*Phytoseiulus persimilis*), compared to plants grown in the shade. With regard to the direct effect of light on pests, the use of light traps for monitoring pest populations has a long history. Moreover, as stated before, several studies have reported a strongly reduced immigration of thrips, whiteflies, aphids and leafminers into UV-deficient greenhouses compared to greenhouses with high UV-transmitting (plastic) cladding materials, as well as a reduced incidence of insect vector-transmitted crop viruses and a suppression of several foliar diseases (reviewed by Raviv and Antignus, 2004; Antignus, 2000). Both Johansen et al. (2011) and Diaz and Fereres (2007) have summarized the effect of specific properties of greenhouse cladding materials on several species of pest insects. Application of light in the visible spectrum can also be an important tool for suppressing plant diseases, and especially the timing of the application can be crucial for its effect. Suthaparan et al. (2014) showed that exposure of cucumber plants to additional red light without UV-B at the beginning and the end of the day suppressed the disease intensity of *P. xanthii*, and that a short exposure of rose plants to red light during the dark period suppressed the sporulation of the powdery mildew *Podosphaera pannosa*. Far-red light, on the other hand, had a stimulating effect on the sporulation of this pathogen. With respect to direct effects of light spectrum on BCAs of arthropod pests, only few studies have been published. However, Cochard et al. (2017), as well as Wang et al. (2013), found inhibiting effects of red light on the parasitoid activity and development of Orius predatory bugs, respectively. Not only light spectrum, but also light intensity and daylength, can affect pests, pathogens and natural enemies (Zilahi-Balogh et al., 2006, 2007; Wang et al., 2013). Apart from other factors, such as a change in light spectrum, temperature and humidity levels, this can help explain differences in IPM success between the different seasons, and give hints on how to improve this. Day length can also affect the production and germination of spores of *P. pannosa* in rose, with a stronger inhibiting effect with increasing daylength (Suthaparan et al., 2010). More research is warranted to decipher the multiple direct and indirect effects that light can have on both pests and diseases as well as their BCAs. This type of research will be, among others, important to be able to assess the effects of developments in artificial (LED) light technologies, which make it possible to illuminate crops with varying spectral compositions.

5.5 Fertilization

Mineral nutrition can exert an important effect on pest and disease development, and has long been recognized as an important aspect of disease control in horticultural crops (Hobbs and Waters, 1964; Stall et al., 1965). Knowledge about the effects of mineral nutrition of plant diseases has been summarized in a number of reviews and books (Dordas, 2008; Amtmann et al., 2008; Walters and Bingham, 2007; Engelhard, 1989; Datnoff et al., 2007). In general, disease severity tends to increase with increasing nitrogen (N) supply, while deficiencies of phosphorus (P), potassium (K), sulphur (S) and silicon (Si) enhance disease susceptibility, and the same is true for most of the micronutrients. However, one needs to be careful with these kind of generalizations, as many exceptions exist. For example, Hoffland et al. (2000b) found the influence of nitrogen supply in tomato disease to be pathogen-dependent. No effect of N supply was found on tomato susceptibility to the vascular wilt pathogen *Fusarium oxysporum* f.sp. *lycopersici*, while – in line with the general observed pattern – the susceptibility of tomato plants to the powdery mildew fungus *Oidium lycopersicum* and the bacterial pathogen *Pseudomonas syringae* pv. tomato increased with increasing N supply. On the contrary, however, tomato plants grown under high

N conditions were 2.5 times less susceptible to *Botrytis cinerea* grey mould. One of the reasons why it is difficult to draw general conclusions on the effect of plant nutrition on disease development is that the underlying mechanisms are complex and diverse. Among others, nutrient supply can change (a) the value of the plant tissue as a source of nutrients and energy to the pathogen, (b) the presence of plant defence chemicals and physical barriers that can suppress pathogen infection and development and (c) the canopy structure of the crop producing a microclimate that is more or less conducive to disease development (Walters and Bingham, 2007).

Only a few studies exist that have researched the effect of fertilization on pests in greenhouse crops. Lowering the nitrogen, phosphorus and potassium in cut roses to 33% or to 50% of the recommended fertilization level caused, respectively, a 30% reduction in thrips abundance and a 50% reduction in spider mite abundance without compromising flower production (Chow et al., 2009, 2012). Moreover, lower soil nitrogen levels in a chrysanthemum crop coincided with reduced thrips populations (Schuch et al., 1998) and high phosphorus rates in Impatiens favoured thrips development, although this did not result in increased thrips damage (Chen et al., 2014). Two-spotted spider mites preferred tomato plants that were grown at high nitrogen availability, with a low C/N ratio, both when given a choice between leaf disks and between plant odours (Hoffland et al., 2000a). Moreover, the form in which nitrogen is offered to the plant can co-determine the effect on pests and diseases. For example, oviposition rate of the sweet potato whitefly, *Bemisia tabaci* (biotype B), was higher on NH_4NO_3 treated plants than on non-fertilized or $Ca(NO_3)_2$ treated plants. Moreover, cultivars of four important horticultural crops (lettuce, tomato, pepper and cucumber) with low concentrations of aromatic amino acids compared to total leaf protein were found to have reduced western flower thrips damage (Mollema and Cole, 1996). Close attention to timing and the amount of irrigation in the root medium can result in improved quality and can reduce many problems in the greenhouse such as root rot caused by *Pythium* and *Phytophthora* pathogens, and fungus gnats. Moreover, preventing excessive moisture in the greenhouse can reduce the growth of algae that form an ideal breeding ground for shore flies. Sub-irrigation also has multiple benefits. It avoids the prolonged leaf wetness resulting from overhead irrigation, thus preventing the germination and infectivity of many foliar pathogens. Moreover, it diminishes the spread and dissemination of many foliar and soilborne pathogens through splash dispersal. On the other hand, sprinkler irrigation may help reducing pest populations of whiteflies (Castle et al., 1996), thrips (Schuch et al., 1998) and mites (Opit et al., 2006), but may at the same time cause the washing-off and accelerated spoiling of alternative food that is often applied to the crop for generalist predators.

6 Methods to enhance the establishment of BCAs through the supplementation of resources

Poor establishment and persistence of natural enemies is one of the main problems in BC. Besides the application of toxic pesticides or fungicides, causes for this poor establishment include (a) a lack of sufficient good-quality resources such as alternative prey, plant-provided food, shelter or oviposition sites; (b) unfavourable climatic conditions; (c) the presence of intraguild predation, hyperpredation and/or hyperparasitation; and (d) the presence of crop resistance traits that impede the behaviour or development of BCAs. When the main cause of poor establishment is the lack of sufficient good-quality resources, supplementing the missing resources can boost BC efficacy (Messelink et al., 2014). Inoculative augmentative BC leading to persistent populations of natural enemies in the absence of pests has been coined the 'standing army approach', a term that emphasizes that the 'army' of natural enemies should be ready for the 'battle' before pests arrive. Poor establishment due to missing resources is partly caused by the types of natural enemies used. The selection of natural enemies for augmentative BC traditionally focussed on specialist natural enemies that were released to obtain rapid pest control (van Lenteren and Woets, 1988). Well-known examples are the spider mite predator *Phytoseiulus persimilis*, the whitefly parasitoid *Encarsia formosa* and the aphid parasitoids *Aphidius* spp., which are still successfully used worldwide. However, while these specialists can be very effective as they are well adapted to combat specific pests, they cannot establish in the absence of these pests (or their close relatives), and often disappear after pest densities have been reduced. Therefore, their implementation requires intensive monitoring to assure an accurate timing of intervention. To overcome the problem of establishment and monitoring, in some cases, specialist natural enemies (e.g. aphid- and whitefly parasitoids) are released routinely (e.g. weekly) as an 'insurance policy', but these prophylactic releases are not always economically viable.

Biocontrol scientists were aware of these problems for a long time and developed several methods to support the establishment of specialist natural enemies (Ramakers and Rabasse, 1995). Rather controversial is the pest-in-first method, which has been developed for spider mite control (Markkula and Tiittanen, 1976). This method involves the inoculation of plants with a low level of spider mites early in the growing season and the subsequent release of their predators, thus allowing the predators to establish at a low equilibrium density in the crop and give protection against subsequent spider mite invasions. Currently, this method is mainly used in sweet pepper crops. The method was not adopted immediately, but it promoted another way of thinking about pest control, based on living with the pest rather than trying to eliminate it. Another method, mainly developed to support specialist natural enemies, is the so-called banker plant method (Huang et al., 2011). With this method, alternative prey/host species that are closely related to the target pest species of the specialist natural enemy are made available through the introduction of a non-crop plant harbouring these alternative species. A widely applied system in greenhouse crops has been the use of monocotyledonous plants with cereal aphids that serve as alternative hosts for parasitoids of pest aphids (Bennison, 1992).

The options to support establishment and augment densities of natural enemies increase when generalist predators are used, because their broader diet range enables them to persist or even reproduce on alternative prey or plant-provided food sources in the absence of pest organisms (Symondson et al., 2002). This offers the opportunity to introduce generalist predators before pests invade the crop (standing army) when the crop itself provides such food sources (e.g. Orius predatory bugs in sweet pepper) or when alternative food sources are separately applied to the crop. Various alternative food sources, such as pollen, sterilized moth eggs, shrimp cysts and astigmatid mites, are now commercially produced for this objective (Messelink et al., 2014). However, some of these food sources are also edible for omnivorous thrips species and caution is needed when thrip densities are relatively high compared to predator densities (Leman and Messelink, 2015; Vangansbeke et al., 2016).

Another method to assist establishment of natural enemies is by application of rearing units or the so-called slow-release systems. Most popular for greenhouse application became the sachets for predatory mites containing prey mites as a base for reproduction (Sampson, 1998). The mites reproduce continuously in the sachets over several weeks and progeny subsequently disperse into the crop. Several types of such sachets are nowadays commercially available. Another example of an open rearing unit is the system developed for short-winged rove beetles (Bennison et al., 2008).

Besides food sources, natural enemies also need suitable shelter and oviposition sites to establish populations. Many plants provide typical tufts of hairs and other minute structures in the vein axils, so-called acarodomatia, which offer predatory mites protection from low humidity conditions, and can stimulate their egg production (Walter, 1996; Grostal and O'Dowd, 1994). A lack of such domatia, particularly on many ornamental plants, may be one of the reasons for poor predatory mite establishment. Some studies have shown potential for the enhancement of predatory mite establishment through the provision of artificial acarodomatia (Loughner et al., 2011; Bresch et al., 2019), but so far this has not been applied. Suitable oviposition sites are also important for predatory bugs that lay their eggs into plant tissue, such as anthocorid and mirid predators. Some ornamental plants are too woody and not suitable for oviposition (like roses) or some plants exude toxic latex after an oviposition attempt (like Poinsettia), which hampers population establishment.

This may in some cases be solved by providing banker plants that serve mainly as 'oviposition plants'.

Conservation of microbial BCAs for the control of arthropod pests has, in contrast to arthropod natural enemies, not received much attention, probably because applications were traditionally used as curative rather than preventative control measures. However, in the review paper of Pell et al. (2010), several methods for the conservation of fungal entomopathogens and the stimulation of natural epizootics are suggested. These include (1) preventing intense soil disturbance as many fungal entomopathogens reside at least during part of their life cycle in the soil, (2) preventing frequent use of fungicides, (3) encouraging non-pest alternative hosts and (4) encouraging non-pest arthropods for the dispersal of conidia. Although these practices are not widely used in greenhouse crops, some organic growers are exploring the use of banker plants with non-pest aphids, such as *Sitobion avenae* on wheat plants, infected by entomophthoralean fungi to encourage the induction of natural epizootics among pest aphids.

The composition of the microbiome in the rhizosphere, including the community of antagonistic microbes, can be influenced by multiple factors, among which (1) the exudates excreted by the roots of the crop, (2) the characteristics of the growing medium, (3) soil disinfestation practices and (4) augmentation of the rhizosphere with commercially available antagonistic microbes. Interestingly, even in relatively 'inert' substrates with little available organic carbon, an active microbiome can develop due to the excretion of exudates by plant roots that contain various organic and inorganic components (Koohakan et al., 2004). It is important to bear in mind, however, that disease suppression in soils and other growth media is often not solely based on so-called general suppressiveness, which relies on the competitive activities of the overall antagonistic microbial community. In certain cases, disease suppression can be attributed to the enrichment of specific (subsets of) soil microorganisms (Raaijmakers and Mazzola, 2016). Plants can play an active role in shaping the rhizosphere microbiome through the composition and quantity of exudates excreted from their roots. The studies of Mendes et al. (2018) and Berendsen et al. (2018) have described some of the mechanisms underlying this phenomenon, and nicely illustrate how disease outbreaks - even of foliar pathogens - can aid the build-up of specific disease-suppressive microbiomes. Moreover, it has long been known that the incorporation of certain types of compost can promote the disease-suppressiveness of the soil, although not all types of compost are equally suitable for the promotion of a disease-suppressive microbiome (see Bonanomi et al., 2007: and references therein). Abiotic characteristics of the (amended) substrate, such as the availability of easily degradable carbon (C), micro(nutrients), C/N ratio, pH and EC, can influence the composition and the activity of the microbiome. However, how exactly these abiotic characteristics relate to disease-suppressiveness is thus far not well understood (Bonanomi et al., 2010). Disease-suppressiveness of organic amendments turned out to most often be pathogen-specific, with amendments effective against one pathogen being ineffective or even conducive to other pathogens (Bonanomi et al., 2010). While disease-suppressiveness can result from the action of antagonistic microorganisms that originate from the substrate and/or organic amendments (Montagne et al., 2016; Antoniou et al., 2017), the residential microbiome can also interfere with the establishment of separately added beneficial microbes. Interestingly, several studies have found that the combined application of commercial thermotolerant antagonistic microbes in combination with solarization, or directly following steam disinfestation, resulted in better control of soilborne diseases than the application of either method alone (Sivan and Chet, 1993; Porras et al., 2007; Jayaraj and Radhakrishnan, 2008; Minuto et al., 2006; Luvisi et al., 2006). Future research should shed more light on the way plants respond to biotic and abiotic challenges with the excretion of root exudates, how root exudates and growth media characteristics affect the establishment of functional microbiomes and what their effects are on soilborne pathogens. This knowledge can aid plant breeders to develop crop cultivars that can actively shape a disease-suppressive rhizosphere microbiome against key soil pathogens as an indirect defence strategy as well as manipulate the growth medium with the amendment of compounds that trigger a functional microbiome. Moreover, future tools that are able to accurately assess the disease-suppressiveness of the rhizosphere microbiome will be of great help to decision-making by the grower.

7 Curative control measures

Preventative measures for pest and disease management are not always sufficient to keep densities below damage thresholds and in those cases, additional curative tactics are needed. A first step in curative control is usually the release of specialist natural enemies that that do not establish in absence of their host/prey. Such natural enemies can supplement the control of generalist predators. For example, the control of whiteflies in cucumber and pepper can be based on preventative releases of generalist predatory mites that are supplemented with specialized parasitoids after pest detection (Messelink et al., 2006; Calvo et al., 2009). In this case, both types of natural enemies nicely complement each other by attacking different stages of the pest, which is an effective strategy. Release densities of mass-produced natural enemies may need to be increased when pests get out of control. Extra introductions of high numbers of natural enemies with the aim to outnumber pest densities and achieve a rapid control is defined as 'inundative BC'. Inundative BC is often focussed on pests with a strong patchy distribution, such as aphids, spider mites and mealy bugs, to treat the 'hot spots' with an overkill of natural enemies. For example, some growers release high numbers of lacewing larvae when the control of aphids gets out of balance. Mass releases of *Trichogramma* egg parasitoids to control Lepidopteran pests is another example of inundative control. The strategy of natural enemy releases strongly depends on the pest- and natural enemy species. Specialist parasitoids or midges with a good searching ability do not need to be released close to the pest colonies they attack, whereas the release of commercially produced predatory larvae with limited mobility, such as lacewing or coccinellid larvae, needs to be targeted specifically to the pest-infested plants.

BCAs of arthropod pests that are based on microorganisms (fungi, bacteria and viruses) or entomopathogenic nematodes are mainly developed for curative BC. Because the effects of these agents usually do not persist and can only cause short-term effects, they have also been coined 'biotic insecticides' (Stern et al., 1959). The microbial agents require direct contact with the target pest or pathogen and well-designed spraying techniques are therefore essential to achieve effective control. Moreover, many of these entomopathogenic microorganisms and nematodes only work well under the right climatic conditions, so the timing of spraying and if possible, the temporary modification of the (micro)climate, is also very important (Gonzalez et al., 2016).

A last step in ecologically based IPM is the use of pesticides as a correction tool in case the pest or disease cannot be controlled well with biocontrol agents. However, this should be limited as much as possible, because most pesticides have lethal effects on natural enemies, including some underestimated sub-lethal side-affects (Desneux et al., 2007). In greenhouses, it is particularly important to prevent side-effects that have negative effects on established generalist predators that need a long time to develop population densities that are high enough for effective pest control. Disruption of such systems is not easy to recover and can have tremendous effects on the total biological system. Limiting undesired side-effects on natural enemies can be achieved by selecting pesticides that are compatible, or as close to compatible as possible, with natural enemies. This selectivity is based on their target specificity and residual time rather than on their origin, as many pesticides of biological origin can be disruptive to natural enemies. Several databases exist, mainly provided by biocontrol companies, that offer information on the side-effects of pesticides on commercially available natural enemies. Undesired side-effects of pesticides on natural enemies could be further reduced by adapting the timing, place and mode of application (Croft, 1990). However, caution is needed especially for pesticides with a high level of persistence as these could disrupt natural enemy establishment over long periods.

8 Food web complexities in BC

One of the challenging aspects of BC is the complexity of interactions within communities of pests and natural enemies, as some of these interactions can strongly influence the results of BC. Consumption (i.e. herbivory, predation and parasitism) and competition are considered the two most important interactions determining the structure of communities (Chase et al., 2002). Understanding these interactions is important in order to optimize efficacy of BC systems. Here, we will shortly summarize some of the interactions that can occur when applying BC.

The shift towards a 'standing army approach' in greenhouse BC promoted the use of generalist natural enemies that, because of their generalist feeding habits, not only feed on prey, but also on other natural enemies. This interaction is called intraguild predation when both these natural enemy species compete for the same prey species (Polis et al., 1989). Intraguild predation can be unidirectional with one intraguild predator and one intraguild prey (the other natural enemy), or bidirectional with both predator species preying upon each other. The phenomenon of intraguild predation has been described for many natural enemies that are used for BC in greenhouse crops and can in some cases have negative consequences for BC (Rosenheim et al., 1995; Janssen et al., 2006). The intraguild predator can for example disrupt pest control when the intraguild prey is more effective in controlling the shared pest than the intraguild predator, but such disruptive effects are rare (Janssen et al., 2006).

Predators can also predate on other natural enemies without sharing a pest, which is referred to as hyperpredation. Such interactions can potentially strongly disrupt BC. In greenhouses, it has been observed that generalist predatory mites seriously disrupt the BC of aphids with predatory midges by predating on the eggs of these midges (Messelink et al., 2011). The predatory mites themselves do not predate upon aphids and thereby do not compensate for the reduced efficacy of the specialized aphid predator. Similar trophic disruptions occur with hyperparasitoids that parasitize on primary parasitoids, but not directly on the herbivore (Sullivan and Völkl, 1999). Although hyperparasitoids are not introduced intentionally into greenhouses, they are very common in aphid-parasitoid communities in greenhouse crops and can strongly reduce the efficacy of the primary parasitoids (Bloemhard et al., 2014; Acheampong et al., 2012).

Not only can the diversity of natural enemies affect pest control, but also the diversity of herbivores can affect the control of particular pest species through resource competition or indirect interactions mediated by the shared host plant or natural enemy (Karban and Baldwin, 1997; Holt, 1977). Such interactions will increasingly occur within ecological based IPM programmes with limited use of pesticides, and thus a larger variation of pest species. The term 'apparent competition' was introduced for the interaction between prey species that share a predator, and resembles the dynamics of two species competing for resources (Holt, 1977). Apparent competition is usually positive for BC, because prey diversity commonly results in higher densities of generalist predators that in turn lead to lower equilibrium densities of both prey species (Chailleux et al., 2014). BC of whiteflies and spider mites by generalist predatory mites is, for example, enhanced by the presence of thrips (Messelink et al., 2008, 2010). Alternatively, pests may temporarily escape predation when predators are distracted from predation of a particular prey through the presence of a second prey, which is referred to as 'apparent mutualism' (Abrams and Matsuda, 1996). Such interactions can weaken the efficacy of BC, but these short-term reduced predation effects are usually soon compensated by an increased population growth of the shared predator (van Maanen et al., 2012). Similar effects may occur with the application of alternative food sources such as pollen or moth eggs (Nomikou et al., 2010; Leman and Messelink, 2015). Caution is needed when the predator has a strong preference for the alternative food source and a long generation time, which can result in negative effects on BC (Messelink et al., 2016).

Herbivores can also interact indirectly through induced changes in plant quality. In tomato plants, it has, for example, been demonstrated that whiteflies can induce resistance against leafminers (Inbar et al., 1999). More recently, it was found that plant-feeding predatory mirid bugs that are used in BC can also induce resistance against herbivores (Pappas et al., 2015; Zhang et al., 2018; Perez-Hedo et al., 2015) and can make plants less attractive for pests (Zhang et al., 2019). However, plant-feeding arthropods may also induce susceptibility to (other) pests (Nombela et al., 2009; Sarmento et al., 2011). Such plant-mediated interactions among herbivores, but also among microorganisms, diseases and herbivores, have nowadays caught the interest of many scientists, and the number of publications about this subject is growing fast.

Greenhouse crops can accommodate a diverse community of various pests and natural enemies that are either released or occur naturally. These complexities have increased strongly during the last decades because of the reduced pesticide use, new invasive pest species and the increased use of generalist predators. The myriad food web complexities resulting from this increased diversity is something to be aware of when designing BC programmes, but also offers opportunities to enhance the results of BC by promoting interactions that are positive for pest and disease management.

9 Case study

In this section we will illustrate how an ecologically based IPM system can be implemented in a greenhouse tomato crop, the most cultivated crop in protected horticulture. Tomato plants are well known for their trichome-mediated defence system against pests (Kennedy, 2003). The glandular type VI trichomes are, for example, effective in entrapping small insects such as thrips (Glas et al., 2012). However, many pests and pathogens are adapted to, or even specialized in, feeding on or infecting tomato plants and can cause severe damage when not controlled (Vallad et al., 2018). Hence, growing tomatoes without the application of pesticides soon results in a complex of various pests and diseases that interact either directly through resource competition, or indirectly through shared natural enemies or plant-mediated effects (Fig. 3). Controlling all these various pests and pathogens with an ecologically based IPM system is challenging, but has been well developed (van Lenteren, 2000). The most important natural enemies are the omnivorous predatory bugs of the family Miridae, such as *Macrolophus pygmaeus* (Rambur), *Nesidiocoris tenuis* (Reuter), *Dicyphus errans* and *Dicyphus hesperus*, which perform very well in tomato (Sanchez and Cassis, 2018). These omnivorous predators consume

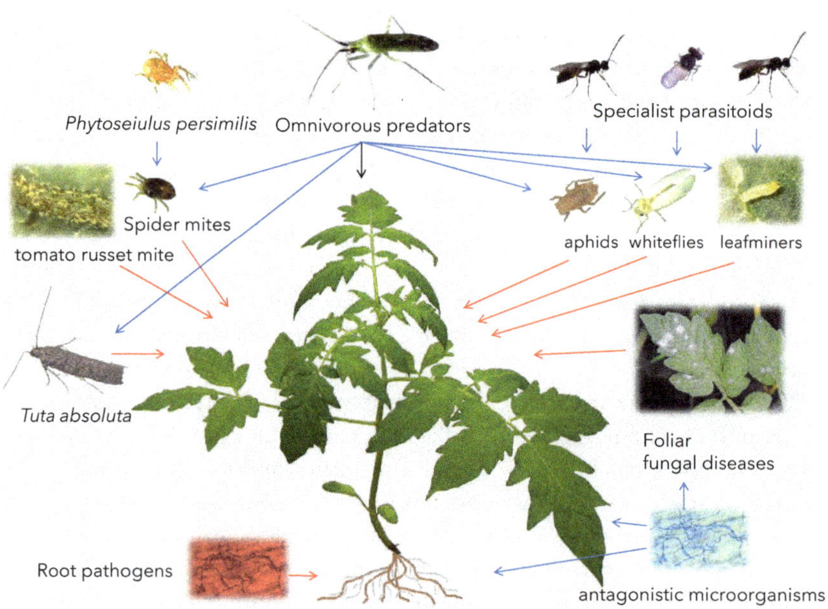

Figure 3 Simplified food web of common pests, pathogens and their natural enemies in a tomato crop. Arrows indicate consumption. Red arrows are negative and blue arrows positive for the plant, black can be both positive and negative.

various important pest species such as whiteflies, aphids (Alvarado et al., 1997), spider mites (Hansen et al., 1999), leafminers (Arnó et al., 2003) and Lepidopterans, including *Tuta absoluta* (Urbaneja et al., 2009). Interestingly, the plant-feeding behaviour of these predators can also induce plant responses that can increase the developmental time and mortality of spider mites (Pappas et al., 2015). Moreover, it can alter the composition of the crop volatile blend resulting in an increased attraction of the whitefly parasitoid *E. formosa* and for some predators a decreased attraction of *T. absoluta* and whiteflies (Perez-Hedo et al., 2015). Inoculating tomato plants with these predators and providing alternative food sources is therefore a very effective strategy to prevent pest infestations later on in the cropping season (Calvo et al., 2012). This strategy can be supported with preventative or curative releases of specialist predators or parasitoids (see Fig. 3). However, this system which is mainly based on pest control by omnivorous mirid predators is also vulnerable when populations of these predators are somehow disrupted and collapse. This can for example happen when pesticide application is needed to control pests that currently cannot be controlled biologically, such as the tomato russet mite *Aculops lycopersici* (Tryon). While generalist predatory mites are in principle suitable predators of this tiny pest mite, their mobility and efficacy on tomato plants is hampered by the presence of glandular trichomes (van Houten et al., 2013), thus currently limiting its control options to chemical control.

Another recent threat is the invasion of the Mediterranean omnivorous predator *N. tenuis*, which outcompetes other mirids such as *M. pygmaeus* (Moreno-Ripoll et al., 2012) and causes serious damage by their typical plant-feeding behaviour (Castañé et al., 2011), although in some Mediterranean areas this predator is still intentionally used because of its control capacity for *T. absoluta*. Growers sometimes try to control this predator with insecticides, which are impossible to apply without negative effects on related mirid predators that are needed for pest control. The integration of pest control with disease management is another challenge in tomato. Sulphur applications against mildew and tomato russet mites can, for example, have strong negative side-effects on parasitoids and predators (Venzon et al., 2013; Zappala et al., 2012). Antagonistic microorganisms such as *Bacillus subtillus* strains offer new opportunities to prevent foliar diseases and root-colonizing antagonists such as *Trichoderma* spp. can be used to prevent infection by fungal soil pathogens. Despite the various complexities and challenges in pest and disease management in tomato, it is in general possible to manage most of them without the application of pesticides. Future research may help to further develop non-chemical control measures for some remaining challenging pests such as the tomato russet mite, *T. absoluta* and *N. tenuis*.

10 Conclusion and future trends

Maintaining access to safe and healthy food for a rapidly growing world population will be a major challenge for the twenty-first century. Intensification of cropping systems is needed to increase production and yield, but at the same time, reduced use of pesticides, chemicals, water and energy is needed to meet the requirements for sustainability. Greenhouse cropping systems offer a great opportunity to take on these challenges. Ecologically based IPM is the only way to ensure sustainable production of food and ornamentals and, as shown in this chapter, requires an ecosystems approach considering all biotic and abiotic aspects that can affect pest and disease management. Within this IPM system, BC is the most important alternative for pesticides and its application is growing fast worldwide (Bale et al., 2008; van Lenteren et al., 2018).

However, there are also factors that may hamper the development and application of BC. Some of them have to do with social aspects such as the attitude of growers and governments (van Lenteren, 2012). Other factors concern the regulation of BCAs. An ongoing discussion in BC is the use of exotic natural enemies that potentially can disrupt natural ecosystems. The most well-known example is the impact of the harlequin ladybird, *Harmonia axyridis*. This natural enemy has been released for BC programmes for aphid control, including greenhouse crops, but appeared to threaten biodiversity by competing with and predating on indigenous aphidophagous insects (Roy et al., 2016). Many countries now implemented regulation with the aim to prevent such effects (van Lenteren et al., 2006). However, models for risk assessments are still in development and there are worries among biocontrol scientists that strict regulation can potentially hamper the implementation and development of BC. This has been the reason for the International Organization for Biological Control (IOBC) to start a special commission for this topic to consult policy makers in Europe about the risks of exotic biocontrol agents (website IOBC-WPRS). A second major change in regulation came with the Convention on Biological Diversity that developed the so-called Nagoya Protocol on access to genetic resources and the sharing of benefits arising from their utilization. The implementation of this new regulation has complicated the collection and export of new natural enemies for BC research in several countries (Cock et al., 2010). A major concern is the strong delay this regulation causes for the development of new BCAs, and in addition, the costs for sharing benefits and regulation of natural enemies may become disproportionally high. All these new developments in regulation for augmentative BC have resulted and will probably result in a stronger focus on indigenous natural enemies per country, particularly when it concerns generalist predators. Registration of microbial organisms is in most countries even more complex than for arthropod natural enemies, because microorganisms often require to have the same registration procedures as synthetic pesticides (e.g. in Europe), which makes the development of products extremely expensive (Ravensberg, 2011). Policy makers are nowadays aware of this problem and new legislation will hopefully speed up the processes of developing new biocontrol products.

Despite these developments that may hamper BC, we strongly believe that BC will remain the most important pillar of IPM systems. Yet, we also realize that strong efforts are required for those countries where pest and disease control is still based on pesticide use. In areas where the introduction of commercially produced natural enemies is not allowed due to legislation, or where it is problematic because of long transport times and high costs, new solutions need to be developed. One opportunity can be to start small rearing units of indigenous BCA species that are locally produced, combined with the application of food sources that increase natural enemy population densities. Such relatively simple systems can help to further increase the application of ecologically based IPM worldwide.

Also, existing ecologically based IPM systems will always require new developments and innovations. *First of all*, there are still several pest species that cannot be controlled effectively with available biocontrol agents. The control of aphids, for example, is rather expensive without the use of insecticides. New strategies for enhancing BC need to be developed. *Secondly*, new invasive pests and pathogens continue to threaten greenhouse production areas and the absence of suitable control agents for these invasive species can completely disrupt the existing BC programmes that target established pest species. Basic epidemiological knowledge is required for the management of new viral diseases, such as host range information, and vector dynamics in the case of vector-transmitted viruses. Also new sources of host-plant resistance that can be combined with other management options are necessary. *Thirdly,* pesticides disappear and new pesticides appear and the types of pesticides used will strongly affect the community structure of pests and diseases. So-called secondary pests may become primary pests when less broad-spectrum pesticides are used, and adequate BC methods for these pests will need to be developed. *Fourthly,* greenhouse structures, climate management, cropping systems and light strategies develop fast, particularly in high-tech glasshouses with artificial light. These changes in the greenhouse environment will also affect the performance of pests, pathogens and their natural enemies. Light, for example, has various direct and indirect effects on both pests and pathogens (Vänninen et al., 2010; Johansen et al., 2011; Raviv and Antignus, 2004; Ballare, 2014; Schumacher, 2017). *Fifthly,* understanding the shaping and the role of the rhizosphere microbiome, and the way in which it determines the level and specificity of disease-suppressiveness of soilborne pathogens as well as the induction of induced systemic resistance against aboveground pathogens merits further investigation, with spin-offs for breeding crop cultivars that can actively shape a disease-suppressive rhizosphere microbiome against key soil pathogens as an indirect defence strategy as well as for the manipulation of the growth medium through the amendment of compounds that trigger a functional microbiome. *Finally*, we believe that plant breeding offers great opportunities to contribute to robust IPM systems for greenhouse crops. Powdery mildew-tolerant cultivars, for example, largely reduce fungicide applications, which increases the use of arthropod natural enemies for pest control. One step further would be to include plant traits in breeding programmes that enhance the performance of natural enemies (Pappas et al., 2017; Cortesero et al., 2000).

In summary, although major challenges to further develop sustainable cropping systems lie ahead of us, greenhouses offer great opportunities to apply ecologically based IPM systems for sustainable food and ornamental production. The combined efforts of policy makers, funding agencies, industry, growers and scientists will be needed to further develop these systems worldwide.

11 Where to look for further information

We hope that reading this chapter made you more interested in IPM and BC of pests and diseases. One single chapter is obviously way too short to discuss all aspects of IPM in greenhouses in detail for the various cropping systems. For further reading, we recommend the recent book of Gullino et al (Gullino et al., 2019), which discusses IPM in detail for greenhouse crops. Those who became very interested in BC in general are recommended to read the recent book of Heimpel and Mills (Heimpel and Mills, 2017). A nice overview of all registered biocontrol agents is presented in the book of Gwynn (Gwynn, 2014).

We also recommend some key papers that discuss the following major topics: (1) the current state of augmentative BC (van Lenteren et al., 2018), (2) methods to support natural enemies (Messelink et al., 2014), (3) effects of food web interactions on pest control (Messelink et al., 2012), (4) effects of artificial light on pest and disease control (Johansen et al., 2011; Vänninen et al., 2010; Schumacher, 2017), (5) management of the greenhouse microclimate in relation to disease control (Jewett and Jarvis, 2001), (6) pesticide side-effects (Desneux et al., 2007), (7) induced plant responses and BC (Pappas et al., 2017), (8) the effects of root exudates and organic amendments on microbiome disease-suppressiveness (Bonanomi et al., 2018; Berendsen et al., 2012) and (9) the integration of entomopathogenic microorganisms with biological pest control (Gonzalez et al., 2016).

Finally, there are a few websites that are of interest for further information:

- The International Organisation for Biological Control: www.iobc-global.org
- The invasive species compendium of CABI: www.cabi.org/isc
- Guidelines for safe food production: www.globalgap.org

12 References

Abrams, P. A. and Matsuda, H. 1996. Positive indirect effects between prey species that share predators. *Ecology* 77(2), 610–6. doi:10.2307/2265634.

Acheampong, S., Gillespie, D. R. and Quiring, D. J. M. 2012. Survey of parasitoids and hyperparasitoids (Hymenoptera) of the green peach aphid, *Myzus persicae* and the foxglove aphid, *Aulacorthum solani* (Hemiptera: Aphididae) in British Columbia. *Journal of the Entomological Society of British Columbia* 109, 12–22.

Alvarado, P., Balta, O. and Alomar, O. 1997. Efficiency of four Heteroptera as predators of *Aphis gossypii* and *Macrosiphum euphorbiae* (Hom.: Aphididae). *Entomophaga* 42(1–2), 215–26. doi:10.1007/BF02769899.

Amtmann, A., Troufflard, S. and Armengaud, P. 2008. The effect of potassium nutrition on pest and disease resistance in plants. *Physiologia Plantarum* 133(4), 682–91. doi:10.1111/j.1399-3054.2008.01075.x.

Anon. 2015. Ecosystem services, agriculture and neonicotinoids. EASAC policy report 26. EASAC – the European Academies' Science Advisory Council, Halle, Germany, p. 61.

Antignus, Y. 2000. Manipulation of wavelength-dependent behaviour of insects: an IPM tool to impede insects and restrict epidemics of insect-borne viruses. *Virus Research* 71(1–2), 213–20. doi:10.1016/S0168-1702(00)00199-4.

Antoniou, A., Tsolakidou, M. D., Stringlis, I. A. and Pantelides, I. S. 2017. Rhizosphere microbiome recruited from a suppressive compost improves plant fitness and increases protection against vascular wilt pathogens of tomato. *Frontiers in Plant Science* 8, 2022. doi:10.3389/fpls.2017.02022.

Arnó, J., Alonso, E. and Gabarra, R. 2003. Role of the parasitoid *Diglyphus isaea* (Walker) and the predator *Macrolophus caliginosus* Wagner in the control of leafminers. *IOBC/WPRS Bulletin* 26, 79–84.

Baker, K. F. 1962. Principles of heat treatment of soil and planting material. *Journal of the Australian Institute of Agricultural Science* 28, 118–26.

Bale, J. S., van Lenteren, J. C. and Bigler, F. 2008. Biological control and sustainable food production. *Philosophical Transactions of the Royal Society of London. Series B, Biological Sciences* 363(1492), 761–76. doi:10.1098/rstb.2007.2182.

Ballare, C. L. 2009. Illuminated behaviour: phytochrome as a key regulator of light foraging and plant anti-herbivore defence. *Plant, Cell and Environment* 32(6), 713–25. doi:10.1111/j.1365-3040.2009.01958.x.

Ballare, C. L. 2014. Light regulation of plant defense. *Annual Review of Plant Biology* 65, 335–63. doi:10.1146/annurev-arplant-050213-040145.

Barzman, M., Barberi, P., Birch, A. N. E., Boonekamp, P., Dachbrodt-Saaydeh, S., Graf, B., Hommel, B., Jensen, J. E., Kiss, J., Kudsk, P., et al. 2015. Eight principles of integrated pest management. *Agronomy for Sustainable Development* 35(4), 1199–215. doi:10.1007/s13593-015-0327-9.

Bayu, M. S. Y. I., Ullah, M. S., Takano, Y. and Gotoh, T. 2017. Impact of constant versus fluctuating temperatures on the development and life history parameters of *Tetranychus urticae* (Acari: Tetranychidae). *Experimental and Applied Acarology* 72(3), 205–27. doi:10.1007/s10493-017-0151-9.

Bennison, J. 1992. Biological control of aphids on cucumbers use of open rearing systems or 'banker plants' to aid establishment of *Aphidius matricariae* and *Aphidoletes aphidimyza*. *Mededelingen van de Faculteit Landbouwwetenschappen, Rijksuniversiteit Gent* 57, 457–66.

Bennison, J., Maulden, K., Maher, H. and Tomiczek, M. 2008. Development of a grower rearing-release system for *Atheta coriaria*, for low cost biological control of ground-dwelling pest life stages. *IOBC/WPRS Bulletin* 32, 21–4.

Ben-Yakir, D., Teitel, M., Tanny, J., Chen, M. and Barak, M. 2008. Optimizing ventilation of protected crops while minimizing invasion by whitefly and thrips. *Acta Horticulturae* 797(797), 217–22. doi:10.17660/ActaHortic.2008.797.29.

Berendsen, R. L., Pieterse, C. M. J. and Bakker, P. A. 2012. The rhizosphere microbiome and plant health. *Trends in Plant Science* 17(8), 478–86. doi:10.1016/j.tplants.2012.04.001.

Berendsen, R. L., Vismans, G., Yu, K., Song, Y., de Jonge, R., Burgman, W. P., Burmølle, M., Herschend, J., Bakker, P. A. H. M. and Pieterse, C. M. J. 2018. Disease-induced assemblage of a plant-beneficial bacterial consortium. *The ISME Journal* 12(6), 1496–507. doi:10.1038/s41396-018-0093-1.

Bethke, J. A. and Paine, T. D. 1991. Screen hole size and barriers for exclusion of insect pests of glasshouse crops. *Journal of Entomological Science* 26(1), 169–77. doi:10.18474/0749-8004-26.1.169.

Blacquiere, T., Aa-Furnée, J. v. d., Cornelissen, B. and Donders, J. 2006. Behaviour of honey bees and bumble bees beneath three different greenhouse claddings. Nederlandse Entomologische Vereniging (NEV).

Bloemhard, C. M. J., van der Wielen, M. and Messelink, G. J. 2014. Seasonal abundance of aphid hyperparasitoids in organic greenhouse crops in the Netherlands. *IOBC-WPRS Bulletin* 102, 15–9.

Bollen, G. J. 1969. The selective effect of heat treatment on the micro flora of a greenhouse soil. *Netherlands Journal of Plant Pathology* 75(1–2), 157–63. doi:10.1007/BF02137211.

Bonanomi, G., Antignani, V., Pane, C. and Scala, F. 2007. Suppression of soilborne fungal diseases with organic amendments. *Journal of Plant Pathology* 89, 311–24.

Bonanomi, G., Antignani, V., Capodilupo, M. and Scala, F. 2010. Identifying the characteristics of organic soil amendments that suppress soilborne plant diseases. *Soil Biology and Biochemistry* 42(2), 136–44. doi:10.1016/j.soilbio.2009.10.012.

Bonanomi, G., Lorito, M., Vinale, F. and Woo, S. L. 2018. Organic amendments, beneficial microbes, and soil microbiota: toward a unified framework for disease suppression. *Annual Review of Phytopathology* 56, 1–20. doi:10.1146/annurev-phyto-080615-100046.

Bravenboer, L. 1959. De chemische en biologische bestrijding van de spintmijt *Tetranychus urticae* Koch Proefstation voor de Groenten- en Fruitteelt onder Glas. Dissertation. Wageningen University, Wageningen.

Bravenboer, L. and Dosse, G. 1962. *Phytoseiulus riegeli* Dosse as a predator of harmful mites of the group of *Tetranychus urticae*. *Entomologia Experimentalis et Applicata* 5(4), 291–304. doi:10.1111/j.1570-7458.1962.tb00594.x.

Bresch, C., Carlesso, L., Suay, R., Van Oudenhove, L., Touzeau, S., Fatnassi, H., Ottenwaelder, L., Paris, B., Poncet, C., Mailleret, L., et al. 2019. In search of artificial domatia for predatory mites. *Biocontrol Science and Technology* 29(2), 131–48. doi:10.1080/09583157.2018.1540030.

Buitenhuis, R., Brownbridge, M., Brommit, A., Saito, T. and Murphy, G. 2016. How to start with a clean crop: biopesticide dips reduce populations of *Bemisia tabaci* (Hemiptera: Aleyrodidae) on greenhouse poinsettia propagative cuttings. *Insects* 7(4), 13. doi:10.3390/insects7040048.

Butler, D. M., Kokalis-Burelle, N., Muramoto, J., Shennan, C., McCollum, T. G. and Rosskopf, E. N. 2012. Impact of anaerobic soil disinfestation combined with soil solarization on plant-parasitic nematodes and introduced inoculum of soilborne plant pathogens in raised-bed vegetable production. *Crop Protection* 39, 33–40. doi:10.1016/j.cropro.2012.03.019.

Calvo, F. J., Bolckmans, K. and Belda, J. E. 2009. Development of a biological control-based Integrated Pest Management method for *Bemisia tabaci* for protected sweet pepper crops. *Entomologia Experimentalis et Applicata* 133(1), 9–18. doi:10.1111/j.1570-7458.2009.00896.x.

Calvo, F. J., Lorente, M. J., Stansly, P. A. and Belda, J. E. 2012. Preplant release of *Nesidiocoris tenuis* and supplementary tactics for control of *Tuta absoluta* and *Bemisa tabaci* in greenhouse tomato. *Entomologia Experimentalis et Applicata* 143(2), 111–9. doi:10.1111/j.1570-7458.2012.01238.x.

Campbell, A., Frazer, B. D., Gilbert, N., Gutierrez, A. P. and Mackauer, M. 1974. Temperature requirements of some aphids and their parasites. *The Journal of Applied Ecology* 11(2), 431–8. doi:10.2307/2402197.

Casals-Casas, C. and Desvergne, B. 2011. Endocrine disruptors: from endocrine to metabolic disruption. In: Julius, D. and Clapham, D. E. (Eds), *Annual Review of Physiology* (vol. 73). Annual Reviews, Palo Alto, pp. 135–62.

Castañé, C., Arnó, J., Gabarra, R. and Alomar, O. 2011. Plant damage to vegetable crops by zoophytophagous mirid predators. *Biological Control* 59(1), 22–9. doi:10.1016/j.biocontrol.2011.03.007.

Castle, S. J., Henneberry, T. J. and Toscano, N. C. 1996. Suppression of *Bemisia tabaci* (Homoptera: Aleyrodidae) infestations in cantaloupe and cotton with sprinkler irrigation. *Crop Protection* 15(7), 657–63. doi:10.1016/0261-2194(96)00037-3.

Chailleux, A., Mohl, E. K., Alves, M. T., Messelink, G. J. and Desneux, N. 2014. Natural enemy-mediated indirect interactions among prey species: potential for enhancing biocontrol services in agroecosystems. *Pest Management Science* 70(12), 1769–79. doi:10.1002/ps.3916.

Chase, J. M., Abrams, P. A., Grover, J. P., Diehl, S., Chesson, P., Holt, R. D., Richards, S. A., Nisbet, R. M. and Case, T. J. 2002. The interaction between predation and competition: a review and synthesis. *Ecology Letters* 5(2), 302–15. doi:10.1046/j.1461-0248.2002.00315.x.

Chen, Y., Story, R. and Samuel-Foo, M. 2014. Effects of nitrogen and phosphorous fertilization on western flower thrips population level and quality of susceptible and resistant Impatiens. *Advances in Crop Science and Technology* 2, 145.

Chow, A., Chau, A. and Heinz, K. M. 2009. Reducing fertilization for cut roses: effect on crop productivity and twospotted spider mite abundance, distribution, and management. *Journal of Economic Entomology* 102(5), 1896–907. doi:10.1603/029.102.0521.

Chow, A., Chau, A. and Heinz, K. M. 2012. Reducing fertilization: a management tactic against western flower thrips on roses. *Journal of Applied Entomology* 136(7), 520–9. doi:10.1111/j.1439-0418.2011.01674.x.

Cochard, P., Galstian, T. and Cloutier, C. 2017. Light environments differently affect parasitoid wasps and their hosts' locomotor activity. *Journal of Insect Behavior* 30(6), 595–611. doi:10.1007/s10905-017-9644-y.

Cock, M. J. W., van Lenteren, J. C., Brodeur, J., Barratt, B. I. P., Bigler, F., Bolckmans, K., Cônsoli, F. L., Haas, F., Mason, P. G. and Parra, J. R. P. 2010. Do new access and benefit sharing procedures under the convention on biological diversity threaten the future of biological control? *BioControl* 55(2), 199–218. doi:10.1007/s10526-009-9234-9.

Coll, M. and Guershon, M. 2002. Omnivory in terrestrial arthropods: mixing plant and prey diets. *Annual Review of Entomology* 47, 267–97. doi:10.1146/annurev.ento.47.091201.145209.

Cortesero, A. M., Stapel, J. O. and Lewis, W. J. 2000. Understanding and manipulating plant attributes to enhance biological control. *Biological Control* 17(1), 35–49. doi:10.1006/bcon.1999.0777.

Croft, B. A. 1990. *Arthropod Biological Control Agents and Pesticides*. Wiley, New York.

D'Addabbo, T., Miccolis, V., Basile, M. and Candido, V. 2010. Soil solarization and sustainable agriculture. In: Lichtfouse, E. (Ed.), *Sociology, Organic Farming, Climate Change and Soil Science*. Springer Netherlands, Dordrecht, pp. 217–74.

Datnoff, L. E., Elmer, W. H. and Huber, D. M. 2007. *Mineral Nutrition and Plant Disease*. American Phytopathological Society, St. Paul, MN.

Denholm, I. and Rowland, M. W. 1992. Tactics for managing pesticide resistance in arthropods: theory and practice. *Annual Review of Entomology* 37, 91–112. doi:10.1146/annurev.en.37.010192.000515.

Desneux, N., Decourtye, A. and Delpuech, J. M. 2007. The sublethal effects of pesticides on beneficial arthropods. *Annual Review of Entomology* 52, 81–106. doi:10.1146/annurev.ento.52.110405.091440.

Diaz, B. M. and Fereres, A. 2007. Ultraviolet-blocking materials as a physical barrier to control insect pests and plant pathogens in protected crops. *Pest Technology* 1, 85–95.

Dixon, A. F. G., Honek, A., Keil, P., Kotela, M. A. A., Šizling, A. L. and Jarošík, V. 2009. Relationship between the minimum and maximum temperature thresholds for development in insects. *Functional Ecology* 23(2), 257–64. doi:10.1111/j.1365-2435.2008.01489.x.

Dordas, C. 2008. Role of nutrients in controlling plant diseases in sustainable agriculture. A review. *Agronomy for Sustainable Development* 28(1), 33–46. doi:10.1051/agro:2007051.

Duso, C., Chiarini, F., Conte, L., Bonora, V., Dalla Montà, L. and Otto, S. 2004. Fogging can control *Tetranychus urticae* on greenhouse cucumbers. *Journal of Pest Science* 77(2), 105–11. doi:10.1007/s10340-003-0037-y.

Dyer, A. G. and Chittka, L. 2004. Bumblebee search time without ultraviolet light. *The Journal of Experimental Biology* 207(10), 1683–8. doi:10.1242/jeb.00941.

EC. 2009. Sustainable Use Directive. European Parliament and of the Council of 21 October 2009 establishing a framework for community action to achieve the sustainable use of pesticides. *Official Journal of the European Union* L309, 71–86.

Eilenberg, J., Hajek, A. and Lomer, C. 2001. Suggestions for unifying the terminology in biological control. *BioControl* 46(4), 387–400. doi:10.1023/A:1014193329979.

Engelhard, A. W. 1989. *Soilborne Plant Pathogens: Management of Diseases with Macro- and Microelements*. APS Press, St. Paul, MN.

Escobar-Bravo, R., Ruijgrok, J., Kim, H. K., Grosser, K., Van Dam, N. M., Klinkhamer, P. G. L. and Leiss, K. A. 2018. Light intensity-mediated induction of trichome-associated allelochemicals increases resistance against thrips in tomato. *Plant and Cell Physiology* 59(12), 2462–75. doi:10.1093/pcp/pcy166.

Georgis, R., Koppenhofer, A. M., Lacey, L. A., Bélair, G., Duncan, L. W., Grewal, P. S., Samish, M., Tan, L., Torr, P. and van Tol, R. W. H. M. 2006. Successes and failures in the use of parasitic nematodes for pest control. *Biological Control* 38(1), 103–23. doi:10.1016/j.biocontrol.2005.11.005.

Glas, J. J., Schimmel, B. C. J., Alba, J. M., Escobar-Bravo, R., Schuurink, R. C. and Kant, M. R. 2012. Plant glandular trichomes as targets for breeding or engineering of resistance to herbivores. *International Journal of Molecular Sciences* 13(12), 17077–103. doi:10.3390/ijms131217077.

Gonzalez, F., Tkaczuk, C., Dinu, M. M., Fiedler, Ż, Vidal, S., Zchori-Fein, E. and Messelink, G. J. 2016. New opportunities for the integration of microorganisms into biological pest control systems in greenhouse crops. *Journal of Pest Science* 89, 295–311. doi:10.1007/s10340-016-0751-x.

Grostal, R. and O'Dowd, D. J. 1994. Plants, mites and mutualism - leaf domatia and the abundance and reproduction of mites on *Viburnum tinus* (Caprifoliaceae). *Oecologia* 97(3), 308–15. doi:10.1007/BF00317319.

Gullino, M. L., Albajes, R., Nicot, P. and van Lenteren, J. C. 2019. *Integrated Pest and Disease Management in Greenhouse Crops* (2nd edn.). Springer (in press).

Gwynn, R. L. 2014. *The Manual of Biocontrol Agents: a World Compendium*. BCPC, Alton, Hampshire.

Hansen, D. L., Brødsgaard, H. F. and Enkegaard, A. 1999. Life table characteristics of *Macrolophus caliginosus* preying upon *Tetranychus urticae*. *Entomologia Experimentalis et Applicata* 93(3), 267–73. doi:10.1046/j.1570-7458.1999.00587.x.

Heimpel, G. E. and Mills, N. J. 2017. *Biological Control - Ecology and Applications*. Cambridge University Press, New York.

Heinz, K. M., Van Driesche, R. G. and Parella, M. P. 2004. *Biocontrol in Protected Culture*. Ball Publishing, Batavia.

Hemming, S. 2004. *Optimaal gebruik van natuurlijk licht in de glastuinbouw*. Agrotechnology & Food Innovations, Wageningen.

Hobbs, E. L. and Waters, W. E. 1964. Influence of nitrogen + potassium on susceptibility of *Chrysanthemum morifolium* to *Botrytis cinerea*. *Phytopathology* 54, 674

Hoffland, E., Dicke, M., Van Tintelen, W. 2000a. Nitrogen availability and defense of tomato against two-spotted spider mite. *Journal of Chemical Ecology* 26(12), 2697–711. doi :10.1023/A:1026477423988.

Hoffland, E., Jeger, M. J. and van Beusichem, M. L. 2000b. Effect of nitrogen supply rate on disease resistance in tomato depends on the pathogen. *Plant and Soil* 218, 239–47.

Holt, R. D. 1977. Predation, apparent competition and structure of prey communities. *Theoretical Population Biology* 12(2), 197–29. doi:10.1016/0040-5809(77)90042-9.

Huang, N. X., Enkegaard, A., Osborne, L. S., Ramakers, P. M. J., Messelink, G. J., Pijnakker, J. and Murphy, G. 2011. The banker plant method in biological control. *Critical Reviews in Plant Sciences* 30(3), 259–78. doi:10.1080/07352689.2011.572055.

Huber, L. and Gillespie, T. J. 1992. Modeling leaf wetness in relation to plant-disease epidemiology. *Annual Review of Phytopathology* 30(1), 553–77. doi:10.1146/annurev.py.30.090192.003005.

Hussey, N. W. and Scopes, N. 1985. *Biological Pest Control, the Glasshouse Experience*. Cornell University Press, Ithaca, NY.

Inbar, M., Doostdar, H., Leibee, G. L. and Mayer, R. T. 1999. The role of plant rapidly induced responses in asymmetric interspecific interactions among insect herbivores. *Journal of Chemical Ecology* 25(8), 1961–79. doi:10.1023/A:1020998219928.

Jalali, M. A., Tirry, L. and De Clercq, P. 2010. Effect of temperature on the functional response of *Adalia bipunctata* to *Myzus persicae*. *BioControl* 55(2), 261–9. doi:10.1007/s10526-009-9237-6.

Janssen, A. and Sabelis, M. W. 2015. Alternative food and biological control by generalist predatory mites: the case of *Amblyseius swirskii*. *Experimental and Applied Acarology* 65(4), 413–8. doi:10.1007/s10493-015-9901-8.

Janssen, A., Montserrat, M., HilleRisLambers, R., Roos, A. M., Pallini, A. and Sabelis, M. W. 2006. Intraguild predation usually does not disrupt biological control. In: Brodeur, J. and Boivin, G. (Eds), *Trophic and Guild Interactions in Biological Control*. Springer Netherlands, Dordrecht, pp. 21–44.

Jayaraj, J. and Radhakrishnan, N. V. 2008. Enhanced activity of introduced biocontrol agents in solarized soils and its implications on the integrated control of tomato damping-off caused by *Pythium* spp. *Plant and Soil* 304(1–2), 189–97. doi:10.1007/s11104-008-9539-y.

Jerbi-Elayed, M., Lebdi-Grissa, K., Le Goff, G. and Hance, T. 2015. Influence of temperature on flight, walking and oviposition capacities of two aphid parasitoid species (Hymenoptera: Aphidiinae). *Journal of Insect Behavior* 28(2), 157–66. doi:10.1007/s10905-015-9490-8.

Jewett, T. J. and Jarvis, W. R. 2001. Management of the greenhouse microclimate in relation to disease control: a review. *Agronomie* 21(4), 351–66. doi:10.1051/agro:2001129.

Johansen, N. S., Vänninen, I., Pinto, D. M., Nissinen, A. I. and Shipp, L. 2011. In the light of new greenhouse technologies: 2. Direct effects of artificial lighting on arthropods and integrated pest management in greenhouse crops. *Annals of Applied Biology* 159(1), 1–27. doi:10.1111/j.1744-7348.2011.00483.x.

Karban, R. and Baldwin, I. T. 1997. *Induced Responses to Herbivory*. University of Chicago Press, Chicago.

Katan, J. 1987. Soil solarization. In: Chet, I. (Ed.), *Innovative Approaches to Plant Disease Control*. Wiley, New York, pp. 77–105.

Katan, J. 1996. Cultural practices and soil borne disease management. In: Utkhde, R. S. and Gupta, V. K. (Eds), *Management of Soil-borne Diseases*. Kalyani Publishers, India, p. 100.

Katan, J. 2000. Physical and cultural methods for the management of soil-borne pathogens. *Crop Protection* 19(8–10), 725–31. doi:10.1016/S0261-2194(00)00096-X.

Kennedy, G. G. 2003. Tomato, pests, parasitoids, and predators: tritrophic interactions involving the genus *Lycopersicon*. *Annual Review of Entomology* 48, 51–72. doi:10.1146/annurev.ento.48.091801.112733.

Kirchner, S. M., Doring, T. F. and Saucke, H. 2005. Evidence for trichromacy in the green peach aphid, *Myzus persicae* (Sulz.) (Hemiptera: Aphididae). *Journal of Insect Physiology* 51(11), 1255–60. doi:10.1016/j.jinsphys.2005.07.002.

Kontodimas, D. C., Eliopoulos, P. A., Stathas, G. J. and Economou, L. P. 2004. Comparative temperature-dependent development of *Nephus includens* (Kirsch) and *Nephus bisignatus* (Boheman) (Coleoptera: Coccinellidae) preying on *Planococcus citri* (Risso) (Homoptera: Pseudococcidae): evaluation of a linear and various nonlinear models using specific criteria. *Environmental Entomology* 33(1), 1-11. doi:10.1603/0046-225X-33.1.1.

Koohakan, P., Ikeda, H., Jeanaksorn, T., Tojo, M., Kusakari, S., Okada, K. and Sato, S. 2004. Evaluation of the indigenous microorganisms in soilless culture: occurrence and quantitative characteristics in the different growing systems. *Scientia Horticulturae* 101(1-2), 179-88. doi:10.1016/j.scienta.2003.09.012.

Körner, O. and Holst, N. 2005. Model based humidity control of *Botrytis* in greenhouse cultivation. *Acta Horticulturae* 691, 141-8.

Kraiselburd, I., Moyano, L., Carrau, A., Tano, J. and Orellano, E. G. 2017. Bacterial photosensory proteins and their role in plant-pathogen interactions. *Photochemistry and Photobiology* 93(3), 666-74. doi:10.1111/php.12754.

Kruidhof, H. M. and Elmer, W. H. 2019. Cultural methods for pest and disease management. In: Gullino, M. L., Albajes, R., Nicot, P. and van Lenteren, J. C. (Eds), *Integrated Pest and Disease Management in Greenhouse Crops* (2nd edn.). Springer (in press).

Le Gall, H., Philippe, F., Domon, J. M., Gillet, F., Pelloux, J. and Rayon, C. 2015. Cell wall metabolism in response to abiotic stress. *Plants* 4(1), 112-66. doi:10.3390/plants4010112.

Leman, A. and Messelink, G. J. 2015. Supplemental food that supports both predator and pest: a risk for biological control? *Experimental and Applied Acarology* 65(4), 511-24. doi:10.1007/s10493-014-9859-y.

Lindquist, R. K. and Short, T. L. 2004. Effects of greenhouse structure and function on biological control. In: Heinz, K. M., Van Driesche, R. G. and Parrella, M. P. (Eds), *Biocontrol in Protected Culture*. Ball Publishing, Batavia, IL, pp. 37-53.

Loughner, R., Nyrop, J., Wentworth, K. and Sanderson, J. 2011. Effects of supplemental pollen and fibers on canopy abundance of *Amblyseius swirskii*. *IOBC/WPRS Bulletin* 68, 105-9.

Louws, F. J., Rivard, C. L. and Kubota, C. 2010. Grafting fruiting vegetables to manage soilborne pathogens, foliar pathogens, arthropods and weeds. *Scientia Horticulturae* 127(2), 127-46. doi:10.1016/j.scienta.2010.09.023.

Luvisi, A., Materazzi, A. and Triolo, E. 2006. Steam and exothermic reactions as alternative techniques to control soil-borne diseases in basil. *Agronomy for Sustainable Development* 26(3), 201-7. doi:10.1051/agro:2006016.

Magarey, R. D., Sutton, T. B. and Thayer, C. L. 2005. A simple generic infection model for foliar fungal plant pathogens. *Phytopathology* 95(1), 92-100. doi:10.1094/PHYTO-95-0092.

Magie, R. O. 1971. Effectiveness of treatments with hot water plus benzimidazoles and ethephon in controlling Fusarium disease of gladiolus. *Plant Disease Reporter* 55, 82-5.

Markkula, M. and Tiittanen, K. 1976. 'Pest-in-First' and 'natural infestation' methods in the control of *Tetranychus urticae* Koch with *Phytoseiulus persimilis* Athias-Henriot on glasshouse cucumbers. *Annales Entomologici Fennici* 15, 81-5.

Matteson, N., Terry, I., Ascoli-Christensen, A. and Gilbert, C. 1992. Spectral efficiency of the western flower thrips, *Frankliniellaoccidentalis*. *Journal of Insect Physiology* 38(6), 453-9. doi:10.1016/0022-1910(92)90122-T.

Mellor, H. E., Bellingham, J. and Anderson, M. 1997. Spectral efficiency of the glasshouse whitefly *Trialeurodes vaporariorum* and *Encarsia formosa* its hymenopteran parasitoid. *Entomologia Experimentalis et Applicata* 83(1), 11–20. doi:10.1046/j.1570-7458.1997.00152.x.

Mendes, L. W., Raaijmakers, J. M., de Hollander, M., Mendes, R. and Tsai, S. M. 2018. Influence of resistance breeding in common bean on rhizosphere microbiome composition and function. *The ISME Journal* 12(1), 212–24. doi:10.1038/ismej.2017.158.

Messelink, G. J., Van Steenpaal, S. E. F. and Ramakers, P. M. J. 2006. Evaluation of phytoseiid predators for control of western flower thrips on greenhouse cucumber. *BioControl* 51(6), 753–68. doi:10.1007/s10526-006-9013-9.

Messelink, G. J., van Maanen, Rv, van Steenpaal, S. E. F. and Janssen, A. 2008. Biological control of thrips and whiteflies by a shared predator: two pests are better than one. *Biological Control* 44(3), 372–9. doi:10.1016/j.biocontrol.2007.10.017.

Messelink, G. J., Van Maanen, R., Van Holstein-Saj, R., Sabelis, M. W. and Janssen, A. 2010. Pest species diversity enhances control of spider mites and whiteflies by a generalist phytoseiid predator. *BioControl* 55, 387–98.

Messelink, G. J., Sabelis, M. W. and Janssen, A. 2012. Generalist predators, food web complexities and biological pest control in greenhouse crops. In: Larramendy, M. L. and Soloneski, S. (Eds), *Integrated Pest Management and Pest Control - Current and Future Tactics*. InTech, Rijeka, pp. 191–214.

Messelink, G. J., Bennison, J., Alomar, O., Ingegno, B. L., Tavella, L., Shipp, L., Palevsky, E. and Wäckers, F. L. 2014. Approaches to conserving natural enemy populations in greenhouse crops: current methods and future prospects. *BioControl* 59(4), 377–93. doi:10.1007/s10526-014-9579-6.

Messelink, G. J., Vijverberg, R., Leman, A. and Janssen, A. 2016. Biological control of mealybugs with lacewing larvae is affected by the presence and type of supplemental prey. *BioControl* 61(5), 555–65. doi:10.1007/s10526-016-9739-y.

Minuto, A., Spadaro, D., Garibaldi, A. and Gullino, M. L. 2006. Control of soilborne pathogens of tomato using a commercial formulation of *Streptomyces griseoviridis* and solarization. *Crop Protection* 25(5), 468–75. doi:10.1016/j.cropro.2005.08.001.

Mollema, C. and Cole, R. A. 1996. Low aromatic amino acid concentrations in leaf proteins determine resistance to *Frankliniella occidentalis* in four vegetable crops. *Entomologia Experimentalis et Applicata* 78(3), 325–33. doi:10.1111/j.1570-7458.1996.tb00797.x.

Montagne, V., Capiaux, H., Cannavo, P., Charpentier, S., Renaud, S., Liatard, E., Grosbellet, C. and Lebeau, T. 2016. Protective effect of organic substrates against soil-borne pathogens in soilless cucumber crops. *Scientia Horticulturae* 206, 62–70. doi:10.1016/j.scienta.2016.04.035.

Moreno-Ripoll, R., Agusti, N., Berruezo, R. and Gabarra, R. 2012. Conspecific and heterospecific interactions between two omnivorous predators on tomato. *Biological Control* 62(3), 189–96. doi:10.1016/j.biocontrol.2012.04.005.

Morra, L., Carrieri, R., Fornasier, F., Mormile, P., Rippa, M., Baiano, S., Cermola, M., Piccirillo, G. and Lahoz, E. 2018. Solarization working like a 'solar hot panel' after compost addition sanitizes soil in thirty days and preserves soil fertility. *Applied Soil Ecology* 126, 65–74. doi:10.1016/j.apsoil.2018.02.018.

Nihoul, P. 1993. Do light intensity, temperature and photoperiod affect the entrapment of mites on glandular hairs of cultivated tomatoes? *Experimental and Applied Acarology* 17(9), 709–18. doi:10.1007/BF00058510.

Nombela, G., Garzo, E., Duque, M. and Muñiz, M. 2009. Preinfestations of tomato plants by whiteflies (*Bemisia tabaci*) or aphids (*Macrosiphum euphorbiae*) induce variable resistance or susceptibility responses. *Bulletin of Entomological Research* 99(2), 183–91. doi:10.1017/S0007485308006214.

Nomikou, M., Sabelis, M. W. and Janssen, A. 2010. Pollen subsidies promote whitefly control through the numerical response of predatory mites. *BioControl* 55(2), 253-60. doi:10.1007/s10526-009-9233-x.

Northfield, T. D., Paini, D. R., Funderburk, J. E. and Reitz, S. R. 2008. Annual cycles of *Frankliniella* spp. (Thysanoptera: Thripidae) thrips abundance on North Florida uncultivated reproductive hosts: predicting possible sources of pest outbreaks. *Annals of the Entomological Society of America* 101, 769–78.

Opit, G. P., Fitch, G. K., Margolies, D. C., Nechols, J. R. and Williams, K. A. 2006. Overhead and drip-tube irrigation affect twospotted spider mites and their biological control by a predatory mite on impatiens. *Hortscience* 41(3), 691–4. doi:10.21273/HORTSCI.41.3.691.

Pappas, M. L., Steppuhn, A., Geuss, D., Topalidou, N., Zografou, A., Sabelis, M. W. and Broufas, G. D. 2015. Beyond predation: the zoophytophagous predator *Macrolophus pygmaeus* induces tomato resistance against spider mites. *PLoS ONE* 10(5), e0127251. doi:10.1371/journal.pone.0127251.

Pappas, M. L., Broekgaarden, C., Broufas, G. D., Kant, M. R., Messelink, G. J., Steppuhn, A., Wäckers, F. and van Dam, N. M. 2017. Induced plant defences in biological control of arthropod pests: a double-edged sword. *Pest Management Science* 73(9), 1780–8. doi:10.1002/ps.4587.

Pell, J. K., Hannam, J. J. and Steinkraus, D. C. 2010. Conservation biological control using fungal entomopathogens. *BioControl* 55(1), 187–98. doi:10.1007/s10526-009-9245-6.

Perdikis, D., Fantinou, A. and Lykouressis, D. 2011. Enhancing pest control in annual crops by conservation of predatory Heteroptera. *Biological Control* 59(1), 13-21. doi:10.1016/j.biocontrol.2011.03.014.

Perez-Hedo, M., Bouagga, S., Jaques, J. A., Flors, V. and Urbaneja, A. 2015. Tomato plant responses to feeding behavior of three zoophytophagous predators (Hemiptera: Miridae). *Biological Control* 86, 46–51. doi:10.1016/j.biocontrol.2015.04.006.

Pilkington, L. J., Messelink, G., van Lenteren, J. C. and Le Mottee, K. 2010. 'Protected Biological Control' – biological pest management in the greenhouse industry. *Biological Control* 52(3), 216–20. doi:10.1016/j.biocontrol.2009.05.022.

Polis, G. A., Myers, C. A. and Holt, R. D. 1989. The ecology and evolution of intraguild predation: potential competitors that eat each other. *Annual Review of Ecology and Systematics* 20(1), 297–330. doi:10.1146/annurev.es.20.110189.001501.

Porras, M., Barrau, C. and Romero, F. 2007. Effects of soil solarization and Trichoderma on strawberry production. *Crop Protection* 26(5), 782–7. doi:10.1016/j.cropro.2006.07.005.

Raaijmakers, J. M. and Mazzola, M. 2016. Soil immune responses. *Science* 352(6292), 1392–3. doi:10.1126/science.aaf3252.

Ramakers, P. M. J. and Rabasse, J.-M. 1995. Integrated pest management in protected cultivation. In: Reuveni, R. (Ed.), *Novel Approaches to Integrated Pest Management*. CRC Press, Florida, pp. 199-229.

Ramakers, P. M. J. and van Lieburg, M. J. 1982. Start of commercial production and introduction of *Amblyseius mckenziei* Sch. & Pr. (Acarina: Phytoseiidae) for the control of *Thrips tabaci* Lind. (Thysanoptera: Thripidae) in glasshouses. *Mededelingen van de Faculteit Landbouwwetenschappen, Rijksuniversiteit Gent* 47, 541-5.

Ravensberg, W. J. 2011. *A Roadmap to the Successful Development and Commercialization of Microbial Pest Control Products for Control of Arthropods*. Springer Science+Business Media B.V, Dordrecht.

Raviv, M. and Antignus, Y. 2004. UV radiation effects on pathogens and insect pests of greenhouse-grown crops. *Photochemistry and Photobiology* 79(3), 219-26. doi:10.1111/j.1751-1097.2004.tb00388.x.

Riddick, E. W. and Simmons, A. M. 2014. Do plant trichomes cause more harm than good to predatory insects? *Pest Management Science* 70(11), 1655-65. doi:10.1002/ps.3772.

Rosenheim, J. A., Kaya, H. K., Ehler, L. E., Marois, J. J. and Jaffee, B. A. 1995. Intraguild predation among biological control agents: theory and evidence. *Biological Control* 5(3), 303-35. doi:10.1006/bcon.1995.1038.

Rowlandson, T., Gleason, M., Sentelhas, P., Gillespie, T., Thomas, C. and Hornbuckle, B. 2015. Reconsideration leaf wetness duration determination for plant disease management. *Plant Disease* 99(3), 310-9. doi:10.1094/PDIS-05-14-0529-FE.

Roy, H. E., Brown, P. M. J., Adriaens, T., Berkvens, N., Borges, I., Clusella-Trullas, S., Comont, R. F., De Clercq, P., Eschen, R., Estoup, A., et al. 2016. The harlequin ladybird, *Harmonia axyridis*: global perspectives on invasion history and ecology. *Biological Invasions* 18(4), 997-1044. doi:10.1007/s10530-016-1077-6.

Runia, W. T. 1983. A recent development in steam sterilization. *Acta Horticulturae* 152, 195-200.

Sampson, C. 1998. The commercial development of an *Amblyseius cucumeris* controlled release method for the control of *Frankliniella occidentalis* in protected crops. In: *The 1998 Brighton Conference - Pests & Diseases*. Brighton, UK, pp. 409-16.

Sanchez, J. A. and Cassis, G. 2018. Towards solving the taxonomic impasse of the biocontrol plant bug subgenus *Dicyphus* (*Dicyphus*) (Insecta: Heteroptera: Miridae) using molecular, morphometric and morphological partitions. *Zoological Journal of the Linnean Society* 184(2), 330-406. doi:10.1093/zoolinnean/zly005.

Sarmento, R. A., Lemos, F., Bleeker, P. M., Schuurink, R. C., Pallini, A., Oliveira, M. G., Lima, E. R., Kant, M., Sabelis, M. W. and Janssen, A. 2011. A herbivore that manipulates plant defence. *Ecology Letters* 14(3), 229-36. doi:10.1111/j.1461-0248.2010.01575.x.

Schuch, U. K., Rdak, R. A. and Behtke, J. A. 1998. Cultivar, fertilizer and irrigation effect vegetative growth and susceptibility of chrysanthemum to western flower thrips. *Journal of the American Society for Horticultural Science* 123(4), 727-33. doi:10.21273/JASHS.123.4.727.

Schumacher, J. 2017. How light affects the life of *Botrytis*. *Fungal Genetics and Biology: FG and B* 106, 26-41. doi:10.1016/j.fgb.2017.06.002.

Sivan, A. and Chet, I. 1993. Integrated control of fusarium crown and root-rot of tomato with *Trichoderma harzianum* in combination with methyl-bromide or soil solarization. *Crop Protection* 12(5), 380-6. doi:10.1016/0261-2194(93)90082-T.

Skirvin, D. J. and Fenlon, J. S. 2003. The effect of temperature on the functional response of *Phytoseiulus persimilis* (Acari : Phytoseiidae). *Experimental and Applied Acarology* 31(1-2), 37-49. doi:10.1023/B:APPA.0000005107.97373.87.

Speyer, E. R. 1927. An important parasite of the greenhouse white-fly (*Trialeurodes vaporariorum*, Westwood). *Bulletin of Entomological Research* 17(3), 301-8. doi:10.1017/S0007485300019374.

Stall, R. E., Hortenst, C. C. and Iley, J. R. 1965. Incidence of Botrytis gray mold of tomato in relation to a calcium-phosphorus balance. *Phytopathology* 55, 447-&.

Stapleton, J. J. and Devay, J. E. 1984. Thermal components of soil solarization as related to changes in soil and root microflora and increased plant-growth response. *Phytopathology* 74(3), 255–9. doi:10.1094/Phyto-74-255.

Stern, V. M., Smith, R. F., van den Bosch, R. and Hagen, K. S. 1959. The integration of chemical and biological control of the spotted alfalfa aphid: the integrated control concept. *Hilgardia* 29(2), 81–101. doi:10.3733/hilg.v29n02p081.

Sullivan, D. J. and Völkl, W. 1999. Hyperparasitism: multitrophic ecology and behavior. *Annual Review of Entomology* 44, 291–315. doi:10.1146/annurev.ento.44.1.291.

Suthaparan, A., Stensvand, A., Torre, S., Herrero, M. L., Pettersen, R. I., Gadoury, D. M. and Gislerød, H. R. 2010. Continuous lighting reduces conidial production and germinability in the rose powdery mildew pathosystem. *Plant Disease* 94(3), 339–44. doi:10.1094/PDIS-94-3-0339.

Suthaparan, A., Stensvand, A., Solhaug, K. A., Torre, S., Telfer, K. H., Ruud, A. K., Mortensen, L. M., Gadoury, D. M., Seem, R. C. and Gislerød, H. R. 2014. Suppression of cucumber powdery mildew by supplemental UV-B radiation in greenhouses can be augmented or reduced by background radiation quality. *Plant Disease* 98, 1349–57.

Symondson, W. O. C., Sunderland, K. D. and Greenstone, M. H. 2002. Can generalist predators be effective biocontrol agents? *Annual Review of Entomology* 47, 561–94. doi:10.1146/annurev.ento.47.091201.145240.

Teitel, M. 2007. The effect of screened openings on greenhouse microclimate. *Agricultural and Forest Meteorology* 143(3–4), 159–75. doi:10.1016/j.agrformet.2007.01.005.

Ullah, M. S. and Lim, U. T. 2015. Life history characteristics of *Frankliniella occidentalis* and *Frankliniella intonsa* (Thysanoptera: Thripidae) in constant and fluctuating temperatures. *Journal of Economic Entomology* 108(3), 1000–9. doi:10.1093/jee/tov035.

Urbaneja, A., Monton, H. and Molla, O. 2009. Suitability of the tomato borer *Tuta absoluta* as prey for *Macrolophus pygmaeus* and *Nesidiocoris tenuis*. *Journal of Applied Entomology* 133(4), 292–6. doi:10.1111/j.1439-0418.2008.01319.x.

Vallad, G. E., Smith, H. A. and Messelink, G. J. 2018. Crop protection: pest and disease management. In: Heuvelink, E. (Ed.), *Tomatoes*. CABI Publishing, Wallingford, UK.

Van Driesche, R. G. and Heinz, K. M. 2004. An overview of biological control in protected culture. In: Heinz, K. M., Van Driesche, R. G. and Parrella, M. P. (Eds), *Biocontrol in Protected Culture*. Ball Publishing, Batavia, IL, pp. 1–25.

Vangansbeke, D., De Schrijver, L., Spranghers, T., Audenaert, J., Verhoeven, R., Nguyen, D. T., Gobin, B., Tirry, L. and De Clercq, P. 2013. Alternating temperatures affect life table parameters of *Phytoseiulus persimilis, Neoseiulus californicus* (Acari: Phytoseiidae) and their prey *Tetranychus urticae* (Acari: Tetranychidae). *Experimental and Applied Acarology* 61(3), 285–98. doi:10.1007/s10493-013-9704-8.

Vangansbeke, D., Audenaert, J., Nguyen, D. T., Verhoeven, R., Gobin, B., Tirry, L. and De Clercq, P. 2015a. Diurnal temperature variations affect development of a herbivorous arthropod pest and its predators. *PLoS ONE* 10(4), e0124898. doi:10.1371/journal.pone.0124898.

Vangansbeke, D., Nguyen, D. T., Audenaert, J., Verhoeven, R., Gobin, B., Tirry, L. and De Clercq, P. 2015b. Prey consumption by phytoseiid spider mite predators as affected by diurnal temperature variations. *BioControl* 60(5), 595–603. doi:10.1007/s10526-015-9677-0.

Vangansbeke, D., Nguyen, D. T., Audenaert, J., Verhoeven, R., Gobin, B., Tirry, L. and De Clercq, P. 2016. Supplemental food for *Amblyseius swirskii* in the control of thrips: feeding friend or foe? *Pest Management Science* 72(3), 466–73. doi:10.1002/ps.4000.

van Houten, Y. M., Glas, J. J., Hoogerbrugge, H., Rothe, J., Bolckmans, K. J., Simoni, S., van Arkel, J., Alba, J. M., Kant, M. R. and Sabelis, M. W. 2013. Herbivory-associated degradation of tomato trichomes and its impact on biological control of *Aculops lycopersici*. *Experimental and Applied Acarology* 60(2), 127–38. doi:10.1007/s10493-012-9638-6.

van Kruistum, G., Verschoor, J. and Hoek, H. 2014. CATT as a non-chemical pest and nematode control method in strawberry mother planting stock. *Journal of Berry Research* 4(1), 29–35. doi:10.3233/JBR-140063.

van Lenteren, J. C. 2000. A greenhouse without pesticides: fact or fantasy? *Crop Protection* 19(6), 375–84. doi:10.1016/S0261-2194(00)00038-7.

van Lenteren, J. C. 2012. The state of commercial augmentative biological control: plenty of natural enemies, but a frustrating lack of uptake. *BioControl* 57(1), 1–20. doi:10.1007/s10526-011-9395-1.

van Lenteren, J. C. and Woets, J. 1988. Biological and integrated pest control in greenhouses. *Annual Review of Entomology* 33(1), 239–69. doi:10.1146/annurev.en.33.010188.001323.

van Lenteren, J. C., Bale, J., Bigler, F., Hokkanen, H. M. and Loomans, A. J. 2006. Assessing risks of releasing exotic biological control agents of arthropod pests. *Annual Review of Entomology* 51, 609–34. doi:10.1146/annurev.ento.51.110104.151129.

van Lenteren, J. C., Bolckmans, K., Kohl, J., Ravensberg, W. J. and Urbaneja, A. 2018. Biological control using invertebrates and microorganisms: plenty of new opportunities. *BioControl* 63(1), 39–59. doi:10.1007/s10526-017-9801-4.

van Maanen, R., Messelink, G. J., van Holstein-Saj, R., Sabelis, M. W. and Janssen, A. 2012. Prey temporarily escape from predation in the presence of a second prey species. *Ecological Entomology* 37(6), 529–35. doi:10.1111/j.1365-2311.2012.01395.x.

Vänninen, I., Pinto, D. M., Nissinen, A. I., Johansen, N. S. and Shipp, L. 2010. In the light of new greenhouse technologies: 1. Plant-mediated effects of artificial lighting on arthropods and tritrophic interactions. *Annals of Applied Biology* 157(3), 393–414. doi:10.1111/j.1744-7348.2010.00438.x.

Venzon, M., Oliveira, R. M., Perez, A. L., Rodríguez-Cruz, F. A. and Martins Filho, S. 2013. Lime sulfur toxicity to broad mite, to its host plants and to natural enemies. *Pest Management Science* 69(6), 738–43. doi:10.1002/ps.3431.

Walter, D. E. 1996. Living on leaves: mites, tomenta, and leaf domatia. *Annual Review of Entomology* 41, 101–14. doi:10.1146/annurev.en.41.010196.000533.

Walters, D. R. and Bingham, I. J. 2007. Influence of nutrition on disease development caused by fungal pathogens: implications for plant disease control. *Annals of Applied Biology* 151(3), 307–24. doi:10.1111/j.1744-7348.2007.00176.x.

Wang, S., Tan, X. L., Michaud, J. P., Zhang, F. and Guo, X. 2013. Light intensity and wavelength influence development, reproduction and locomotor activity in the predatory flower bug *Orius sauteri* (Poppius) (Hemiptera: Anthocoridae). *BioControl* 58(5), 667–74. doi:10.1007/s10526-013-9522-2.

Zappala, L., Siscaro, G., Biondi, A., et al. 2012. Efficacy of sulphur on *Tuta absoluta* and its side effects on the predator *Nesidiocoris tenuis*. *Journal of Applied Entomology* 136, 401–9.

Zhang, Y. and Shipp, J. L. 1998. Effect of temperature and vapor pressure deficit on the flight activity of *Orius insidiosus* (Hemiptera : Anthocoridae). *Environmental Entomology* 27(3), 736–42. doi:10.1093/ee/27.3.736.

Zhang, N. X., Messelink, G. J., Alba, J. M., Schuurink, R. C., Kant, M. R. and Janssen, A. 2018. Phytophagy of omnivorous predator *Macrolophus pygmaeus* affects performance of herbivores through induced plant defences. *Oecologia* 186(1), 101–13. doi:10.1007/s00442-017-4000-7.

Zhang, N. X., van Wieringen, D., Messelink, G. J. and Janssen, A. 2019. Herbivores avoid host plants previously exposed to their omnivorous predator *Macrolophus pygmaeus*. *Journal of Pest Science* 92(2), 737–45. doi:10.1007/s10340-018-1036-3.

Zilahi-Balogh, G. M. G., Shipp, J. L., Cloutier, C. and Brodeur, J. 2006. Influence of light intensity, photoperiod, and temperature on the efficacy of two aphelinid parasitoids of the greenhouse whitefly. *Environmental Entomology* 35(3), 581–9. doi:10.1603/0046-225X-35.3.581.

Zilahi-Balogh, G. M. G., Shipp, J. L., Cloutier, C. and Brodeur, J. 2007. Predation by Neoseiulus cucumeris on western flower thrips, and its oviposition on greenhouse cucumber under winter vs. summer conditions in a temperate climate. *Biological Control* 40(2), 160–7. doi:10.1016/j.biocontrol.2006.10.011.

Part 3

System management

Chapter 12

Automation and robotics in greenhouses

E. J. van Henten, Wageningen University and Research, The Netherlands

1 Introduction

Protected cultivation is a very effective way of growing crops generating high yields and high-quality products. Covers made of screening materials, plastic or glass protect the crop from unfavourable outdoor climate conditions and pests. When the environment is sufficiently closed and proper technology is installed, protected cultivation offers the opportunity to modify the indoor climate to create an environment that is optimal for crop growth and production, both in terms of quantity and of quality. Depending on the technical infrastructure, protected cultivation is an intensive production method with high investment and operational costs, that permits only the production of high-value fruit and vegetable crops like tomatoes, sweet pepper, cucumber, strawberries, flowers like roses, chrysanthemums and gerberas and many types of ornamentals. Currently, protected cultivation is a cultivation method commonly adopted around the globe with implementations ranging from relatively low-tech, low-cost shelters to high-tech and expensive greenhouse systems.

Capital, energy and labour are the main cost factors in a cost breakdown of a high-tech greenhouse production system in the Netherlands. Since labour costs constitute a significant cost factor, labour efficiency and labour cost

http://dx.doi.org/10.19103/AS.2019.0052.14

reduction by means of automation and robotics have received, still receive and will continue to receive considerable attention in protected cultivation. Competition on the national and international markets demands extreme vigilance from horticultural entrepreneurs regarding the cost-effectiveness of their production systems.

Beyond labour costs, some additional issues and trends stimulate this focus on supporting and replacing human labour in protected cultivation, amongst others (based on Henten, 2006):

- Increasing size of production facilities:
  In 1990, the greenhouse crop production area in the Netherlands amounted to 10 000 ha, with more than 14 000 companies. Until 2012 the production area remained around 10 000 ha and then started to decline gradually to around 9 000 ha in 2015. In that same year, the number of companies amounted to around 4500. The average size of a commercial greenhouse facility rapidly increased from an average of 0.7 ha in 1990 to around 2 ha in 2015 and this trend will continue. The number of large companies has grown rapidly. Around half of the total production area in the Netherlands is operated by companies having an acreage of 5 ha or more. Scale enlargement of greenhouse crop production facilities is a worldwide trend.
- Increasing problems with the availability of sufficiently skilled labour:
  A job in greenhouse crop production is not considered to be very prestigious due to heavy and repetitive tasks under sometimes very uncomfortable climate conditions, and the wages are relatively low compared to many other types of available jobs, so growers have considerable problems acquiring skilled labourers.
- Employees encountering health problems due to heavy and repetitive tasks:
  Compared with standard industrial practice, the working environment in greenhouses is not very comfortable. Both temperature and humidity are high, and repetition of tasks performed while the worker has an unfavourable posture, results in musculoskeletal disorders among which back injuries and repetitive strain injury (RSI) are the most common.
- Food quality and food safety:
  Consumers wish to eat food that meets high quality and safety standards. Equipped with objective measuring devices, robotic systems, not subject to fatigue, offer the opportunity to consistently produce uniform product batches. Also replacing humans by machines that can easily be cleaned reduces the risk of contamination.
- Precision treatment of individual plants or products for improved and more uniform product quality according to customer specifications:

> There is a distinct trend of specialization of crop production and individualization of the treatment of crops. Companies tend to shift from bulk production to the production of high-quality produce or specialties. Growth and development of individual plants is monitored and action is taken to meet the requirements imposed by individual consumer groups. As a consequence greenhouse crop production becomes more labour intensive.

In the face of international competition and to meet the consumer's demand for low-cost, high-quality and safe horticultural products, growers wish to produce as efficiently and effectively as possible. As a matter of fact, these requirements are not new in industrial production (e.g. car industry, consumer electronics etc.). And, in a way, greenhouse crop production has become an industrial production process. For more than a century, industrial production has gone through a process of alleviating or even replacing human labour with machines. Also, product(ion) quality management and the improvement of the overall efficiency and effectiveness of the production process have received considerable attention in industrial production. It is not surprising that the greenhouse industry follows similar trends in mitigating the aforementioned challenges, for instance, by means of automation and robotics.

It should be noted that this analysis only relates to handling and physical operations performed on the crop and harvested products. Control of the greenhouse climate in terms of light intensity, temperature, humidity, carbon dioxide and the root environment of the crop in terms of water and nutrient content and supply, albeit being extremely important in this domain, are outside the scope of this chapter.

It is fair to state that the descriptions and analyses contained in this chapter are largely based on the situation in the Dutch high-tech greenhouse industry. Depending on local conditions throughout the world the challenges to be addressed might be (significantly) different. And clearly, the technology that can be deployed depends on these local conditions as well. However, there is a generic core in the material presented in this chapter that hopefully serves as a reference for analysing local conditions in greenhouse crop production systems and can be used to direct research and development for potential solutions and innovations.

2 A crop production cycle

To be able to analyse the achieved or the required level of automation in horticultural crop production, we first briefly describe the various steps in horticultural crop production. A generic description of the production cycle covering most crops in protected cultivation consists of the steps or operations provided in Table 1 (Henten, 2006; Henten and Bontsema, 2010; Henten et al., 2013).

It all starts with the production of seedlings or cuttings. Seeds are sown in substrate or soil plugs and may be transplanted once or even twice before being planted into or onto the final growing substrate (e.g. soil, rockwool, perlite, coco fibre etc.). Cuttings are directly planted in substrate. Once they have reached a certain size, seedlings and cuttings are transplanted into the final container or rockwool pot. For certain crops, grafting is a necessary step to assure production quality and quantity or to prevent diseases. Depending on the crop and local production habits, the production of seedlings, cuttings and grafts is done by companies specialized in effectively producing high-quality starting material for a crop production cycle. It also happens frequently however that starting material is produced on site or at least within the company at a different location, sometimes even abroad.

Before crop production in the production greenhouse actually starts, the greenhouse is cleaned or even disinfected. For crops grown in or on the soil, the soil is disinfected and in some cases a soil profile is prepared for drainage. In the Netherlands vegetables like tomatoes, cucumbers, sweet pepper and eggplant are mostly grown on substrate slabs in a soilless culture either on the soil or on gullies elevated from the ground. In both cases the soil underneath the substrate slabs and gullies is covered with plastic. In cases where crops are grown on concrete floors (e.g. ornamentals), only the floors have to be cleaned. Then, the heating system and irrigation system are put in place and the crop is planted.

During crop growth and production, plant maintenance operations (PMOs) are performed on the crop. PMOs differ per crop. Some crops require physical support from sticks or support wires and mounting the sticks to the plants or attaching plants to the support wire is one of the operations needed on a regular basis. Removal of side shoots and thinning of flowers and fruit are quite common in most fruits, vegetables and flowering crops. In the high-wire cultivation system of tomatoes and cucumbers leaf picking, that is removal of old leaves, is a regular operation, as are the lowering of the plants and plant protection.

Clearly, all crops have one PMO in common, harvesting. Harvesting is implemented either as a full-field operation when all plants are ready at almost the same time or on a selective basis, when readiness of individual plants, flowers or fruits differs among the crop and when crops are grown for a prolonged period of time.

Harvested products are transported to a processing area where they are graded and packed before shipment to the auction or retailer. Depending on the product and the (range of) customer(s), products are graded and packed according to customer specifications. The crop production cycle ends with the

Table 1 A generic description of production cycle including individual steps or operations in the production with indication of availability of equipment to support or replace human labour or of ongoing research and development aiming at supporting or replacing human labour in that task

Step No.	Operation		Commercially available	Research and development
1	Seeding		X	
	Grafting		X	
	Production of cuttings (taking and sticking of cuttings in soil)		X	
2	Transplanting		X	
3	Greenhouse cleaning and preparation			
4	Internal transport of plants		X	
5	Planting		X	
6	Plant maintenance operations (PMOs)			
	a	Sticking support sticks or mounting wires		
	b	Attaching plants to support sticks and knotting wires		
	c	Side shoot removal		
	d	Weeding/plant thinning		
	e	Fruit/flower thinning		
	f	Leaf picking		X
	g	Lowering plants		
	h	Crop protection/spraying	X	X
	i	Harvesting		X
7	Internal transport of plants or harvested products		X	
8	Grading		X	
9	Packing		X	
10	Crop removal			

removal of plants from the greenhouse and implementing preparations for the new production cycle.

Besides details of operations that depend on the particularities of each crop, crops also differ in the extent to which this generic production cycle is repeated in time. Single harvest crops like lettuce and chrysanthemum go through steps 1–9 once, starting with seeding or planting and ending with grading and packing. The production is finished at harvest time and then the cycle is repeated with a new crop. For multiple harvest crops like tomatoes and roses, crop maintenance, harvest and sorting and packing are repetitive tasks. These crops go through steps 6–9 several times over a prolonged period of time. The frequency at which steps 6–9 are implemented depends on the growth and production rate. The production rate, in turn, depends on cultivation measures including climate control, water and nutrient supply and on the weather conditions outside the greenhouse. Though the growth environment is much more controlled as opposed to the uncontrolled conditions in the open field, seasonal influences, especially light conditions, strongly affect crop growth and production.

3 State of the art in automation and robotics

The following review of state of the art in automation and robotics in protected cultivation should be considered as being more indicative than exhaustive.

Seeding is a highly mechanized process now. A fully automated seeding line consists of a tray washer, a tray de-stacker, a machine that fills trays with substrate, a seeding machine, a machine covering the trays with a thin layer of substrate, a watering machine and a tray stacking machine. For special purposes seed sorting machines are available. To improve product quality and uniformity as well as for efficient use of the production area, camera-assisted tray inspection and filling machines are used to achieve 100% filling with good quality plant material.

Contrary to seeding, the production of cuttings requires a considerable amount of expertise and skill. Grafting is a delicate process which requires a high degree of skill and the operation is physically and mentally demanding. Both the production of cuttings and grafts still rely heavily on manual labour though technical development is progressing in both operations (van Hoogstraten and van de Zedde, 2014).

In early 2000, the RomboMatic appeared on the Dutch market (Rombouts and Rombouts, 2002). This cutting production robot for potted roses used camera vision and industrial manipulators and produced rose cuttings from stock material. Stock material was manually fed into the machine and then processed by four parallel operating camera vision-controlled industrial manipulators. Cuttings were taken from the stock material, dipped in root growth stimulating powder and stuck into the potting soil substrate. The machine could also be used to produce cuttings of Euonymus, Lavender and Hedera. The ISO Cutting and Planting 1800 of the IsoGroup seems to continue this strand of technical innovation (IsoGroup, 2018a).

A grafting robot was developed by Nishiura et al. (1996). The grafting robot achieved a success rate of 97% at a speed ten times faster than human workers (Kondo and Ting, 1998). This strand of technical innovation has matured into a commercial product (IsoGroup, 2018b). This machine can be used for grafting of cucumber, tomato and sweet pepper.

Sticking of cuttings into the potting soil has been a human operation for a long time. The development of geranium and chrysanthemum cutting-sticking robots were reported by Simonton (1990) and Kondo and Monta (1999), respectively. Being commercially available, this kind of machinery is able to process large amounts of cuttings (IsoGroup, 2018c). Also sorting and transplanting of seedlings fits into this line of technology (IsoGroup, 2018d).

Internal transport in greenhouses can be categorized into two groups: (1) internal transport of plants and harvested products in a soil-based or stationary growing system and (2) mobile growing systems in which the crop circulates through the greenhouse.

Stationary cultivation systems are common practice in vegetable and flower production. Then plants have to be transported to the greenhouse and harvested plants and products need to be transported from the greenhouse to the central processing area. A wide range of equipment is available for internal transport of plants and harvested products inside the greenhouse. Partially, this is dedicated equipment specifically developed for high-tech greenhouse production environments (Berg Hortimotive, 2018). For crops grown in soils, systems for soil preparation and transport systems taken from outdoor cultivation practices have been adopted and sometimes modified for this particular operation.

Mobile growing systems are primarily used in ornamentals, though these systems are used in rose flower production as well. Aluminium benches are central to these growing systems. Using suitable infrastructure, these benches circulate through the greenhouse. Bench transportation systems are used to transport benches between greenhouse and processing area. A bench filling machine automatically puts the plant on the bench. Systems are available to reduce plant density during the growing period. Plants can be collected automatically using an automatic collection system after which products can be graded according to customer's specification. An automatic cleaning system cleans the benches at the end of the production cycle (Logiqs, 2018).

Fully automatic in-row pesticide sprayers are commonly used in greenhouse vegetable production (e.g. Micothon, 2018). Automatic overhead pesticide sprayers are used during the production of seedlings, flowers and potted plants.

Grading and packing of products usually takes place in a separate area at the production site. Commercial systems for grading and packing of fruits, flowers and vegetables have been around for two decades or more. Companies involved include GREEFA, Aweta, MAF RODA, TOMRA and Sermac. Sorting parameters are colour, weight and diameter for tomatoes and weight for cucumber and sweet pepper. A typical sorting line consists of buffered feeders, the sorting machine with multiple sorting lines operating in parallel, followed by automated box fillers and box stacking machines or robots. Sorting lines for flowers consist of buffered feeders, maturity measurement, measurement of length and thickness of stems, and automatic bunching and sealing machines are used for sorting roses and other flowers. In recent years, more advanced camera vision-based assessment of ripeness was implemented. Still human labour commonly plays a role during grading and packing.

4 Analysis of state of the art in automation and robotics

Reflecting on the contents of Table 1, it appears there is some technology available in the first and final phases of crop production, that is in steps 1–5 and 7–9. Available technology at least supports human labour, but in quite a number of operations technology replaces human labour. This observation should not be interpreted as a general conclusion, since availability of technology may differ among different crops. Most technology essentially addresses the 'big crops', vegetables like tomatoes, cucumbers, sweet peppers, and flowers like roses and cut chrysanthemum. The small crops may have to do with adapted versions of existing equipment.

Interestingly enough, especially in the PMOs (step 6) human labour is still key.

From a technological perspective, the state-of-the-art analysis of automation and robotics in greenhouses can be summarized quite effectively as follows. Operations in greenhouse production that can relatively easily be automated have been automated. Or, the other way around, operations that are still primarily the domain of human labour are apparently by their very nature difficult to automate.

What makes some operations easier to automate than others? And what kind of technology suffices for automation of the 'simple' operations and what technology is required for the 'challenging' operations?

Though hard to provide a thorough proof, it seems that in the 'simple' operations location, shape and size of objects to be treated, be it seeds, seedlings or potted ornamentals, are quite well known and uniform. The kind of automation technology able to perform these operations has the following characteristics:

- It replaces a human being or human activity in relatively simple tasks.
- It does not have much flexibility with respect to the task, the objects it has to handle and the working environment in which the machine operates.
- It does not contain many sensors.
- It does not contain much 'intelligence'.
- It is usually a purely mechanical system.

Compared to technology as previously characterized, humans excel in their perceptive and cognitive skills, their very-fast and effective eye-hand coordination, their excellent dexterity and soft gripping skills, their flexibility when it comes to task switching as well as their flexibility and robustness in the face of (natural) variation in the working environment or unfavourable conditions like varying light levels. These are the kind of capabilities needed to perform a flower thinning operation, a side shoot removal operation or selective harvesting. Current technology, essentially based on mechanical engineering solutions, is not able to meet that level of specification. More advanced technology is needed to support or replace human labour in the more challenging tasks. Such technology will have the following characteristics:

- It shows flexibility in task switching.
- It is flexible with respect to the natural variation in the objects it has to handle and the working environment in which the machine operates.
- It uses many sensors and concepts like sensor fusion.
- It contains a considerable amount of software-based 'intelligence'.
- It is based on a mix of mechanics and electronics, or so-called mechatronics or robotics.

5 Some examples of research and development in automation and robotics

In the past three decades research on robotics and high-tech mechanization mainly focused on harvesting high-value crops in greenhouses and examples addressing other complex and challenging crop maintenance operations are very limited. Bac et al. (2014) analysed in detail 50 initiatives on robotic harvesting of high-value crops of the previous thirty years. In the domain of protected cultivation this included robots for harvesting cucumber, tomato, eggplant, sweet pepper, strawberry, roses and gerbera. It is not surprising that tomato received most attention, since in protected cultivation this is one of the key fruit-vegetable crops. Hereafter some examples of past and current research and product development will be presented in a bit more detail.

5.1 Sweet pepper harvesting

The EU project CROPS (Crops-robots, 2018) focused among others on a robot for autonomous harvesting of sweet peppers (see Fig. 1a). The developed prototype sweet pepper harvester consists of a platform which uses the existing pipe-rail system (heating pipes) in the greenhouse to move along the plants. The robot is equipped with a nine degrees-of-freedom manipulator and two different types of end-effectors for grasping and detaching of the fruits. The system is set-up in a modular way concerning both hardware and software. Sensor modules are mounted on the robot platform and the end-effector. The colour camera mounted on the platform is used to locate the harvestable fruit. Based on that information, the manipulator positions the end-effector in the neighbourhood of the fruit. Then the camera on the end-effector is used to make a close-up image of the fruit and its environment, information that is used to re-plan the motion to fruit. Finally, the gripper detaches the fruit from the plant.

Figure 1 Sweet pepper harvesting robots, (a) the EU CROPS robot, (b) the Australian robot and (c, d) the EU Sweeper robot.

The harvesting module is mounted onto the platform and comprises a robotic arm and end-effector to grip and detach the fruit. A sensor module is also located on the platform and acquires data using a colour camera and a time-of-flight (TOF) camera. While dealing with limited visibility of the targets due to occlusion by leaves or other objects the module obtains information to measure the fruit ripeness and the position and orientation of fruits, leaves, stems, other obstacles using advanced image analysis methods and sensor-fusion techniques. With this information a feasible free path towards a fruit and an optimal grasp pose for the manipulator to grasp the fruit is calculated. On the end-effector a second camera module is located. This module is used to correct the target position and orientation in the last phase of approaching.

Figure 1b shows a parallel development from Australia (Lehnert et al., 2017). In this system, harvested ripe fruits were detected using colour information. The robot then actively viewed the fruit from different angles and created a 3D representation of the scene using its RGB-D camera. A geometric model of a bell pepper was fitted to the point cloud in order to determine the pose of the fruit (Lehnert et al., 2016). Next, an end-effector combining gripping and cutting attaches to the fruit and once in contact cuts the peduncle using an oscillating blade. Harvesting results were improved by explicitly detecting the peduncle (Sa et al., 2017). The overall system was able to harvest approximately 75% of the bell peppers in modified crop conditions at an average harvest time of 20 s/fruit.

A further development of the CROP sweet pepper harvesting robot was continued in the EU project Sweeper (Sweeper, 2018). The prototype is shown in Fig. 1c and d. Various aspects of the vision and control modules of this robot have been addressed in Barth et al. (2016, 2017, 2018).

5.2 Leaf removal in high-wire grown cucumbers and tomatoes

Leaf removal in high-wire cucumber cultivation was addressed by Henten et al. (2005). A stereo vision system identified and located the main stem of the cucumber plant. This information was used to mount an end-effector around the stem. The robotic manipulator then moved the end-effector upwards along the cucumber stem. Leaf petioles were detected using the thermal cutting device, similar to the one used for the cucumber harvester reported by Henten et al. (2002). It is worth noting that this deleafing robot was built on the same hardware and software basis as the cucumber harvesting robot, demonstrated some years before (Henten et al., 2005). Though in a very simplistic way, it was the first robotic platform in agriculture that was able to switch between tasks by replacing some software modules and the end-effector module.

In recent years the company Priva has developed a leaf removal robot for tomato, the Kompano (Priva, 2018) as shown in Fig. 2. The robot uses the pipe-rail system to move through the greenhouse. A stereo camera is used to observe the plant. The lowest leaf is detected and located and a four degrees-of-freedom manipulator navigates the cutting tool to the leaf petiole to cut the leaf as close to the plant as possible.

5.3 Strawberry harvesting

Strawberry harvesting has received considerable interest in research in the past decades leading to prototypes demonstrated in a greenhouse environment (Hayashi et al., 2010, 2012). A recent development in Europe is the prototype robotic strawberry harvester of the company Octinion from Belgium (Octinion, 2018). The system navigates through the greenhouse using local beacons with centimetre precision. A 3D vision system is used for detection and localization of ripe strawberries. A soft touch gripper picks the strawberries without using any other tools like scissors or other cutting devices. The system includes a quality monitoring device to harvest and sorts fruits according to size and quality (Fig. 3).

Figure 2 The Priva Kompano leaf removal robot for tomato (www.priva.com).

Figure 3 The prototype strawberry-harvesting robot of Octinion (www.octinion.com).

5.4 Crop monitoring and crop protection

A modular agricultural robot, Thorvald, allows easy adaptation to different greenhouse layouts and cultivation systems (Grimstad and From, 2017). Applications include crop protection. A similar strand of work is an autonomous crop monitoring robot on an Iseki platform (Fig. 4) as reported by Takayama et al. (2011). The robot travels on the pipe-rail system and deploys a chlorophyll imaging system to evaluate the photosynthetic functioning of the plant and, by doing so, to identify local stress in the crop. With greenhouse facilities growing in size we may expect more of these systems to appear in the future.

5.5 Sorting and packing

Technology exists for grading and packing of fruit and flowers, yet more advanced operations on individual fruits and flowers are still largely the subject of human labour. In the EU project PicknPack a flexible robotized production line (Fig. 5) was developed to assess quality and pack fresh and processed food products. The focus of the project was aiming for more flexibility and short changeover times to change from one product to another within seconds without loss of packaging material and spillage of food. With these technologies it should be easier to adapt to customer demands. Secondly, it

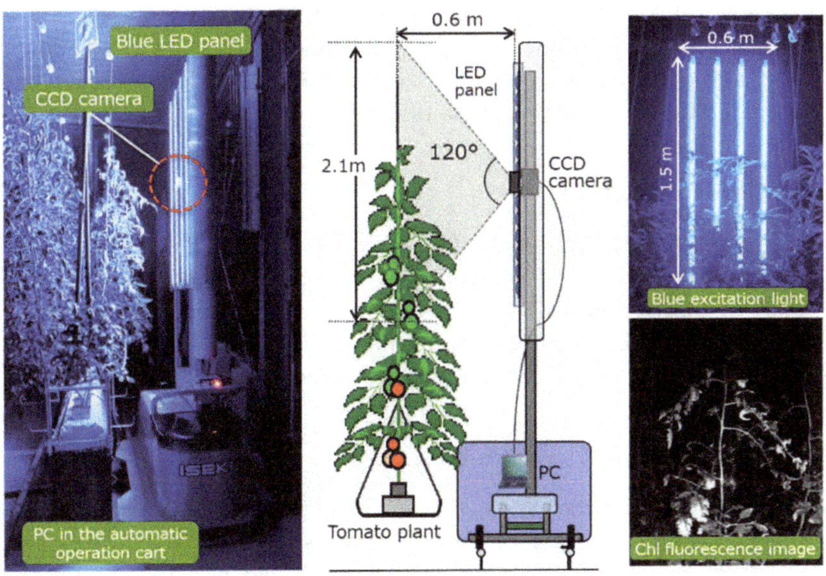

Figure 4 The crop-monitoring robot of Iseki. Source: adapted from Takayama et al. (2011).

Figure 5 Bin picking, quality control and packing of truss tomatoes, a case study in the PicknPack project (www.picknpack.eu).

should be possible to use the line for many different food products to adapt to a larger market, so that robotized lines become interesting for smaller and seasonal products. Robotic technology is entering the domain of sorting and packing as illustrated by the SmartPackr technology provided by Greefa (Greefa, 2018). The system is able to put fruits in the tray according to customer specifications.

6 Challenges and future trends

6.1 Challenges

The analysis of the state of art in automation and robotics revealed that a number of operations still strongly rely on human labour. What are the key reasons why these operations are so hard to automate? Essentially these operations are so challenging for robotics due to different forms of variation and uncertainty encountered while performing these tasks.

6.1.1 Object variation

Natural products have an inherent variation. Every plant, flower or fruit is in a way unique and appearance, geometry and mechanical properties vary from instance to instance. While this is clearly the case between different crops, it is also the case between different cultivars of a crop and even within the same cultivar of a crop, not least depending on growth stage.

6.1.2 Environmental variation

In industry robots operate in a highly controlled environment. Such conditions cannot be realized in a greenhouse. So, robots must be robust to changes in environmental conditions like illumination. While humans are relatively insensitive to large fluctuations in illumination, dealing with these conditions still proves to be a challenge for most computer vision-based sensing techniques embedded in most robotic systems.

6.1.3 Variation in cultivation systems

Depending on local conditions and preferences of the farmer, cultivation systems used to grow crops can differ significantly. For example, a tomato crop can be grown in soil while in high-tech hydroponic farming the plants are grown in substrate on gullies elevated around 80 cm above the soil. Though relatively trivial, this requires a different layout of automated crop maintenance and harvesting system. Even more subtle differences do exist within one growing system when looking at the implementation of, for instance, a high-wire cultivation system at different farms. Robots need to cater for these differences in the implementations of growing systems.

6.1.4 Task variation

As discussed in Section 2, many different processes take place in the production chain. Robots that have been designed for one specific task run the risk of being idle for a period of time. Robots that can perform multiple tasks have an increased economic viability.

6.1.5 Incomplete information

Most agricultural environments are complex with a high level of clutter and many occlusions. This requires that robots can deal with incomplete information and uncertainty. Similarly, uncertainty about future events such as weather, crop performance and the onset of a disease limits the amount of detailed information that can be used for planning.

6.2 Future trends

When evaluating current technology and providing guidance for future research and development it is worth reflecting on human capabilities when dealing with more complex tasks. As indicated earlier, perceptive and cognitive skills, fast eye-hand coordination, dexterity and soft gripping skills and flexibility in the face of (natural) variation in the shape of objects, working environment and tasks are key capabilities needed to deal with more complex tasks, capabilities in which humans excel. This is not a plea in support of humanoid robotics, a dominant line of research in the field of robotics. This list indicates key capabilities or functionalities that require attention when dealing with more complex operations in greenhouse cultivation. Future research should be directed along the the shape.

6.2.1 Modification of cultivation systems

Since variation and uncertainty are the main challenges for automation in agriculture, reduction of variation and uncertainty might support the development of robotic technology in this domain. For example, compared to the traditional cucumber growing system, the high-wire cultivation system significantly simplified the task of robotic harvesting (Henten, 2002, 2003) by increasing visibility and accessibility of the crop. Modification of the cultivation system is a strategy also implemented in orchards (Bac et al., 2014).

6.2.2 Sensing, multi-modal sensing, active perception

Current robot technology in agriculture strongly relies on computer vision-based sensing in the visible and near-infrared domain. Such technology might be sufficient for a range of operations in horticulture with low levels of complexity; some of the more difficult tasks may require fusion of multiple sensor modalities and novel sensors. For instance tactile sensing has not received much attention yet in this application domain. Current robotic systems in agriculture are largely based on the traditional sense-plan-act cycle, where perception and action are viewed as individual and sequential processes. This scheme cannot deal with the challenges that the agro-food environment poses. Active perception, in which the robot resolves uncertainty in the environment by actively gathering new sensory input by changing perspective or by manipulating objects, is a potential direction when dealing with more complex environments and tasks.

6.2.3 Compliant actuators and end-effectors

Current robotic gripping technology in agri-food is able to handle relatively well-defined, rigid objects. However, grippers for harvesting and packaging have to perform their tasks under demanding requirements, such as robustness to variability in product size, shape and softness and fast operation to reduce cycle times. Current robotic technology does not address these challenges. This large variation demands different grasp types, with different force distributions and different grasp stiffness with improved tactile sensing and control.

6.2.4 Including domain knowledge in the design and operation of robotic capabilities

Autonomous systems currently do not reasonably exploit prior knowledge about the structure of the working environment, the potential presence or absence of objects and the evolution of the object characteristics such as growth and ripeness in time.

6.2.5 Human-robot co-working

Variation in agriculture might prohibit near-term adoption of autonomous agricultural systems. A viable alternative to accelerate innovations and adoption may be mixed human-robot systems working cooperatively (Murakami et al., 2008; Berenstein et al., 2012; Cheein, 2015).

7 Conclusion

This chapter provided an overview of the state of the art in automation and mechanization for protected cultivation as well as ongoing robotics research in that domain. The chapter started with outlining reasons for automation and mechanization. In the face of a growing world population that needs to be supported by sustainable production of high-quality food, it is expected that technology will keep playing an important role in agriculture. An overview of available technology confirmed the presence of a wide range of technology commercially available in this domain. Yet, the overview indicated as well that in some phases of the production cycle human labour still dominates. The operations involved in these production phases are still too challenging for machines to successfully take over due to the inherent natural variability of the products, the poorly structured working environment and varying lighting conditions. With their sensory and cognitive skills as well as their highly effective eye-hand coordination, humans excel in the more complex operations. The chapter concluded with directions for future research on technology that is more flexible and robust and able to deal with complex operations in challenging working environments.

8 Where to look for further information

A concise and complete overview of the state of the art in robotics including applications in various domains such as the agri-food domain is provided in Siciliano and Katib (2016). The chapter of Bergerman et al. (2016) in that compendium covers agricultural robotics in general and contains some examples of robots for protected cultivation as well.

Further to that it is worth browsing journals in the field of robotics, such as the *Journal of Field Robotics* that regularly hosts special issues on agricultural robotics including protected cultivation (e.g. Ball et al., 2017). Also journals in the domain of agricultural and horticultural engineering like *Biosystems Engineering*, *Computers and Electronics in Agriculture* and *Transactions of the ASABE* are valuable sources of information on this topic.

Results of robotic research in protected cultivation are also communicated through conferences and workshops of the International Society of Horticultural

Science (ISHS), the European Society of Agricultural Engineers (EurAgEng) and the American Society of Agricultural and Biological Engineers (ASABE). More fundamental aspects of (agricultural) robotics are communicated through conferences and workshops like the annual IEEE International Conference on Intelligent Robots and Systems (IROS), the annual IEEE International Conference on Robots and Automation (ICRA) and the conferences of the International Federation of Automatic Control (IFAC). The IFAC regularly hosts a conference on Biorobotics specifically addressing fundamentals and applications of agricultural robotics.

Societies like ISHS, IEEE and IFAC also host technical committees or special interest groups addressing agricultural robotics and automation.

First initiatives in agricultural robotics date back to the late 1970s and early 1980s. In recent years both progress in the domain of robotics and artificial intelligence and a growing awareness of the need of supporting a growing world population with more sustainable food production has led to a sharp increase in interest from universities and research institutes worldwide on the topic of agricultural robotics and automation in general. Research and development of robotics and automation technology in protected cultivation specifically is mainly found in countries and regions with a strong position in protected cultivation itself. These regions usually also host a range of technology providers for that branch of agriculture.

9 References

Bac, C. W., Henten, E. J. Van., Hemming, J. and Edan, Y. 2014. Harvesting robots for high-value crops: state-of-the-art review and challenges ahead. *Journal of Field Robotics* 31(6), 888–911. doi:10.1002/rob.21525.

Ball, D., Upcroft, B., Van Henten, E., van den Hengel, A., Tokekar, P. and Das, J. 2017. *Journal of Field Robotics* - Special Issue on Agricultural Robotics. *Journal of Field Robotics* 34(6), 1037–38.

Barth, R., Hemming, J. and Henten, E. J. van. 2016. Design of an eye-in-hand sensing and servo control framework for harvesting robotics in dense vegetation. *Biosystems Engineering* 146, 71–84. doi:10.1016/j.biosystemseng.2015.12.001.

Barth, R., IJsselmuiden, J., Hemming, J. and Henten, E. J. van. 2017. Synthetic bootstrapping of convolutional neural networks for semantic plant part segmentation. *Computers and Electronics in Agriculture* (in press). doi:10.1016/j.compag.2017.11.040.

Barth, R., IJsselmuiden, J., Hemming, J. and Henten, E. J. van. 2018. Data synthesis methods for semantic segmentation in agriculture: a *Capsicum annuum* dataset. *Computers and Electronics in Agriculture* 144, 284–96. doi:10.1016/j.compag.2017.12.001.

Berenstein, R., Ben Halevi, I. and Edan, Y. 2012. A remote interface for a human-robot cooperative sprayer. Paper presented at the *Proceedings of the 11th International Conference on Precision Agriculture*, Indianapolis, IN.

Berg Hortimotive. 2018. Crop protection. Available at: https://www.berghortimotive.com/en/crop-protection/(Accessed on July 2018).

Bergerman, M., Billingsly, J., Reid, J. and Henten, E. J. van. 2016. Robotics in agriculture and forestry. In: Siciliano, B. and Khatib, O. (Eds), *Springer Handbook of Robotics*. Springer-Verlag, Berlin, Germany, pp. 1463-92. Chapter 56.

Cheein, F. A., Herrera, D., Gimenez, J., Carelli, R., Torres-Torriti, M., Rosell-Polo, J. R., Escolà, A. and Arnó, J. 2015. Human-robot interaction in precision agriculture: sharing the workspace with service units. *2015 IEEE International Conference on Industrial Technology (ICIT)*, Seville, pp. 289–95.

Crops-robots. 2018. Clever robots for crops. Available at: www.crops-robots.eu (Accessed on July 2018).

Greefa. 2018. SmartPackr™. Available at: https://www.greefa.com/product/packing/ (Accessed on August 2018).

Grimstad, L. and From, P. 2017 The Thorvald II agricultural robotic system. *Robotics* 6(4), 24. doi:10.3390/robotics6040024.

Hayashi, S., Shigematsu, K., Yamamoto, S., Kobayashi, K., Kohno, Y., Kamata, J. and Kurita, M. 2010. Evaluation of a strawberry-harvesting robot in a field test. *Biosystems Engineering* 105(2), 160–71. doi:10.1016/j.biosystemseng.2009.09.011.

Hayashi, S., Yamamoto, S., Saito, S., Ochiai, Y., Kohno, Y., Yamamoto, K., Kamata, J. and Kurita, M. 2012. Development of a movable strawberry-harvesting robot using a travelling platform. In: *Proceedings of the International Conference of Agricultural Engineering CIGR-AgEng 2012*, 8–12 July, Valencia, Spain.

Henten, E. J. van. 2006. Greenhouse mechanization: state of the art and future perspective. *Acta Horticulturae* 710(710), 55–70. doi:10.17660/ActaHortic.2006.710.3.

Henten, E. J. van and Bontsema, J. 2010. *Vision on Agricultural Robots; In Dutch: Visie* op agro-robotica. Wageningen UR, Farm Technology Group and WUR Greenhouse Horticulture, Wageningen, The Netherlands.

Henten, E. J. van, Hemming, J., Tuijl, B. A. J. van, Kornet, J. G., Meuleman, J., Bontsema, J. and van Os, E. A. 2002. An autonomous robot for harvesting cucumbers in greenhouses. *Autonomous Robots* 13(3), 241–58. doi:10.1023/A:1020568125418.

Henten, E. J. van, Van Tuijl, B. A. J., Hemming, J., Kornet, J. G., Bontsema, J. and Van Os, E. A. 2003. Field test of an autonomous cucumber picking robot. *Biosystems Engineering* 86(3), 305–13. doi:10.1016/j.biosystemseng.2003.08.002.

Henten, E. J. van, Tuijl, B. A. J. van, Hoogakker, G.-J., Weerd, M. J. van der, Hemming, J., Kornet, J. G. and Bontsema, J. 2005. An autonomous robot for de-leafing cucumber plants grown in a high wire cultivation system. *Biosystems Engineering* 94, 317–23.

Henten, E. J. van, Bac, C. W., Hemming, J. and Edan, Y. 2013. Robotics in protected cultivation. In: *4th IFAC Conference on Modelling and Control in Agriculture, Horticulture and Post Harvest Industry*, Finland, pp. 170–87. doi:10.3182/20130828-2-SF-3019.00070.

Hoogstraten, K. Van and Zedde, H. J. van de 2014. Robot makes grafting and rose cutting faster and easier: three party cooperation delivers innovations. *Greenhouses: the International Magazine for Greenhouse Growers* 3, 38–9.

IsoGroup. 2018a. ISO cutting and planting 1800. Available at: http://www.iso-group.nl/en/machines/iso-cutting-and-planting-1800 (Accessed on July 2018).

IsoGroup. 2018b. ISO graft 1100. Available at: https://www.iso-group.nl/en/machines/iso-graft-1100 (Accessed on July 2018).

IsoGroup. 2018c. ISO vision planter. Available at: http://www.iso-group.nl/en/machines/iso-visionplanter (Accessed on July 2018).

IsoGroup. 2018d. ISO robot plug planting machine. Available at: https://www.iso-group.nl/en/machines/iso-robot-plug-planting-machine (Accessed on July 2018).

Kondo, N. and Monta, M. 1999. Chrysanthemum cutting sticking robot system. *Journal of Robotics and Mechatronics* 11(3), 220–4. doi:10.20965/jrm.1999.p0220.

Kondo, N. and Ting, K. C. 1998. *Robotics for Bioproduction Systems*. ASAE, St. Joseph, MO, 325pp.

Lehnert, C., Sa, I., McCool, C., Upcroft, B. and Perez, T. 2016. Sweet pepper pose detection and grasping for automated crop harvesting. In: *Proceedings of the IEEE International Conference on Robotics and Automation*, 16–21 May 2016, Stockholm.

Lehnert, C., English, A., McCool, C., Tow, A. W. and Perez, T. 2017. Autonomous sweet pepper harvesting for protected cropping systems. *IEEE Robotics and Automation Letters* 2(2), 872–9. doi:10.1109/LRA.2017.2655622.

Logiqs. 2018. Greenhouse automation. Available at: https://www.logiqs.nl/en/greenhouse-automation/(Accessed on July 2018).

Micothon. 2018. Micothon tube rail spraying robot. Available at: https://micothon.nl/portfolio-items/micothon-tube-rail-spraying-robot/(Accessed on July 2018).

Murakami, N., Ito, A., Will, J. D., Steffen, M., Inoue, K., Kita, K. and Miyaura, S. 2008. Development of a teleoperation system for agricultural vehicles. *Computers and Electronics in Agriculture* 63(1), 81–8. doi:10.1016/j.compag.2008.01.015.

Nishiura, Y., Murase, H., Honami, N., Taira, T. and Wadano, A. 1996. Development of a gripper for a plug-in grafting robot system. *Acta Horticulturae* 440(440), 475–80. doi:10.17660/ActaHortic.1996.440.83.

Octinion. 2018. Strawberry harvesting robot. Available at: http://www.octinion.com/products/harvesting-series/strawberry-picking-robot (Accessed on September 2018).

Priva. 2018. Priva Kompano deleafing robot. Available at: https://www.priva.com/discover-priva/news-and-stories/priva-kompano-deleaf-line (Accessed on July 2018).

Rombouts, N. J. L. and Rombouts, P. K. M. 2002. Inrichting voor het machinaal afscheiden van stekken van een plantentak. Patent NL1017794C, 29pp.

Sa, I., Lehnert, C., English, A., McCool, C., Dayoub, F., Upcroft, B. and Perez, T. 2017. Peduncle detection of sweet pepper for autonomous crop harvesting–combined color and 3-D information. *IEEE Robotics and Automation Letters* 2(2), 765–72. doi:10.1109/LRA.2017.2651952.

Siciliano, B. and Khatib, O. 2016. *Springer Handbook of Robotics*. Springer-Verlag, Berlin, Germany.

Simonton, W. 1990. Automatic geranium stock processing in a robotic workcell. *Transactions of the ASAE* 33(6), 1274–80. doi:10.13031/2013.31579.

Sweeper. 2018. Sweet pepper harvesting robot. Available at: http://www.sweeper-robot.eu/(Accessed on July 2018).

Takayama, K., Nishina, H., Iyoki, S., Arima, S., Hatou, K., Ueka, Y. and Miyoshi, Y. 2011. Early detection of drought stress in tomato plants with chlorophyll fluorescence imaging -practical application of the speaking plant approach in a greenhouse. *IFAC Proceedings Volumes*, 44(1), 1785–90. doi:10.3182/20110828-6-IT-1002.01490.

Chapter 13

Models, sensors and decision support systems in greenhouse cultivation

Oliver Körner, Leibniz Institute of Vegetable and Ornamental Crops (IGZ), Germany

1 Introduction

Model-based decision support and automation, accompanied by the development of Industry 4.0, is progressively digitising the horticultural business. The most automated area is the production of plants in controlled environment agriculture (CEA), utilising greenhouses. However, not all greenhouse nurseries are ready for Industry 4.0. On one side of the scale, simple tunnel systems cover the soil; on the other are highly technological, fully equipped plant factories. In between these two extremes is a complete range of intermediate solutions. Nonetheless, all greenhouse technological systems evolved from simple structures without automatised climate control and sensors (and thus no option for model-based climate control or decision support systems (DSS)). Greenhouses used for the protected cultivation of horticultural crops have, over the centuries, developed from simple structures using solar radiation to increase indoor temperature and thereby extend the production season, to advanced production systems with the possibility to control the indoor climate (and with that, the crop), very accurately. For this reason, model-based controllers, sensors and DSS were developed. This chapter discusses the potential for sensors and models for growing automation in greenhouses.

http://dx.doi.org/10.19103/AS.2019.0052.15

2 Greenhouses as controllable units for decision making

2.1 Structures and equipment

Today, the typical modern commercial greenhouse is a structure of steel and aluminium, covered with either glass, highly insulating sheets of multi-layer polycarbonate or plastic, or foil (Ponce et al., 2015). Various types of greenhouse structures exist, each developed for the needs of the crop, the outside climate and the method of production. Inside, it is usually fitted with, amongst other things, movable screens to save energy during the night and provide shade during the day, fixtures for supplemental lighting, ventilation windows, air conditioning, pipe or air-heating systems and an installation to supply extra carbon dioxide for increased photosynthesis and thus growth (Fig. 1).

As greenhouses were made taller in order to increase the buffering of climate dynamics to create a constant and homogeneous climate, all activities are carried out to create perfect microclimate conditions (climate at and within the canopy) for the crop (Körner et al., 2007; Singh et al., 2016; Stanghellini and Van Meurs, 1989). The level of (micro)climate control varies greatly, from the basic shelter type greenhouse to the fully computerised actively conditioned greenhouse (Vox et al., 2010). In the latter, all installations are controlled by advanced climate computers. These systems also monitor the climate in the different sections of the greenhouse and provide alarms when key variables are outside their predefined limits (Ding et al., 2013; Wang and Tao, 2015).

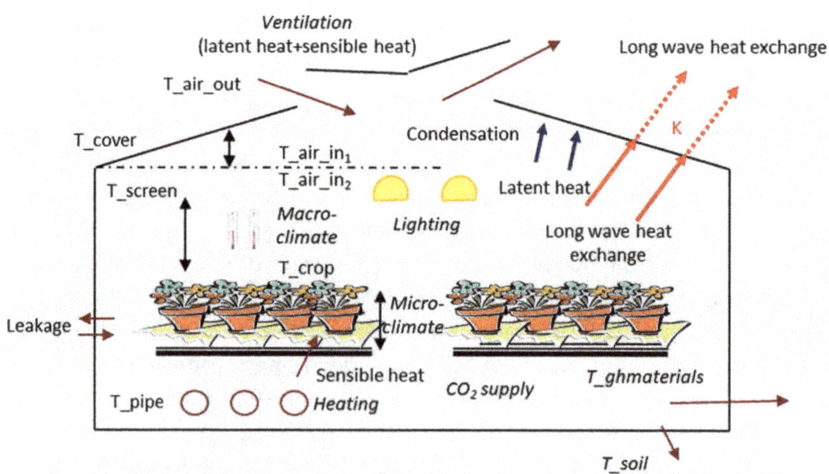

Figure 1 Example of a greenhouse with different physical processes, actuators and states. With temperature (T), greenhouse (gh), inside above screen (in_1), inside below screen (in_2), outside (out).

2.2 Energy management

While the energy efficiency of greenhouses has improved gradually since the 1970s energy crisis, energy remains a major expenditure in greenhouse production (Cuce et al., 2016). Since then, the high amount of energy used for climate control has led to research into resource-friendly concepts of greenhouse production in temperate and/or warm climate zones, while energy-saving climate regimes and strategies with reduced energy loss were established many years ago (e.g. Langhans et al., 1981; Meyer, 1983). Today, technical solutions exist for greenhouse crop production for almost all climatic environments.

Updating greenhouses with high-technological equipment, such as combined heat and power units, heat pumps, underground seasonal and daytime energy storage systems, and air treatment units as used in the closed greenhouse concept (Gelder et al., 2012; Opdam et al., 2005), has reduced energy consumption significantly (Cuce et al., 2016). In these closed environments, temperature and humidity control remain the two main challenges (Körner and Challa, 2003; Vadiee and Martin, 2012), and so optimising investment and climate control have led to limitations of (semi-) closed greenhouse systems.

2.3 Climate control

In many respects, greenhouse climate control is a trade-off between the desire to create optimal conditions for the crop (e.g. avoiding plant stress) and attempting to save energy and resources, as well as of course ROI (return on investment). Modern greenhouses are equipped with automated computer-based climate control technology, a long-existing technology (Udink ten Cate, 1983). In terms of climate control, the greenhouse is a black box with inputs and outputs, with outside climate as the *disturbing factor* (Fig. 2). The climate is controlled by an 'intelligent' algorithm steering the actuators of the greenhouse as heating system, air conditioning (cooling/dehumidification), ventilation, venting, screening, CO_2 dosage or lighting. In theory, the climate control unit should be capable of realising the targeted crop's (micro)climate, in any environment. Ideally, minimal action is then left to the grower (concerning climate control). However, in daylight situations in greenhouses (in contrast to closed environments without daylight) with fast dynamic adjustments to outside conditions and slow changes between hours, days or seasons, the creation of a homogeneous climate (both vertical and horizontal) is perfect. Large temperature gradients exist even in well-designed greenhouses, with local temperature differences of up to 5°C (Bojacá et al., 2009). This heterogeneity is caused by fluctuations

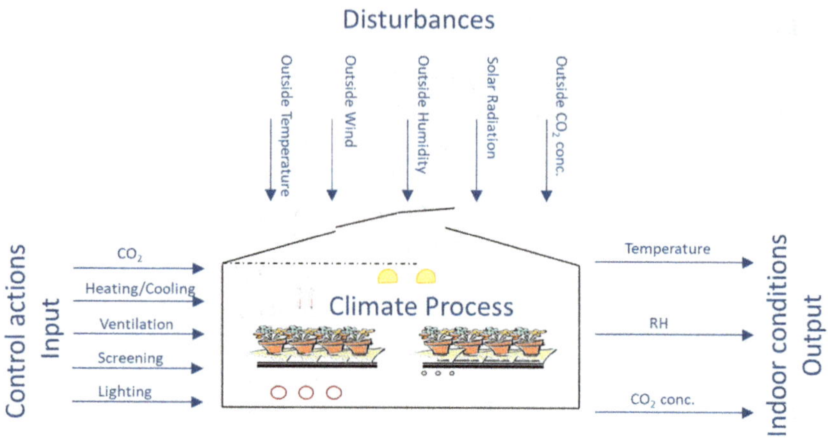

Figure 2 The greenhouse as input-output box with simplified inputs, outputs and disturbances.

in outside weather and the responsive dynamics of greenhouse climate control, actuated by, for example, heating and ventilation systems (Bot, 2001). Unwanted consequences include uneven production and product quality, and local build-ups of pests and diseases (Fargues et al., 2003). Moreover, a non-uniform distribution of air temperature and humidity leads to higher energy consumption as the microclimate must be controlled globally to maintain crop production locally. Structures therefore increase in height.

As greenhouse climate is controlled within larger (control) areas, plant- or spot-specific treatments cannot usually be carried out. Innovative use of wireless sensor networks could make distributed climate control possible (Gonda and Cugnasca, 2006), although current site-specific measurements cannot be transferred to climate control. However, the mapping of climate/crop spatial variability provides a way to adjust production inputs in different greenhouse compartments according to local demands. Such precision horticulture could reduce inputs while improving crop quality. The ability to predict the three-dimensional microclimate within the crop is therefore essential for the proper control of greenhouse actuators.

3 The decision-making process

3.1 Greenhouse management systems

One of the main aims of a greenhouse management system is to avoid growth-inhibiting stress and crop damage – i.e. maintaining high quality production while keeping resource consumption in mind. The decision-making process is ongoing and can be separated in an endless loop of planning and the cultivation phase (Fig. 3).

In Fig. 3 the process of decision making with a management system is illustrated as follows:

1 Before starting the cultivation phase the first planning phase is performed, in which the experience of the grower or inputs from planning tools - such as greenhouse investments tools - are used. In most situations combinations of these two are employed.
2 The planning phase is followed by a cultivation phase, in which experience and judgement is backed up by information from sensors, models, climate control and DSS software. During this phase, errors or valuable improvements will become evident. This is followed by another planning phase building up on the experiences of the preceding phases.

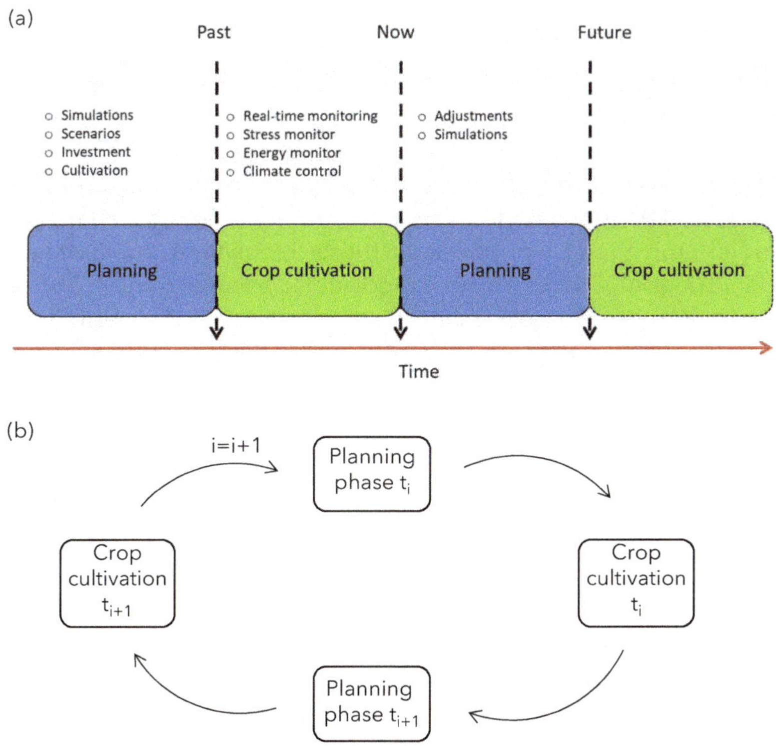

Figure 3 Decision-making process with decision support systems (DSS) in greenhouse cultivation. With process in a timely sequence (a) and in a repeating loop (b).

Over time both the grower and the mathematical models will increase their level of experience. The grower will naturally become an expert in using the system, while models can be improved depending on the model type: deterministic models can be improved in terms of parameter quality (auto-calibration) and data-based models by training (Fitz-Rodríguez and Giacomelli, 2009). At the same time, growers learn from decisions suggested by self-improving algorithm and model-based DSS. Eventually, this can lead to growers understanding the DSS logics and adapting them to the point where the DSS itself becomes redundant, which happened especially in early DSS. Due to the programming nature of these early DSS versions (simple non self-learning or calibrating algorithms), decisions tended to have a repetitive pattern. These DSS served the horticultural business for a limited period (e.g. CARE (Fisher and Heins, 1996), IntelliGrow (Aaslyng et al., 2003), PREDICT (Jensen et al., 2009)), but nevertheless contributed to the development of greenhouse horticulture and the formation of new generations of support systems.

3.2 Decision support targets

More sophisticated DSS, with embedded application of current scientific knowledge, are now in common use in plant production: mostly in open-field crops, such as, cereals (Rydahl, 2003), open-field horticultural crops (Gallardo et al., 2011; Sánchez-Molina et al., 2015; Van Loon et al., 2011) and fruits (Jones et al., 2010). The targets can differ, while having the common aim of optimised crop production management. In greenhouse horticulture, DSS exist for various subcategories in the production and trading chain. DSS can be used both for planning (e.g. technology investments, comparison of climate control strategies) and for real-time greenhouse decision support (based on sensor input and model predictions integrated in the DSS). This chapter only discusses systems for the crop production unit. The reason for using DSS in any cropping system is the expectation of improved results, in every possible sense.

Every decision process begins with a target, or by stating what it aims to achieve. The actual targets could be the following:

- The economic result,
- Plant growth, that is, measured as increase in dry/fresh matter,
- Harvest scheduling,
- Minimising the use of natural resources, that is, environmentally friendly production,
- Buyer's satisfaction with the product or the product quality.

The climate control system is one of several possible methods for realising chosen targets, that is, to get the best possible plant growth, yield and quality amongst other things. It is used, in particular, to control the climate's effect on crop growth, plant development and morphological aspects, and, aside from economic gains, significant factors to be optimised include keeping to the production plan, environmental friendliness and improving product quality (Buwalda et al., 2006). While growth is particularly important with regard to vegetable production, the situation is more complex when it comes to ornamental crops. In these products, aesthetic values play a larger role than weight and size. For decision making in crop cultivation, information regarding plant cultivation processes is essential. The individual situation determines which method to use.

3.3 Decisions by the climate controller

Alongside DSS, the climate controller makes decisions, supervised or non-supervised. In order to optimise the various often-conflicting aims in greenhouse climate control, non-supervised optimal control algorithms can be used (Van Ooteghem et al., 2005; Van Straten et al., 2002). Using this concept, goal functions are defined, often in economic terms, and further defined by sub-targets (Rytter et al., 2012; Van Ooteghem, 2007). Optimal greenhouse control uses models to predict future conditions and real-time sensor information combined with, among others, climate modelling (Seginer, 1999). This concept, although mainly based on mathematical algorithms, can be seen as one further development of the speaking-plant approach, yet combining forecast models and soft sensors (i.e. mathematical models using real-time sensor data). In this concept, there are two fundamentally different factors to consider: the highest possible income and the lowest possible expenditure. However, in different locations and climate zones, decisions need to be adjusted as conditions vary tremendously between locations (see Fig. 4); in Spain, strong radiation and heat in summer demands a completely different strategy to crop cultivation in cooler regions at higher latitudes. Thus, models need to be adjusted to the circumstances. A system employing self-learning models that is trained to, for example, Spanish conditions would fail in the Netherlands.

4 Growers' decision tools

4.1 Tool types

To assist growers in decision making and to give them tools to handle the increasing complexity of crop cultivation, a large number of model-based

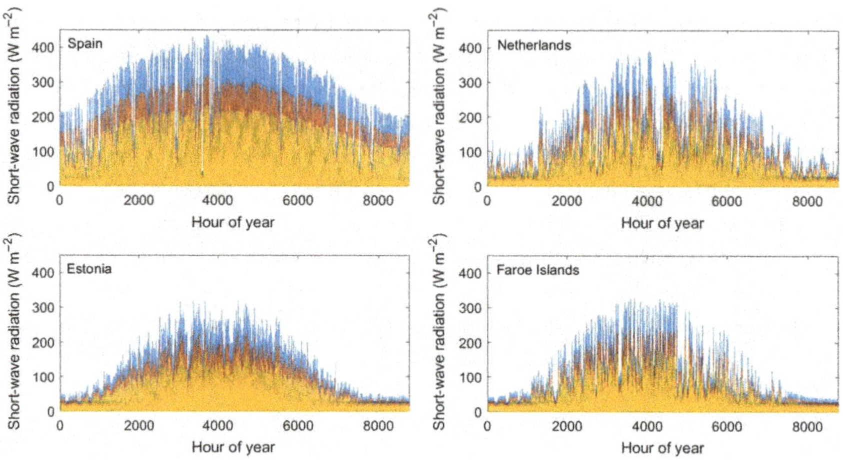

Figure 4 Short-wave radiation inside a standard greenhouse with three vertical crop levels in a tomato canopy with LAI 3 (blue = top LAI 2.66, red = middle LAI 1.5, yellow = bottom LAI 0.34) in four European climate regions calculated with the virtual greenhouse simulator tool (Körner, 2013; Körner and Hansen, 2012; Körner and Holst, 2017).

DSS have been developed for greenhouse growers as, for example, energy-optimising tools (Körner et al., 2008; Watson et al., 2018). All types of DSS can be summarised into two groups: 'planning tools' and 'real-time support'. Planning tools are used for decision making at strategic level (e.g. greenhouse simulators), which (among others) can simulate the effects of the planned investment for the specific greenhouse in terms of energy use, climate and crop production. Real-time support tools use real-time data from measurements and/or simulation for decision support, for example early stress warning. An example of a stand-alone real-time support tool is a soft sensor, which combines sensor data with intelligent algorithms (De Koning, 2006), and is often embedded in climate control systems.

In most DSS, soft sensors have great advantages. The model quality, however, can vary a great deal, from simple empirical regression models (e.g. Boote and Jones, 1987) via deterministic models (e.g. Heuvelink, 1996) to functional structural plant models (FSPM) (e.g. Buck-Sorlin et al., 2011). As current crop growth and development models are inaccurate and have limited predictive power (Poorter et al., 2013), models are rarely employed in crop management alone, but are combined with sensors or used for planning issues in greenhouse simulators (e.g. Körner and Hansen, 2012; Vanthoor, 2011).

4.2 Soft sensors and real-time support tools

In cultivation, decisions need to be made instantly. The greenhouse grower must decide upon important issues by analysing an increasing amount of data. Modern greenhouses, plant factories or vertical farming systems are complex production units. The complexity of the system requires many decisions and many of them need to be made on the hoof. Greenhouse climate controllers certainly do large parts of the job in steering actuators. Today, decisions on actuator setting and command signals are usually based on models, where advanced systems use highly detailed deterministic models.

Although optimal control algorithms are capable of controlling the optimized climate and theoretically do not require human-machine interaction, some cultivation-related decisions are left to the daily business of the grower. Alongside automated control systems, real-time model-based DSS are developed to support the grower in these decisions. A typical real-time DSS consists of an internet-based server platform that is connected to the greenhouse and also acts as a personal database for greenhouse nurseries. As well as the actual climate control software, using a real-time internet-enabled DSS allows easy access to the services provided by the DSS. This type of DSS is based on plant and crop models and sometimes include models of the physical parameters of the individual grower's greenhouse, as well as ecological and financial evaluations of the nurseries' use of resources. Recently a real-time DSS was incorporated into a greenhouse simulator (Kjaer et al., 2019), enabling the use of features close to the optimal climate control concept.

In real-time mode, DSS are fed with data from physical sensors, which can be of various types. As models are simply a mathematical description of a simplified reality, their predictions will never completely match reality. Physical sensors provide (allowing for technical errors) another type of information. Measurement signals alone, however, are of limited importance for controlling crop growth (or crop performance) and only have an additional value when combined with models, eventually forming soft sensors. This technology quantifies the consequences of the value of the measured parameter (Marcelis et al., 2000). One of the first soft sensor systems was developed for hydroponic nutrient uptake: A mechanistic model for tomato was created simulating dry and fresh weight growth of different plant organs, transpiration, water use and demand and uptake of the different macro nutrients. This model was then combined with physical sensors to create a tool for online optimal control of nutrient uptake (Marcelis et al., 2000). Following this development, sensors for total nutrient uptake and ion-specific types were developed. The increasing interest in hydroponics from several movements, such as vertical farming and aquaponics, has accelerated demand and thus the development of these systems (Eck et al., 2019; Keesman et al., 2019).

However, one example of a real-time DSS technology was suggested more than 30 years ago: the above-mentioned speaking-plant concept combined models and sensors (Hashimoto et al., 1985; Udink ten Cate, 1983). In this system, information concerning the physiological status of individual or several plants within a larger canopy is monitored (Ehret et al., 2001). Individual plants are equipped with all kinds of available sensors and the information is upscaled to the canopy. Nowadays the concept is used for smart and intelligent greenhouse applications (Nishina, 2015), using multi-sensor networks and models (Aiello et al., 2018) for various applications such as, for example, fertigation (Ahmad et al., 2011). Its monitoring application can be categorised as soft sensors.

Previous processes that drew attention in modelling for soft sensors and cultivation monitoring included photosynthesis, respiration and transpiration. Today, additional processes related to plant microclimate, such as, for example, stomata behaviour (Giday, 2014; Voogt and Van Weel, 2008) are used. Common to all endeavours is that, compared to laboratory conditions, it is harder to measure these processes under greenhouse conditions, especially with the different types of cultivation measures performed. For practical and economic reasons, most applied lab methods are not directly transferable to a greenhouse canopy. Therefore, using robust and simple sensors in cultivation and combining them with data parameterised models gathered from thorough laboratory experiments is the most suitable method (Janka et al., 2013). However, the choice of model largely determines its usability. The choice of model type (family) already points in the direction of prediction quality and transferability and is thus one of the most important decisions when creating or using a DSS.

The use of technology in greenhouse climate control and decision support is increasing. While modern greenhouse systems are essentially plant factories, equipped with various commercially available sensors, high-tech sensors used in plant phenotyping are mainly applicable to movable plant units, as for example single pots, movable benches or racks (Moriyuki and Fukuda, 2016). Only in rare cases do sensors move around in the greenhouse. This technology is usually not practical for commercial greenhouse crop production where plants are usually fixed. Regular sensors alone, on the other hand, often produce little data about early plant physiological processes such as stress, while stress avoidance is often carried out via highly energy consuming rigid climate strategies (Körner and Challa, 2003). This means that climate control in a greenhouse (or plant factory) is either carried out preventatively under large security margins, or it runs behind and attempts to mitigate any damage. A control system and/or a DSS using soft sensors, combined with the ideas and developments of Industry 4.0 in protected cultivation systems, will be the next step forward. Part of this development is that internet-enabled technology is used to provide the nurseries with intelligent DSS or to control Internet-of-Things (IoT) capable devices and facilities (Liu, 2016). Thus, this kind of DSS relies on the interaction of sensors, models and data. For this purpose, data from sensors (and/or other devices) is logged, stored, categorised and linked with mathematical or statistical models to perform action supporting various crop management related processes. Regardless of the choice of models and sensors, the principle can be simply explained in five steps:

1 Sensors acquire measured data (e.g. leaf temperature, soil moisture, EC).
2 Data is stored in a database (e.g. a cloud server).
3 Information from existing earlier measurements and/or other measured data are combined with new data.
4 The information is processed into usable information and restored in new form in the database.
5 Information is wrapped and made ready for presentation to a user friendly interface and sent to the end user (e.g. early stress warning system app on the smartphone or for direct control to the IoT ready irrigation system).

When defining a sensor model-based DSS one needs to keep in mind that the importance or the balance between sensors or models need to be defined and this leads to two points to consider:

1 With improving quality of the measured data, fewer and probably simpler models can be applied, while data from simple measurements need strong and probably more complex models.
2 Model type can vary between simple regression models, artificial intelligence (AI) and neural networks and explanatory (deterministic) models; or any combination of these types (*further explained below*). Differences in model choice affect the amount and quality of data needed, and the transferability of the system.

Although sensor monitoring and model development have greatly improved online alerting and control systems, and real-time DSS have been developed (Körner and Hansen, 2012), they are integrated into existing large-scale greenhouses with static plants (immobile). While only a few high-tech solutions such as online chlorophyll-fluorescence measurements have been introduced to greenhouses, many solutions are designed for smaller areas (Tschiersch et al., 2017) or for plant stress detection on single leaves, with most employing optical measurement methods (Mahlein, 2016). The realisation of soft sensors is thus still under development, while information is increasingly being handled with stochastic (non-explanatory) models (Aiello et al., 2018). Explanatory models that are necessary to understand the processes and interactions are due to the development of database-technology for storing and processing data on their way out. Hence, more and more black-box modelling is introduced, accompanied by a loss in deep process information understanding.

4.3 Greenhouse planning tools

All greenhouse model simulators are compilations of sub-models that integrate plants, greenhouse structure, equipment and actuators. Models were compiled as planning tools into complete simulation systems. The main driver in the development of planning tools was energy consumption modelling. This, however, is the result of all underlying processes. One of the main variables in greenhouse energy household is humidity and here crop microclimate is a key issue: that is, energy household of the greenhouse is strongly influenced by crop evapotranspiration. Thus, in order to calculate the water and nutrient uptake of crops, the microclimate needs to be known (Challa and Bakker, 1999). Therefore, prediction of microclimate is a central aspect of greenhouse simulation and is an ongoing issue in greenhouse modelling and the implementation of models in simulators.

A DSS as planning tool can be used to estimate the consequences of planned actions, such as installing greenhouse facilities in different climatic regions (e.g. Faroe Islands in Fig. 4), adjusting existing greenhouses, or just changing the cultivation system (Frantz et al., 2010; Körner and Hansen, 2012; Vanthoor, 2011). Complete systems include all models related to crop cultivation in protected environments combining dead and living environments and their highly dynamic interactions (and sub-model interactions) combining models from physical greenhouse structure and equipment, physical microclimate, plant physiology and pest and diseases (Körner et al., 2014). Rather complex and differentiated model approaches have been carried out for the greenhouse climate (Bot, 1993; De Zwart, 1996), and greenhouse crop growth was intensively modelled in the 1990s, for example on major greenhouse crops such as tomatoes (Heuvelink, 1996), cucumber (Marcelis, 1994) and lettuce (Liebig and Alscher, 1993).

Greenhouse physics were modelled, starting with the general greenhouse physical simulation model carried out by Bot (1983), the expert system HORTEX (Rath, 1992), and the early simulation model KASPRO (De Zwart, 1996). Later, these mainly physical model compilations where combined with detailed and often deterministic crop microclimate models (e.g. Janka et al., 2016; Kim and Lieth, 2003; Körner et al., 2007; Stanghellini, 1987), deterministic crop growth

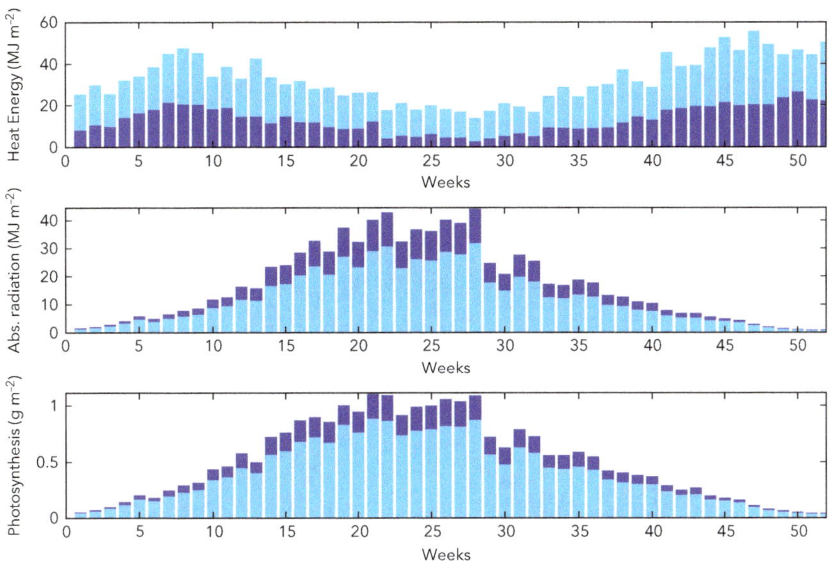

Figure 5 Example of a simulated data with a greenhouse simulator (Körner, 2013; Körner and Hansen, 2012; Körner and Holst, 2017) analysing heat energy, absorbed short-wave radiation and crop photosynthesis per week during a standard year in Denmark (Lund, 1995) for a standard greenhouse (dark blue) vs. a high insulated greenhouse (light blue).

models (e.g. Heuvelink, 1996; Lee, 2002; Marcelis, 1994; Marcelis et al., 2006) and pest and disease models (Körner et al., 2014; Körner and Jakobsen, 2006). These agglomerations of mathematical physical, biological and control models in greenhouses in the form of software to calculate the complex interactive processes in a greenhouse are collectively known as greenhouse simulators. These model compilations can be used to analyse adjustments to the cropping environment (in line with Fig. 3) such as increased insulation (Fig. 5) or screening (Fig. 6).

Recently, some product prototypes and/or business based consultancy tools have been developed based on greenhouse simulation models: The Virtual Grower (Frantz et al., 2010), Greenergy Energy Audit Tool (Körner et al., 2008), The Adaptive Greenhouse (Vanthoor, 2011), The Virtual Greenhouse (Körner and Hansen, 2012) or Greenmod (Rodríguez et al., 2015). At research level, some model systems (i.e. simulation models in combination with certain greenhouse technologies) have been developed that potentially can be used to optimise investments and structural modifications on the production unit and production process - for example, electricity consumption with LED lighting (Watson et al., 2018). However, most systems entail closed software environments that can only be used by the developers, and many only exist in research mode and lack further

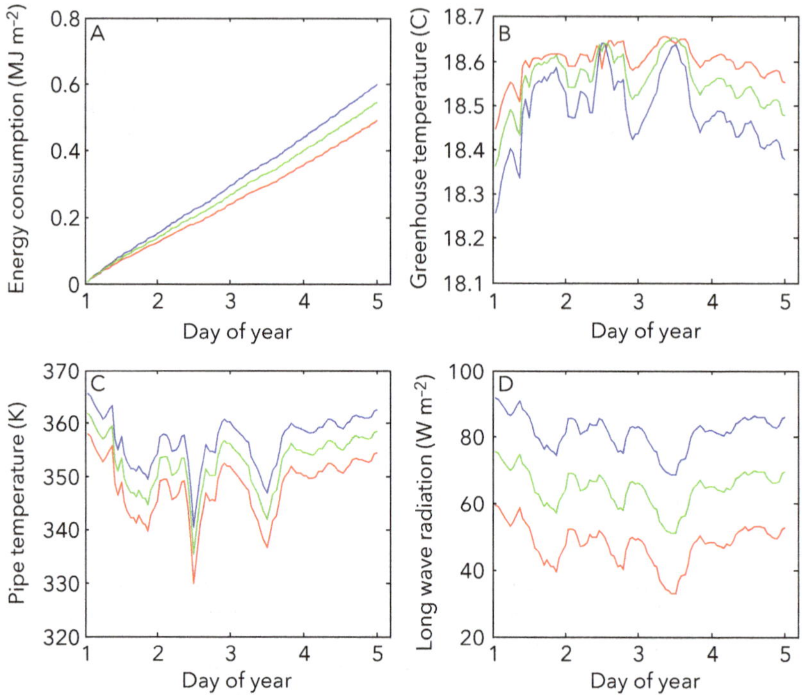

Figure 6 Example of energy consumption (a), inside greenhouse temperature (b), heating pipe temperature (c) and net long-wave radiation loss (d) simulated with a greenhouse simulator analysing the installation of screens with low (blue), medium (green) and high (red) energy saving properties in a standard greenhouse from 1 January to 5 January in Odense (Denmark) using a standard year in Denmark (Lund, 1995). Source: taken from Körner and Hansen (2012).

development and acceptance from the industry (Körner and Holst, 2017). As yet there is no common basis for model sharing and collaborative model development, and as a result, most modellers and modelling teams work in isolation when developing their own models and codes. A shortcoming of that procedure is that greenhouse simulation models are developed in parallel in disparate research environments, leading to little cooperative growth and development.

5 Models for decision support

5.1 Models and their application

Models are schematic representations of systems, which in turn comprise a part of reality, defined by its boundaries (Wit, 1969). Models are thus simplified systems from reality, in which some aspects are overlooked, whilst retaining the essential elements are used (De Wit and Van Keulen, 1987). All models used in decision support have to be incorporated in a larger system. Figure 7 shows the three main roles when creating models for a decision support system: the engineer, the modeller and the user.

Prediction accuracy is jeopardised by many sources of uncertainty, such as model errors, variability between plants, variability between greenhouses and an uncertain climate. As mentioned above, the latter is the disturbing factor in greenhouse control terminology (Fig. 2) and varies strongly between geographical regions, seasons and days (Fig. 4). Thus, prediction accuracy can vary strongly depending on which situation it is in. One solution is using self-learning models, that is models - online feeding of sensor information for auto-calibration of a soft sensor can make plant model predictions considerably more reliable and useful to the grower in the monitoring and decision process.

In greenhouse and crop growth simulation systems, two families of models can be distinguished - data-based models and mechanistic explanatory models. Mechanistic explanatory models have the advantage of transparency of processes and offer the possibility to consider sub-processes and thus a deeper understanding of the respective plant culture. Data-based models are black-box models. In contrast to descriptive models, the latter type shows only the main result, and deeper analysis and consideration of the sub-processes is not possible. However, there are good reasons for using one or the other model family. Each type has its pros and cons.

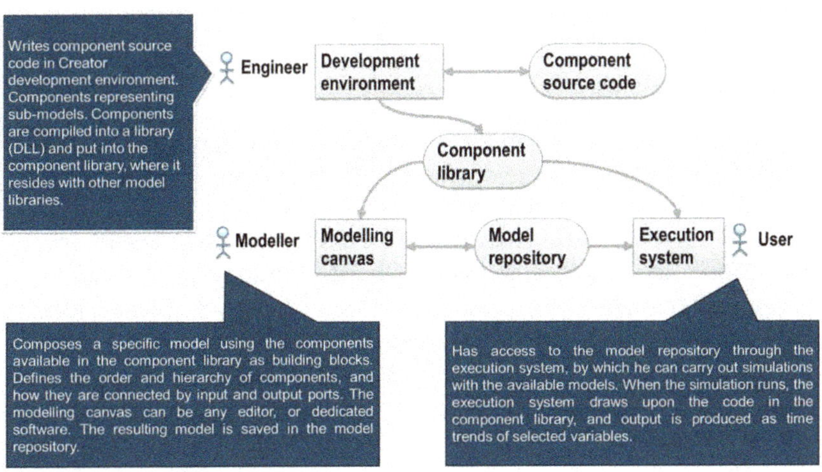

Figure 7 Example of model development with various actors. Source: modified from Körner and Holst (2017).

5.2 Deterministic models

As explained above, a major challenge for growers is controlling the microclimate of the crop to optimise growth, whilst avoiding physiological damage and disease through stress (Janka et al., 2016). Mechanistic-deterministic crop models can solve this, and are powerful tools in research and teaching – they are the growers' preferred choice for applications in DSS (Lentz, 1998). Mathematical-deterministic models, descriptive or explanatory, usually consist of a series of mathematical equations that describe the individual processes of the crop cultivation process, often down to cell level (Thornley, 1976). Deterministic models can be used to explore phenomena in cultivation. The complex world of a greenhouse consists of various climate factors – greenhouse physics and the living crop (as well as other living organisms in the greenhouse) interacts on a highly complex level. To understand these relationships and to achieve insight into the combined processes, deterministic models are the preferred choice. Here, some of the central models calculate on the basis of greenhouse climate the photosynthesis and respiration (Gijzen, 1992) and thus indirectly the growth and yield (Challa, 1998; Lee, 1998; Marcelis et al., 1998; Rodríguez et al., 2015). In order to make these often universal mathematical equations suitable for different cultures, a number of culture-specific parameters are used and eventually build in crop-specific models as for tomatoes or cucumbers (Heuvelink, 1996; Marcelis, 1994). Parameterising – the qualified assessment of the parameters – is a key but time-consuming task. However, metrics can help and be used for automatic model calibration (estimation of parameters). Then, measured data can be combined with models and finally to IoT enabled facilities and DSS.

5.3 Data-based models

Data-based models all have their origin in mathematical fitting of data. The opposite are deterministic explanatory models that are collections of theoretical equations describing processes. Often, models in greenhouse horticultural applications are a combination of these two families. The simplest data-based models are regression models. Here, data of experimental measurements of a certain type are fitted with regression, where linear, non-linear and multiple regression models exist. Early modellers used the regression technique for all kinds of greenhouse crop processes as the basic processes of photosynthesis or transpiration in crop stands (e.g. Hanson et al., 1987; Heissner, 1997a; Heissner, 1997b) or plant morphology and development in ornamentals (Hopper et al., 1994). Today, methodologies such as support vector machine algorithms as a further development of simple regressions are sometimes applied, for example, to a photosynthesis model for climate control

(Li et al., 2017). Although prediction accuracy might be acceptable, regression model types are only valid within the boundaries of the underlying data as e.g. specific cultivar, climate condition, greenhouse set-up and so on. These models are thus not transferrable to other conditions. In some specific areas, however, regression-type models make sense, such as in the estimation of leaf area in canopies from simple measurements for easily determining the leaf area index during cultivation (Blanco and Folegati, 2005; Dannehl et al., 2015). In these situations, however, the same rules apply: The model estimations cannot be transferred to other cultivars or cultivation conditions.

Whereas in many advanced DSS (in contrast to the simple and very limited (data-based) regression models) purely mathematical-deterministic models (descriptive) based on generally valid but parameterisable physical and biological processes are used, today raw data is increasingly used in stochastic and machine learning models, such as neural networks (Ehret et al., 2011). The amount of data available through original databases and data-sharing possibilities (starting with cloud storage and computing) enables the use of big data in ecological modelling (Hampton et al., 2013) and related fields such as agriculture (Wolfert et al., 2017). In horticulture, and especially in high-tech protected cultivation systems or CEA, data-based modelling approaches are increasingly used, while due to a lack of data, only a few approaches have been done that mainly deal with specific neural networks only using relatively low amounts of data (e.g. Kücükönder et al., 2016; Miranda, 2017; Moon et al., 2018; Salazar et al., 2015). This is basically another further development of the early regression-type models.

Recently, phenotyping and machine learning has been combined, as for example in plant water status (Guo et al., 2017). Models based on machine learning – methods that use large amounts of data to train algorithms – have been introduced into precision agriculture and several methods have been utilized, such as support vector machines and neural networks for classification (supervised learning); k-means and self-organising maps for clustering (unsupervised learning) (Behmann et al., 2015). Successful applications include the early detection of plant diseases based on spectral features using supervised or unsupervised learning methods (Behmann et al., 2015) – i.e. data are collected from phenotyping machines and fed into mathematical algorithms that can eventually predict the risk of plant stress or disease (Mahlein, 2016). The more data the better the quality of this AI – the models learn from mistakes, deviations and successes. Therefore, the more data that is collected and fed into the system, the smarter the model and more timely and accurate the warning and/or the IoT systems become. Of course, the amount of data multiplied by the connection of additional operations results in a higher precision of IoT connected systems. With data-based models such as stochastically big data types, the crux is often the exploitation of large amounts of data for a more detailed analysis and thus the improved predictability and control of the actuators, for example the IoT systems.

6 Case studies

To illustrate the theoretical description above, three case studies have been designed based on the (1) importance of microclimate for greenhouse models, (2) application of models in a planning tool and (3) application of models as real-time early warning tool.

6.1 Case study 1: model prediction of boundary layer and greenhouse microclimate

Microclimate is a key issue for almost all processes related to crop cultivation. As such, prediction quality of microclimate is important for all DSS related to CEA cultivation. Air movement affects the crop by changing the thickness of the boundary layer around the leaves (Moreshet et al., 1969): a layer of heavy humid air acting as a barrier for leaf gas exchange. The boundary layer affects all processes, and despite their importance, optimal air movement conditions for crops in protected cultivation are neither known nor have been studied thoroughly. Only simple set-ups were tested for laminar air flows (Yokoi et al., 2008) resulting in a single wind speed optimum value for a certain crop, 0.4 m/s for protected rice (Goto et al., 2014). The thickness of the boundary layer, however, has major consequences, for example on energy consumption (Körner and Holst, 2017). A thicker boundary layer due to low wind speed results in reduction in convective heat exchange, photosynthesis, leaf evaporation and CO_2 uptake (Dubey, 1997), increased respiration, directly followed by increased heat stress (Janka et al., 2013), disease risk (Grange and Hand, 1987; Tantau and Lange, 2003) and physiological disorders (Guichard et al., 2001; Ho et al., 1993), and thus reduced plant growth and yield (Sato et al., 2000; Weis and Berry, 1988).

The crop microclimate on the leaf boundary layer, however, is difficult to measure and control (Körner et al., 2007) and thus reliable models would be useful. Model estimations have been done to predict the boundary layer conductance near leaves and in canopies (Jones, 1992; Monteith and Unsworth, 1990), or to approximate the outside wind speed influence on the climate conditions inside a naturally ventilated greenhouse (Wang et al., 2000). In these simplified models the influence of low wind speed (below 2 m/s) and its nature within the leaf canopy is ignored, leading to the most commonly used assumptions of a constant wind speed (Bailey, 1985) and boundary layer in greenhouse canopies (Stanghellini, 1995). At low wind speed, however, existing empirical models approximate the boundary layer under mixed convection regimes (natural and forced), because regular models for gas exchange fail to predict theoretically the boundary layer resistances (Jones, 1992; Stanghellini and De Jong, 1995). To illustrate, three different models of horizontal plate approaches (M1, M2 and M3) as described in Stanghellini (1995) or Jones (1992) (M1, M2), and an explicit tomato model (Stanghellini, 1987) (M3) on boundary layer resistance were compared. The three models were implemented in a greenhouse simulator (Körner and Hansen, 2012) and simulations were carried out with (a) wind speed calculated as logarithmic function of canopy depths (Jones, 1992) and (b) constant wind speed.

Regardless of the fast development in greenhouse 3-D crop modelling research (Buck-Sorlin et al., 2011) incorporating both architectural and deterministic approaches (Kahlen, 2015), crop microclimate prediction based on calculating the boundary layers has been neglected. The difference between existing models of boundary layer and wind speed is illustrated in Fig. 8 and Table 1. The existing models based on rough estimations or simple empirical data are not feasible to be used in protected cultivation as no data exists. Neither calculation for growth planning nor crop scheduling or model-based climate control can be reliable under these circumstances. Thus, one of the most central issues in greenhouse horticultural modelling is yet to be explored.

6.2 Case study 2: model-based planning tools for energy management

Energy efficiency (i.e. energy consumption per unit produced) is the first important parameter in the design of a sustainable crop production system, and is one of the current main targets of the horticultural industry in Europe, combined with the use of sustainable energy sources (EPSO, 2014). Improved energy efficiency for heating was initiated in the early 2000s by the introduction of dynamic climate control in greenhouses. To reduce the energy consumption of supplemental lighting, the DynaLight (Kjær et al., 2012) and SmartGrid approach (Kläring and Ramírez, 2018; Ramírez et al., 2019) (a dynamic light controller based on the combination of photosynthesis models, light integration and the dynamic electricity price structure) is a solution that can be used to optimally control light with integral daily light (Fig. 9). This approach can be used for greenhouse construction planning. Additionally, planning tools can be used as decision support on energy consumption depending on location. Increasing greenhouse insulation with sealing (Fig. 10) or screens (Fig. 11), for example, can have a strong energy saving effect on energy saving in high latitudes as shown in simulation with the Greenhouse Energy Auditing Tool (Körner et al., 2008), whereas in warmer regions this can have negative consequences with too high temperatures and humidity problems (not shown). Additional DSS sub-models for humidity controlled related issues, as for example fungal attacks (Baptista et al., 2012; Körner et al., 2014) can be added.

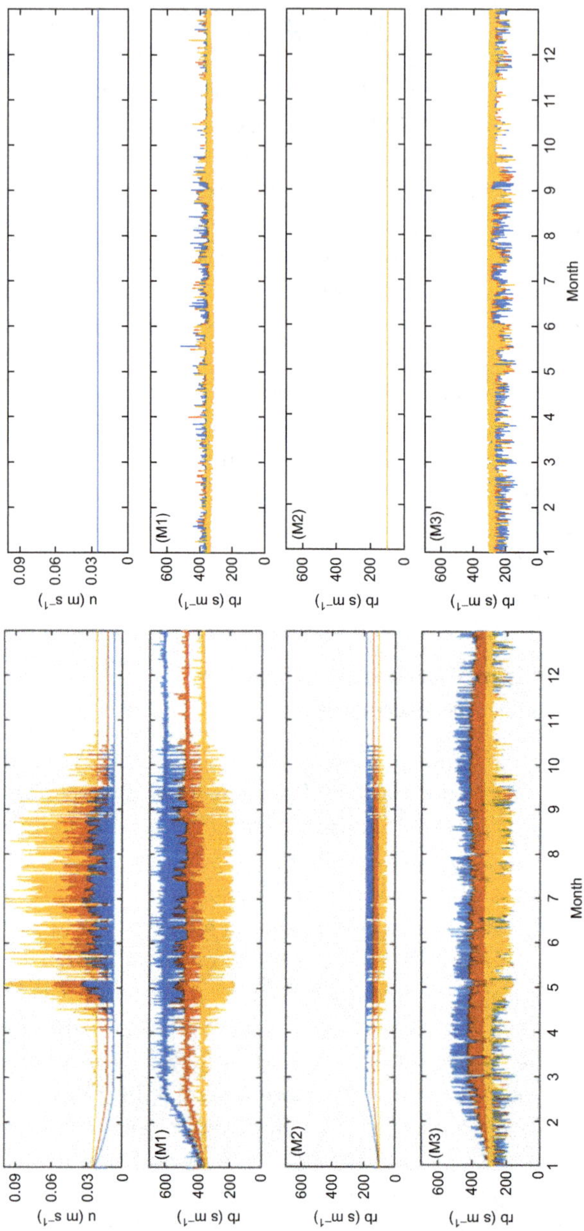

Figure 8 Simulated wind speed (u) and boundary layer resistances (rb) with simulated (top figures) and constant wind speed (bottom figures) in three canopy heights (top, yellow; middle, red; bottom, blue) and three boundary layer models (M1 (Stanghellini, 1995), M2 (Jones, 1992), M3 (Stanghellini, 1987)). Source: taken from Körner and Holst (2017).

Table 1 Simulated yearly heat energy consumption (GJ m^{-2}) for three different sub-models of boundary layer resistance, rb (M1 (Stanghellini, 1995), M2 (Jones, 1992), M3 (Stanghellini, 1987)), of boundary layer resistance and two wind models in a tomato crop in a standard greenhouse in Denmark

	M1	M2	M3
Calculated windspeed	1.2379	1.3436	1.2774
Constant windspeed	1.2645	1.3572	1.2863

Source: taken from Körner and Holst (2017).

6.3 Case study 3: using models for early detection/control of stress

Plants can cope with various types of stresses (biological and physical) and are capable of overcoming stress without further damage or reduction in quality or yield. However, the stress dosage (intensity and duration) is of utmost importance for its reversibility (Weis and Berry, 1988). Damage to the physiological apparatus caused by plant stress are known to be reversible up to a certain point (Ferris et al., 1998; Larcher and Bodner, 1980), while mild stress can have positive effects on plants in terms of harvest product composition and other quality parameters. A too high dosage would result in non-reversible

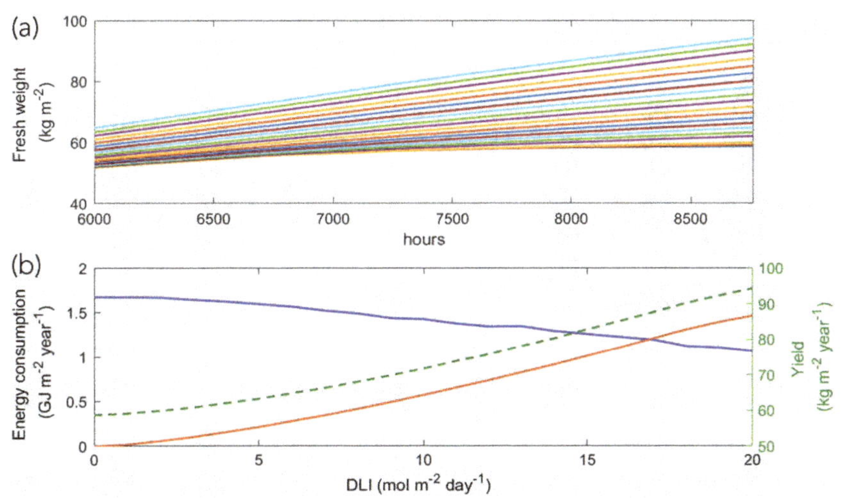

Figure 9 Simulations with different set point levels of daily light integral, DLI, in a tomato greenhouse cultivation in a standard greenhouse with (a) fresh weight yield for different levels of supplementary lighting between light set point of 0 and 20 mol m^{-2} day^{-1} (each line represents one light level) shown for hour 6000–8760 of the year and (b) annual energy consumption (heat, red; light, blue). Simulations were performed with a greenhouse simulator (Körner, 2013; Körner and Hansen, 2012; Körner and Holst, 2017).

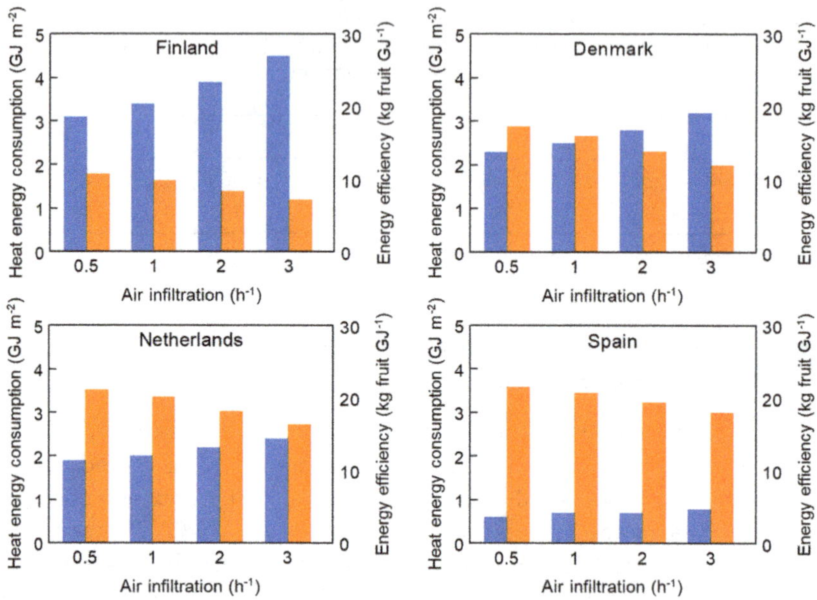

Figure 10 Simulated influence of greenhouse air infiltration rate on energy consumption (blue) and energy use efficiency (red). Source: data taken from Körner et al. (2008).

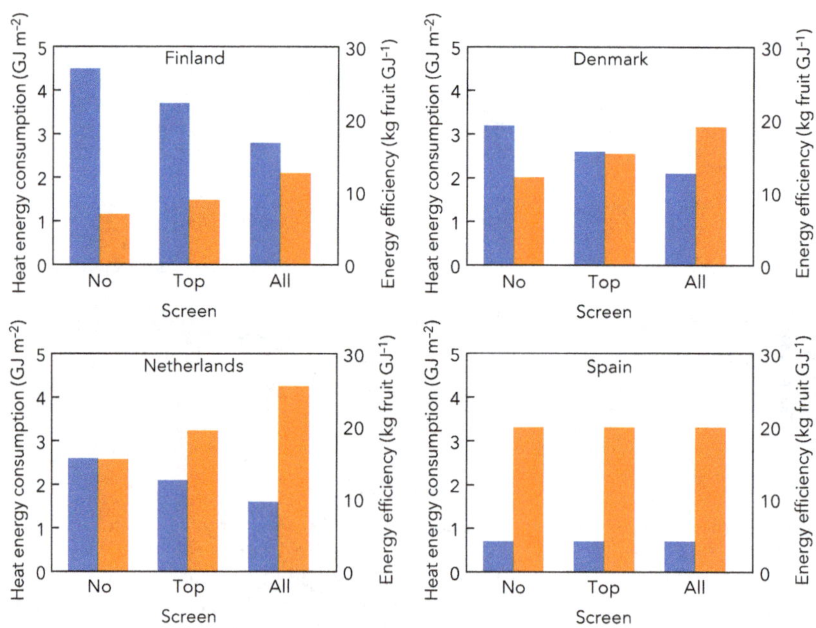

Figure 11 Simulated influence of standard energy screens on energy consumption (blue) and energy use efficiency (red). Source: data taken from Körner et al. (2008).

damage to the crop and thus lead to negative consequences, such as lower yield and/or quality (Sato et al., 2000). Often, the (negative) result of stress is only visible a long time after the stress actually happened. On the other hand, a carefully selected dosage of stress can yield in planned changes to the crop and its produce (e.g. tomato fruits) such as for example, better keeping quality, better taste or increased high-value compounds such as lycopene in tomato.

Understanding the dosage and the functional relation between stress intensity and stress duration means the positive effects of stress can be achieved while minimising the risk of negative consequences. Stress can therefore be induced with resulting higher lycopene content in tomatoes used as an additional cultivation target. However, stress often leads to multiple reactions and thus the complete system or matrix of processes needs to be known. If, for example, desiccation stress is induced in tomato with the aim of a higher lycopene content, the crop can face problems such as blossom end. Thus, DSS that incorporate the knowledge of physiological processes can be used to induce stress, while detection of negative stress is basically carried out by the same intelligent models and sensors. Only the decision algorithm needs to be adjusted to the target.

7 Future trends and conclusion

Despite the technological advances implemented in greenhouse crop production, greenhouse operation relies on human expertise to decide on the optimum values of each environmental control parameter (Fitz-Rodríguez and Giacomelli, 2009). Therefore, creating DSS on AI by collecting large data-sets and using neural networks algorithms with fuzzy logic (i.e. black-box models) or other machine learning algorithms have been developed (Linker et al., 1998) and is growing in its usage, for climate (Zou et al., 2017), leaf area prediction for transpiration (Wang et al., 2017) or crop growth (Fitz-Rodríguez and Giacomelli, 2009). Despite the deeper understanding of deterministic models and their advantage in system analysis, data-based approaches will probably increase and the development of AI and IoT technologies in CEA will continue to grow. Thus, the combination of models and sensors to soft sensors will increase, while the possibilities of cloud computing and big data enable the usage of data-based models. As data-based models strongly depend on the availability of data, data sharing is needed. Although this may partly be hampered by the new EU data policy, in many regions data sharing is not a new concept: for example in the 1990s production of potted plants in greenhouses became an industry in Denmark and have since then dominated the European market, and for a long time the Dutch have dominated the global development of glass greenhouse production. The few remaining Danish potted plants production companies (a 62% decrease between 2002 and 2014) can only keep up with niche and specialty crops or extreme specialisation in the globalised market. In the 1990s, the strength in Denmark was clustering and the common goal of being stronger in the network and, above all, the lively exchange of experiences and data sharing. The same is true today in the highly organised Netherlands. A synergetic effect through cooperation is, of course, only possible if there are also several producers with the same or at least similar products and the openness to share internal practices with their competitors. The more that likeminded companies provide data for shared use, the better the quality of today's IoT systems and the higher the optimisation levels. Image analysis (which has not been discussed in this chapter) is another connected field that will increase in importance and be further incorporated in DSS. Commercial companies already use image recognition for pest and disease detection, as discussed in Mahlein (2016).

In conclusion, data-based models will grow in importance and algorithms will improve prediction quality. However, as data-based models lack the system understanding, a combination of mathematical-deterministic models and big data might be the best solution. Science will continue working with explanatory models, as the advantage of system understanding is a superior gain in research.

8 Where to look for further information

The rapid development of the greenhouse horticultural industry towards the use of knowledge and data-based models in combination with various sensor types as described above is an ongoing process. The literature on sensor-based climate control in greenhouse horticulture from the early 1980s (e.g. Hashimoto et al., 1985; Udink ten Cate, 1983; Leung et al., 1985) was the foundation of later developments towards soft sensors. Nowadays, with the increased capability of computers to handle large data sets and the development of microsensors, and the combination of these with the application of IoT devices is leading to a revolution in the crop production industry. For many years greenhouse horticulture has been one of the main innovation drivers for new developments in agricultural crop production. Today, sensors and advanced technological solutions are available for all parts of the agricultural industry. This role as innovation driver is now accelerating due to the shift from pure greenhouse horticulture (with plastic or glass shelters) to production in closed rooms and vertical horticulture using only artificial light sources. It is noticeable that, since 2013, the title of the world's largest (biannual) scientific conference on protected cultivation technology, GreenSys, now includes the term 'plant factory' alongside the traditional 'greenhouse'. In general, the GreenSys conferences with their proceedings (as Acta Horticulturae) are probably the richest sources of information on development in this field.

It is important to note that innovation in this sector comes from a range of industries. As an example, the development of now commonly used LED lamps was mainly driven by the car and other industries and then applied to crop production. Ecological and health movements, as well as urban planners, are influencing developments in horticultural crop production too. Plant production using hydroponics is now common and synergistic co-cultivation systems with closed nutrient cycling, as well as modern de-coupled aquaponics systems, are possible solutions to more sustainable production in such areas as urban agriculture (Goddek and Körner, 2019). Currently various research projects are being funded, such as CityFood under the European 'Sustainable Urbanisation Global Initiative'. However, there are also global developments in synergetic co-cultivation systems with sensor model-based controllers and/or DSS. Production systems in the Namib Desert (desertfoods-international.com) or in Saudi Arabia (redseafarms.com) are current examples. The transition from pure plant production to co-cultivation and high-tech equipped plant factories are accompanied by developments in AI. Rapid advances in AI are influencing the crop production industry, especially CEA. Worldwide, increasing numbers of AI conferences as well as the World Intelligence Congress (Tianjin, China) are being held on various topics such as deep learning and big data. Greenhouse horticulture and CEA will be recipients of developments in this field, and such conferences will be an important source of information on key developments.

9 References

Aaslyng, J. M., Lund, J. B., Ehler, N. and Rosenqvist, E. 2003. Intelligrow: a greenhouse component-based climate system. *Environmental Modelling and Software* 18(7), 657–66. doi:10.1016/S1364-8152(03)00052-5.

Ahmad, U., Subrata, D. M. and Arif, C. 2011. Speaking plant approach for automatic fertigation system in greenhouse. *International Journal of Signal Processing, Image Processing and Pattern Recognition* 4, 95–106.

Aiello, G., Giovino, I., Vallone, M., Catania, P. and Argento, A. 2018. A decision support system based on multisensor data fusion for sustainable greenhouse management. *Journal of Cleaner Production* 172, 4057–65. doi:10.1016/j.jclepro.2017.02.197.

Bailey, B. J. 1985. Wind dependent control of greenhouse temperature. *Acta Horticulturae* 174(174), 381–6. doi:10.17660/ActaHortic.1985.174.50.

Baptista, F. J., Bailey, B. J. and Meneses, J. F. 2012. Effect of nocturnal ventilation on the occurrence of *Botrytis cinerea* in Mediterranean unheated tomato greenhouses. *Crop Protection* 32, 144–9.

Behmann, J., Mahlein, A. K., Rumpf, T., Römer, C. and Plümer, L. 2015. A review of advanced machine learning methods for the detection of biotic stress in precision crop protection. *Precision Agriculture* 16(3), 239–60. doi:10.1007/s11119-014-9372-7.

Blanco, F. F. and Folegati, M. V. 2005. Estimation of leaf area for greenhouse cucumber by linear measurements under salinity and grafting. *Scientia Agricola* 62(4), 305–9. doi:10.1590/S0103-90162005000400001.

Bojacá, C. R., Gil, R. and Cooman, A. 2009. Use of geostatistical and crop growth modeling to assess the variability of greenhouse tomato yield caused by spatial temperature variations. *Computers and Electronics in Agriculture* 65(2), 219–27. doi:10.1016/j.compag.2008.10.001.

Boote, K. J. and Jones, J. W. 1987. Equations to define canopy photosynthesis from given quantum efficiency, maximum leaf rate, light extinction, leaf area index, and photon flux density. In: Biggins, J. (Ed.), *Progress in Photosynthesis Research*. Martinus Nijhoff, Dordrecht, pp. 415–8.

Bot, G. P. A. 1983. Greenhouse climate: from physical processes to a dynamic model. PhD thesis. Wageningen Agricultural University, Wageningen, 240p.

Bot, G. P. A. 1993. Physical modelling of greenhouse climate. In: Hashimoto, Y., Bot, G. P. A., Day, W., Tantau, H. J. and Nonami, H. (Eds), *The Computerized Greenhouse*. Academic Press, Inc., San Diego, pp. 51–74.

Bot, G. P. A. 2001. Developments in indoor sustainable plant production with emphasis on energy saving. *Computers and Electronics in Agriculture* 30(1–3), 151–65. doi:10.1016/S0168-1699(00)00162-9.

Buck-Sorlin, G., de Visser, P. H., Henke, M., Sarlikioti, V., van der Heijden, G. W., Marcelis, L. F. and Vos, J. 2011. Towards a functional-structural plant model of cut-rose: simulation of light environment, light absorption, photosynthesis and interference with the plant structure. *Annals of Botany* 108(6), 1121–34. doi:10.1093/aob/mcr190.

Buwalda, F., Van Henten, E. J., De Gelder, A., Bontsema, J. and Hemming, J. 2006. Towards an optimal control strategy for sweet pepper cultivation – 1. A dynamic crop model. *Acta Horticulturae* 718, 367–74.

Challa, H. 1998. Greenhouse crop growth models, from description to application. In: Grossbeeren, I. f. G. u. Z. (Ed.), *Models and Strategies to Control Growth and Yield of Greenhouse Vegetables*. Institut fuer Gemuese und Zierpflanzenbau, Grossbeeren, pp. 9–15.

Challa, H. and Bakker, M. 1999. Potential production within the greenhouse environment. In: Stanhill, G. and Enoch, H. Z. (Eds), *Ecosystems of the World 20 – Greenhouse Ecosystems*. Elsevier, Amsterdam, pp. 333–47.

Cuce, E., Harjunowibowo, D. and Cuce, P. M. 2016. Renewable and sustainable energy saving strategies for greenhouse systems: a comprehensive review. *Renewable and Sustainable Energy Reviews* 64, 34–59. doi:10.1016/j.rser.2016.05.077.

Dannehl, D., Rocksch, T. and Schmidt, U. 2015. Modelling to estimate the specific leaf area of tomato leaves ('Pannovy'). *Acta Horticulturae* 1099, 79–86.

De Koning, A. N. M. 2006. Models and sensors in greenhouse control. *Acta Horticulturae* 718(718), 175–82. doi:10.17660/ActaHortic.2006.718.19.

De Wit, C. T. and Van Keulen, H. 1987. Modelling production of field crops and its requirements. *Geoderma* 40(3–4), 253–65. doi:10.1016/0016-7061(87)90036-X.

De Zwart, H. F. 1996. Analyzing energy-saving options in greenhouse cultivation using a simulation model. PhD thesis. Wageningen Agricultural University, Wageningen, The Netherlands, 236p.

Ding, X., Xiong, G., Hu, B., Xie, L. and Zhou, S. 2013. Environment monitoring and early warning system of facility agriculture based on heterogeneous wireless networks. In: *Proceedings of 2013 IEEE International Conference on Service Operations and Logistics, and Informatics*, pp. 307–10.

Dubey, R. S. 1997. Photosynthesis in plants under stressful conditions. In: Pessarakli, M. (Ed.), *Handbook of Photosynthesis. Books in Soils, Plants, and the Environment*. Marcel Dekker, New York, pp. 859–75.

Eck, M., Schmautz, Y., Junge, R., Körner, O. and Jijakli, M. H. 2019. Nutrient cycling. In: Goddek, S. and Kotzen, B. (Eds), *Aquaponics Food Production*.

Ehret, D. L., Lau, A., Bittman, S., Lin, W. and Shelford, T. 2001. Automated monitoring of greenhouse crops. *Agronomie* 21(4), 403–14. doi:10.1051/agro:2001133.

Ehret, D. L., Hill, B. D., Helmer, T. and Edwards, D. R. 2011. Neural network modeling of greenhouse tomato yield, growth and water use from automated crop monitoring data. *Computers and Electronics in Agriculture* 79(1), 82–9. doi:10.1016/j.compag.2011.07.013.

EPSO. 2014. *Horticulture Research in Europe - to 2020 and Beyond*. EPSO, Berlin, Germany.

Fargues, J., Vidal, C., Smits, N., Rougier, M., Boulard, T., Mermier, M., Nicot, P., Reich, P., Jeannequin, B., Ridray, G., et al. 2003. Climatic factors on entomopathogenic hyphomycetes infection of *Trialeurodes vaporariorum* in Mediterranean glasshouse tomato. *Biological Control* 28, 320–31.

Ferris, R., Wheeler, T. R., Hadley, P. and Ellis, R. H. 1998. Recovery of photosynthesis after environmental stress in soybean grown under elevated CO_2. *Crop Science* 38(4), 948–55. doi:10.2135/cropsci1998.0011183X003800040012x.

Fisher, P. R. and Heins, R. D. 1996. The *greenhouse care system*: a decision-support system for height control and scheduling of potted flowering plants. *Acta Horticulturae* 417(417), 41–6. doi:10.17660/ActaHortic.1996.417.4.

Fitz-Rodríguez, E. and Giacomelli, G. A. 2009. Yield prediction and growth mode characterization of greenhouse tomatoes with neural networks and fuzzy logic. *Transactions of the ASABE* 52(6), 2115–28. doi:10.13031/2013.29200.

Frantz, J. M., Hand, B., Buckingham, L. and Ghose, S. 2010. Virtual grower: software to calculate heating costs of greenhouse production in the United States. *HortTechnology* 20, 778–85.

Gallardo, M., Giménez, C., Martínez-Gaitán, C., Stöckle, C. O., Thompson, R. B. and Granados, M. R. 2011. Evaluation of the VegSyst model with muskmelon to simulate crop growth, nitrogen uptake and evapotranspiration. *Agricultural Water Management* 101(1), 107–17. doi:10.1016/j.agwat.2011.09.008.

Gelder, A. d., Dieleman, J. A., Bot, G. P. A. and Marcelis, L. F. M. 2012. An overview of climate and crop yield in closed greenhouses. *The Journal of Horticultural Science and Biotechnology* 87(3), 193–202. doi:10.1080/14620316.2012.11512852.

Giday, H. 2014. *On the Role of Abscisic Acid in the Regulation of Phenotypic and Genetc Variation in Stomatal Closing Ability*. Aarhus University, Denmark, Aarhus, 129pp.

Gijzen, H. 1992. Simulation of photosynthesis and dry matter production of greenhouse crops. CABO-TT nr. 28. CABO-DLO, Wageningen, The Netherlands.

Goddek, S. and Körner, O. 2019. A fully integrated simulation model of multi-loop aquaponics: a case study for system sizing in different environments. *Agricultural Systems* 171, 143–54.

Gonda, L. and Cugnasca, C. E. 2006. A proposal of greenhouse control using wireless sensor networks. In: *4th World Congress Conference on Computers in Agriculture and Natural Resources*, Orlando, FL.

Goto, E., Ishigami, Y., Kimura, K., Arai, K., Kashima, K., Nakajima, H. and Maruyama, S. 2014. Optimum air current speed for rice plant canopy in a closed plant-production system. *Acta Horticulturae* 1037(1037), 305–10. doi:10.17660/ActaHortic.2014.1037.36.

Grange, R. I. and Hand, D. W. 1987. A review of the effects of atmospheric humidity on the growth of horticultural crops. *Journal of Horticultural Science* 62(2), 125–34. doi:10.1080/14620316.1987.11515760.

Guichard, S., Bertin, N., Leonardi, C. and Gary, C. 2001. Tomato fruit quality in relation to water and carbon fluxes. *Agronomie* 21(4), 385–92. doi:10.1051/agro:2001131.

Guo, D., Juan, J., Chang, L., Zhang, J. and Huang, D. 2017. Discrimination of plant root zone water status in greenhouse production based on phenotyping and machine learning techniques. *Scientific Reports* 7(1), 1–12. doi:10.1038/s41598-017-08235-z.

Hampton, S. E., Strasser, C. A., Tewksbury, J. J., Gram, W. K., Budden, A. E., Batcheller, A. L., Duke, C. S. and Porter, J. H. 2013. Big data and the future of ecology. *Frontiers in Ecology and the Environment* 11(3), 156–62. doi:10.1890/120103.

Hanson, P. J., McRoberts, R. E., Isebrands, J. G. and Dixon, R. K. 1987. An optimal sampling strategy for determining CO_2 exchange rate as a function of photosynthetic photon flux density. *Photosynthetica* 21(1), 98–101.

Hashimoto, Y., Morimoto, T. and Fukuyama, T. 1985. Some speaking plant approach to the synthesis of control system in the greenhouse. *Acta Horticulturae* 174(174), 219–26. doi:10.17660/ActaHortic.1985.174.26.

Heissner, A. 1997a. Der CO_2-Gaswechsel von Paprikapflanzen in Abhängigkeit von der Bestrahlungsstärke, der CO_2-Konzentration, der Lufttemperatur und dem Dampfdrucksättigungsdefizit der Luft: Messungen und Modell. (CO_2 gas-exchange of sweet-pepper plants depending on irradiation, CO_2 concentration, air temperature and vapour pressure deficit: measurements and model). *Gartenbauwissenschaft* 62(2), 78–90.

Heissner, A. 1997b. Die Transpiration von Paprika im Gewaechshaus: Messungen und Modellierung. *Gartenbauwissenschaft* 62(5), 228–36.

Heuvelink, E. 1996. Tomato growth and yield: quantitative analysis and synthesis. PhD thesis. Wageningen Agricultural University, Wageningen, The Netherlands, 326p.

Ho, L. C., Belda, R., Brown, M., Andrews, J. and Adams, P. 1993. Uptake and transport of calcium and the possible causes of blossom-end rot in tomato. *Journal of Experimental Botany* 44(2), 509–18. doi:10.1093/jxb/44.2.509.

Hopper, D. A., Hammer, P. A. and Wilson, J. R. 1994. A simulation model of Rosa hybrida growth response to constant irradiance and day and night temperatures. *Journal of the American Society for Horticultural Science* 119(5), 903–14. doi:10.21273/JASHS.119.5.903.

Janka, E., Körner, O., Rosenqvist, E. and Ottosen, C. O. 2013. High temperature stress monitoring and detection using chlorophyll a fluorescence and infrared thermography in chrysanthemum (*Dendranthema grandiflora*). *Plant Physiology and Biochemistry* 67, 87–94. doi:10.1016/j.plaphy.2013.02.025.

Janka, E., Körner, O., Rosenqvist, E. and Ottosen, C. O. 2016. A coupled model of leaf photosynthesis, stomatal conductance, and leaf energy balance for chrysanthemum (*Dendranthema grandiflora*). *Computers and Electronics in Agriculture* 123, 264–74. doi:10.1016/j.compag.2016.02.022.

Jensen, M. R. L., Jørgensen, B. N., Körner, O. and Ottosen, C. O. 2009. PREDICT - A component-based software platform for dynamic climate control. In: AgroTech (Ed.), *Intelligent Use of Energy in Greenhouses*. AgroTech, Odense, Denmark, pp. 43–4.

Jones, H. G. 1992. *Plants and Microclimate*. Cambridge University Press, Cambridge.

Jones, V. P., Brunner, J. F., Grove, G. G., Petit, B., Tangren, G. V. and Jones, W. E. 2010. A web-based decision support system to enhance IPM programs in Washington tree fruit. *Pest Management Science* 66, 587–95. doi:10.1002/ps.1913.

Kahlen, K. 2015. Predicting crop productivity in changing environments. In: Marcelis, L. F. M. and Schouten, R. (Eds), *Model-IT*, Wageningen, The Netherlands.

Keesman, K. J., et al. 2019. System modelling. In: Goddek, S. and Kotzen, B. (Eds), *Aquaponics Food Production* (in press).

Kim, S. H. and Lieth, J. H. 2003. A coupled model of photosynthesis, stomatal conductance and transpiration for a rose leaf (*Rosa hybrida* L.). *Annals of Botany* 91(7), 771–81. doi:10.1093/aob/mcg080.

Kjær, K. H., Ottosen, C. O. and Jørgensen, B. N. 2012. Timing growth and development of *Campanula* by daily light integral and supplemental light level in a cost-efficient light control system. *Scientia Horticulturae* 143, 189–96.

Kjær, K. H., et al. 2019. An integrated simulation and decision support system for greenhouse climate control (itGrows 2.0) based on an open source greenhouse modelling platform (HortSim V0.0). *Acta Horticulturae* (submitted).

Kläring, H. -P. and Ramírez, T. 2018. Interrupted dark phase does not affect greenhouse tomato growth and yield. *Scientia Horticulturae* 240, 221–3. doi:10.1016/j.scienta.2018.06.023.

Körner, O. 2013. An online greenhouse simulator *The Virtual Greenhouse*. In: Schamp, B. (Ed.), *The Future Low Carbon Greenhouse Production*. PCS Ornamental Plant Research and PCS Ornamental Plant Research, Lokeren, Belgium.

Körner, O. and Challa, H. 2003. Process-based humidity control regime for greenhouse crops. *Computers and Electronics in Agriculture* 39(3), 173–92. doi:10.1016/S0168-1699(03)00079-6.

Körner, O. and Hansen, J. B. 2012. An on-line tool for optimising greenhouse crop production. *Acta Horticulturae* 957(957), 147–54. doi:10.17660/ActaHortic.2012.957.16.

Körner, O. and Holst, N. 2017. An open source greenhouse modelling platform. *Acta Horticulturae* 1154(1154), 241–8. doi:10.17660/ActaHortic.2017.1154.32.

Körner, O. and Jakobsen, L. 2006. A thrips pest pressure model for greenhouse climate control. *Acta Horticulturae* 718(718), 407–14. doi:10.17660/ActaHortic.2006.718.47.

Körner, O., Aaslyng, J. M., Andreassen, A. U. and Holst, N. 2007. Modelling microclimate for dynamic greenhouse climate control. *HortScience* 42(2), 272–9. doi:10.21273/HORTSCI.42.2.272.

Körner, O., Warner, D., Tzilivakis, J., Eveleens-Clark, B. and Heuvelink, E. 2008. Decision support for optimising energy consumption in European greenhouses. *Acta Horticulturae* 801(801), 803–10. doi:10.17660/ActaHortic.2008.801.94.

Körner, O., Holst, N. and De Visser, P. 2014. A model-based decision support tool for grey mould prediction. *Acta Horticulturae* 1037(1037), 569–74. doi:10.17660/ActaHortic.2014.1037.71.

Kücükönder, H., Boyaci, S. and Akyüz, A. 2016. A modeling study with an artificial neural network: developing estimation models for the tomato plant leaf area. *Turkish Journal of Agriculture and Forestry* 40, 203–12. doi:10.3906/tar-1408-28.

Langhans, R. W., Wolfe, M. and Albright, L. D. 1981. Use of average night temperatures for plant growth for potential energy savings. *Acta Horticulturae* 115(115), 31–8. doi:10.17660/ActaHortic.1981.115.3.

Larcher, W. and Bodner, M. 1980. Dosisletalität-Nomogramm zur Charakterisierung der Erkältungsempfindlichkeit tropischer Pflanzen. (Dose-lethality nomogram for evaluating susceptibility to chilling in tropical plants). *Angewandte Botanik* 54, 273–8.

Lee, J. H. 1998. *Validation of a Crop Growth Model for Cut Chrysanthemum, cv. Cassa (Dendranthema grandiflorum (Ramat.) Kitamura)*. Wageningen Agricultural University, Wageningen, The Netherlands.

Lee, J. H. 2002. Analysis and simulation of growth and yield of cut chrysanthemum. PhD thesis. Wageningen University, Wageningen, The Netherlands, 120pp.

Lentz, W. 1998. Model applications in horticulture: a review. *Scientia Horticulturae* 74(1–2), 151–74. doi:10.1016/S0304-4238(98)00085-5.

Leung, C. M., Wilson, D. R., Marquand, C. J. and Tassou, S. A. 1985. A knowledge based system for the control of a greenhouse environment. *Acta Horticulturae* 174, 425–31.

Li, T., Ji, Y. H., Zhang, M., Sha, S. and Li, M. Z. 2017. Universality of an improved photosynthesis prediction model based on PSO-SVM at all growth stages of tomato. *International Journal of Agricultural and Biological Engineering* 10(2), 63–73.

Liebig, H. P. and Alscher, G. 1993. Combination of growth models for optimized CO_2- and temperature-control of lettuce. *Acta Horticulturae* 328(328), 155–62. doi:10.17660/ActaHortic.1993.328.13.

Linker, R., Seginer, I. and Gutman, P. O. 1998. Optimal CO_2 control in a greenhouse modelled with neural networks. *Computers and Electronics in Agriculture* 19(3), 289–310. doi:10.1016/S0168-1699(98)00008-8.

Liu, J. 2016. Design and implementation of an intelligent environmental-control system: perception, network, and application with fused data collected from multiple sensors in a greenhouse at Jiangsu, China. *International Journal of Distributed Sensor Networks* 12(7), 1–10. doi:10.1177/155014775056460.

Lund, H. 1995. The design reference year user manual, a report of Task 9: solar radiation and pyranometer studies. Report No. IEA-SHCP-9E-1; Report No. 274, Solar Materials Research and Development, International Energy Agency Solar Heating and Cooling Programme, Thermal Insulation Laboratory, Technical University of Denmark, Denmark.

Mahlein, A.-K. 2016. Plant disease detection by imaging sensors – parallels and specific demands for precision agriculture and plant phenotyping. *Plant Disease* 100(2), 241–51. doi:10.1094/PDIS-03-15-0340-FE.

Marcelis, L. F. M. 1994. Fruit growth and dry matter partitioning in cucumber. PhD thesis. Wageningen Agricultural University, Wageningen, 173p.

Marcelis, L. F. M., Heuvelink, E. and Goudriaan, J. 1998. Modelling biomass production and yield of horticultural crops: a review. *Scientia Horticulturae* 74(1–2), 83–111. doi:10.1016/S0304-4238(98)00083-1.

Marcelis, L. F. M., van den Boogaard, R. and Meinen, E. 2000. Control of crop growth and nutrient supply by the combined use of crop models and plant sensors. *IFAC Proceedings Volumes* 33(19), 161–6. doi:10.1016/S1474-6670(17)40906-2.

Marcelis, L. F. M., Elings, A., Bakker, M. J., Brajeul, E., Dieleman, J. A., de Visser, P. H. B. and Heuvelink, E. 2006. Modelling dry matter production and partitioning in sweet pepper. *Acta Horticulturae* 718(718), 121–8. doi:10.17660/ActaHortic.2006.718.13.

Meyer, J. 1983. The influence of thermal screens on energy consumption of greenhouses. *Gartenbauwissenschaft* 49(2), 74–6.

Miranda, L., Lara, B., Rocksch, T., Dannehl, D. and Schmidt, U. 2017. Neural networks for simulation of transpiration and photosynthesis in a semi-closed greenhouse. *Acta Horticulturae* 1170(1170), 193–200. doi:10.17660/ActaHortic.2017.1170.22.

Monteith, J. L. and Unsworth, M. H. 1990. *Principles of Environmental Physics*. Edward Arnold, London, 291p.

Moon, T. W., Jung, D. H., Chang, S. H. and Son, J. E. 2018. Estimation of greenhouse CO_2 concentration via an artificial neural network that uses environmental factors. *Horticulture, Environment and Biotechnology* 59(1), 45–50. doi:10.1007/s13580-018-0015-1.

Moreshet, S., Koller, D. and Stanhill, D. 1969. The partitioning of resistances to gaseous diffusion in the leaf epidermis and the boundary layer. *Annals of Botany* 32(4), 695–701.

Moriyuki, S. and Fukuda, H. 2016. High-throughput growth prediction for *Lactuca sativa* L. seedlings using chlorophyll fluorescence in a plant factory with artificial lighting. *Frontiers in Plant Science* 7, 394. doi:10.3389/fpls.2016.00394.

Nishina, H. 2015. Development of speaking plant approach technique for intelligent greenhouse. *Agriculture and Agricultural Science Procedia* 3, 9–13. doi:10.1016/j.aaspro.2015.01.004.

Opdam, J. J. G., Schoonderbeek, G. G., Heller, E. M. B. and De Gelder, A. 2005. Closed greenhouse: a starting point for sustainable entrepreneurship in horticulture. *Acta Horticulturae* 691(691), 517–24. doi:10.17660/ActaHortic.2005.691.61.

Ponce, P., Molina, A., Cepeda, P., Lugo, E. and MacCleery, B. 2015. *Greenhouse Design and Control*. CRC Press, London.

Poorter, H., Anten, N. P. R. and Marcelis, L. F. M. 2013. Physiological mechanisms in plant growth models: do we need a supra-cellular systems biology approach? *Plant, Cell and Environment* 36(9), 1673–90. doi:10.1111/pce.12123.

Ramírez, T., Kläring, H. P. and Körner, O. 2019. Long term measurements of carbon dioxide gas exchange under interrupted supplemental lighting on greenhouse tomato crops (submitted).

Rath, T. 1992. *Einsatz wissensbasierter Systeme zur Modellierung und Darstellung von gartenbautechnischem Fachwissen am Beispiel des hybriden Expertensystems HORTEX*. University of Hannover, Germany.

Rodríguez, F., Berenguel, M., Guzmán, J. L. and Ramírez, A. 2015. *Modelling and Control for Greenhouse Crop Growth*. Springer, London, UK.

Rydahl, P. 2003. A web-based decision support system for integrated management of weeds in cereals and sugarbeet. In: York, E. B. (Ed.), *EPPO Conference on Computer Aids for Plant Protection*, pp. 455–460.

Rytter, M., Sørensen, J. C., Sørensen, B. N. and Körner, O. 2012. Advanced model-based greenhouse climate control using multi-objective optimization. *Acta Horticulturae* 957(957), 29–35. doi:10.17660/ActaHortic.2012.957.2.

Salazar, R., López, I., Rojano, A., Schmidt, U. and Dannehl, D. 2015. Tomato yield prediction in a semi-closed greenhouse. *Acta Horticulturae* 1107(1107), 263–70. doi:10.17660/ActaHortic.2015.1107.36.

Sánchez-Molina, J. A., Pérez, N., Rodríguez, F., Guzmán, J. L. and López, J. C. 2015. Support system for decision making in the management of the greenhouse environmental based on growth model for sweet pepper. *Agricultural Systems* 139, 144-52. doi:10.1016/j.agsy.2015.06.009.

Sato, S., Peet, M. M. and Thomas, J. F. 2000. Physiological factors limit fruit set of tomato (*Lycopersicon esculentum* Mill.) under chronic, mild heat stress. *Plant, Cell and Environment* 23(7), 719–26. doi:10.1046/j.1365-3040.2000.00589.x.

Seginer, I. 1999. Optimal dynamic management of the greenhouse climate. In: Stanhill, G. and Enoch, H. Z. (Eds), *Ecosystems of the world 20 – Greenhouse Ecosystems*. Elsevier, Amsterdam, pp. 349–83.

Singh, M. C., Yousuf, A. and Singh, J. P. 2016. Greenhouse microclimate modeling under cropped conditions – a review. *Research in Environment and Life Sciences* 9, 1552–7.

Stanghellini, C. 1987. Transpiration of greenhouse crops. PhD thesis. Wageningen Agricultural University, Wageningen, The Netherlands, 150pp.

Stanghellini, C. 1995. Vapour balance. In: Bakker, J. C., Bot, G. P. A., Challa, H. and Van der Braak, N. J. (Eds), *Greenhouse Climate Control*. Wageningen Pers, Wageningen, The Netherlands, pp. 141–50.

Stanghellini, C. and De Jong, T. 1995. A model of humidity and its applications in a greenhouse. *Agricultural and Forest Meteorology* 76(2), 129–48. doi:10.1016/0168-1923(95)02220-R.

Stanghellini, C. and Van Meurs, W. T. M. 1989. Crop transpiration: a greenhouse climate control parameter. *Acta Horticulturae* 245(245), 384–8. doi:10.17660/ActaHortic.1989.245.51.

Tantau, H. J. and Lange, D. 2003. Greenhouse climate control: an approach for integrated pest management. *Computers and Electronics in Agriculture* 40(1–3), 141–52. doi:10.1016/S0168-1699(03)00017-6.

Thornley, J. H. M. 1976. *Mathematical Models in Plant Physiology*. Academic Press (Inc.), London, UK, 318pp.

Tschiersch, H., Junker, A., Meyer, R. C. and Altmann, T. 2017. Establishment of integrated protocols for automated high throughput kinetic chlorophyll fluorescence analyses. *Plant Methods* 13, 54. doi:10.1186/s13007-017-0204-4.

Udink ten Cate, A. J. 1983. Modeling and (adaptive) control of greenhouse climates. PhD thesis. Wageningen Agricultural University, Wageningen, The Netherlands, 159pp.

Vadiee, A. and Martin, V. 2012. Energy management in horticultural applications through the closed greenhouse concept, state of the art. *Renewable and Sustainable Energy Reviews* 16, 5087–100.

Van Loon, J., Vensteenkiste, J., Diels, J. and Schrevens, E. 2011. Developing and testing a model for open field horticultural crops to enable use of a 'just-in-time' fertilization management. In: *19th International Congress on Modelling and Simulation*, Perth, Australia.

Van Ooteghem, R. J. C. 2007. Optimal control design for a solar greenhouse. PhD thesis. Wageningen University, Wageningen, The Netherlands, 304pp.

Van Ooteghem, R. J. C., Van Willigenburg, L. G. and Van Straten, G. 2005. Receding horizon optimal control of a solar greenhouse. *Acta Horticulturae* 691(691), 797–806. doi:10.17660/ActaHortic.2005.691.98.

Van Straten, Gv, Van Willigenburg, L. G. and Tap, R. F. 2002. The significance of crop co-states for receding horizon optimal control of greenhouse climate. *Control Engineering Practice* 10(6), 625–32. doi:10.1016/S0967-0661(02)00023-0.

Vanthoor, B. 2011. A model-based greenhouse design method. PhD thesis. Wageningen University, Wageningen, The Netherlands, 307pp.

Voogt, J. and Van Weel, P. 2008. Climate control based on stomatal behavior in a semi-closed greenhouse system 'Aircokas'. *Acta Horticulturae* 797(797), 151-6. doi:10.17660/ActaHortic.2008.797.19.

Vox, G., Teitel, M., Pardossi, A., Minuto, A., Tinivella, F. and Schettini, E. 2010. Sustainable greenhouse systems. In: Salazar, A. and Rios, I. (Eds), *Sustainable Agriculture*. Nova Science Publishers, Inc., New York, p. 79.

Wang, J. and Tao, Y. 2015. Design and realization of temperature monitoring and warning system under temperature greenhouse. *The Open Automation and Control Systems Journal* 7(1), 929–33. doi:10.2174/1874444301507010929.

Wang, S., Boulard, T. and Haxaire, R. 2000. Measurements and analysis of air speed distribution in a naturally ventilated greenhouse. *Acta Horticulturae* 534(534), 277–84. doi:10.17660/ActaHortic.2000.534.32.

Wang, H., Sánchez-Molina, J. A., Li, M., Berenguel, M., Yang, X. T. and Bienvenido, J. F. 2017. Leaf area index estimation for a greenhouse transpiration model using external climate conditions based on genetics algorithms, back-propagation neural networks and nonlinear autoregressive exogenous models. *Agricultural Water Management* 183, 107–15. doi:10.1016/j.agwat.2016.11.021.

Watson, R. T., Boudreau, M.-C. and Van Iersel, M. W. 2018. Simulation of greenhouse energy use: an application of energy informatics. *Energy Informatics* 1(1), 1-14. doi:10.1007/s42162-018-0005-7.

Weis, E. and Berry, J. 1988. Plants and high temperature stress. In: Long, S. P. and Woodward, F. I. (Eds), *Plants and Temperature*. The Company of Biologists Ltd., Cambridge, pp. 329–46.

Wit, D. 1969. Dynamic concepts in biology, Prediction and measurement of photosynthetic productivity. In: *Proceedings of the IBP/PP Technical Meeting*, Trebon, pp. 17–23.

Wolfert, S., Ge, L., Verdouw, C. and Bogaardt, M. J. 2017. Big data in smart farming – a review. *Agricultural Systems* 153, 69–80. doi:10.1016/j.agsy.2017.01.023.

Yokoi, S., Goto, E., Kozai, T., Nishimura, M., Taguchi, K. and Ishigami, Y. 2008. Effects of planting density and air current speed on the growth and uniformity of qing-geng-cai and spinach plug seedlings in a closed transplant production system. *Environmental Control in Biology* 46(2), 103–14. doi:10.2525/ecb.46.103.

Zou, W., Yao, F., Zhang, B., He, C. and Guan, Z. 2017. Verification and predicting temperature and humidity in a solar greenhouse based on convex bidirectional extreme learning machine algorithm. *Neurocomputing* 249, 72–85. doi:10.1016/j.neucom.2017.03.023.

Chapter 14

Assessing the impact of environmental factors on the quality of greenhouse produce

Nazim Gruda, University of Bonn, Germany

1 Introduction

Product quality is growing in importance, due to consumer concerns. Protected cultivation not only protects crops from external natural hazards, but also manipulates the micro-environment to facilitate optimal plant performance, extends production duration, increases the variety of produce throughout the year, induces early flowering and improves the quality of produce (Gruda and Tanny, 2014).

The primary crops cultivated in greenhouses are vegetables, followed by ornamentals (both cut and pot plants), and exotic fruits, such as bananas and strawberries. The impact of horticultural plants on human nutrition and wellbeing greatly depends on produce quality in terms of visual, organoleptic and/or nutritional properties. Ornamental plants brighten our lives and increase the feeling of being closer to nature, even in urban areas, whereas vegetables and fruits make a significant contribution to our nutritional health, including the prevention of chronic diseases, such as coronary heart disease and cancer.

http://dx.doi.org/10.19103/AS.2019.0052.16

The positive effects on health result from the diversity of biological substances, such as dietary fibres, minerals, vitamins, secondary metabolites and so on, contained in these products and less from the intake of individual ingredients. Thus, the phrase of the ancient Greek philosopher, Aristotle 'The whole is more than the sum of its parts' is appropriate in this context.

Most studies on vegetable and fruit quality consider the influence of post-harvest and/or processing factors. Good post-harvest conditions can retain the quality characteristics defined by pre-harvest factors but cannot improve the quality of produce. This is true for ornamentals as well. In addition, consumers eat greenhouse vegetables fresh and not processed. Thus, it is imperative to maximize the quality of greenhouse products through appropriate pre-harvest growing techniques (Gruda et al., 2018).

Product quality is a complex, multidimensional and multi-factorial issue that depends on both pre- and post-harvest factors and includes objective, measurable quality traits as well as subjective characteristics (Gruda, 2005; Gruda et al., 2018). This chapter outlines the objective and measurable criteria related to plant reaction to various environmental and pre-harvest growing factors. In addition, we present their effects on visual, organoleptic and nutritional properties, referring to studies undertaken in the last 20 years. Before this, some basic knowledge needs to be explained, starting with the definition of product quality.

1.1 Product quality: a definition

The definition of quality for horticultural products is not as straightforward as it may appear, due to the variety of different stakeholders, namely breeders, producers, traders, processors and consumers involved in the horticultural value chain (Fig. 1). Each has different perspectives, views and expectations on quality characteristics of horticultural produce. For instance, while for a tomato breeder and seed producer, selling high vigour seeds is important, a farmer aims to achieve high marketable yields, with good commercial characteristics. For a consumer, tomatoes with high nutritional value is the most important. With the exception of the breeder and end consumer, each stakeholder acts as both a producer and a consumer for the next and the preceding chain participant, respectively. Therefore, Shewfelt (1999) has suggested a product-oriented and a consumer-oriented approach to define the quality of horticultural produce.

Whereas from the perspective of product orientation, quality attributes are inherent in a product and can be readily and objectively quantified, a consumer orientation, which defines quality in terms of consumer satisfaction, is a much less tangible and quantifiable concept, based mainly on subjective criteria.

For ornamental plants, there are no quality standards, while market standards based on objective criteria for the quality of fresh fruits and

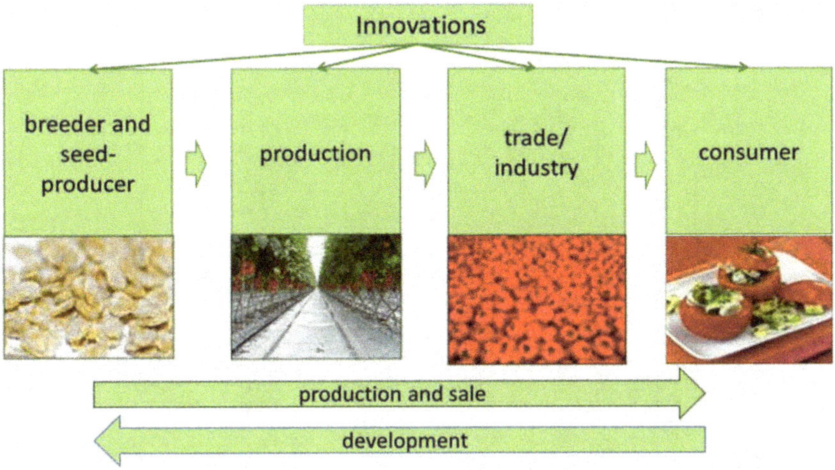

Figure 1 Value chain, using tomatoes as an example. Whilst the direction for production and sales generally goes from left to right, current development is now an interactive approach with the 'consumer' as a starting point. To improve the quality of produce, innovative approaches, including all components of the chain as well as the interactions between them should be applied.

vegetables are well established (UNECE Commercial Agricultural Quality Standards – FAO; OECD Scheme for the Application of International Standards for Fresh Fruits and Vegetables). These quality standards, as well as being minimum requirements, are objective quality measures for the market value of vegetables, aimed at preventing inferior quality products, bringing production in line with consumer requirements, facilitating trade based on fair competition and improving profitability. However, they are generally based on appearance and visual quality characteristics, such as size and shape, colour, form and freshness (Table 1). Whereas critical concentrations of pesticides and nitrates in plant tissues are included in food standards, important quality attributes, such as flavours and nutraceutical properties, are not considered. Consequently, positive changes in organoleptic characteristics, such as sugar, aroma compounds or organic acid content are not reflected in terms of price. Recently, consumer demand for these components has greatly increased, together with requirements for higher nutritional value (Gruda, 2005). In addition to health awareness, the environmental consciousness of consumers has increased enormously (Table 1).

In this chapter, we will discuss several quality characteristics for greenhouse vegetables as well as potential pre-harvest improvement measures, focused on product-oriented quality and objective criteria. However, first let us examine consumer preferences and differences between intrinsic and extrinsic quality characteristics.

Table 1 Quality values of greenhouse vegetables and their quality criteria determined by the customer groups of the supply chain

External quality		Internal and nutritional quality		Cultivation methods	
Market value	Utilization value	Sensory value	Nutritional and health value	Ecological value	Imaginary value
Shape	Shelf life	Taste	*Desired ingredients:*	Integrated production	Prestige
Colour, shine	Transportability	Smell	Vitamins	Organic production	Image
Size	Storability	Flavour	Minerals	Soilless culture system	Religion
Freshness	Processing	Textural properties	Flavourings	Resource conservation cultivation	
Consistency	Suitability to be packed	Secondary metabolites	Dietary fibres		
Integrity	Kitchen characteristics	Mouthfeel	Bioactive substances etc.	Use of renewable raw materials	
Cleanliness			*Undesired ingredients:*		
Free from foreign odour			Endogenous substances, (oxalic acids, solanine, allergens, heavy metals)	Sustainable cultivation	
				Local and regional production and marketing, short transportation distances	
			Residues of pesticides		
			Nitrate		
			Mycotoxins etc.		

Source: Adapted from Huyskens-Keil and Schreiner (2003), Wonneberger and Keller (2004) and Gruda (2005).

1.2 Two words to consumer preferences: Intrinsic and extrinsic quality characteristics

According to Schreiner et al. (2013), a typical consumer with consistent preferences does not exist. In addition, individual consumers often change their lifestyle according to personal preferences. Consumer requirements are in a constant state of flux, not only due to changes in personal preferences, but also social value systems. Even in the case of ornamental plants, demand is also influenced by lifestyle.

Intrinsic and extrinsic quality features characterize horticultural products. In contrast to intrinsic quality characteristics, which are directly derived from the natural product, extrinsic quality characteristics are not directly related to product performance. They contribute to the general benefit of the product and are therefore a means of adding value to a core product (Grunert et al., 2004; Schreiner et al., 2013). Extrinsic quality parameters in the broader sense are product characteristics, conveyed through marketing instruments such as distribution, pricing, communication policies and advertising (Schreiner et al., 2013). According to Schnitzler and Gruda (2002), consumers form their expectations on intrinsic quality parameters based on cues from extrinsic variables when lacking immediate insight into intrinsic product quality parameters. Consider the label of provenance. Although the differences for vegetable greenhouse produce in two neighbouring countries are marginal (both use greenhouses, i.e. controlled production conditions), the end consumer is more likely to choose 'local' produce, for example from Germany, when the product is placed on the German market, over a product from just a few km away from the border, for example from the Netherlands.

In addition, branding allows production companies to distinguish their horticultural products from comparable ones. Schreiner et al. (2013) has shown that an additional value of more than 100% can be achieved in German retail outlets when increasing the extrinsic product quality, for example with the wording 'Tasty Tom', known as one of the premier tomato brands in Europe. A protected geographical indication registered by the EU also falls into this category.

Quality evaluation is based on market, utilization, sensory, nutritional and health value of horticultural products (Table 1). The influence of extrinsic factors and the use of subjective criteria are excluded from the remit of this chapter.

2 Greenhouse production and produce quality

Protected cultivation represents different advantages compared to outdoor production with regard to quality assurance (Bot, 2003). First, controlled environments offer a way to optimize environmental parameters, as the product is not suddenly exposed to changes in climate conditions. Besides environmental factors, cultivation factors such as water and fertilizer supply can be better regulated under protected cultivation. Furthermore, by using soilless culture techniques the crop can be grown independently from, for example, unfavourable soil conditions (Gruda, 2005; Gruda et al., 2018). As a result, production duration is extended and variety increased, achieving high quality produce even in sub-optimal climates and geographies. Demand for healthy products all year round is therefore much better supported.

On the other hand, the content of several secondary metabolites, such as flavonoid or glucosinolate content, is much higher in field-grown plants than greenhouse-grown plants, indicating increased metabolite synthesis due to UV exposure (Schirrmacher et al., 2004; He, 1999). The physical, chemical, organoleptic, nutritional and functional quality properties of horticultural produce are primarily dependent on internal factors, such as the genetic constitution and composition, as well as external factors, such as environment and practical agronomic practices.

The first factor having an influence on produce quality is the genotype. According to Hanna (2009), the choice of an adequate-yielding cultivar with desired fruit qualities and longer shelf life is a vital decision for growers. In general, there are some differences between the cultivars used for indoor and outdoor cultivation. For example, greenhouse tomatoes are indeterminate plants, as opposed to determinate field and processing tomato plants (Dorais et al., 2001), while greenhouse cucumbers are generally tall with large leaves, whose fruits are used for fresh consumption, as opposed to short field plants with relatively small leaves whose fruits are generally used in the pickling industry (Gruda, 2005). Several studies have shown cultivar differences in the susceptibility of greenhouse crops to different physiological disorders. Physiological behaviour, tolerance against different stress reactions and the susceptibility of greenhouse vegetable cultivars to these conditions is not only expressed by a variation in terms of visual characteristics, but also by the modification of organoleptic and nutritional quality, for example for vegetables (Gruda et al., 2018). Nowadays, plant breeding programmes are used to increase the concentrations of bioactive compounds with antioxidant activities, including health-promoting metabolites such as lycopene and other carotenoids in new tomato cultivars, together with other desirable characteristics, such as high yield and resistance to pests and diseases.

The expression of genotype characteristics depends on several external factors, including pre- and post-harvest measures that influence primary and secondary metabolites, that is the composition of intrinsic values. Recently, this practice has been used to increase the intake, bioavailability and occurrence of plant metabolites in fresh vegetables. Usually, suboptimal biotic and abiotic stress conditions are applied for a short period. Under such conditions, the plant generally increases its antioxidant metabolism to counteract oxidative stress, thus improving nutritional quality from a human point of view (Pérez-López et al., 2015).

The main factors influencing the quality of produce are presented below. In addition, methods of implementing these tasks to achieve the desired result are also shown. For didactic and practical reasons, we present the reactions of protected crops to singular environmental variables. Furthermore, with some examples, we present the reaction of crops to real-time conditions within protected cultivations, giving an overview of the whole concept.

3 Assessing the impact of light on the quality of greenhouse produce

Protected cultivation is always a compromise between the required protection and the need to provide optimal diffuse radiation (Gruda and Tanny, 2014). Greenhouse structure and cover reduce the amount of daylight that greenhouse crops receive, even if transmissivity of greenhouses has been improved over the last 25 years. In addition, photo-selective film covers modify the light spectrum by filtering solar radiation in plastic houses. Moreover, the structure of the roof, the geometrical properties of the crop and the artificial light application, if used, also affects the radiation reaching the canopy and strongly influences the quality of horticultural produce.

Both low and high light intensity greatly influence, and usually reduce, the product quality of horticultural products. At low light intensities, plants are etiolated, elongated and can have fewer and longer internodes. For example, ornamental potted plants might not be compact and therefore not meet a particular market demand, whereas leaf vegetables may build up high levels of nitrite. Long photoperiods of supplemental lighting can cause photo-injury such as leaf chlorosis and necrotic spotting in greenhouse fruit vegetables. In year-round greenhouse vegetable production, photo-injury caused by supplemental lighting could limit the application for a long time (Hao et al., 2017a). Extreme light intensity combined with high temperatures can adversely affect quality leading to disorders in appearance and development, such as uneven ripening, the occurrence of green shoulder, degradation of pigmentation, sunscald, and blossom end rot (BER) in tomato as well as cracking in tomatoes and peppers fruits (Gruda, 2005; Gruda and Tanny, 2014). Moreover, excessive amounts of light are associated with damages in photosynthetic system II and photoinhibition (Ksas et al., 2015). However, Frantz and Ling (2011) reported an increased number of flowering shoots and inflorescence size of Kalanchoe (*Kalanchoe* spp.) by increasing photosynthetic photon flux, and Gaudreau et al. (1994) reported increased head firmness of lettuce as a result of supplemental light (Gruda and Tanny, 2014).

Although light is not essential for the synthesis of ascorbic acid, it is a regulatory factor for L-ascorbate (vitamin C), an essential antioxidant for plants, animals and human beings. The ascorbic acid synthesis and its concentration in plant tissues accelerates and increases with increasing irradiance (Shinohara, 1987; Ntagkas et al., 2018). A correlation between light conditions and the ascorbic acid content of different vegetables such as tomato, lettuce, sweet pepper and strawberry has been made (Shinohara, 1987). For instance, the ascorbic acid content of tomatoes, was low in lower clusters and remarkably high in the uppermost fourth cluster of tomato plants (Shinohara et al., 1982; Gruda, 2005). Therefore, in greenhouse crops with high plant density, supplemental lighting to the lower understory leaves of the canopy can increase year-round productivity and improve the quality of produce (Fig. 2). For instance, Dorais and Gosselin (2002) reported a higher sugar content and ascorbic acid concentration in tomato by application of additional light, while Tewolde et al. (2016) found that night-time inter-lighting in winter significantly increased the total soluble solids (TSS) and ascorbic acid content of the tomatoes, by 20% and 25%, respectively. According to Ntagkas et al. (2018), respiration and photosynthesis interact in light regulation of ascorbic acid biosynthesis via carbohydrate availability. However, according to Gautier et al. (2009), Massot et al. (2010) and Bertin and Génard (2018), ascorbic acid accumulation is not limited by leaf photosynthesis or sugar substrate, but strongly depends on the fruit irradiance itself.

Figure 2 LED inter-lighting in production of cherry tomatoes in a horticultural company in North Rhine-Westphalia, Germany. Since LEDs are not only able to be applied at specific wave lengths, but in addition, generate less heat, they can be used directly near the plants. White screens are used under the glass to prevent external light emissions when it gets dark. Photo: Gruda (2017), private collection.

Recent developments in the field of light source technologies have opened new perspectives for sustainable and highly efficient light sources in the form of light-emitting diodes (LED) (Singh et al., 2015), which provide ideal light quality, intensity, spectrum and photoperiod for protected crops. Bian et al. (2015) reviewed the effects of light quality regulation in controlled environments, highlighting the research progress and advantages of LED technology as a regulation tool for decreasing the amounts of harmful substances while modifying and increasing the amounts of beneficial phytochemicals in vegetables. LED lamps with red and blue spectrums are applied in horticultural production (Fig. 3), ensuring higher yields and higher external and nutritional quality of produce in comparison to monochromatic light (Bantis et al., 2018). Thus, the biosynthesis of secondary metabolites of vegetables can be manipulated and improved. Several studies have shown positive effects of LED light spectra with increased nutritional value and enhanced antioxidant levels in green vegetables, such as carotenoids (Lefsrud et al., 2008), vitamin C (Li et al., 2012), anthocyanins (Mizuno et al., 2011) and polyphenols (Li and Kubota, 2009; Johkan et al., 2012; Samuolienė et al., 2012). Meanwhile, application of supplemental LED lighting to dill and parsley results in the accumulation of vitamin C and carbohydrates and in the enhancement of free radical-binding activity and the activity of nitrate-reducing enzymes (Bliznikas et al., 2012). Piovene et al. (2015) reported that LED lighting enabled an increase in antioxidant compounds and a decrease in nitrate content in basil leaves.

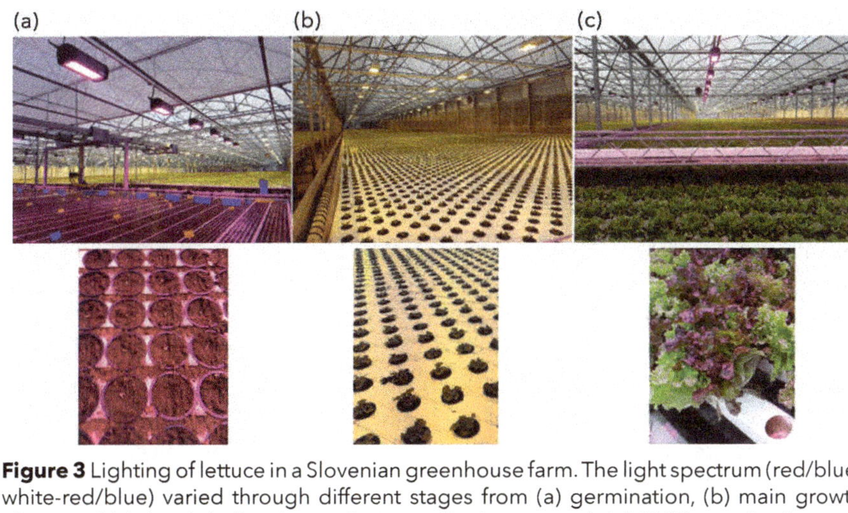

Figure 3 Lighting of lettuce in a Slovenian greenhouse farm. The light spectrum (red/blue-white-red/blue) varied through different stages from (a) germination, (b) main growth phase and (c) shortly before harvesting, contributing to nearly (a) 100% germination rate, (b) a quick growth speed and (c) a compact and coloured final product with a reduced biomass accumulation and higher number of carotenoids and chlorophyll, respectively. Photo: Gruda (2017), private collection.

According to Mizuno et al. (2011) and Liu et al. (2018), anthocyanin biosynthesis has been shown to be light-dependent. Environmental stimuli such as high light intensity, blue/UV light and low temperature are recommended to be applied during the cultivation period to promote anthocyanin production, stabilize the structure and reduce their degradation. For instance, the surface of the tomato fruit directly exposed to the light showed a more intense anthocyanin pigmentation compared to the shaded parts (Mazzucato et al., 2013). In addition to intensity, light quality also affects anthocyanin biosynthesis. However, sometimes contradictory results are obtained. For instance, Stutte et al. (2009) reported a total content of anthocyanin produced in red leaf lettuce that was twice as high in the presence of blue light than with red alone. The addition of green wavelengths to the spectra had no additional effect on anthocyanin concentration or content in the leaf, whereas addition of far red resulted in a slight, but not statistically significant, reduction in the anthocyanin content in the leaf. Ouzounis et al. (2016), investigated the effect of red and blue LED lighting on physiological, morphological, developmental and chemical parameters of nine tomato genotypes. The combination of blue and red LED lighting increased total dry matter in seven of the nine genotypes compared to just red LED light. Upward or downward leaf curling was observed in all genotypes of the 100% red treatment. In addition, blue light increased chlorophyll and flavonol contents in three genotypes.

Several studies show a correlation between the presence of secondary metabolites, decreasing temperatures and global radiation. The exposure of plants to global radiation fosters an elevated synthesis and accumulation of shielding compounds such as flavonoids (Schreiner et al., 2013). However, Taulavuori et al. (2016) found that the synthesis of flavonoids and phenolic acids under increasing periods of enhanced blue light was species-specific. The most abundant compound in red leaf lettuce was cichoric acid (a dicaffeoyltartaric acid) while rosmaric acid (an ester of caffeic acid and 3.4-dihydroxyphenyllactic acid) dominated in basil. These results tally with results from Ouzounis et al. (2015) with *Lactuca sativa*. The quantification of secondary metabolites showed that all treatments with additional blue light had a higher amount of pigments, phenolic acids and flavonoids compared to the control; however, the plant responses to light were cultivar-dependent.

It is well known that ultraviolet B (UV-B) radiation induces changes in secondary metabolites such as phenolic compounds of vegetables and herbs (Ghasemzadeh et al., 2016; Mosadegh et al., 2018). However, when these crops are cultivated in protected production systems, for example in a glasshouse, UV-B radiation is blocked. Therefore, cover films that allow some of the ultraviolet light energy spectrum to pass through or the installation of UV-B lights in the greenhouse are recommended. On the other hand, the use of UV lamps has to consider the overall photosynthetic performance. Mosadegh et al. (2018) therefore suggested discontinuing UV-B irradiation after six days, without affecting the efficiency of photosynthetic apparatus of sweet basil (*Ocimum basilicum*).

Photo-selective cover films and nets of different colours can also influence the product quality of vegetables. Furthermore, NIR-blocking and UV-blocking covers have been shown to have some beneficial effects (Kittas et al., 2006; Lamnatou and Chemisana, 2013). In addition, anti-fog, anti-drip, anti-condensation and anti-dust additives (Gruda and Tanny, 2014) are used to improve the light transmittance in greenhouses that in turn have an influence on the quality of horticultural products.

According to Bian et al. (2015), light quality combination strategies and modulation methods in plant species and plant development stages should be carefully evaluated when selecting a supplemental light source for vegetables, because the effect of light parameters on the response of phytochemical accumulation in plants varies among species and cultivars.

Nowadays, optimized vertical light spectral distribution profiles are tested to improve not only the vegetative and fruit growth, but also the fruit quality in tall greenhouse crops (Hao et al., 2017b). Although some of the effects of light on phytochemicals in plants are known, the biochemical and molecular mechanisms are still not clear. Therefore, detailed scientific studies are necessary to understand the effect of different light spectra on physiology and produce quality in different specific plant species (Singh et al., 2015) and cultivars. In the future, attention should be paid to red, blue, white and green light, but also to other photosynthetically non-effective light qualities, such as UV and far-red light (Bian et al., 2015).

4 Assessing the impact of temperature on the quality of greenhouse produce

Temperature is one of the most important environmental factors influencing the quality of produce. In general, the speed of all biochemical processes is temperature-dependent, and the reaction rate of different biological processes increases with rising temperature. With enhanced exposition duration and temperature intensity above the optimum level, the reaction rate decreases, because most enzymes lose their effectiveness or have been damaged, resulting in reduced plant growth and development and, in turn, impaired quality (Gruda, 2005; Gruda and Tanny, 2014). For instance, Sublett et al. (2018) reported that when looking at relevant variations in practice, temperature is an important regulatory factor in the quality of dark red 'Lollo Rosso' lettuce.

In heated greenhouses, temperature management is directly related to energy consumption. Elings et al. (2005) calculated that lowering day and night temperature set points for tomato by 1°C led to a reduced energy consumption of 8% per year. As we reported before dynamic temperature integration with temperature drop can promote the translocation of photo-assimilates from leaf to fruit, having the potential to reduce photo-injury at long periods of supplemental lighting (Hao et al., 2017a). Hao et al. (2017a) found this to be an energy-efficient climate control strategy for improving the response of greenhouse tomatoes and sweet peppers to long photoperiods of supplemental lighting in year-round greenhouse fruit and vegetable production.

Lowering temperatures can adversely affect the production and the quality of produce (Gruda and Tanny, 2014). For instance, low temperatures extend plant development time. This could be a serious problem especially for seasonal crops, for example poinsettia (*Euphorbia pulcherrima*) that should be colourful and ready to enjoy over the Christmas period, as well as for earlier protected vegetable crops, because of high market prices. Sánchez-González et al. (2015) reported that lower temperatures, associated with lower values of radiation, lengthen the time of tomato fruit development, increasing in this way the firmness of tomatoes at the same colour stage. According to Kläring et al. (2015), each degree reduction in greenhouse temperature increases the time required for tomato fruit development by three-and-a-half days, leading to considerable delay of the first harvest. Moccaldi and Runkle (2007) and Blanchard et al. (2011) reported that the flowering time of salvia (*Salvia splendens*), marigold (*Tagetes patula*) and petunia (*Petunia* × *hybrida*) decreased as the temperature increased.

Low temperatures can cause fruit malformation and distortion, seedlessness, pericarp cracking and pigmentation formation in various fruits and vegetables (Gruda 2005). Moreover, low night temperatures can reduce the number of pollen grains per flower and impair the germination ability of vegetables such as tomatoes and peppers with a tendency to develop parthenocarpic fruits (Gruda and Tanny, 2014). Temperature may, in addition, indirectly influence the colour intensity of most fruits and vegetables, for example in greenhouse-grown tomatoes (Dumas et al., 2003), or eggplants grown under cool winter conditions in unheated polyethylene covered greenhouses (Zipelevish et al., 2000).

Since both water and nutrient uptake can be inhibited at low temperatures, these conditions may influence the organoleptic properties of vegetables. For instance, low temperatures will produce less juicy tomatoes with low acidity content and a mealy taste (Brückner et al., 2004), whereas Kano and Goto (2003) reported a higher occurrence of bitter fruits in cucumber (*Cucumis sativus*), when grown under lower temperatures rather than higher temperatures (Gruda and Tanny, 2014). Increasing temperature in a given range can improve fruit taste through an increased sugars/acids ratio (Bertin and Génard, 2018). Rosales et al. (2007) reported that Mediterranean summer conditions with a high temperature of around 35°C and solar radiation of 7 MJm^{-2} day^{-1} increased fruit hexose and starch content in cherry tomatoes, on a fresh weight basis, due to increased sucrose synthase activity. Moreover, according to Gautier et al. (2008) an increase in fruit temperature from 21°C to 26°C during fruit ripening strongly reduces the titratable acidity, possibly due to a decreased vacuolar malic acid accumulation (Bertin and Génard, 2018). However, good eating quality of vegetables depends on species and cultivar-specific temperatures. For instance, temperatures of 20°C and higher are desirable for tomatoes; but not for swede roots. Johansen et al. (2016) found that growth temperatures of 21°C enhanced the intensities of sensory attributes such as pungent odour, bitterness, astringency and fibrousness, while low temperature (9°C) was associated with acidic odour, sweet taste, crispiness and juiciness of swede roots.

Pollination, fruit set formation and horticultural product quality are also influenced by extreme high temperatures. Deleterious effects of high temperature can be direct or indirect. Temperature can directly damage cellular membranes, proteins and nucleic acids, for example by BER in tomatoes, pepper and eggplant and tipburn of Chinese cabbage and lettuce. Indirect temperature effects can include inhibited pigment synthesis and thermal degradation of existing pigments, for example by sunscalds and/or desiccate tissue and plant organs induced by water stress. In this case, internal fruit and vegetable temperature is more important than air temperature (Gruda, 2005).

Gruda (2005) and Castilla and Hernandez (2007) reported that high temperatures can limit tomato fruit acidity, negatively influence taste and flavour, lead to poor development of colour and exhibit low lycopene content. According to Gautier et al. (2008), temperatures of 27–32°C had a detrimental effect on lycopene content and its precursors, whereas temperatures within the range of 21°C and 26°C decreased total carotenes. High temperatures are the reason that most tomato cultivars cultivated in tropical areas with high temperatures are yellow coloured or show colour disturbances. Pardossi et al. (2000), Islam and Khan (2001) and Kano (2004) reported that sugar accumulation could be suppressed by high air temperatures in melons, cherry tomatoes and watermelon. Dorais et al. (2001) reviewed the results of different authors, reporting a maximum dry matter and sugar content for tomatoes at the beginning of summer, when solar radiation and photosynthesis reach their maximum.

As well as air temperature, temperature in the root zone is important for the quality of horticultural produce. The results very much depend on aboveground conditions and season, as well as on temperatures in the root area. For instance, Kafkafi (2001) stated that increasing the temperature of the nutrient solution from 12°C to 20°C increased water flow through the stems of tomato plants by 2.5 times, which consequently improved nutrient uptake and fruit quality. Urrestarazu et al. (2008) reported that heating of nutrient solution in soilless cultivation systems at 18–22°C increased the soluble sugar content of greenhouse melon. Sometimes root cooling is beneficial, especially in warm regions, where reducing air temperature to an optimal crop growth level requires a lot of energy. According to Thompson et al. (1998), keeping water temperature at 24°C maintains the market quality of lettuce heads even at 31°C air temperature. Benoit and Ceustermans (2001) and Li et al. (2002) reported a much lower rate of BER of soilless culture sweet pepper plants grown on cooled slabs than on non-cooled slabs, possibly due to a higher root activity from higher oxygen content in the root environment. Masaru et al. (2016) reported an increase in reproductive organs biomass, such as inflorescences and fruits of strawberries, when treated with a root zone temperature of 10°C by an air temperature of 20°C. However, the cooling of the root zone did not significantly influence the contents of ascorbic acid and sugar in fruits, although the fruit maturation period was significantly prolonged.

In the future, climate change will generally induce a temperature and CO_2 increase that in turn will directly or indirectly influence plant growth, physiological and metabolic processes, and the nutritional and flavour quality of vegetables. Gruda et al. (2019) have shown that these influences will affected protected cultivation to a lesser degree compared to open field crops. However, the interaction between climate change and other environmental greenhouse growth stress factors will additionally impact the quality of produce.

5 Assessing the impact of air humidity on the quality of greenhouse produce

Greenhouse humidity is one of the most important and difficult environmental factors to be controlled in greenhouses. The main challenge with humidity control is the interaction with temperature, both air and crop, air movement and evapotranspiration. For instance, warm air can hold more water vapour and is more difficult to be saturated than cold air.

In the scientific literature, apart from relative humidity (RH) in per cent, very often the term vapour pressure deficit (VPD) in kPa (kilo-Pascal) is used to characterize greenhouse humidity. VPD is defined as the difference between the amount of water in the air of current air humidity and the amount by saturation at the same temperature. High or low VPD will adversely influence plant growth, yield and product quality, depending on the availability of water in the root zone, because in the end it is a question of plant water balance (Gruda and Tanny, 2014).

At high VPD (low air humidity), transpiration is too high. To minimize water loss, plants close their stomata with negative effects on photosynthesis, plant growth and quality. High humidity or low VPD may result in heat damage due to a reduction in evapotranspiration rates, the main cooling mechanism of plants (Gruda, 2005). From the view of product quality, the management of humidity has two main purposes: maintaining crop transpiration within boundaries and preventing condensation on the crop. Excessively high or low rates of transpiration may result in local calcium deficiencies, loss of turgor, partial stomatal closure and loss of assimilation. In addition, condensation increases the incidence of diseases (Gruda and Tanny, 2014).

Gruda (2005) has shown that controlling VPD could influence the excess or deficiency of calcium content in the fruits and consequently the occurrence of related physiological disorders. Vanhassel et al. (2015) reported a reduction of tipburn in hydroponically cultivated butterhead lettuce due to an improved Ca transport to the leaf margins of young leaves by increasing nightly air humidity. On the other hand, low VPD can cause other disorders, such as cracking and russeting of vegetable fruits. According to Demers et al. (2007), low VPD most probably decreases leaf transpiration, but increases root pressure, which in turn increases fruit water supply and turgor pressure. Under such conditions, a greater stress level would be applied to the fruit skin and cuticle, which increases the likelihood of the development of cuticular and fruit cracking. More information regarding crop responses to controlled climates is provided elsewhere in this book.

By contrast with light and temperature, data concerning the influence of air humidity on nutritional greenhouse vegetable quality are generally scarce except for tomato fruits (Gruda, 2005). Under Mediterranean summer conditions, high VPD could have negative effects on yield and nutrient transport to the tomato fruit. In contrast, the accumulation of sugar contents (on a fresh weight basis), phytonutrients such as phenolic compounds and ascorbic acid, as well as the antioxidant capacity can be increased (Bertin et al., 2000; Rosales et al., 2011). Moreover, Rosales et al. (2011) found a reduction in lycopene content at low VPD. These results are in line with earlier results from Dorais et al. (2001). The authors demonstrated that higher VPD brought out an increase in colour intensity of tomato fruits. On the other hand, low levels of VPD and high temperatures in combination with high CO_2 concentrations promoted the secondary metabolism of tomatoes. For instance, Dannehl et al. (2012) reported an increase in lycopene, β-carotene and phenolic compounds as well as associated antioxidant activity of tomatoes, grown in semi-closed greenhouses.

According to Torre and Fjeld (2001), Torre et al. (2003) and Mortensen and Gislerød (2005), relative air humidity between 85% and 90% during active growth is a critical environmental factor that reduces the post-harvest life of cut roses. Fanourakis et al. (2013, 2016) also found that cut flowers grown at high RH showed poorer keeping quality than plants cultivated at moderate RH, mainly due to uncontrolled water loss from the cut shoot.

6 Assessing the impact of CO_2 enrichment on the quality of greenhouse produce

CO_2 is a crucial component of photosynthesis and CO_2 concentration in the greenhouse can be reduced during the day due to plant consumption. Therefore, elevated CO_2 (eCO_2) is a measure commonly used in the greenhouse industry to increase the photosynthetic rate, growth and plant yield. As with high light intensities, elevated CO_2 may lead to short internodes, thick leaves, increased growth of tubers, beets and roots and delays in the ageing of leaves. For instance, Mortensen (1987) and Mortensen and Moe (1995) reported that horticultural greenhouse plants demonstrated positive effects due to eCO_2. Frantz and Ling (2011) also observed a positive effect of eCO_2 on leaf and flower mass after five weeks on the growth of *Petunia* × *hybrida*.

Rapid plant growth and yield increase are generally accompanied by changes in quality of produce. According to Loladze (2014) these conditions promote a dilution in mineral, protein content and other nutritional quality plant parameters. Giri et al. (2016) reported that CO_2 of 700 ppm decreased the concentration of many important nutrients including nitrogen (protein), potassium and phosphorus in the edible parts of both lettuce and spinach. The total nitrogen concentration in lettuce shoots was reduced by more than 30% at eCO_2 compared to the plants grown at ambient levels of CO_2 (400 ppm). Similarly, the concentration of several micronutrients including S, Z, Cu and Mg was depressed in lettuce shoots. However, other authors have not always detected changes or reductions, or they have actually detected increases in minerals (Baslam et al., 2012). On the other hand, the responses detected are species- and cultivar-specific (Pérez-López et al., 2015). Baslam et al. (2012), for instance, detected that lettuce green-leaf cultivar accumulated higher levels of the most part of macro- and micronutrients when cultivated under eCO_2, while in the red leaf cultivar only the Fe concentration was increased by eCO_2. According to Pérez-López et al. (2015), the concentration of B and Mn increased in green-pigmented lettuce cultivar, whereas the concentration of K, Cu and Mn increased in red-pigmented lettuce cultivar by eCO_2 in comparison to the control treatment. On the other hand, both cultivars decreased the concentration of Mg and Fe, whereas the concentrations of the other minerals remained constant. In a recent review, Dong et al. (2018) reported the greatest decrease in Fe concentration in leafy vegetables, followed by fruit and root vegetables, whereas the decrease in Zn concentration was similar in fruit and root vegetables followed by stem vegetables.

Generally, eCO_2 promotes soluble sugar accumulation in edible parts of vegetables. This effect was stronger in leafy vegetables than in fruit and root vegetables (Dong et al., 2018). It is likely that the synthesized carbohydrates in leaves cannot be fully translocated to fruits or roots (Gruda, 2005).

Increased soluble sugar as precursors can increase the synthesis and accumulation of antioxidants (Becker and Kläring, 2016; Dong et al., 2018) and consequently enhance the nutraceutical value of vegetables (Pérez-López et al., 2015). The contents of health-promoting compounds, including lycopene, β-carotene and ascorbic acid, as well as the flavour, indicated by sugars, titrable acidity and sugar/acid ratio of tomatoes, were markedly increased in eCO_2-enriched tomato fruits (Zhang et al., 2014). According to the authors, CO_2 enrichment significantly enhanced other organoleptic characteristics, including firmness, aroma and sensory attributes in tomato fruits. Becker and Kläring (2016) showed that sugars, flavonoids and caffeic acid derivatives increased in red leaf lettuce grown at 1000 ppm CO_2. In broccoli, eCO_2 concentrations increased the total glucosinolate content; however the indole glucosinolates declined (Schreiner et al., 2013).

Technically, when greenhouse windows are open, the CO_2 concentration is dramatically reduced in ventilated greenhouses with CO_2 enrichment. This makes the CO_2 enrichment inefficient and expensive. The development of closed- and semi-closed greenhouses has helped to increase the efficacy of eCO_2. In such greenhouses, window ventilation is usually reduced or replaced by an active cooling system. In addition, energy saving measures have been implemented where excess solar energy is collected and stored, in order to be reused at night or in periods in which the solar radiation is limited, for example on cloudy days or in winter. Under these conditions, it is possible and preferable to maintain high CO_2 concentrations, even at high light levels (Gruda and Tanny, 2014).

On the other hand, CO_2 is expected to accumulate in the atmosphere and will increase in the future due to climate change. The increased atmospheric concentrations of CO_2 may be advantageous for crop production; however, the quality of produce does not always correlate with increased yields (Gruda et al., 2019). Dong et al. (2018) conducted a meta-analysis for vegetables, cultivated in outdoor and indoor conditions, and found that elevated CO_2 increased the concentration of fructose, glucose, total soluble sugar, total antioxidant capacity, total phenols, total flavonoids, ascorbic acid and calcium in the edible part of vegetables, but decreased the concentrations of protein, nitrate, magnesium, iron and zinc. However, even with the results from the meta-analysis, it is difficult to generalize because the effect of elevated CO_2 is not only dependent on CO_2 concentrations, but also on the interaction with other environmental factors.

The negative effects of low light conditions (Fierro et al., 1994), low temperatures (Frantz and Ling 2011), high salinity levels in irrigation water available in Mediterranean countries (Romero-Aranda et al., 2002) and/or high electric conductivity (EC) levels of nutrient solutions (Li et al., 1999) can be diminished by elevated CO_2 (Gruda and Tanny, 2014). For instance, according to Sánchez-González et al. (2015), CO_2 enrichment increased the production of a high value commercial crop grown under saline conditions by reducing the time needed for complete fruit development of tomatoes without compromising organoleptic quality. Dong et al. (2018) analysed within the framework of climate change the interaction of plant cultivar, CO_2 level, growth stage, light, O_3, stress, nutrient and salinity with elevated CO_2 and recommended different ways to optimise elevated CO_2 benefits in combination with other factors, in order to enhance vegetable nutritional quality. This includes (1) plant breeding; (2) optimisation of other environmental factors and nutrient availability; (3) improvement of elevated CO_2 efficacy and reduced vegetative growth; and (4) combining elevated CO_2 with mild environmental stress (e.g. ultraviolet B radiation or salinity). In addition, temperature, atmospheric humidity and water supply are very important. All these measures will enhance vegetable quality, which might counteract dilution effect or direct metabolic pathways towards synthesis of health-beneficial compounds (Dong et al., 2018). Finally, as shown before, the quality of produce is species- and cultivar-dependent and elevated CO_2 is only applied if the market value of the produce can compensate for the higher production costs.

7 Assessing the impact of other factors on greenhouse product quality

7.1 Impact of the root environment on the quality of greenhouse produce

Here we address the impact of cultural measures and practices related to root environment, such as plant nutrition and water supply on vegetable quality.

According to Wang et al. (2008), fertilizers increase the concentration of essential amino acids and vitamins in vegetables. However, excessive fertilizer application, especially N fertilizer, can result in undesirable changes such as increases in nitrate, titratable acidity and acid:sugar ratio, as well as decreases in vitamin C, soluble sugar, soluble solids, Mg and Ca in some. Therefore, a balanced nutrient supply is recommended. Tomasi et al. (2015) reported that a balanced nutrient ratio might help to reduce the accumulation of undesired compounds, such as nitrate, and enhance nutritional value for human consumption. One easy way to reduce the accumulation of nitrate in vegetables is by short nitrate starvation, by replacing the nutrient solution with water some days before harvest (Santamaria et al., 2001). Gonnella et al. (2004)

reported that the replacement of the nutrient solution with rainwater three days before harvesting resulted in one-third nitrate reduction in lamb lettuce leaves. Nowadays, microgreens seem to be an interesting and innovative low-nitrate product, requiring low fertilizer levels. Bulgari et al. (2016) reported that the nitrate content of basil, Swiss chard and rocket microgreens grown in soilless culture systems was lower than usually measured in baby leaf or adult vegetables of the same species.

Apart from a faster crop growth, year-round production and an improvement in water use efficiency (WUE) and nutrient use efficiency, soilless culture is an appropriate tool for targeted management of crop mineral nutrients and enhancement of product quality (Gruda, 2009). Numerous studies confirm that this method enables growers to produce crops without quality loss compared to soil cultivation and with even better quality when all other plant biological and product technological requirements are met. Quantity and time of absorbed mineral nutrients as well as the way they are utilized in plant metabolism are better managed in hydroponics (Gruda, 2009). Through fertigation management it is possible to modulate the plant's secondary metabolism and increase the healthy compounds, for example in cultivated vegetables. Di Gioia et al. (2018) found that three species of rocket grown in soilless culture systems synthesized 36%, 51% and 41% more glucosinolates than plants grown in a conventional soil system. More information on water and nutrient management is given elsewhere in this book.

Regarding different soilless culture systems, Bonasia et al. (2017) reported that wild rocket genotypes cultivated in a floating system improved leaf firmness, visual quality, total antioxidant capacity (+11.6%) and reduced nitrate leaf accumulation (−37%) in comparison to an ebb-and-flow system. In addition, the authors found a high chloride leaf accumulation in the ebb-and-flow system. Soilless culture can also be effective in the biofortification of edible plant portions with essential and/or beneficial micronutrients to human health (Tomasi et al., 2015; Rouphael et al., 2018).

Changes caused by Na and Cl in some ion activities in the root may cause changes in mineral, carbohydrate and protein concentrations (Martínez-Ballesta et al., 2010), finally inducing changes in plant growth and quality. Therefore, moderate salinity is widely used to improve product quality and enhance health-promoting substances of different vegetables (Gruda 2009; Gruda et al. 2018). The favourable effects of salinity on the quality of tomatoes are mainly increasing dry matter content, and higher sugar and titratable acid concentrations in tomato fruit juice. Krauss et al. (2006) reported that rising EC values of the nutrient solution from 3 to 10 dS m^{-1} increased vitamin C, lycopene and ß-carotene (the precursor to vitamin A) in fresh tomato fruits by up to 35%. In addition, the phenols' antioxidative capacity was increased on a fresh weight basis. Wu et al. (2004) reported a higher increase of lycopene in comparison to TSS for five cultivars. Ehret et al. (2013) also reported an increase in antioxidants, such as oxygen radical absorbance capacity, β-carotene, lycopene and vitamin C concentrations in greenhouse tomatoes treated continuously with nutrient solution electrical conductivities of 2, 4 or 6 dS m^{-1}. However, in other studies, such as from Krumbein et al. (2006), there were no differences in lycopene content. The reason for different results could be the time of analysis. According to Wu and Kubota (2008), lycopene analysis should be done throughout the fruit ripening process rather than at the last stage of ripeness.

Pérez-López et al. (2015) reported that salt stress might increase the antioxidant capacity of lettuce. Under combined conditions of salt stress and elevated CO_2, total antioxidant capacity increased by 61% in green-pigmented lettuce cultivar and by 109% in red-pigmented lettuce cultivar compared to salt stress and ambient CO_2 (Pérez-López et al., 2015).

Moreover, the presence of plant biostimulants, such as humic substances, protein hydrolysates, seaweed extracts and beneficial microbials, such as mycorrhizal fungi and N_2 fixation bacteria, can improve the product quality, for example by inducing the accumulation of secondary metabolites, vitamins, antioxidants and minerals (Baum et al., 2015; Rouphael et al., 2018).

More specific details concerning the impact of individual nutrients on the quality characteristics of greenhouse products can be found in the work of Schnitzler and Gruda (2002), Passam et al. (2007) and Gruda et al. (2018).

Producing high-quality horticultural products requires an efficient water supply and irrigation management (Gruda and Tanny, 2014). Moderate water stress conditions can significantly enhance the dry matter, TSS and sugar content in tomatoes. Furthermore, similar to the salt stress effect, water shortage might increase the content of nutraceutical or health-promoting substances. For instance, the ascorbic acid content of many crops can be increased with less frequent irrigation (Lee and Kader, 2000). Dumas et al. (2003) reported that, depending on cultivars, water shortage generally tends to increase the ascorbic acid content of tomato fruits. Deficit irrigation strategies, such as regulated deficit irrigation (RDI) or partial root drying (PRD) are potential ways to improve WUE by allowing crops to withstand mild water stress with no or only marginal decreases of yield and quality, although some vegetables tend to not cope so well with any stress situation (Costa et al., 2007). Mitchell et al. (1991) found that deficit irrigation reduced water accumulation and fruit yield of tomatoes, but increased TSS in fruit. Campos et al. (2009) reported that yield, number of fruits and TSS content of tomatoes were similar among the control and PRD treatments. However, an increase of 25% in the fruit titratable acidity was reached in one PRD treatment. In addition, PRD treatments allowed a water irrigation saving of up to 46% and fruit firmness increased by up to 31% in PRD treatments.

7.2 Assessing the impact of growing techniques and cultural practices on the quality of greenhouse produce

Cultural practices and growing techniques are important to maintaining the balance between plant growth, plant development, expected marketable yield and produce quality (Gruda et al., 2018). For instance, an increase in plant density correlates negatively with the proportion of the marketable yield (Papadopoulos and Pararajasingham, 1997). The size reduction of the individual fruits and an insufficient supply of photo-assimilates mainly caused by a competition for light interception influence the photosynthetic rate and carbohydrate distribution per plant (Gruda and Tanny, 2014). Therefore, Heuvelink and González-Real (2008) recommend that plant densities not only relate to cultivars used, but also to cultivation seasons.

Training and pruning are crucial to optimize light distribution and directly contribute to the size, colour and quality of horticultural produce, as well as the nutritional value of greenhouse vegetables. Pruning is used to support training, with the aim of improving light relationships, equilibrating plant growth and development, providing better control of diseases, minimizing yield losses and improving the quality of produce (Gruda and Tanny, 2014). Incorrect pruning, such as a higher number of cucumber fruits, will lead to a pushing-off of young fruits, even with well-supplied nutrients. According to Prudent et al. (2009), the effect of pruning on other quality traits of the tomato fruit depends on many factors including the sink developmental stage, fruit to leaf ratio, fruit position and genetic background.

The positive effects of vigorous interspecific rootstocks on scion performance are often reflected in (1) fruit size, particularly in crops such as watermelon, cucumber and tomato; (2) fruit texture, for example in watermelon grafted on interspecific cucurbit rootstocks; and (3) sensorial attributes, such as fruit sweetness, elicited by soluble carbohydrates (Kyriacou et al., 2017). We predict that this method will continue to contribute to further improving fruit sensorial and nutritional as well as post-harvest quality of vegetables in the future.

7.3 Assessing the impact of maturity and ripeness stage at harvest on the quality of greenhouse produce

Kader (2008) classified maturity stage at harvest as the second most important factor, after genotype, influencing flavour quality. Most vegetables are non-climacteric and not associated with increases in ethylene production in the ripening process. Therefore, the highest quality level of, for example, leafy vegetables is achieved by harvest, afterwards the quality will decline. In contrast, some fruit vegetables are climacteric and can be harvested before the ripening stage. This makes it possible to assure a sufficient firmness, to withstand transport and handling procedures and to maximize their storage potential (Kader, 2008). However, even fruit vegetables taste best when harvested fully ripe and as stated by Schnitzler and Gruda (2002), the physiological status is very important to analyse product quality parameters. Therefore, in order to maximize nutritional value, it is recommended to harvest fruit vegetables near to fully ripe stage. Hence, tomatoes picked at the vine-ripe red stage are sweeter, less sour and have more aroma than tomatoes picked at earlier stages (Gruda et al., 2018). On the other hand, β-carotene, lycopene, vitamin C and phenolic compounds increase during the ripening stages of tomatoes and sweet pepper, for example (Dumas et al., 2003; Pérez-López et al., 2007). Freshness, short transport distances and a full aroma and flavour are the reasons why demand for local products has permanently increased and why farm gate sales and weekly markets have become so popular in some European cities.

8 Conclusion

Product quality is a complex and multidimensional issue that relates to different factors. The quality of produce follows the same patterns as yield – low and high intensities negatively influence the quality and for each parameter an optimum level of the implied factor is required. However, keeping in mind the complexity of horticultural plants as biological organisms, the optimums are different for quality parameters and for yield. On the other hand, the optimum does not depend on just one factor but the interaction of all factors together. Therefore, compromises have to be made in order to realize high-quality, year-round, environmentally-friendly high yields in greenhouse production.

Improvements in quality will only be achieved by respecting both intrinsic and extrinsic factors. In order to do so, improvements in plant breeding, greenhouse engineering, cultivation systems and post-harvest management are required, in addition to improved legislation and marketing standards. On the other hand, high-quality greenhouse products have to meet the expectations of producers, traders and consumers, balanced against environmental and sustainability goals and the demand of future generations. With regard to potential climate change, adaptive strategies have to be applied for sustainable production and high product quality.

9 Future trends

The examples mentioned above clearly highlight the strong influence of cultivar choice and crop management strategies on environmental conditions and plant growth that in turn can improve the visual, organoleptic and nutritional quality of greenhouse produce. Future researchers should focus on different approaches that include the use of high-yield and high-quality greenhouse genotypes, well-designed environments and sustainable management practices (Fig. 4).

Tieman et al. (2017) studied whole-genome sequencing and genome association and identified the genetic loci that affect most of the target flavour chemicals, including sugars, acids and volatiles of tomatoes. These results provide the necessary information and a genetic roadmap to improved flavour. Breeding new cultivars of vegetables with better flavour, high healthy compounds and nutritional quality through new, modern methods will be a long-term task to improve the biosynthesis and the level of different nutritional benefit elements and compounds as well as to reduce the negative and undesirable ones.

In addition, well-designed environments and well-performed sustainable management practices should be further developed in the short-term. Future research should be focused on the interaction of genotypes, greenhouse engineering and cultivation systems in order to select the best combinations, not only from an economical point of view, for example energy saving and high-yield production methods, but also from the point of view of produce quality (Fig. 4). Here belongs the development of new quality-oriented cultivars, modern close and semi-close greenhouses, use of LED lighting, intelligent use

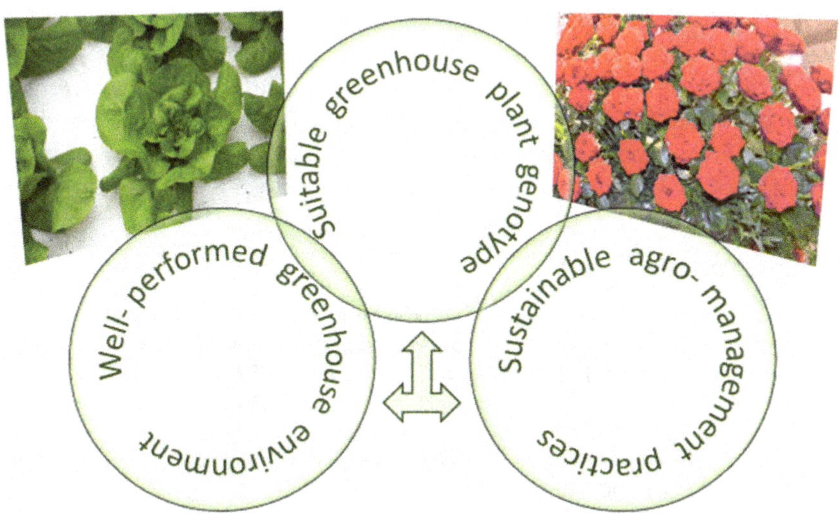

Figure 4 The quality of greenhouse produce depends on plant genotypes, the performance of environmental conditions and applied agro-management practices. High quality of horticultural greenhouse plants can only be achieved when all factors and the interactions between them are well performed. Sometimes compromises have to be reached, related to yield as well as specific quality parameters. Photos: Gruda, private collection.

of elevated CO_2 and air and root temperatures as well as a precise dosage of water, fertigation and plant biostimulants.

10 Where to look for further information

In this chapter, we have described the impact of different environmental factors on the quality of greenhouse produce. In order to be comprehensive, we presented the impact of cultivar choice and some agrotechnical measures and cultural practices. This is a very large area to cover, and therefore, we have provided some additional sources for further consideration.

Standard introductions to the subject are contributions by R.L. Shewfelt, 'What is quality?' and Schreiner et al., 'Current understanding and use of quality characteristics of horticulture products' (see Shewfelt (1999) and Schreiner et al. (2013) in the References for full details). Other publications, such as Kader (2008) add to these recommendations with respect to favours and quality of fruits and vegetables.

Gruda (2005), 'Impact of environmental factors on product quality of greenhouse vegetables for fresh consumption' and Gruda and Tanny (2014), 'Protected Crops' further present how to improve the quality of greenhouse products and Dorais et al. (2001) is an example for tomatoes, the most consumed vegetable worldwide (see these contributions in the References for full details).

Concerning the nutritional value of vegetables (not only from greenhouses), individual countries issue their own guidance, based usually on societies for human nutrition. As an example, more information can be found in several opinions of the German Nutrition Society (https://www.dge.de/en/). A prominent health initiative in many countries is '5 A Day', which recommends the consumption of at least five portions of fruit and vegetables every day. Protected cultivation plays an important role in the production of year-round fresh produce.

11 References

Bantis, F., Smirnakou, S., Ouzounis, T., Koukounaras, A., Ntagkas, N. and Radoglou, K. 2018. Current status and recent achievements in the field of horticulture with the use of light-emitting diodes (LEDs). *Sci. Hort.*, 235, 437–51.

Baslam, M., Garmendia, I. and Goicoechea, N. 2012. Elevated CO_2 may impair the beneficial effect of arbuscular mycorrhizal fungi on the mineral and phytochemical quality of lettuce. *Ann. Appl. Biol.*, 161, 180–91. doi:10.1111/j1744-7348.2012.00563.x.

Baum, C., El-Tohamy, W. and Gruda, N. 2015. Increasing the productivity and product quality of vegetable crops using arbuscular mycorrhizal fungi: A review. *Sci. Hortic.*, 187, 131–41. doi:10.1016/j.scienta.2015.03.002.

Becker, C. and Kläring, H. P. 2016. CO_2 enrichment can produce high red leaf lettuce yield while increasing most flavonoid glycoside and some caffeic acid derivative concentrations. *Food Chem.*, 199, 736–45.

Benoit, F. and Ceustermans, N. 2001. Impact of root cooling on blossom end rot in soilless paprika. *Acta Hortic.*, 548, 319-25.

Bertin, N. and Génard, M. 2018. Tomato quality as influenced by preharvest factors. *Sci. Hortic.*, 233, 264-76. doi:10.1016/j.scienta.2018.01.056.

Bertin, N., Guichard, S., Leonardi, C., Longuenesse, J. J, Langlois, D. and Navez, B. 2000. Seasonal evolution of the quality of fresh glasshouse tomatoes under Mediterranean conditions, as affected by air vapour pressure deficit and plant fruit load. *Ann. Bot.*, 85, 741-50.

Bian, Z. H., Yanga, Q. C. and Liu, W. K. 2015. Effects of light quality on the accumulation of phytochemicals in vegetables produced in controlled environments: A review. *J. Sci. Food Agric.*, 95(5), 869-77. doi:10.1002/jsfa.6789.

Blanchard, M. G., Runkle, E. S. and Fisher, P. R. 2011. Modeling plant morphology and development of petunia in response to temperature and photosynthetic daily light integral. *Sci. Hortic.*, 129, 313-20.

Bliznikas, Z., Žukauskas, A., Samuolienė, G., Viršilė, A., Brazaitytė, A., Jankauskienė, J., Duchovskis, P. and Novičkovas, A. 2012. Effect of supplementary pre-harvest LED lighting on the antioxidant and nutritional properties of green vegetables. *Acta Hortic.*, 939, 85-91. doi:10.17660/ActaHortic.2012.939.10.

Bonasia, A., Lazzizera, C., Elia, A. and Conversa, G. 2017. Nutritional, biophysical and physiological characteristics of wild rocket genotypes as affected by soilless cultivation system, salinity level of nutrient solution and growing period. *Front. Plant Sci.* doi:10.3389/fpls.2017.00300.

Bot, G. P. A. 2003. The solar greenhouse: technology for low energy consumption. *Acta Hortic.*, 611, 61-9.

Brückner, B., Krumbein, A., Schwarz, D. and Kläring, P. 2004. Temperature effects on tomato quality. *IX International Symposium on Soilless Culture and Hydroponics*, Almeria, Spain, 14-19 November 2004. Book of summary, 173.

Bulgari, R., Baldi, A., Ferrante, A. and Lenzi, A. 2016. Yield and quality of basil, Swiss chard and rocket microgreens grown in a hydroponic system. *N. Z. J. Crop Hortic. Sci.*, 45(2), 119-29. doi:10.1080/01140671.2016.1259642.

Campos, H., Trejo, C., Pena-Valdivia, C. B., Ramírez-Ayala, C. and Sánchez-García, P. 2009. Effect of partial rootzone drying on growth, gas exchange and yield of tomato (*Solanum lycopersicum* L.). *Sci. Hortic.*, 20(4), 493-9. doi:10.1016/j.scienta.2008.12.014.

Castilla, N. and Hernandez, J. 2007. Greenhouse technological packages for high-quality crop production. *Acta Hortic.*, 761, 285-97.

Costa, J. M., Ortuño, M. F. and Chaves, M. M. 2007. Deficit irrigation as a strategy to save water: Physiology and potential application to horticulture. *J. Integr. Plant Biol.*, 49(10), 1421-34. doi:10.1111/j.1672-9072.2007.00556.x.

Dannehl, D., Huber, C., Rocksch, T., Huyskens-Keil, S. and Schmidt, U. 2012. Interactions between changing climate conditions in a semi-closed greenhouse and plant development, fruit yield, and health-promoting plant compounds of tomatoes. *Sci. Hortic.*, 138, 235-43. doi:10.1016/j.scienta.2012.02.022.

Demers, D. A., Dorais, M. and Papadopoulo, A. P. 2007. Yield and russeting of greenhouse tomato as influenced by leaf-to-fruit ratio and relative humidity. *HortSci.*, 42(3), 503-7.

Di Gioia, F., Avato, P., Serio, F. and Argentieri, M. P. 2018. Glucosinolate profile of *Eruca sativa*, *Diplotaxis tenuifolia* and *Diplotaxis erucoides* grown in soil and soilless systems. *J. Food Compost. Anal.*, 69, 197-204. doi:10.1016/j.jfca.2018.01.022.

Dong, J., Gruda, N., Li, X. and Duan, Z., 2018. Effects of elevated CO_2 on nutritional quality of vegetables - A review. *Front. Plant Sci.* doi:10.3389/fpls.2018.00924.

Dorais, M. and Gosselin, A. 2002. Physiological response of greenhouse vegetable crops to supplemental lighting. *Acta Hortic.*, 580, 59-67.

Dorais, M., Papadopoulos, A. and Gosselin, A. 2001. Greenhouse tomato fruit quality. *Hortic. Rev.*, 26, 239-319.

Dumas, Y., Dadomo, M., Di Lucca, G. and Grolier, P. 2003. Effects of environmental factors and agricultural techniques on antioxidant content of tomatoes. *J. Sci. Food Agric.*, 83, 369-82.

Ehret, D. L., Usher, K., Helmer, T., Block, G., Steinke, D., Frey, B., Kuang, T. and Diarra, M. 2013. Tomato fruit antioxidants in relation to salinity and greenhouse climate. *J. Agric. Food Chem.*, 61(5), 1138-45. doi:10.1021/jf304660d.

Elings, A., Kempkes, F. L. K., Kaarsemaker, R. C., Ruijs, M. N. A., van de Braak, N. J. and Dueck, T. A. 2005. The energy balance and energy-saving measures in greenhouse tomato cultivation. *Acta Hortic.*, 691, 67-74.

Fanourakis, D., Pieruschka, R., Savvides, A., Macnish, A. J., Sarlikioti, V. and Woltering, E. J. 2013. Sources of vase life variation in cut roses: A review. *Postharvest Biol. Technol.*, 78, 1-15. doi:10.1016/j.postharvbio.2012.12.001.

Fanourakis, D., Bouranis, D., Giday, H., Carvalho, D. R. A., Nejad, A. R. and Ottosen, C.-O. 2016. Improving stomatal functioning at elevated growth air humidity: A review. *J. Plant Physiol.*, 207, 51-60. doi:10.1016/j.jplph.2016.10.003.

Fierro, A., Tremblay, N. and Gosselin, A. 1994. Supplemental carbon dioxide and light improved tomato and pepper seedling growth and yield. *HortSci.*, 29(3), 152-4.

Frantz, J. M. and Ling, P. 2011. Growth, partitioning, and nutrient and carbohydrate concentration of *Petunia* × *hybrida* Vilm. are influenced by altering light, CO_2, and fertility. *HortSci.*, 46(2), 228-35.

Gaudreau, L., Charbonneau, J., Vézina, L. P. and Gosselin, A. 1994. Photoperiod and photosynthetic photon flux influence growth and quality of greenhouse grown lettuce. *HortSci.*, 29, 1285-9.

Gautier, H., Diakou-Verdin, V., Bénard, C., Reich, M., Buret, M., Bourgaud, F., Poëssel, J. L., Caris-Veyrat, C. and Génard, M. 2008. How does tomato quality (sugar, acid, and nutritional quality) vary with ripening stage, temperature, and irradiance? *J. Agric. Food Chem.*, 56, 1241-50.

Gautier, H., Massot, C., Stevens, R., Sérino, S. and Génard, M. 2009. Regulation of tomato fruit ascorbate content is more highly dependent on fruit irradiance than leaf irradiance. *Ann. Bot.*, 103, 495-504.

Ghasemzadeh, A., Ashkani, S., Baghdadi, A., Pazoki, A., Jaafar, H. Z. E. and Rahmat, A. 2016. Improvement in flavonoids and phenolic acids production and pharmaceutical quality of sweet basil (*Ocimum basilicum* L.) by ultraviolet-B irradiation. *Molecules*, 21(9), 1203. doi:10.3390/molecules21091203.

Giri, A., Armstrong, B. and Rajashekar, C. B. 2016. Elevated carbon dioxide level suppresses nutritional quality of lettuce and spinach. *Am. J. Plant Sci.*, 7(1), 246-58. doi:10.4236/ajps.2016.71024.

Gonnella, M., Serio, F., Conversa, G. and Santamaria, P. 2004. Production and nitrate content in lamb's lettuce grown in floating system. *Acta Hortic.*, 644, 61-8. doi:10.17660/ActaHortic.2004.644.5.

Gruda, N. 2005. Impact of environmental factors on product quality of green house vegetables for fresh consumption. *Crit. Rev. Plant Sci.*, 24, 227-47. doi:10.1080/07352680591008628

Gruda, N. 2009. Do soilless culture systems have an influence on product quality of vegetables? *J. Appl. Bot. Food Qual.*, 82, 141-7.

Gruda, N. and Tanny, J. 2014. Protected crops. In: Dixon, G. R. and Aldous, D. E. (Eds), *Horticulture - Plants for People and Places, Volume 1: Production Horticulture*. Springer Science+Business Media, Dordrecht, The Netherlands, pp. 327–405. doi:10.1007/978-94-017-8578-5_10.

Gruda, N., Savvas, D., Colla, G. and Rouphael, Y. 2018. Impacts of genetic material and current technologies on product quality of selected greenhouse vegetables - A review. *Eur. J. Hortic. Sci.*, 83, 319–28. doi:10.17660/eJHS.2018/83.5.5.

Gruda, N., Bisbis, M. B. and Tanny, J. 2019. Influence of climate change on protected cultivation: Impacts and sustainable adaptation strategies. *J. Clean. Prod.* 225, 481-95. doi: 10.1016/j.jclepro.2019.03.210.

Grunert, K., Bredahl, L. and Brunsø, K. 2004. Consumer perception of meat quality and implications for product development in the meat sector - a review. *Meat Sci.*, 66, 259–72. doi:10.1016/S0309-1740(03)00130-X.

Hanna, H. Y. 2009. Influence of cultivar, growing media, and cluster pruning on greenhouse tomato yield and fruit quality. *HortTechnology*, 19, 395–9.

Hao, X., Zhang, Y., Guo, X., Little, C., Zheng, J. M. and Khosla, S. 2017a. Temperature drop improved responses of greenhouse fruit vegetables to long photoperiod of supplemental lighting. *Acta Hortic.*, 1182, 185–92. doi:10.17660/ActaHortic.2017.1182.22.

Hao, X., Guo, X., Zheng, J., Celeste, L., Kholsa, S. and Chen, X. 2017b. Response of greenhouse tomato to different vertical spectra of LED lighting under overhead high-pressure sodium and plasma lighting. *Acta Hortic.*, 1170, 1003–110. doi:10.17660/ActaHortic.2017.1170.129.

He, H. 1999. Studies growth adaptation and identification of glucosinolates on Chinese brassicas. Thesis, Herbert Utz Verlag, Munich-Germany.

Heuvelink, E. and González-Real, M. M. 2008. Innovation in plant-greenhouse interactions and crop management. *Acta Hortic.*, 801, 63–74. doi:10.17660/actahortic.2008.801.2.

Huyskens-Keil, S. and Schreiner, M. 2003. Quality of fruits and vegetables. *J. Appl. Bot.*, 77, 147-51.

Islam, M. S. and Khan, S. 2001. Seasonal fluctuations of carbohydrate accumulation and metabolism of three tomato (*Lycopersicon esculentum* Mill.) cultivars grown at seven sowing times. *J. Hortic. Sci. Biotech.*, 76(6), 764–70.

Johansen, T. J., Hagen, S. F., Bengtsson, G. B. and Mølmann, J. A. B. 2016. Growth temperature affects sensory quality and contents of glucosinolates, vitamin C and sugars in swede roots (*Brassica napus* L. ssp. *rapifera* Metzg.). *Food Chem.*, 196, 228–35.

Johkan, M., Shoji, K., Goto, F., Hahida, S. and Yoshihara, T. 2012. Effect of green light wavelength and intensity on photomorphogenesis and photosynthesis in *Lactuca sativa*. *Environ. Exp. Bot.*, 75, 128–33.

Kader, A. A. 2008. Flavor quality of fruits and vegetables. *J. Sci. Food Agric.*, 88(11), 1863-8. doi.org/10.1002/jsfa.3293.

Kafkafi, U. 2001. Root zone parameters controlling plant growth in soilless culture. *Acta Hortic.*, 554, 27–38.

Kano, Y. 2004. Effects of summer day-time temperature on sugar content in several portions of watermelon fruit (*Citrullus lanatus*). *J. Hortic. Sci. Biotech.*, 79(1), 142–5.

Kano, Y. and Goto, H. 2003. Relationship between the occurrence of bitter fruit in cucumber (*Cucumis sativus* L.) and the contents of total nitrogen, amino acid nitrogen, protein and HMG-CoA reductase activity. *Sci. Hortic.*, 98(1), 1–8.

Kittas, C., Tchamitchian, M., Katsoulas, N., Karaiskou, P. and Papaioannou, C. 2006. Effect of two UV-absorbing greenhouse-covering films on growth and yield of an eggplant soilless crop. *Sci. Hortic.*, 110, 30–7.

Kläring, H-P., Klopotek, Y., Krumbein, A. and Schwarz, D. 2015. The effect of reducing the heating set point on the photosynthesis, growth, yield and fruit quality in greenhouse tomato production. *Agric. Forest Meteorol.*, 214–215, 178–88. doi:10.1016/j.agrformet.2015.08.250.

Krauss, S., Schnitzler, W. H, Grassmann, J. and Woitke, M. 2006. The influence of different electrical conductivity values in a simplified recirculating soilless system on inner and outer fruit quality characteristics of tomato. *J. Agric. Food Chem.*, 54, 441–8.

Krumbein, A., Schwarz, D. and Kläring, H. P. 2006. Effects of environmental factors on carotenoid content in tomato (*Lycopersicon esculentum* (L.) Mill.) grown in a greenhouse. *J. Appl. Bot. Food Qual.*, 80, 160–4.

Ksas, B., Becuwe, N., Chevalier, A. and Havaux, M. 2015. Plant tolerance to excess light energy and photooxidative damage relies on plastoquinone biosynthesis. *Sci. Rep.*, 5, 10919. doi:10.1038/srep10919.

Kyriacou, MC., Rouphael, Y., Colla, G., Zrenner, R. and Schwarz, D. 2017. Vegetable grafting: The implications of a growing agronomic imperative for vegetable fruit quality and nutritive value. *Front. Plant Sci.* doi:10.3389/fpls.2017.00741.

Lamnatou, C. and Chemisana, D. 2013. Solar radiation manipulations and their role in greenhouse claddings: Fresnel lenses, NIR- and UV-blocking materials. *Renew. Sustain. Energy Rev.*, 18, 271–87.

Lee, S. K. and Kader, A. A. 2000. Preharvest and postharvest factors influencing vitamin C content of horticultural crops. *Postharvest Biol. Technol.*, 20(3), 207–20. doi:10.1016/S0925-5214(00)00133-2.

Lefsrud, M. G., Kopsell, D. A. and Sams, C. E. 2008. Irradiance from distinct wavelength light-emitting diodes affect secondary metabolites in kale. *HortSci.*, 43, 2243–4.

Li, Q. and Kubota, C. 2009. Effects of supplemental light quality on growth and phytochemicals of baby leaf lettuce. *Environ. Exp. Bot.*, 67, 59–64.

Li, J. H, Safi, M., Gale, J., Volokita, M. and Novoplansky, A. 1999. Response of tomato plants to saline water as affected by carbon dioxide supplementation. I. Growth, yield and fruit quality. *J. Hortic. Sci. Biotechnol.*, 74, 132–237.

Li, G., Zhang, W., Benoit, F. and Ceustermans, N. 2002. Effects of environment factors on the growth and incidence of blossom-end rot in soilless *Capsicum fructescens* var. *grossum*. *Acta Hortic.*, 633, 382–9.

Li, H., Tang, C., Xu, Z., Liu, X. and Han, X. 2012. Effects of different light sources on the growth of non-heading Chinese cabbage (*Brassica campestris* L.). *J. Agric. Sci.*, 4, 262–73.

Liu, Y., Tikunov, Y., Schouten, R. E., Marcelis, L. F. M., Visser, R. G. F. and Bovy, A. 2018. Anthocyanin biosynthesis and degradation mechanisms in solanaceous vegetables: A Review. *Front. Chem.* doi:10.3389/fchem.2018.00052.

Loladze, I. 2014. Hidden shift of the ionome of plants exposed to elevated CO_2 depletes minerals at the base of human nutrition. *eLife*, 3, e02245. doi:org/10.7554/eLife.02245.

Martínez-Ballesta, M. C., Dominguez-Perles, R., Moreno, D. A., Muries, B., Alcaraz-López, C., Bastías, E., García-Viguera, C. and Carvajal, M. 2010. Minerals in plant food: Effect of agricultural practices and role in human health: A review. *Agron. Sustain. Dev.*, 30, 295–309. doi:10.1051/agro/2009022.

Masaru, S., Uenishi, M., Miyamoto, K. and Suzuki, T. 2016. Effect of root-zone temperature on the growth and fruit quality of hydroponically grown strawberry plants. *J. Agric. Sci.* 8(5). doi:10.5539/jas.v8n5p122.

Massot, C., Génard, M., Stevens, R. and Gautier, H. 2010. Fluctuations in sugar content are not determinant in explaining variations in vitamin C in tomato fruit. *Plant Physiol. Biochem.*, 48, 751–7. doi:10.1016/j.plaphy.2010.06.001.

Mazzucato, A., Willems, D., Bernini, R., Picarella, M. E., Santangelo, E., Ruiu, F., Tilesi, F. and Soressi, G. P. 2013. Novel phenotypes related to the breeding of purple-fruited tomatoes and effect of peel extracts on human cancer cell proliferation. *Plant Physiol. Biochem.*, 72, 125–33. doi:10.1016/j.plaphy.2013.05.012.

Mitchell, J. P., Shennan, C., Grattan, S. R. and May, D. M. 1991. Tomato fruit yields and quality under water deficit and salinity. *J. Am. Soc. Hortic. Sci.*, 116(2), 215–21.

Mizuno, T., Amaki, W. and Watanabe, H. 2011. Effects of monochromatic light irradiation by LED on the growth and anthocyanin contents in leaves of cabbage seedlings. *Acta Hortic.*, 907, 179–84.

Moccaldi, L. A. and Runkle, E. S. 2007. Modeling the effects of temperature and photosynthetic daily light integral on growth and flowering of *Salvia splendens* and *Tagetes patula*. *J. Am. Soc. Hortic. Sci.*, 132(3), 283–8.

Mortensen, L. V. 1987. CO_2 enrichment in greenhouses. Crop responses. *Sci. Hortic.*, 33(1–2), 1–25.

Mortensen, L. M. and Gislerød, H. R. 2005. Effect of air humidity variation on powdery mildew and keeping quality of cut roses. *Sci. Hortic.*, 104(1), 49–55.

Mortensen, L. M. and Moe, R. 1995. Effects of temperature, carbon dioxide concentration, daylength, and photon flux density on growth, morphogenesis, and flowering of miniature roses. *Acta Hortic.*, 378, 63–70.

Mosadegh, H., Trivellini, A., Ferrante, A., Lucchesini, M., Vernieri, P. and Mensuali, A. 2018. Applications of UV-B lighting to enhance phenolic accumulation of sweet basil. *Sci. Hortic.*, 229, 107–16. doi:10.1016/j.scienta.2017.10.043.

Ntagkas, N., Woltering, E. J. and Marcelis, L. F. M. 2018. Light regulates ascorbate in plants: An integrated view on physiology and biochemistry. *Environ. Exp. Bot.*, 147, 271–80. doi:10.1016/j.envexpbot.2017.10.009.

Ouzounis, T., Parjikolaei, B. R., Fretté, X. and Rosenqvist, E. 2015. Predawn and high intensity application of supplemental blue light decreases the quantum yield of PSII and enhances the amount of phenolic acids, flavonoids, and pigments in *Lactuca sativa*. *Front. Plant Sci.*, 6, 19. doi:10.3389/fpls.2015.00019.

Ouzounis, T., Heuvelink, E., Ji, Y., Schouten, H. J., Visser, R. G. F. and Marcelis, L. F. M. 2016. Blue and red LED lighting effects on plant biomass, stomatal conductance, and metabolite content in nine tomato genotypes. *Acta Hortic.*, 1134, 251–8. doi:10.17660/ActaHortic.2016.1134.34.

Papadopoulos, A. P. and Pararajasingham, S. 1997. The influence of plant spacing on light interception and use in greenhouse tomato (*Lycopersicon eculentum* Mill.): A review. *Sci. Hortic.*, 69, 1–29. doi:10.1016/s0304-4238(96)00983-1.

Pardossi, A., Giacomet, P., Malorgio, F., Albini, F. M., Murelli, C., Serra, G., Vernieri, P. and Tongnoni, F. 2000. The influence of growing season on fruit yield and quality of greenhouse melon (*Cucumis melo* L.) grown in nutrient film technique in a Mediterranean climate. *J. Hortic. Sci. Biotech.*, 75(4), 488–93.

Passam, H. C., Karapanos, I. C., Bebeli, P. J. and Savvas, D. 2007. A review of recent research on tomato nutrition, breeding and post-harvest technology with reference to fruit quality. *Eur. J. Plant Sci. Biotechnol.*, 1, 1–21.

Pérez-López, A. J., del Amor, F. M., Serrano-Martínez, A., Fortea, M. I. and Núñez-Delicado, E. 2007. Influence of agricultural practices on the quality of sweet pepper fruits as affected by the maturity stage. *J. Sci. Food Agric.*, 87, 2075–80. doi:org/10.1002/jsfa.2966.

Pérez-López, U., Miranda-Apodaca, J., Lacuesta, M., Mena-Petite, A. and Mûnoz-Rueda, A. 2015. Growth and nutritional quality improvement in two differently pigmented lettuce cultivars grown under elevated CO_2 and/or salinity. *Sci. Hortic.*, 195, 56–66.

Piovene, C., Orsini, F., Bosi, S., Sanoubar, R., Bregola, V., Dinelli, G. and Gianquinto, G. 2015. Optimal red: Blue ratio in led lighting for nutraceutical indoor horticulture. *Sci. Hortic.*, 193, 202–8. doi:10.1016/j.scienta.2015.07.015.

Prudent, M., Causse, M., Génard, M., Tripodi, P., Grandillo, S. and Bertin, N. 2009. Genetic and physiological analysis of tomato fruit weight and composition: Influence of carbon availability on QTL detection. *J. Exp. Bot.*, 60(3), 923–37.

Romero-Aranda, R., Soria, T. and Cuartero, J. 2002. Greenhouse mist improves yield of tomato plants grown under saline conditions. *J. Am. Soc. Hortic. Sci.*, 127(4), 644–8.

Rosales, M., Rubio-Wilhelmi, M. M., Castellano, R. and Romero, L., 2007. Sucrolytic activities in cherry tomato fruits in relation to temperature and solar radiation. *Sci. Hortic.*, 113, 244. doi:10.1016/j.scienta.2007.03.015.

Rosales, M. A., Cervilla, L. M., Sánchez-Rodríguez, E., Rubio-Wilhelmi, M. M., Blasco, B., Ríos, J. J., Soriano, T., Castilla, N., Romero, L. and Ruiz, J. M. 2011. The effect of environmental conditions on nutritional quality of cherry tomato fruits: Evaluation of two experimental Mediterranean greenhouses. *J. Sci. Food Agric.*, 91(1), 152–62. doi.org/10.1002/jsfa.4166.

Rouphael, Y., Kyriacou, M. C., Petropoulos, S. A., De Pascale, S. and Colla, G. 2018. Improving vegetable quality in controlled environments. *Sci. Hortic.*, 234, 275–89. doi:10.1016/j.scienta.2018.02.033.

Samuolienė, G., Sirtautas, R., Brazaitytė, A., Viršilė, A. and Duchovskis, P. 2012. Supplementary red-LED lighting and the changes in phytochemical content of two baby leaf lettuce varieties during three seasons. *J. Food Agric. Environ.*, 10, 701–6.

Sánchez-González, M. J., Sánchez-Guerrero, M. C., Medrano, E., Porras, M. E., Baeza, E. J. and Lorenzo, P. 2015. Influence of pre-harvest factors on quality of a winter cycle, high commercial value, tomato cultivar. *Sci. Hortic.*, 189, 104–11.doi:10.1016/j.scienta.2015.03.044.

Santamaria, P., Gonnella, M., Elia, A., Parente, A. and Serio, F. 2001. Ways of reducing rocket salad nitrate content. *Acta Hortic.*, 548, 529–36. doi:10.17660/ActaHortic.2001.548.64.

Schirrmacher, G., Schnitzler, W. H. and Graßmann, J. 2004. Determination of secondary plant metabolites and antioxidative capacity as new parameter for quality evaluation - indicated by the new Asia salad Gynura bicolour. *J. Appl. Bot.*, 78, 133–4.

Schnitzler, W. H. and Gruda, N. 2002. Hydroponics and product quality. In: Savvas, D., Passam, H. (Eds), *Hydroponic Production of Vegetables and Ornamentals*. Embrio publications, Athens, pp. 373–411.

Schreiner, M., Korn, M., Stenger, M., Holzgreve, L. and Altmann, M. 2013. Current understanding and use of quality characteristics of horticulture products. *Sci. Hortic.*, 163(5), 63–9. doi:10.1016/j.scienta.2013.09.027.

Shewfelt, R. L. 1999. What is quality? *Postharvest Biol. Technol.*, 15, 197–200.

Shinohara, Y. 1987. Growing conditions and quality of vegetables – Effect of light and fertlizer conditions on the ascorbic acid content of vegetables. *Mem. Inst. Agr. & For. Univ. Tsukuba – Agr. & For. Sci.*, 3, 61–156.

Shinohara, Y., Suzuki, Y. and Shibuya, M. 1982. Effects of cultivations method, growing season and cultivar on the ascorbic acid content of tomato fruits. *J. Japan Soc. Hortic. Sci.*, 51(3), 338–43.

Singh, D., Basu, C., Meinhardt-Wollweber, M. and Roth, B. 2015. LEDs for energy efficient greenhouse lighting. *Renew. Sustain. Energ. Rev.*, 49, 139–47. doi:10.1016/j.rser.2015.04.117.

Stutte, G. W., Edney, S. and Skerritt, T. 2009. Photoregulation of bioprotectant content of red leaf lettuce with light-emitting diodes. *HortSci.*, 44, 79–82.

Sublett, W. L., Barickman, T. C. and Sams, C. E. 2018. Effects of elevated temperature and potassium on biomass and quality of dark red 'Lollo Rosso' lettuce. *Horticulturae*, 4(2), 11. doi:10.3390/horticulturae4020011.

Taulavuori, K., Hyöky, V., Oksanen, J., Taulavuori, E. and Julkunen-Tiitto, R. 2016. Species-specific differences in synthesis of flavonoids and phenolic acids under increasing periods of enhanced blue light. *Environ. Exp. Bot.*, 121, 145–50. doi:10.1016/j.envexpbot.2015.04.002.

Tewolde, F. T., Lu, N., Shiina, K., Maruo, T., Takagaki, M., Kozai, T. and Yamori, W. 2016. Nighttime supplemental LED inter-lighting improves growth and yield of single-truss tomatoes by enhancing photosynthesis in both winter and summer. *Front. Plant Sci.*, 7, 448. doi:10.3389/fpls.2016.00448.

Thompson, H. C., Langhans, R. W., Both, A. J. and Albright, L. D. 1998. Shot and root temperature effects on lettuce growth in a floating hydroponic system. *J. Am. Soc. Hortic. Sci.*, 123, 361–4.

Tieman, D., Zhu, G., Resende Jr., M. F., Lin, T., Nguyen, C., Bies, D., Rambla, J. L., Beltran, K. S., Taylor, M., Zhang, B., Ikeda, H., Liu, Z., Fisher, J., Zemach, I., Monforte, A., Zamir, D., Granell, A., Kirst, M., Huang, S. and Klee, H. 2017. A chemical genetic roadmap to improved tomato flavor. *Science*, 355(6323), 391–4. doi:10.1126/science.aal1556.

Tomasi, N., Pinton, R., Dalla, L., Cortella, G., Terzano, R., Mimmo, T., Scampicchio, M. and Cesco, S. 2015. New 'solutions' for floating cultivation system of ready-to-eat salad: A review. *Trends Food Sci. Technol.*, 46(2), 267–76. doi:10.1016/j.tifs.2015.08.004.

Torre, S. and Fjeld, T. 2001. Water loss and postharvest characteristics of cut roses grown at high or moderate relative air humidity. *Sci. Hortic.*, 89(3), 217–26.

Torre, S., Fjeld, T., Gislerød, H. R and Moe, R. 2003. Leaf anatomy and stomatal morphology of greenhouse roses grown at moderate or high air humidity. *J. Am. Soc. Hortic. Sci.*, 128(4), 598–602.

Urrestarazu, M., Salas, C., Valera, D., Gómez, A., Mazuela, P. C., Urrestarazu, M., Salas, C., Valera, D., Gómez, A., Urrestarazu, M., Salas, C., Valera, D. and Mazuela, P. C. 2008. Effects of heating nutrient solution on water and mineral uptake and early yield of two cucurbits under soilless culture. *J. Plant Nutr.*, 31, 527–38.

Vanhassel, P., Bleyaert, P., Van Lommel, J., Vandevelde, I., Crappé, S., Van Hese, N., Hanssens, J., Steppe, K. and Van Labeke, M. C. 2015. Rise of nightly air humidity as a measure for tipburn prevention in hydroponic cultivation of butterhead lettuce. *Acta Hortic.*, 1107, 195–202. doi:10.17660/ActaHortic.2015.1107.26.

Wang, Z.-H., Li, S.-X. and Malhi, S. 2008. Effects of fertilization and other agronomic measures on nutritional quality of crops. *J. Sci. Food Agric.*, 88(1), 7–23. doi:10.1002/jsfa.3084.

Wonneberger, C. and Keller, F. 2004. *Gemüsebau*. Verlag: Ulmer, E; Auflage: 1., Aufl., ISBN: 9783800139859, 384 Seiten.

Wu, M. and Kubota, C. 2008. Effects of high electrical conductivity of nutrient solution and its application timing on lycopene, chlorophyll and sugar concentrations of hydroponic tomatoes during ripening. *Sci. Hortic.*, 116, 122–9.

Wu, M., Buck, J. S. and Kubota, C. 2004. Effects of nutrient solution EC, plant microclimate and cultivars on fruit quality and yield of hydroponics tomatoes (*Lycopersicon esculentum*). *Acta Hortic.*, 659, 541–7.

Zhang, Z., Liu, L., Zhang, M., Zhang, Y. and Wang, Q. 2014. Effect of carbon dioxide enrichment on health-promoting compounds and organoleptic properties of tomato fruits grown in greenhouse. *Food Chem.*, 153, 157–63. doi:10.1016/j.foodchem.2013.12.052.

Zipelevish, E., Grinberge, A., Amar, S., Gilbo, Y. and Kafkafi, U. 2000. Eggplant dry matter composition fruit yield and quality as affected by phosphate and total salinity caused by potassium fertilizers in the irrigation solution. *J. Plant Nutr.*, 23(4), 431–42. doi:10.1080/01904160009382030.

Chapter 15

Sustainable use of energy in greenhouses

S. Hemming, J. C. Bakker, J. B. Campen and F. L. K. Kempkes, Wageningen University and Research, The Netherlands

1 Introduction

Energy use in greenhouse production is essential for reaching optimum crop production. Energy is mainly used for temperature control (heating and cooling), humidity control (mainly dehumidification), and artificial lighting in high-tech greenhouses. Some energy might be used for CO_2 production. Only small amounts of energy (on average 5–10 kWh m^{-2}), are used for operating other greenhouse equipment. This includes ventilation motors, screens, and pumps for irrigation and water treatments. The amount of energy required for greenhouse production depends on crop demands, as well as the desired production levels and the optimum climate conditions for crop growth. Energy usage also depends on outside weather conditions and greenhouse location.

Optimum crop growth conditions can usually be characterized with temperatures of 15–30°C, humidity levels of 70–90% and CO_2 levels of 400–1000 ppm. If one of the growth conditions is outside these boundaries, suboptimal crop yields and product quality must be expected. Energy is used in high-tech greenhouses for active climate control, to continuously keep growth conditions as close as possible to optimum levels. Heating or cooling are applied to keep temperature at optimum levels while, (de)humidification is applied to keep humidity levels in an acceptable range. Light is one of the

http://dx.doi.org/10.19103/AS.2019.0052.07

most important crop growth factors, alongside temperature, humidity and CO_2. Optimum light levels largely depend on crop stage, species and varieties. In general, fruit vegetable crops, such as tomato, cucumber and sweet pepper, have high light demands, for example 90 mol of PAR is needed to produce 1 kg of tomato once the plant is in production (Hemming et al., 2014a). In case of low outside light conditions due to location or season, energy is used to operate artificial lighting to increase production speed or product quality.

The absolute use of energy for greenhouse production is dependent on outside climate conditions and therefore differs between specific locations as described by Bakker et al. (2008). For example, for Finland the total energy consumption has typically been estimated at 1900 MJ m^{-2} $year^{-1}$ (Olofsson et al., 2006), for the Netherlands at 1100 MJ m^{-2} (van der Velden and Smit, 2017) and for Southern France 500–1600 MJ m^{-2} (Grisey et al., 2014). While in Finland the majority of energy consumption is due to the use of artificial light and heating, it is heating, dehumidification and artificial light in the Netherlands and cooling and heating in France. The need for a reduction of energy costs is important for all growers. For Italy, it has been calculated that energy use for greenhouse climate control is about 20–30% of the total production costs (De Pascale and Maggio, 2005), comparable to about 20–25% in the Netherlands (Van der Knijff et al., 2004), whilst in France it is about 25–40% (Grisey et al., 2014).

Since the use of energy not only affects the operational costs, but also the income the crop produces, researchers and growers focus on the improvement of energy (cost) efficiency, measured as the energy usage (cost) per unit produced. In the Netherlands, energy efficiency has increased over the last decades, while production is more intensified. Since 1990, the CO_2 emission of Dutch greenhouse horticulture was decreased from ca. 7.5 to 4.3 Mt used for crop production (van der Velden and Smit, 2017). Bakker et al. (2008) describe that in Northern Europe more artificial light is used at very high levels (e.g. over 100 W_e m^{-2} in the Netherlands and 200 W_e m^{-2} in Finland), to increase production and supply products year-round. Also, in Mediterranean areas the improved environmental control (e.g. application of heating, more CO_2 supply, additional lighting) leads to intensified production schemes to expand the production period, and create a more constant quantitative-qualitative yield. This leads on the one hand to a higher energy efficiency (yield per unit of energy input), but at the same time also to a higher absolute energy use per unit of floor area (De Pascale and Maggio, 2005; Grisey et al., 2014).

In recent years, in many countries more political focus has been aimed at climate change, and with it the reduction of the use of fossil fuels and CO_2 emission. The use of sustainable energy sources for greenhouse production became an important issue (Bot, 2001). For many years, the Netherlands has focussed on the maximum use of solar energy for greenhouse production (Bakker et al., 2008). Alongside this, the use of other sustainable energy sources is increasing, such as geothermal energy or biofuels for heating, green electricity from wind or water, and solar sources for artificial lighting (van der Velden and Smit, 2017).

In general, research concerning the usage of energy for greenhouses focuses on two principles (Hemming et al., 2017a): total energy reduction, and sustainability, both without reduction of crop production. The principles for these are as follows: maximum use of natural sunlight (free solar energy input to greenhouse, free light for crop growth and production), maximum insulation (to prevent energy losses through greenhouse shell). Also, the efficient use of energy by different technologies and control strategies (e.g. temperature integration strategies, efficient heating and cooling systems, high humidity levels, fogging, mechanical dehumidification systems, optimum CO_2 concentration), and sustainable energy sources to replace fossil fuels (e.g. geothermal, biofuels, wind). In the following sections these principles are described in more detail.

2 Maximizing natural solar energy use

A greenhouse can be considered as an energy-efficient solar collector. The transparent greenhouse covering influences the greenhouse energy balance, as well as crop photosynthesis and development. The entry of sunlight (300–2500 nm) into the greenhouse accounts for free solar energy for heating in cold periods, and free light for crop growth and production, and must therefore be maximized for sustainable energy usage. Only if natural sunlight is insufficient to reach (economic) optimum crop production or product quality, should supplementary artificial lighting be applied.

2.1 Capturing solar energy

2.1.1 Light transmission

To make maximum use of natural solar energy, the light transmission of the greenhouse should be as high as possible, at least during part of the production season. In cold periods, every Joule of natural sunlight reduces the use of (fossil) fuels, and every mol contributes to crop production. For higher latitudes a general rule suggests that 1% more light leads to a 0.7–1% increase in production (Cockshull et al., 1992; Marcelis et al., 2006).

Photosynthetic active radiation (PAR light, 400–700 nm) accounts for ca. 50% of the solar radiation, and is the driving force for crop photosynthesis (McCree, 1971). A higher PAR light transmission in a greenhouse system can be reached by optimizing the greenhouse structure. This includes minimizing reflection losses through a modification of roof slope and orientation (e.g. Janssen and Hart, 2006; Soriano et al., 2004; Swinkels et al., 2017) or minimizing the number and size of construction elements while maintaining enough mechanical strength against design loads (Janssen and Hart, 2006). The use of white-coated greenhouse construction elements and white-coated interior equipment increases light reflection towards the crop. Today, high-tech greenhouses are optimized for these factors. Further increase of light transmission by adaptation of the greenhouse structure is limited by up to 1–2% in high-tech greenhouses (Kempkes et al., 2017a).

Additional improvements in light transmission can be achieved by choosing a specific type of covering material (Baeza and López, 2012; Tantau et al., 2012). The majority of low-tech and mid-tech greenhouses worldwide are covered with polyethylene-based plastic films, with an average hemispherical light transmission of 80–82%. Different polymers, such as ETFE or PVDF (Waaijenberg et al., 2005) show 2–4% higher hemispherical light transmission values, but are used only occasionally. A popular material for high-tech greenhouses is glass with a hemispherical light transmission of 82–84%. Here, improvements have been made during the last decade by introducing anti-reflection (AR) coatings on one or two sides of the glass (Hemming et al., 2006a, 2011, 2012) resulting in up to 8% higher hemispherical light transmission.

Another important factor influencing the amount of light, and thus solar energy, entering the greenhouse, is a favourable condensation behaviour of coverings. If condensation occurs in the form of a water film instead of water droplets, light transmission of plastic films is increased by up to 20% (Gbiorczyk, 1997; Gbiorczyk et al., 2004; Geoola et al., 1994; Pieters et al., 1997; Pollet and Pieters, 2000; Schultz, 1997), and light transmission of glass by up to 4% (Stanghellini et al., 2012). Also, regular cleaning of the greenhouse roof increases light transmission. Depending on the surroundings of the greenhouse, pollution can occur by salt in areas close to the sea, dust in desert areas, oily components in areas close to airports, or algae growth in humid conditions. An increase of light transmission by 10–20% has been observed in moderate desert as well as in humid areas due to cleaning (Hemming and Campen, *unpublished data*; Waaijenberg, 2006; Geoola et al., 1998).

To estimate the amount of light entering the greenhouse, Stoffers (1967) developed the first transmission model, an idea that was later followed by different research groups, as summarized by Baeza and López (2012). They range from simple to more complex models (e.g. Baille et al., 1990; Bot, 1983; Takakura et al., 1971), with some of them developed and validated for Mediterranean greenhouse types (Castilla et al., 2001; Soriano, 2002), others for tropical environments (Impron, 2011) or cold climates (Rath, 1992). Some models can cope with the spatial distribution of direct and diffuse radiation inside the greenhouse (De Zwart, 1996; Kurata et al., 1991; Vanthoor, 2011; Wang and Boulard, 2000), while advanced models make use of ray-tracing techniques (Swinkels et al., 2001, 2017). It has been shown recently that the integration of all of the above factors used to increase light transmission, can lead to a substantial increase of solar energy, and thus sustainable energy and photosynthetic light in greenhouses, by using a targeted modelling approach (Kempkes et al., 2017a; Swinkels et al., 2017). This example is described in more detail in Section 6.1.

2.1.2 Diffuse light

An important component of the light transmission of a greenhouse is the geometrical distribution of light, the scattering angles of light entering the greenhouse, often summarized as diffuse light. During the last decade, diffuse light has been shown to be favourable for the production of different greenhouse crops (e.g. Dueck et al., 2012a; Hemming et al., 2008; Li et al., 2014a; Victoria-Garcia et al., 2012). Diffuse light enhances crop photosynthesis. Horizontal light distribution was shown to be increased, even in greenhouses covered with diffuse covering materials, and to be the most important factor for positive effects of diffuse light on crop yield (Li et al., 2014b). Vertical light penetration in high-wire-grown crops is improved (Hemming et al., 2008; Jongschaap et al., 2006). As a consequence, a larger percentage of diffuse light is absorbed by middle leaf layers, leading to increased photosynthesis in those leaf layers (Dueck et al., 2012a; Hemming et al., 2008; Jongschaap et al., 2006; Li et al., 2014a,b). Additionally, Dueck et al. (2012a) and Li et al. (2014b) have shown that those leaves have a larger dry matter content. It has been shown in several experiments that crop morphology and development are influenced by diffuse light. Dueck et al. (2012a) have observed a more generative growth and faster fruit development. Yield increase was mainly due to higher average fruit weight. In potted plants, faster growth is reported him diffuse light, also, higher light intensity can be tolerated by shade-tolerant pot plants resulting in even larger yield increases (Li et al., 2014a). In the Netherlands, yield increase of tomato and cucumber has been shown to be up to 10%, due to the application of structured glass transforming direct outside radiation into diffuse light inside greenhouses.

The development of newer materials and their optical properties are described by Hemming et al. (2014b). Research should focus further on the development of materials with high scattering angles, and thus high diffusion, combined with high hemispherical light transmission. A preliminary rule of thumb can be created based on the research described above, in which 10% more light diffusion (measured as Hortiscatter, NEN 2675 (NEN, 2018)) leads to approximately 2.5–3.5% more tomato production (Hemming, *unpublished*). With that, the energy efficiency per unit crop produced is increased, thus diffuse light contributes to sustainable energy use in greenhouses.

2.1.3 NIR transmission

Near-infrared radiation (NIR, 800–2500 nm) accounts for ca. 40% of solar radiation (Blanchard and Runkle, 2010) and is of significance mainly for the transfer of thermal energy into the greenhouse. In temperate climates, NIR contributes to warming of the air inside greenhouses, and thus reduces heating requirements during cold periods (Max et al., 2012a). Therefore, it helps to reduce the need for additional heating. During summer months, and in warmer climates, the application of NIR-selective filters is considered to be advantageous, (García-Alonso et al., 2006; Impron et al., 2008; Morris et al., 1958; Sonneveld et al., 2012a; Tanaka et al., 1997), because by applying NIR-selective filters the energy demand for cooling can be reduced.

The potential of NIR-filtering greenhouse covers has been known for many years. The first detailed investigations of the cooling effect of a water film with NIR-filtering additives were carried out by Morris et al. (1958). Different materials have been investigated since then, such as plastic films or coated glass for greenhouses covers (e.g. Abdel-Ghany et al., 2001; Hemming et al., 2006b; Verlodt and Verschaeren, 1997), movable screens (Runkle et al., 2002a,b; Stanghellini et al., 2011) or NIR-filtering temporary coatings (Blanchard and Runkle, 2010; Von Elsner and Xie, 2003).

A NIR filter works either by absorbing or reflecting NIR radiation. The principle of absorption was used by different authors in the past (Chiapale et al., 1983; Gale et al., 1996; Sadler and Van Bavel, 1984). A NIR-absorbing liquid was pumped inside the space of a multiwall polycarbonate sheet whenever cooling was needed, and the warm liquid was then stored and reused for heating. In more recent times, solid NIR filters have been proposed. In this case, the filter uses the principle of NIR absorption, the temperature of the filter material (covering or screen) increases and a large fraction of the energy enters the greenhouse as thermal radiation. Alternatively, the filter uses the principle of NIR reflection, and the the energy is directly kept out of the greenhouse (Kempkes and Hemming, 2012). So far, no perfect NIR filters - defined as those capable of cutting out all NIR wavelengths- are available.

Stanghellini et al. (2011) reported that some crops have a high reflectivity for NIR radiation (about 50%), so that NIR radiation transmitted by partial reflection (using an imperfect NIR filter), will end up being trapped between two reflective surfaces (the greenhouse covering and the crop). This leads to a lower effect of cooling energy saving, than could be expected from the NIR-reflecting properties of the filter. The authors investigated the effect of a NIR-reflective screen material in a high-tech greenhouse on ventilation requirement, crop transpiration, and water use efficiency of a rose crop. The NIR filter reduced the energy load of the greenhouse by 12%. The high reflectivity of the canopy in the NIR range caused the energy input of the greenhouse to be reduced by less than the properties of the material would suggest (42% NIR reflection). Perfect NIR-reflecting filters are desired for the future.

Kempkes et al. (2008) point out the importance of seasonal energy demand, especially in unheated greenhouses. While a NIR-reflecting filter is desired for cooling in warm seasons, the effect of such a filter is not desired during cold seasons. Switchable NIR-reflecting filters would be a desirable option for the future.

2.2 Conversion of solar energy

The amount of solar energy entering the greenhouse is dependent on latitude, season, time and weather conditions. In many greenhouses, periods with a surplus of solar energy, and other periods with insufficient solar energy alternate. Energy consumption is related to the amount of available solar energy. In periods with high availability of sunlight, the energy must be used to cool the greenhouse (see Section 4.1.3), whilst in periods with low availability of sunlight the energy must be used to heat the greenhouse (see Section 4.1.2). Solar radiation can be considered as the most important renewable energy source in greenhouses (see Section 5), and must be converted efficiently.

2.2.1 Passive solar energy systems

In small-scale and low-tech greenhouse production systems, solar energy can be considered as the most important renewable energy source to overcome diurnal differences in the availability of energy, and the energy demand of a greenhouse crop. Cuce et al. (2016) give an overview of different ways that solar energy can be used for heating greenhouses. Diurnal energy storage systems are used in low-tech greenhouses to capture solar radiation during daytime, and reuse it for heating during the night-time. Different systems have been described in literature (e.g. Sethi and Sharma, 2007; Von Zabeltitz, 1987), using water (Baeza et al., 2017; Grafiadellis, 1985; Ntinas et al., 2015; Sørensen, 1989), rock or soil (e.g. Arizov and Niyazov, 1980; Öztürk and Bascetincelik, 2003) or phase change material (e.g. Boulard et al., 1990; Kurklu, 1998) as heat storage. Heat storage can either be applied in the greenhouse floor area, the north wall (Chinese Solar greenhouse, Tong et al., 2013) or next to the greenhouse (e.g. Baille et al., 2006). These systems are mainly applied in climates with mild winters. The possibilities of solar energy usage for greenhouse cooling applications are also described (Sethi and Sharma, 2008). This can be done by using ground heat exchangers with horizontal earth tubes, and such systems are mainly applied in hot climates.

Both heating and cooling concepts based on the conversion and diurnal storage of solar energy, can be considered as simple sustainable energy systems for small-scale and low-tech greenhouses.

2.2.2 Closed greenhouses

In large-scale high-tech greenhouses, one of the principles for sustainable energy usage is the efficient conversion of solar energy by the use of heat exchangers, heat pumps, and heat buffers. This helps to overcome seasonal differences in the availability of the solar energy, and the energy demand of a greenhouse crop. The principle of a 'closed greenhouse' has been suggested as a possible solution (Bot, 1992). Cooling by natural ventilation is replaced partly (in semi-closed greenhouses) or completely (in closed greenhouses) by mechanical cooling. Excess solar energy is collected and stored during the warm season and reused to heat the greenhouse during the cold season. In temperate climates, this concept combines improved crop production with energy savings (De Gelder et al., 2012a; Katsoulas et al., 2015).

In such a concept, the greenhouse growing area is equipped with heat exchangers, in which cold water is circulated during warm periods. The cold water (7–11°C) is taken from a cold water storage (e.g. a cold well of the aquifer). It absorbs excess solar energy inside the greenhouse, warms up and is pumped into a warm water storage (e.g. a warm well of the aquifer) for seasonal storage. The warm water (18–20°C) is circulated again inside the greenhouse during cold periods. A heat pump extracts the heat from the water to make it useable for the greenhouse; the cold water is stored in the cold well of the aquifer for the next warm season (Bot, 2001; Courtois et al., 2008; De Zwart et al., 2008; Grisey et al., 2014; Opdam et al., 2005; Wong et al., 2011). The principle is described by De Gelder et al. (2012a) (Fig. 1). The heat pump is usually driven by electrical energy. Other electricity consumers in such a system are the heat exchangers, heat storage pumps, other circulations pumps, greenhouse motor and valves. Therefore, it might make sense to generate electricity with a combined heat and power (CHP) generator, which also generates heat to be used directly in the greenhouse or to be pumped into a heat storage, allowing CO_2 to be used for crop production.

De Gelder et al. (2012a) quantified the energy saving of heating with heat pumps using electricity generated by a CHP, and by exploiting surplus summer solar energy, at 30–40% compared to conventional heating (Bot et al., 2005; Opdam et al., 2005). Considering all electricity consumption of the system, energy savings of 25–35% can be reached (Bakker et al., 2006; De Zwart et al., 2008; De Zwart, 2011). Apart from energy savings, other potential benefits of closed greenhouses have been reported (Dieleman and Hemming, 2011), such as increases in crop production due to higher CO_2 levels (De Gelder et al., 2012a; Heuvelink and González-Real, 2008; Qian et al., 2012), reduction of CO_2 emissions (Qian et al., 2011), water savings (Katsoulas et al., 2015) and improved crop protection. In particular, temperature and humidity control (see Sections 4.1.1 and 4.2.1), combined with high light intensity and CO_2 concentration in a closed greenhouse, was found to boost crop production (see Section 2.2.2).

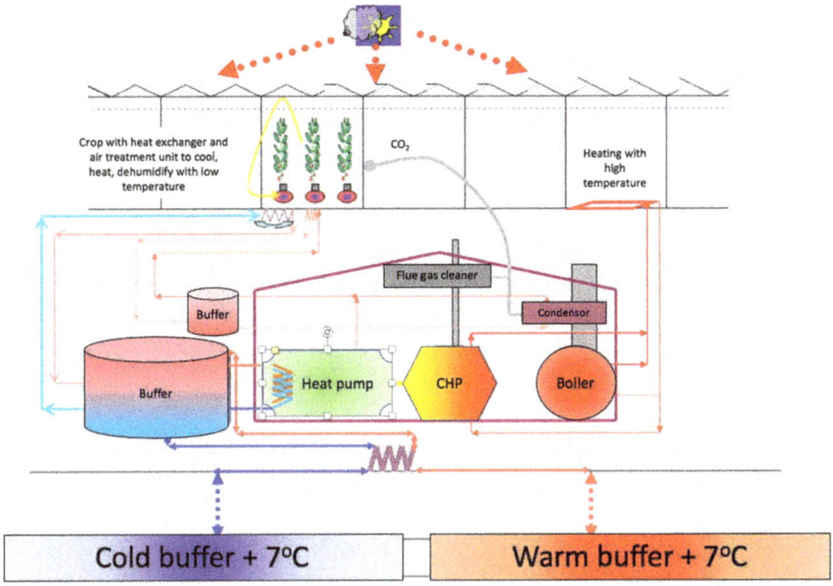

Figure 1 Overview of the principle of a closed greenhouse during cooling mode in warm periods and during heating mode in cold periods including heat exchangers inside the greenhouse, heat buffer (aquifer and heat pump) (after De Gelder and Kipp, 2005).

On the other hand, higher investments (Ruijs et al., 2010) must be balanced by energy prices and crop benefits.

A number of concepts were described for closed and semi-closed greenhouses in different countries, such as research in the Netherlands (e.g. Bakker et al., 2006, 2008; Bot, 1992; Bot et al., 2005; De Zwart, 2011; Hoes et al., 2008; Opdam et al., 2005), Germany (e.g. Buchholz et al., 2005; Tantau et al., 2011), France (e.g. Grisey et al., 2011), Finland (Särkkä et al., 2006) or the United States (e.g. Yildiz and Stombaugh, 2008). Regional differences in climate are important and the equipment should adapt to deliver the optimal climate conditions (De Pascale and Stanghellini, 2011; Vanthoor, 2011). Crop demands should be considered; for example, for high value crops such as *Phalaenopsis*, closed greenhouses has been established widely in the Netherlands (De Gelder et al., 2012a). For vegetable crops only semi-closed systems are currently utilized. An example of such a system is given in more detail in Section 6.3.

2.2.3 Artificial lighting

In cases of low outside light conditions due to location or season, and high crop demand for light, energy is used to operate supplemental artificial lighting to increase production speed or product quality. The traditional way of using supplemental artificial lighting in greenhouses in higher latitudes is the use of high-pressure sodium (HPS) lamps. Since they have an acceptable spectrum, a good energy efficiency and reasonable price, they are widely used in greenhouse horticulture (Hemming, 2011). Current efficiencies can be quantified to be ca. 2.1 μmol J^{-1} for a 1000 W fixture (e.g. Philips, 2018a). Other lighting sources include mercury lamps, which are used for more blue light in the spectrum, plasma lamps with full sunlight spectrums, or fluorescent tubes, which are mainly used under fully artificial light conditions (Hemming, 2011). Modern lighting sources include light emitting diodes (LED). Current efficiencies can be quantified to be up to 3.2 μmol J^{-1} (e.g. Philips, 2018b). Increases in the light output per energy input will lead to higher energy savings in the future.

Apart from technical developments of the lamp, fixture energy savings can be realized by adaptations of the technical layout of systems, like the position of lights in toplighting or interlighting positions, or a combination of these (e.g. Dueck et al., 2012b; Gómez and Mitchell, 2016; Gunnlaugsson and Adalsteinsson, 2006; Guo et al., 2016; Hovi et al., 2004; Hovi-Pekkanen and Tahvonen, 2008; Kumar et al., 2016; Trouwborst et al., 2011). Further energy savings are possible during greenhouse application by increasing the light use efficiency of the crop. This can be done by optimizing light duration and intensity (e.g. Dieleman et al., 2015; Dueck et al., 2007; Stadler, 2012), changes in plant densities (e.g. Näkkilä et al., 2006), changes in LAI or shape of the cropping system (Verheul, 2012), changes in planting date, and even more importantly, selecting more efficient genotypes (Higashide and Heuvelink, 2009; Ouzounis et al., 2016).

Adaptations of the light spectrum might lead to further increase in efficiency of light usage for specific crops, (Gómez and Mitchell, 2016; Hernández and Kubota, 2012; Hogewoning et al., 2012; Van Ieperen, 2016), or increased product quality (Labrie and Verkerke, 2014; Li and Kubota, 2009; Ouzounis et al., 2014). The effects of a LED light spectrum on photosynthetic (Paradiso et al., 2011) and photomorphogenetic (Hao et al., 2016; Van Ieperen, 2012) responses have been investigated. Secondary effects of LED, which do not emit NIR, on crop temperature (Dueck et al., 2012b; Runkle et al., 2011), transpiration (Van Ieperen et al., 2012), water and nutrient uptake or plant protection (Vänninen et al., 2012) have also been studied.

In general, it can be concluded that supplemental artificial light increases energy efficiency, but at the same time also leads to a higher absolute energy use. It certainly involves higher investment and energy costs (Marcelis et al., 2014), which must be balanced with higher crop production and quality. Since for sustainable usage of energy in greenhouses, the maximum use of solar energy is required, only supplemental lighting was discussed here.

3 Maximizing insulation

To prevent energy losses from the greenhouse through the greenhouse shell, maximum insulation must be realized, either permanently (using covering materials) or temporarily (using thermal screens). In both cases, insulation measures must not interfere with solar energy and light entry to the greenhouse. Radiative energy losses can be prevented by the design and choice of the properties of a screen or covering (such as thermal infrared transmissivity and emissivity). Convective energy losses are determined by outside wind speed, material U-value, and materials' air permeability or leakage (involving latent and sensitive heat exchanges).

3.1 Insulating coverings

During cold periods, energy losses through a single-layered greenhouse covering are high, and heating is applied to meet crop temperature demands. To reduce energy usage different types of thermal screens can be used (see also Section 3.2) for temporary insulation. Permanent insulation can be reached with double-layered coverings. The energy-saving potentials of double-layered greenhouse covering materials have been researched in several studies (e.g. Andersson and Nielsen, 2000; Bakker et al., 2008; Bot, 2001; Villeneuve et al., 2005; Zhang et al., 1996). Another way in which radiative convective energy losses can be reduced is via careful choice of covering materials. However, special attention must be paid to total light transmission.

Glass is often used in heated greenhouses in different areas where high-tech greenhouses are being used (Papadakis et al., 2000). From a worldwide perspective, plastic films are the most widely used greenhouse covering material (Von Zabeltitz, 2011). Only in limited areas are single rigid plastic materials or double rigid plastic materials used. A summary of greenhouse covering materials, their properties and usage is given by different authors (e.g. Kittas and Baille, 1998; Max et al., 2012a; Papadakis et al., 2000; Waaijenberg, 1984).

Covering materials have different thermal properties depending on the type and thickness of the material. Both thermal infrared (FIR, >2500 nm) transmissivity and emissivity determine the radiative energy transfer through the material (Papadakis et al., 2000; Silva and Rosa, 1987). If the covering material is transparent for FIR and/or it has a high emissivity, energy is lost. Glass and all wet materials have a low or non-existent permeability to FIR. Dry materials such as PE films have a high FIR transmissivity, which can be reduced by adding special IR absorbing pigments. Also adding other polymers in co-extruded films (e.g. EVA) reduce the FIR transmissivity (Papadakis et al., 2000). The thermal material properties result in different potential energy savings.

Materials with low emissivity lead to energy savings. The emissivity of glass is ca. 0.9. Special low-emission coatings can be applied to reduce the emissivity to 0.3. Due to their electrical behaviour, the transmittance of low-emissivity coatings within the NIR (800–2500 nm) and FIR (>2500 nm) spectral range is very low. Therefore, the energy flux in the direction opposite to which the coating is oriented is distinctly reduced (Max et al., 2012a). Such low-emissivity coatings have also been used on single glazing in the past (e.g. Hortiplus or Agriplus glass). This results in an increase in material temperature which, in turn, results in an increase of the convective heat losses. The energy savings are reported to be ca. 40% with clear skies and low wind speeds, but 5% under cloudy skies and windy conditions (Papadakis et al., 2000). Nowadays, low-emissivity coatings are the focus of much new research (e.g. Hemming et al., 2011, 2012; Kempkes et al., 2014a).

The overall heat transfer through a material is described by the thermal transfer coefficient (U-value, formerly referred to as k-value). The U-value of standard single-pane glazing is around 6 W m^{-2} K^{-1} (e.g. Nijskens et al., 1984; Papadakis et al., 2000). It can be reduced to around 3 W m^{-2} K^{-1} by double glazing or to around 1.1 W m^{-2} K^{-1} for low-emissivity coated double glazing. The U-value of most PE- or EVA-based plastic films is reported to be around 9 W m^{-2} K^{-1} depending on the exact material composition, while the U-value of special plastics like ETFE is substantially lower. This can be explained by the lower transmissivity for FIR. Double plastic sheets typically made from PC or PMMA have a U-value of around 3 W m^{-2} K^{-1} (Papadakis et al., 2000) due to both low FIR transmissivity and lower convection losses. An overview of the U-values of different greenhouse systems is given by Van de Braak (1995).

Double-layer covering materials increase the insulation value of the greenhouse surface permanently. However, at higher latitudes special attention has to be paid to light transmission. The hemispherical light transmission of double plastics can drop down to 60–70%. Hemming et al. (2011) showed that it is possible to develop covering materials combining high insulation with high light transmission. They used two glass panes in a double glass system filled with gas in between. Light transmission was increased by adding AR coatings on three of the four surfaces of the two glass panes. Radiation losses were decreased by adding a low-emissivity coating on the fourth side for low emissivity. Insulation values were increased by a gas-filled space between the two glass panes, resulting in low conduction and convection losses (Hemming et al., 2012; Kempkes et al., 2014a). The details are described in Section 6.2.

Other insulating greenhouse concepts have been reported using insulating greenhouse coverings in different countries (e.g. Andersson and Nielsen, 2000; Bot, 2001; Villeneuve et al., 2005; Zhang et al., 1996). Greenhouse constructions with a combination of glass and plastic film have been investigated by Max et al. (2012b) and Kempkes et al. (2017b). In general, energy savings up to 50% were reported (Hemming et al., 2017a).

3.2 Thermal screens

Thermal screens add an additional barrier between the greenhouse environment and its surroundings to avoid energy losses from the greenhouses (Bailey, 1981). Thermal screens have been studied by several research groups (e.g. Bailey, 1981; Cohen and Fuchs, 1999; Huang and Hanan, 1976; Teitel et al., 1996). When movable, a screen has less impact on the light transmission compared to fixed screens or double coverings (see Section 3.1). If used almost permanently, screens can reduce the energy use by more than 35%, depending on the material (Bakker and Van Holsteijn, 1995). Other researchers quantified the energy saving of different materials in various greenhouse types, and under different climate conditions (e.g. Chandra and Albright, 1980; Meyer, 1984; Teitel et al., 2009). In practice, movable screens are closed only during part of the cropping season. Due to restrictions for closing, for example, too high humidity and light loss by the screen application, in commercial practice the energy saving by thermal screens is restricted to 20% (Bakker et al., 2006).

The total energy saving of screen materials can be determined by the radiation exchange (transmissivity and emissivity of heat radiation), and the air permeability through the screen (convection at low wind speeds). The humidity transport through the material (air and with that humidity and humidity transport through threads or plastics by diffusion, absorption or adsorption) must also be taken into account. Furthermore, the total energy saving of a screen is largely determined by screen management (Hemming et al., 2017b).

Previously, De la Faille et al. (2009) developed a method and measurement equipment to determine the emissivity and transmissivity for thermal infrared radiation of screens. Geoola et al. (2009) determined the U-value of different screens using a hot box method. They found that the U-value depends on wind velocity and temperature differences, and wet or dry conditions. Miguel et al. (1997) developed a method and equipment to determine the air permeability of screens. Results of laboratory experiments were compared with measurements in practice (Miguel, 1998; Miguel et al., 1998). Dijke (2016) has developed a test equipment and protocol to quantify the humidity transport through screens. Hemming et al. (2017b) recently developed a complete, objective method to measure material properties and compare different screen materials by quantifying the energy use under standardized conditions. With this method developed, they found average savings of typical energy screens, in general, varied between 20% and 30% during night-time. They concluded that impermeable screens, and screens with low emissivity and low thermal infrared transmissivity, give highest energy saving (Hemming et al., 2017b). However, very few commercial screens were found to combine these properties.

As well as the properties of the materials, energy savings are also affected by screen management. Transparent screens allow their usage during daytime hours. Efficient screening strategies can save energy while maintaining crop production levels. This was shown by Dieleman and Kempkes (2006), who reached additional energy savings up to 4% without affecting production, when delaying the screen opening to radiation levels above 50–150 W m^{-2}. Currently, screen opening is even more delayed to decrease radiation losses from the crop to the cold sky during the daytime in winter. Experiments by De Gelder et al. (2017) have shown that lower light levels inside the greenhouse due to permanent screen usage, were apparently compensated by higher crop temperatures. Overall, this had no negative influence on yearly crop production. Airtight screens were limited in their use in the past due to increased humidity levels. However, currently mechanical dehumidification systems can be used to remove humidity. Hemming et al. (2017b) have shown that impermeable screens lead to the highest energy savings for highly transpiring crops. In cases where the number of hours the screen is used is increased, more energy can be saved.

Another factor influencing the total energy saving using screens, is their positioning inside the greenhouse. An overview of different ways of screening are given by Bakker and Van Holsteijn (1995). Depending on the crop and outside climate conditions, growers today use one screen for multiple purposes. For example, a thermal screen is also applied as a shading screen especially when the plants are young. Nowadays, as well as positioning, growers use multiple screens with different properties. Often double movable screens are used, in which materials with the highest energy saving are combined with transparent materials for flexible usage during dull days and night-time. Sometimes growers also use fixed plastic foils in addition during the first week of production in winter, as a second or third insulation layer. In cases where the U-value of the screen installation is increased, more energy can be saved.

4 Maximizing energy efficiency by climate control

Energy is used in greenhouse production for reaching optimum crop production. Energy in high-tech greenhouses is mainly used for temperature control (heating and cooling), and humidity control (mainly dehumidification). In some greenhouses, energy is also used for artificial lighting and CO_2 production. The potential of a crop to handle flexible temperature regimes provides opportunities for reducing energy usage during both heating and cooling (see Section 4.1.1). Increasing the humidity set points and reducing crop transpiration give possibilities to reduce energy usage for humidity control (see Section 4.2.1). Energy efficiency can also be improved by the design of the heating (see Section 4.1.2), cooling (see Section 4.1.3) or dehumidification equipment (see Section 4.2.2).

4.1 Temperature control

4.1.1 Temperature strategies

New temperature control strategies can lead to energy saving in greenhouses (Tantau, 1998). Greenhouse climate is traditionally controlled by rather rigid set temperature points for heating. Energy consumption strongly decreases with decreasing set points for heating temperature (Dieleman and Hemming, 2011). Reducing the average 24 h temperature by 2°C in tomato was calculated to reduce the energy consumption of a year-round tomato crop by 16% (Elings et al., 2005). However, lower average temperatures might reduce leaf area development, light interception and therefore production.

Another possibility is the application of temperature integration. Temperatures are then allowed to fluctuate within certain bandwidths across a defined period. Several studies have shown that most horticultural crops tolerate these fluctuations as long as the average temperature over 24 h (Bakker and Van Uffelen, 1988) or several days (De Koning, 1994; Hurd and Graves, 1984) is kept constant. This may result in energy savings of 4% at a bandwidth of 2°C in 24 or 72 h (Elings et al., 2005) in tomato, to 13% at a bandwidth of 10°C (Buwalda et al., 1999) in rose crops. Temperature control regimes must take into consideration the greenhouse location, and therefore climate conditions such as external temperatures and wind speeds. In the case of cold winter climates, which usually can be found in higher latitudes, allowing lower temperatures during heating periods leads to substantial energy savings. These lower temperatures can be compensated with higher temperatures during daytime, using free natural solar energy following temperature integration strategies. In the case of hot climates, which usually can be found in lower latitudes, high radiation levels lead to high cooling loads. Allowing higher temperatures during these periods can lead to substantial energy savings. These higher temperatures should then be compensated with lower temperatures during the cooler night-time, following temperature integration strategies.

Further energy savings can be achieved by analyzing the details of dynamic temperature control. In particular, peaks in energy use that often occur around sunrise in cold climates, are costly and efforts should be made to reduce these (Dieleman and Hemming, 2011). Dieleman et al. (2005) found that a temporarily low temperature drop in a fruit-bearing tomato crop did not affect stem length, plant development or fruit production. Energy saving can be obtained when external temperatures are low, without negative effects for the crop. However, in some young plants or ornamental crops (e.g. potted plants,) a DROP might reduce plant length (Sysoeva et al., 1999), especially if the DROP is applied in the first part of the day (Grimstad, 1993; Myster and Moe, 1995). Crop-dependent dynamic temperature control should be applied for energy saving.

4.1.2 Energy efficient heating

The primary purpose of the heating system in a greenhouse is to raise the temperature of the crop. Different heating systems have been described in literature (Van de Braak, 1995; Sethi and Sharma, 2008). Passive heating systems are described in Section 2.2.1. Active heating can be introduced into the greenhouse through central or local heating systems. Local direct air heating systems are mainly found in small or low-tech greenhouses. They have the disadvantage of uneven temperature distribution (Teitel et al., 1999). In large-scale high-tech greenhouses, a central hot water boiler is traditionally used, and the hot water is generated and circulated through a closed circuit of steel pipes to distribute the heat in the greenhouse (Van de Braak, 1995). The central boiler can be operated with different energy sources. Replacing fossil fuels with renewable sources (see Section 5) will help to reduce CO_2 emissions.

Decreasing the heating pipe temperatures can reduce energy consumption. To compensate for the lower temperature the heat-exchanging surface must be enlarged (such as more pipes, pipes with larger surface area, implementing pipes in the floor or growing benches) (e.g. Van de Braak, 1995; Vogelezang and Van Weel, 1988). Alternatively, the heat transfer coefficient must be improved, (including increased heat radiation by choosing a pipe material with high conductivity such as aluminium, or increasing heat convection by forced instead of free air movement) (e.g. Blom and Ingratta, 1981). Forced air heating systems in general have a higher energy consumption. Air is forced along a heating element, and the heated air can be discharged freely into the greenhouse, but is most effective when distributed evenly by a system of ducts. Due to the additional consumption of electrical power by the fan, 10% higher energy usage for heating has been reported by Van de Braak (1995). This might be the main reason why such systems are not used anymore in modern high-tech greenhouses.

Alternative heating sources such as CHP generators are widely used in greenhouses in higher latitudes. CHP is used on ca. 7000 ha (ca. 70% of the total greenhouse area) in the Netherlands (Van der Velden and Smit, 2017). A CHP generates both electricity, which can be used for artificial lighting, and waste heat, which can be used for heating. Nowadays, electricity is also generated for the public grid. CHP is a suitable instrument to increase the energy efficiency. At the same time CO_2 (after cleaning, mainly NOx) can be supplied for the enrichment of the greenhouse air, where previously the growers would have burnt natural gases for this purpose (Bot, 2001). Nowadays some CHPs utilizing biofuels are used in commercial practice, but further research focused on CO_2 quality is still needed.

Furthermore, heat pumps (see Section 2.2.2) are used in greenhouse for heating purposes. Also, the application of artificial lighting (see Section 2.2.3) provides heating for the crop. With the increase of supplemental artificial light the need for additional heating, in general, decreases.

4.1.3 Energy efficient cooling

Different cooling systems have been described in literature (e.g. Breuer and Knies, 1995; Sethi and Sharma, 2007). An efficient way of cooling, as long as outside temperatures are at acceptable levels for crop growth, is the use of natural ventilation. In the case of natural ventilation, the pressure difference over the openings, caused by wind and differences in air temperature between inside and outside air, results in air exchange. If designed and operated effectively, natural ventilation is able to decrease greenhouse air temperature down to, or even below, outside levels without the use of energy. The natural ventilation process has been studied by many research groups (e.g. Boulard et al., 1996; De Jong, 1990; Mistriotis et al., 1997; Sase et al., 1984). In cases of forced ventilation, additional energy is needed to cause an air exchange between internal and external air.

Forced ventilation was studied by several researchers (e.g. Goodhind, 1965; Kittas et al., 2005; Papadakis et al., 1992). Calculations carried out by Van de Braak and Breuer (1991) show that, in the Netherlands, a ventilation capacity of about 120 m^3 m^{-2} h^{-1} is necessary to achieve desirable climate conditions, in a greenhouse with only mechanical ventilation. The yearly number of ventilation hours has been estimated to be between 1900 and 6000, depending on the method of control and the desired temperatures (Breuer and Knies, 1995).

The application of shading is able to reduce crop temperatures without the use of energy (e.g. Baille et al., 2001; Kittas et al., 1999). However, shading may reduce crop growth and production due to the reduction of PAR light (see Section 2.1.1). Selective shading by only excluding NIR radiation from the greenhouse has already been described in Section 2.1.3.

Different evaporative cooling systems have been studied, such as pad-and-fan, fogging or roof cooling (e.g. Giacomelli et al., 1985; Katsoulas et al., 2001; Montero et al., 1990; Morris et al., 1958; Reisinger et al., 1987; Willits, 2003). Evaporation removes energy (2260 kJ kg^{-1} water) from greenhouse air. This will decrease air temperature, but it will increase air humidity levels. The cooling effect will largely depend on air humidity levels (Breuer and Knies, 1995). Evaporative cooling works well in areas with low outside humidity levels (e.g. desert areas), whereas it is ineffective in areas with high outside humidity (e.g. tropical areas). The cheapest way of evaporative cooling is the transpiration process of the crop itself. Impron (2011) showed that crop transpiration can have a significant effect on greenhouse cooling.

Active cooling systems are used in some (semi-)closed greenhouse systems (see Sections 2.2.2 and 6.3). Electricity usage of such systems largely depends on operation hours and strategy.

4.2 Humidity control

4.2.1 Humidity strategies

In general, humidity in the greenhouse is high due to crop transpiration. The transpiration of a crop depends on solar radiation, air temperature and the humidity of the greenhouse air (Stanghellini, 1987). The most important reason for humidity control in a greenhouse is the risk of fungal diseases and physiological disorders (Grange and Hand, 1987). Bakker et al. (2008) quantified the amount of energy use for humidity control in Dutch greenhouses to be 10–20% of the total energy use.

The energy use involved in dehumidification can be reduced by increasing the humidity set points and reducing crop transpiration. If crop transpiration is reduced, humidity in the greenhouse will be lowered. Li et al. (2001) showed that transpiration in tomato can be reduced by 35% without affecting fruit production. One of these measures may be the removal of leaves from crops with high LAI (Dueck et al., 2006). Simulations showed that reducing the LAI from 6 to 3 in a sweet pepper crop in August, resulted in a 10% reduction in transpiration and 5% energy conservation, without affecting light interception or production. In tomato, 50% of leaf area reduction by removing old leaves was shown to reduce transpiration by 30%, without having a detrimental effect on crop yield (Adams et al., 2002).

Increasing the set point for relative humidity from 85% to 90% in tomato was calculated to reduce the energy consumption by 4% (Elings et al., 2005). Further increases to 95% can be tolerated by most vegetable crops and save more energy, if a homogeneous temperature distribution, and therefore relative humidity distribution in the greenhouse, can be guaranteed. In modern greenhouses this is done by air circulation. If temperature distribution is homogeneous, and always above dew point at all crop parts, an increase of relative humidity levels by RH = 5% can save up to 10% of energy (Balendonck et al., 2014; Balendonck and Janssen, 2015). In ornamental crops, the application of higher humidity set points is often restricted due to higher risks of quality reduction.

4.2.2 Energy efficient dehumidification

Air humidity in a greenhouse is reduced by condensation on the inner side of the roof, by leakage through the greenhouse construction, and by replacement of humid internal air with dryer external air when ventilation windows are opened. Condensation is an efficient way of removing humidity (latent heat) from the greenhouse. Insulated energy saving greenhouse, in particular those with double coverings and reduced leakages, have condensation at 10% of the level found in non-insulated greenhouses. This leads to a higher air humidity. Removing humidity through the ventilation windows becomes more important. Traditionally, this is done by simultaneously opening the ventilation and using the heating system. However, this leads to high energy usage since as well as humidity (latent heat), desirable heat is also removed. In traditional greenhouses at Northern latitudes 140–200 MJ m^{-2} $year^{-1}$ of heating energy is removed by sensible heat losses through ventilation windows for dehumidification purposes (De Halleux and Gauthier, 1998). Depending on location and outside conditions, values can differ.

Alternatively, active control of humidity is possible, leading to energy savings compared to traditional control through ventilation windows. Campen et al. (2003) describe a method using an air-to-air heat exchanger in a balance ventilation system to reduce the losses of sensible heat (Fig. 2 left). In such a heat exchanger, the cold and dry outside air entering the greenhouse passes the warm and humid greenhouse air separated with a thin heat-exchanging surface. De Zwart (2014) estimates that this leads to an energy saving of 80–140 MJ m^{-2} $year^{-1}$ in greenhouses at Northern latitudes. This principle only works effectively if absolute outside humidity is lower than absolute inside humidity. Therefore, in warm humid climates other working principles must be considered.

These alternative approaches include cold surfaces or phase change materials. If the latent heat associated with crop evaporation can also be collected at a cold surface of the heat exchanger, energy savings are even higher (Campen and Bot, 2002) (Fig. 2 middle). De Zwart (2014) describes a system in which harvesting of energy during dehumidification is separated from the application of energy for greenhouse heating. He shows figures of energy saving by closed loop dehumidification of ca. 200 MJ m^{-2} $year^{-1}$ in greenhouses at Northern latitudes. This system is considered to be a good

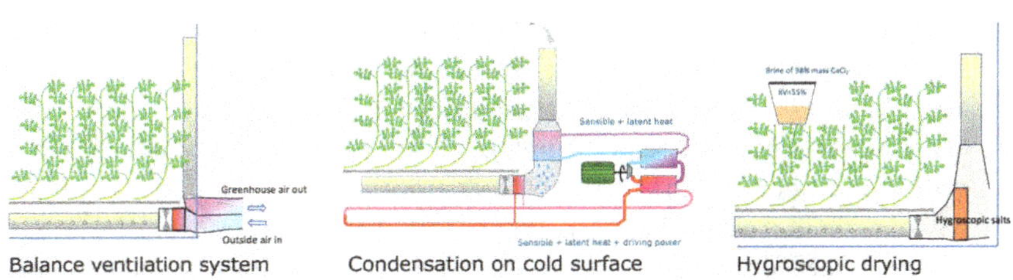

Figure 2 Different principles of mechanical dehumidification of greenhouses. On the left the principle of a balance ventilation system is shown, in the middle the principle of condensation on a cold surface, and on the right the principle of hygroscopic drying.

solution for fossil-free greenhouses of the future. Since the heat collected by the dehumidification is sufficient to cover the heating demand of a well-insulated greenhouse, no additional heating is required.

Dehumidification by hygroscopic drying is another principle by which latent heat can be removed from the greenhouse (Fig. 2 right), and an energy-saving potential up to 240 MJ m^{-2} $year^{-1}$ can be reached (De Zwart and Speentjes, 2013).

Today, mechanical dehumidification is applied in commercial greenhouses in Northern latitudes. Depending on the cropping system, the technical implementation of the three principles described above (a balance ventilation system, condensation on a cold surface, or hygroscopic drying) can vary.

4.3 Integration: 'next-generation cultivation strategies'

The concept of 'Het Nieuwe Telen', translates into English as 'next-generation cultivation' or 'new way of growing' and has been developed based on the following aspects: temperature integration depending on available solar radiation levels, higher humidity set points, and intensified use of thermal screens in combination with mechanical dehumidification. The concept has been summarized by Marcelis et al. (2014).

Insulation of the greenhouse is realized by the use of thermal screens (see Section 3.2). Under conditions of relatively low solar radiation a transparent thermal screen is closed, while at night a highly insulating non-transparent screen is also closed. At the start of the crop cycle a foil screen (perforated or non-perforated) is used for several weeks for additional insulation. To be able to keep the screens closed for prolonged periods of time, an accurate control of air humidity is required. Therefore, air humidity is controlled by a mechanical dehumidification system that injects relatively dry external air into the greenhouse (see Section 4.2.2). This results in an accurate control with a more uniformly distributed air humidity. The combined use of the intensified screening and humidity control reduced energy demand by 220 MJ m^{-2} $year^{-1}$ compared to a reference use of 1300 MJ m^{-2} $year^{-1}$, as estimated by model calculations of De Gelder et al. (2012b).

Temperature integration (see Section 4.1.1) over several days is applied, and temperature is made strongly dependent on radiation, since crops demand higher temperatures at higher light levels. This reduces energy use by 130 MJ m^{-2} $year^{-1}$. For year-round tomato cultivation, young plants are planted in December when it is cold and dark. Postponing the planting date hardly affects yield, while it saves 75 MJ m^{-2} $year^{-1}$ (De Gelder et al., 2012b).

The outlined temperature control in combination with a fogging system allows the windows to remain closed more often, which contributes to higher CO_2 concentrations in the greenhouse. This counteracts potential adverse effects of light loss due to intensified use of thermal screens on crop production. Subsequently, this concept has been tested on a large number of crops, including tomato, sweet pepper, *Alstroemeria*, *Gerbera*, lisianthus, *Chrysanthemum* and *Phalaenopsis* (De Gelder et al., 2014; De Gelder and Dieleman, 2012; Labrie and De Zwart, 2012). In most cases energy savings of about 30% were observed, while production and quality were maintained. The basic elements of 'next-generation cultivation strategies' are also applicable in other countries, though details of the strategy must be adapted.

5 Renewable energy sources

Renewable energy is derived from natural processes that are replenished constantly. Varga and Horváth (2014) categorize renewable energy sources and technologies as follows: (1) solar electricity and heat, (2) wind energy, (3) ocean energy, thermal, tidal and wave energy, (4) dynamic hydropower, (5) biomass energy, (6) geothermal resources, heat pumps, deep geothermal systems and enhanced geothermal systems, (7) bio fuels, and (8) hydrogen from renewable sources. However, not all technologies have the potential to be directly applied for greenhouse production systems. In this section, only a quick overview is given of the most important renewable energy sources for greenhouse application.

Solar energy can be considered as the most important renewable source for greenhouse production (see also Section 2). In low-tech greenhouse production systems, solar energy is used in different passive heat collection and diurnal storage systems (Cuce et al., 2016; Sethi and Sharma, 2008). Van der Velden and Smit (2017) describe the use of solar energy and seasonal storage for high-tech greenhouse cultivation in the Netherlands. Solar energy is collected in semi-closed greenhouses on 293 ha.

Some authors discuss the use of PV modules in novel greenhouse applications to supply heating or electricity (Nayak and Tiwari, 2008; Sonneveld et al., 2010a,b; Yano et al., 2010). PV modules placed next to a greenhouse (e.g. roofs of packaging area, water basins, typically ca. 10% of the area) can provide the greenhouse with renewable electricity. However, dynamic demand and supply rates have to be balanced. The application of PV modules on the roof of a greenhouse has been considered, but the modules must avoid interfering with the light transmission of the greenhouse. A possible greenhouse concept using PV in the roof area is described in Section 6.4.

Deep geothermal heat can be used for greenhouse heating in several countries. Knies and Bakker (2005) have explored the possibilities for the Netherlands, and De Zwart and Ruijs (2009) for Turkey. Also, in Eastern Europe and Iceland, geothermal heat is applied to greenhouses. A general overview of the worldwide availability of geothermal energy and its suitability for either electricity production or direct use is provided by Fridleifsson (2001). In the Netherlands deep geothermal heat is applied on 503 ha (Van der Velden and Smit, 2017).

Other renewable energy sources for heating are biomass or biofuels, which are applied in different heating systems (boiler, CHP). In the Netherlands biofuels are applied on 149 ha (Van der Velden and Smit, 2017). The majority of the renewable electricity used in the greenhouse is mainly obtained from the grid. The use of renewable sources for electricity production depends on the national energy mix provided, and can contain solar, wind, water, geothermal, biomass or biofuel.

Further options to make greenhouse production more sustainable involve the use of thermal energy and CO_2 generated as waste products from other industries, via regional heat and CO_2 grids. Initial uses of these materials have been shown in the Netherlands (KasAlsEnergiebron, 2018).

6 Case studies: integral design of new energy-saving greenhouse systems

6.1 Winterlight Greenhouse: maximum use of solar radiation

The transparent greenhouse covering influences the greenhouse energy balance as well as crop photosynthesis and development. To make maximum use of free natural sunlight for greenhouse heating and crop production in higher latitudes in winter, the light transmission of the greenhouse should be as high as possible (Hemming et al., 2017a) (see Section 2.1). In winter every Joule of natural solar radiation contributes to energy saving of (fossil) fuels. Next to that, research by Marcelis et al. (2014) has shown that especially in higher latitudes, 1% more mol of light gives 0.7–1% increased production of vegetable crops.

The concept of the 'Winterlight Greenhouse' aims at maximum transmission of natural sunlight especially for Dutch winter months (October–March). A methodological study was carried out taking into consideration the following elements. These included the type of greenhouse covering material (basic glass, type of diffuse structure, type of AR coating, condensation behaviour), greenhouse roof construction (roof shape, angle and dimensions, type and dimensions materials used for construction, coatings), screen (basic material used for transparent screen variables, installation) and crop (variety, cropping system, density, crop factors management) (Hemming et al., 2017a). The effects of stepwise and combined changes of the greenhouse system on the greenhouse transmission, was simulated (for every hour of the year) by the ray-tracing model RAYPRO (Swinkels et al., 2001, 2017). Inputs for the ray-tracing model were the hourly outside direct and diffuse radiation, a 3D model of the greenhouse construction and equipment, and the optical properties of all greenhouse construction and covering materials used. Optical properties of covering materials were quantified based on NEN 2675 (NEN, 2018) and used as input for the model. Changes in the crop and cropping system were calculated with a 3D crop model (e.g. Vos et al., 2010). Interactions of single elements were investigated to maximize this use of natural sunlight for crop production. The methodology used in the study could also be used for any other greenhouse type and location.

Swinkels et al. (2017) showed in a feasibility study that an east-west roof orientation of a Venlo-type greenhouse, provided 2–4% more light compared to a north-south orientation during winter months. In an east-west orientation a roof angle between 20° and 25° was optimal, both for clear and diffuse glass coverings. Changes in the greenhouse construction were possible and could lead to a maximum of 2–4% light increase. Highly reflective structural profiles used in the greenhouse roof gave 4–6% light increase. Current AR coatings on glass enabled an increase in light transmission by 4–8%, whereas with an angle-dependent optimization of AR coatings even more light could be gained. Hydrophilic glass surfaces gave favourable condensation properties and were able to increase the amount of light inside the greenhouse. The chosen diffuse AR glass realized 6–7% more transmissivity compared to the reference material if both materials were wet. A new transparent screen material gave 2–4% extra light gain. Interactions of the single effects occurred.

Based on the results of the modelling study and practical design considerations by industrial partners (e.g. legislation, loads, costs), a greenhouse concept of 500 m^2 was designed, built and tested (Fig. 3, Kempkes et al., 2017a). This concept can be characterized as follows.

The greenhouse gutter distance was scaled up from 4 m (reference) to 5.6 m to maximize the glass panel size to approximately 3 m × 1.67 m. For this the glass thickness was increased from 4 mm to 5 mm. To reduce the energy consumption of the 'Winterlight Greenhouse' a double independent controllable screen with a 5 cm split in between was used. This screen was installed under an angle equal to the roof slope, to increase transmissivity when

Figure 3 Photos of the 'Winterlight Greenhouse' with new roof dimensions, new glass and screen materials, a partly closed V-shaped double screen and white powder-coated roof construction materials for cucumber production (Kempkes et al., 2017a).

closed during daytime during winter months. In the reference greenhouse this is a standard double horizontal screen (Kempkes et al., 2017a). R&D samples of glass and screen materials with improved transmissivity under dry and wet conditions, and improved powder coating on all construction elements in order to enhance the reflectivity have been provided. All properties were measured at WUR LightLab. The reference materials were standard glass with 82.9% hemispherical transmissivity, a standard Luxous 1347 screen (Ludvig Svensson) with 72% hemispherical transmissivity, a white powder-coated steel construction with a reflectivity of 82%, and aluminium profiles in the roof with a reflectivity of 65%. The supplied new materials showed a hemispherical transmissivity of 91% for the glass, 76% for the screen, and a reflectivity of 87% for the powder-coated construction elements.

Before construction, the overall transmissivity of the 'Winterlight Greenhouse' was estimated to be 11.8% higher in winter than that of the reference greenhouse (Table 1). In the final process of upscaling the R&D samples of glass and screen materials from the lab size to the production, about 1% of transmissivity in screen cloth and glass, and 2% in reflectivity of the construction were lost. However, the final result was still a 10.5% light gain in winter when the cover was wet and both screens were closed. Table 1 also

Table 1 Overview of the simulated light transmissivity (%) of the reference and 'Winterlight Greenhouse' (WLG) with standard, improved and final materials during winter (October-March), summer (April-September) and January (Kempkes et al., 2017a)

Scenario	Winter	January	Summer
Reference, standard glass (82.9%), screen open	72.6	70.3	73.6
WLG, standard glass (82.9%), screen open	73.6 + 1.0	71.7 + 1.4	74.4 + 0.8
WLG, new glass (91%), screen open	79.9 + 7.2	77.7 + 7.4	81.0 + 7.5
Reference, standard glass (82.9%) - double Luxous 1347 (72%), screen closed	41.2	39.1	43.5
WLG, standard glass (82.9%) - double Luxous 1347 (72%), screen closed	44.8 + 3.6	43.5 + 4.4	45.8 + 2.2
WLG, standard glass (82.9%) - double new screen (76%), screen closed	49.0 + 7.8	47.6 + 8.5	50.0 + 6.5
Reference, standard glass (82.9%), condensation including screen open	68.4	66.3	70.4
WLG, new glass (91%), condensation including screen open	78.6 + 10.1	76.7 + 10.4	80.0 + 9.6
Reference, standard glass (83%), condensation including double Luxous 1347 (72%), screen closed	39.6	37.7	42.2
WLG, new glass (91%), double new screen (76%), condensation including screen closed	51.4 + 11.8	50.0 + 12.4	52.9 + 10.7
WLG, final glass (90%), double final screen (75%), condensation including screen closed	50.1 + 10.5	48.7 + 11.0	51.5 + 9.3

shows that light gains in winter are also positive in summer (Kempkes et al., 2017a). A greenhouse experiment in 2017 confirmed the theoretical results. Cucumber crop production was 9% higher compared to reference growers, (data not published yet).

6.2 VenlowEnergy Greenhouse: a greenhouse with high insulation

Insulation of the greenhouse roof and sidewalls prevents energy losses from the greenhouse by minimizing radiation and convection losses. Temporary insulation can be created by closing energy-saving screens during cold periods (winter, night-time, cold days). Permanent insulation can be created by double-layer covering materials as described by Hemming et al. (2017a) (see Section 3.1). At higher latitudes special attention has to be paid to light transmission. If screens are closed during daytime or a double covering is applied, light reduction easily occurs. Hemming et al. (2011) showed that it is possible to develop new covering materials without light reduction, using double glass with AR coatings on three sides, and realizing high insulation values by a low-emission coating on the fourth side. Based on that study a new greenhouse concept, the 'VenlowEnergy Greenhouse', has been developed and investigated (Hemming et al., 2012; Kempkes et al., 2014a).

A greenhouse concept of 500 m^2 was built and tested (Hemming et al., 2017a). The greenhouse had a double glass covering material (VenloW-e®) for high insulation. Since insulating coverings lead to less condensation and therefore higher humidity levels, special attention was paid to dehumidification (see Section 4.2). The greenhouse was equipped with a central dehumidification system. This involved fans in the greenhouse sidewall blowing in dry outside air into a plastic air duct distribution system, below the cropping gutter in the greenhouse if necessary, to avoid high humidity levels. This system included a heat regain unit, which was able to regain 80% of the sensible heat of the air extracted from the greenhouse by the dehumidification system. This heat exchanger heats up incoming outside cold air, before distributing it inside by exchanging heat with the outflow of warm and humid air from the greenhouse at a capacity of 7 m^3 m^{-2} h^{-1}. The greenhouse was equipped with a transparent energy screen Luxous 1243D (Ludvig Svensson) for further insulation (see Section 3.2). The double glass covering used was based on an earlier study by Hemming et al. (2011) and consisted of one glass pane with two AR coatings, and a second glass pane with one low-emission coating and one AR coating. The double glass had a PAR light transmission of 88% perpendicular (τ_p) and 79% hemispherical (τ_h) and a U-value of 1.2 W m^{-2} K^{-1} (compared to τ_p = 90%, τ_h = 82%, U = 6.7 W m^{-2} K^{-1} of single glass).

As described by Hemming et al. (2017a), an intensive modelling study preceded the greenhouse realization and testing. An energy-saving greenhouse climate strategy (temperature, humidity, CO_2) was developed with the validated dynamic climate model KASPRO (De Zwart, 1996). The model was also used to predict year-round energy consumptions and effects on cucumber production (dry matter), to develop the optimum growing strategy in terms of low energy consumption combined with a high cucumber yield (Hemming et al., 2012). Model calculations showed that a cucumber production of 75 kg m^{-2} $year^{-1}$ should be possible with a gas consumption of 12 m^3 gas m^{-2} $year^{-1}$ (=380 MJ), an electricity consumption of 8 kWh m^{-2} $year^{-1}$ and a CO_2 consumption of 43 kg m^{-2} $year^{-1}$, while reducing the theoretical energy consumption per kg cucumber. An experiment in 2010 with an autumn crop of cucumbers confirmed the theoretical results (Hemming et al., 2012).

More recently, modelling studies and practical experiments were carried out with tomato (Kempkes et al., 2014a). In an average year (SEL2000 year, Fig. 4) an energy consumption of 15.8 m^3 m^{-2} $year^{-1}$ of gas was predicted (=500 MJ). Three years of tomato cultivation in 2011–13 have shown that energy savings up to 50% compared to commercial growers are possible without affecting the production level. Measured energy consumptions were 17.9, 15.8 and 16.5 m^3 gas equivalents m^{-2} $year^{-1}$ in 2011, 2012 and 2013, respectively (=566, 500 and 522 MJ) (Fig. 4), with a tomato production (cv. 'Komeett') of 67, 72 and 71 kg m^{-2} $year^{-1}$ in 2011, 2012 and 2013, respectively. In 2012, electricity necessary to run the dehumidification system was 6 kWh m^{-2} (Kempkes et al., 2014a,b). To reduce the energy consumption in this insulated

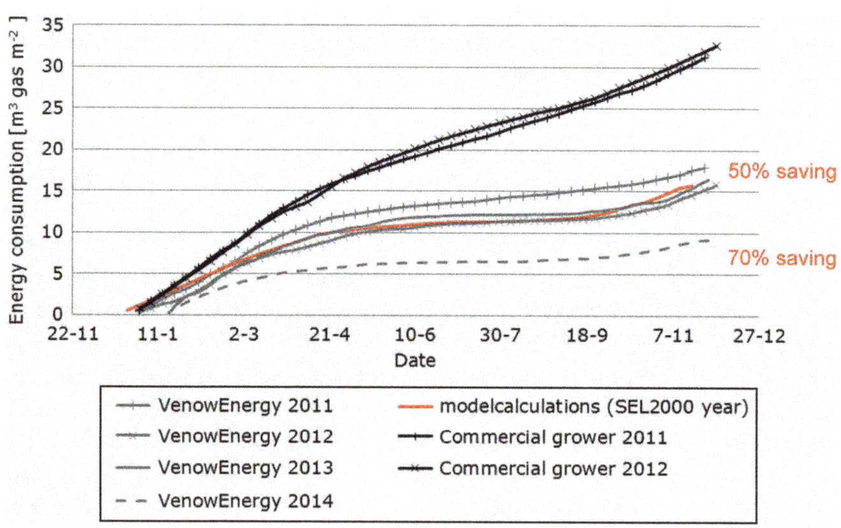

Figure 4 Energy saving in 4 years (2011–14) of tomato production cv. 'Komeett', in a highly insulated greenhouse concept with double glass with modern coatings in the 'VenlowEnergy Greenhouse', compared to commercial growers (e.g. Kempkes et al., 2014a).

greenhouse even further beyond 50% of the original level, crop transpiration had to be limited, especially during night-time (Hemming et al., 2017a). Therefore, higher humidity set points had to be used. An experiment by De Gelder et al. (*unpublished*) in 2013 and 2014 showed that with higher humidity set points, crop transpiration in the night and during periods with heating could be limited, and latent heat losses could be reduced. Using a decreased level of vapour deficit in the night from 3.4 g m^{-3} in 2013 to 2 g m^{-3} in 2014, the night transpiration was reduced by 25% without affecting the crop quality. Blossom end rot and shortage of calcium uptake did not occur. The tomato crop production was 71.2 kg m^{-2} $year^{-1}$ in 2013 and 75.4 kg m^{-2} $year^{-1}$ in 2014. In these experiments the energy consumption of the 'VenlowEnergy Greenhouse' had been reduced from 16.5 m^3 m^{-2} $year^{-1}$ in 2013 (=522 MJ) to 9.2 m^3 m^{-2} $year^{-1}$ in 2014 (=291 MJ) (Fig. 4).

6.3 Sunergy Greenhouse: a semi-closed greenhouse

One of the principles for sustainable energy usage in greenhouses, is the efficient conversion of solar energy by the use of heat exchangers/coolers, heat pumps and heat storage, in order to overcome seasonal differences in the availability of the solar energy and the energy demand of a greenhouse crop production (see Section 2.2). The closed or semi-closed greenhouse concept aims at covering heating, cooling and dehumidification needs of the greenhouse with minimal use of resources (De Gelder and Dieleman, 2012). Cooling by ventilation is replaced partly (in semi-closed greenhouses) or completely (in closed greenhouses) by mechanical cooling. Excess solar energy is collected in summer by heat exchangers/coolers and seasonally stored to be reused to heat the greenhouse in winter. In temperate climates, this concept combines improved crop production with energy and water savings (De Gelder et al., 2012a; Katsoulas et al., 2015).

The 'Sunergy Greenhouse' is a semi-closed greenhouse. An first demonstration greenhouse has been developed and tested on a scale of 500 m^2 by De Zwart (2010). Ventilation windows are fully closed during periods with high solar radiation, to enable harvesting of solar energy at a relatively high temperature by heat exchangers. This energy is stored in an aquifer and reused in cold periods for heating purposes. However, air exchange with outside air is allowed during dull days and during the night for dehumidification purposes. This way of dehumidification lowers the investment costs of the air treatment unit and, moreover, lowers the electricity demand compared to a completely closed greenhouse with mechanical dehumidification (De Zwart, 2011). To be able to keep the greenhouse closed in high radiation periods, the greenhouse is equipped with overhead heat exchange and cooling units with a capacity of 40 m^3 m^{-2} h^{-1} (Fig. 5 right). The cooling power is controlled by the flow of the

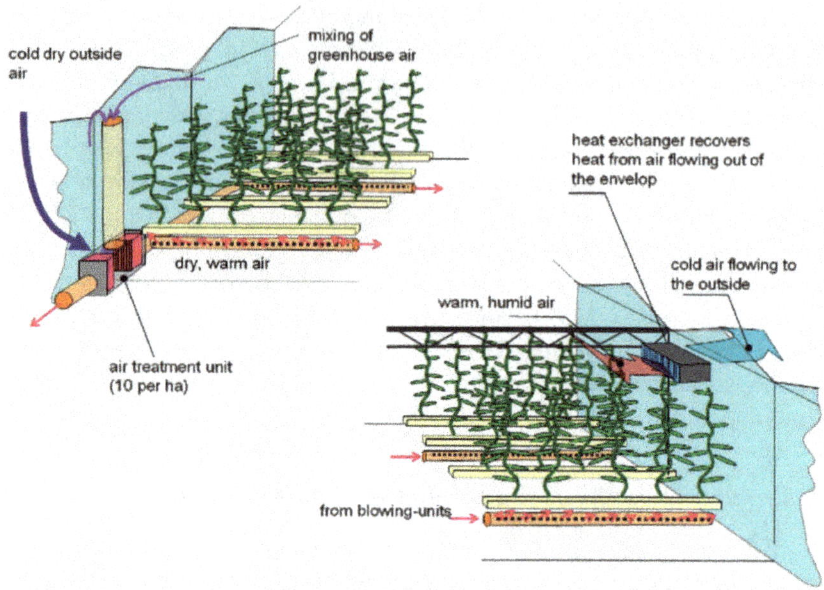

Figure 5 Overview of the air treatment unit with tubes under the crop gutters (left), and the overhead heat exchanger and cooling unit above the crop (right) inside the 'Sunergy Greenhouse', and their working principle (De Zwart, 2010).

cooling water. The temperature of the water supplied to the coolers is kept at 11°C, a temperature chosen to keep the seasonal heat storage system thermally neutral on an annual base (De Zwart, 2011). An important component of the 'Sunergy Greenhouse' is the air treatment unit. This unit blows air via tubes underneath the crop gutters into the greenhouse with a controllable flow rate of 5–15 $m^3 m^{-2} h^{-1}$ (Fig. 5 left). The air can be obtained from the outside and from the inside of the greenhouse with a controllable ratio. Outside air is taken in when the greenhouse becomes too humid. When the air is taken from the inside, the air treatment unit can be used to improve the homogeneity. Other elements of the 'Sunergy Greenhouse' concept are an electrical heat pump with a COP of 1.75, a CHP unit with a capacity of 185 $kW_e ha^{-1}$ and an electrical efficiency of 43% to generate electricity (96 kWh m^{-2}), heat and CO_2 (ca. 70 kg $ha^{-1} h^{-1}$). Both were not practically realized for demonstration purposes but simulated with a model (De Zwart, 2011). Also, for the heat storage a simple simulation model was used, which had earlier been proven to be able to accurately calculate the dynamic thermal behaviour of such a storage system (Campen and De Zwart, 2005).

Testing results were summarized by De Zwart (2010, 2011). Due to the intensive use of two thermal screens and the acceptance of high humidity set points, the energy consumption for heating the greenhouse was reduced by 25% compared to common practice in the Netherlands. Due to the fact that

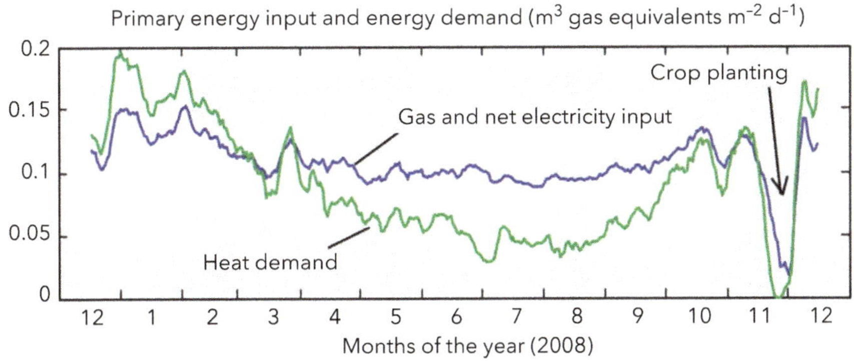

Figure 6 Overview of the energy input (gas and electricity) and the energy need for a tomato crop in the 'Sunergy Greenhouse' during 1 year (2008), a total of 2 m^3 a.e. m^2 $year^{-1}$ resulted (De Zwart, 2010).

the greenhouse remained closed during sunny and warm days, the inside CO_2 concentration could be kept between 900 and 1000 ppm with a supply rate of only 100 kg ha^{-1} h^{-1}. This was around 40% of the supply rate used in the average vegetable greenhouse in the Netherlands. In particular, the high CO_2 levels led to high production rates of cucumber (30 g cucumber per MJ outside radiation) and tomato (20 g of tomato per MJ outside radiation) (De Zwart, 2011). In total the following electricity balance was realized (in kWh m^{-2} y^{-1}): generation by CHP unit was 96 kWh, buying from grid 46, selling to grid 47, consumption by heat pump 44, heat exchangers/coolers 16, circulation pumps 11, air treatment units 11, heat storage pumps 10, greenhouse motors, valves and pumps 3 electricity units (De Zwart, 2010). The following total energy balance was recorded: consumption by CHP 27 m^3 natural gas per m^2 per year, and availability due to energy harvest and storage 29 m^3 gas equivalents per m^2 per year. In total, such a system was able to collect a surplus energy of 2 m^3 m^{-2} $year^{-1}$ (Fig. 6), and therefore can be considered to be more or less energy neutral.

6.4 Daylight Greenhouse: a greenhouse with electricity and heat production

One of the principles for sustainable energy usage in greenhouses is to make maximum use of natural sunlight (providing free energy input to the greenhouse and free light for crop growth and production). In cases where the sun provides more light than is favourable for crop production, part of the solar energy must be rejected. During the production of shade-tolerant pot plants light levels are usually limited by growers to a daily maximum of 8–10 mol of PAR light per m^2. Depending on the outside sunlight conditions, the surplus of energy has to be removed. For example, in the Netherlands, up to 45 mol of PAR is available per m^2 inside a greenhouse that is not shaded, therefore about 35 mol of PAR has to be rejected (Hemming et al., 2014a). This surplus of energy can potentially be transformed into electricity and/or heat and leads to the concept of an electricity-delivering greenhouse. Solar energy is used as a renewable source (see Section 5). However, in such a greenhouse, crop production always remains the primary function.

The idea of electricity-delivering greenhouses is not new. An overview of solar energy use in greenhouses is, for example, given by Lamnatou and Chemisana (2013). In 1991, Kurata described a greenhouse with a Fresnel prism as a covering. Later, other authors explored the idea of using Fresnel lenses as light regulation in greenhouses (Souliotis et al., 2006; Tripanagnostopoulos et al., 2005) and collecting solar energy for thermal energy production (Jirka et al., 1999). The use of Fresnel lenses in greenhouses for the production of concentrated power with CPV collectors is a more recent development (Sonneveld et al., 2010a,b, 2011a,b, 2012b; Korecko et al., 2010). Sonneveld et al. (2012a) investigated the principle of NIR reflection by a greenhouse covering for electricity production, and tested a prototype greenhouse. Luminescent solar concentrators are a newer concept for use on greenhouse roofs (Corrado et al., 2016; Hammam et al., 2007; Sol et al., 2018a,b).

In the concept of the 'Daylight Greenhouse' part of the sunlight is used when it is temporarily not necessary for crop production, as summarized by Hemming et al. (2014a). The direct part of the sunlight was concentrated by linear Fresnel lenses in the greenhouse roof (Sonneveld et al., 2011b, 2012b). The concentrated sunlight is trapped by a CPV collector (Sonneveld et al., 2011a), which moves automatically with a sun-tracking software depending on the position of the sun (Fig. 7 right). The diffuse part of the sunlight is used for crop production. The light level in the greenhouse can be controlled by moving the collector in or out of the focal line. An advanced prototype of that principle has been developed and investigated on a 500 m^2 scale (Fig. 7 left).

Figure 7 Fresnel lens greenhouse (daylight greenhouse), demonstration of 500 m^2 with tropical pot plant production built at Wageningen UR Greenhouse Horticulture in Bleiswijk (Hemming et al., 2014a).

The test results of the 'Daylight Greenhouse' were described by Hemming et al. (2014a). A production of 17 kWh m^{-2} $year^{-1}$ electricity was measured in 2012. Assuming an average conversion efficiency of 43% in the public grid, this electricity production is equivalent to 4.3 m^3 natural gas. Also, a production of 440 MJ m^{-2} $year^{-1}$ of thermal energy was measured in the same period, equivalent to 14 m^3 natural gas. In total 18.3 m^3 gas equivalents were produced. At the same time 16 m^3 gas was used to heat the greenhouse for tropical pot plant production. After accounting for energy losses and electricity demand for the heat pump, the net energy consumption of the Fresnel greenhouse had been 6 m^3 gas m^{-2} $year^{-1}$ in 2012. A commercial tropical pot plant greenhouse (with no artificial lighting) consumes 25 m^3 gas m^{-2} $year^{-1}$ on average. In terms of energy saving it can be stated that the 'Daylight Greenhouse' had a net energy demand 75% less than commercial practice. The Fresnel greenhouse produced in total 500 MJ $m^{-2}$$year^{-1}$ of energy (electricity and heat). Considering that in the Netherlands 1800 MJ m^{-2} $year^{-1}$ direct radiation is available, it can be concluded that it was possible to collect 28% of that potential energy in the Fresnel greenhouse. Comparable results were obtained in a larger greenhouse by a commercial grower (data unpublished).

7 Conclusion and future trends

Energy use in greenhouse production is essential to reach optimum crop production. Energy is mainly used for temperature control (heating and cooling), humidity control (mainly dehumidification) and artificial lighting in high-tech greenhouses. Some of the energy is used for CO_2 production. The absolute use of energy for greenhouse production depends on the outside climate conditions and crop needs. Intensified production schemes to expand the production period and reach more constant quantitative-qualitative yields, often lead to a higher absolute energy use. Therefore, both the reduction of absolute energy use, and the increase of energy efficiency per amount of produce, lead to sustainability.

Four principles for sustainable energy use in greenhouse horticulture can be identified: (1) maximum use of natural sunlight (free solar energy input to greenhouse, free light for crop growth and production), (2) maximum insulation (prevent energy losses through greenhouse shell), (3) efficient use of energy by different technologies and control strategies (e.g. temperature integration strategies, efficient heating and cooling systems, high humidity levels, fogging, mechanical dehumidification systems, optimum CO_2), and (4) sustainable energy sources replacing fossil fuels (e.g. geothermal, biofuels, wind).

In the pursuit of absolute carbon emission reduction, carbon-neutral greenhouse horticulture must be developed in the future, focusing on a greenhouse horticulture fully independent of fossil fuels. For that a more integrated approach to designing carbon-neutral greenhouse production systems is necessary. Combining different energy-saving principles can lead to new integral greenhouse designs. In particular, increased insulation and the use of sustainable energy sources will become more important. Free solar energy collection will play an important role in greenhouse production in the future. As well as this, deep geothermal heat or extraction of heat from soil or water, might play a more important role us greenhouse heat sources. Green electricity from the grid helps to intensify production with artificial lighting. Further future options to make greenhouse production more sustainable include the use of thermal energy and CO_2 sourced as waste products from other industries, via regional heat and CO_2 grids.

The optimal solution for sustainable energy use in greenhouse production will depend on local conditions, national policy and economic prices of products and resources.

8 References

Abdel-Ghany, A. M., Kozai, T. and Chun, C. 2001. Evaluation of selected greenhouse covers for use in regions with a hot climate. *Japanese Journal of Tropical Agriculture* 45, 242–50.

Adams, S. R., Woodward, G. C. and Valdes, V. M. 2002. The effects of leaf removal and of modifying temperature set-points with solar radiation on tomato yields. *Journal of Horticultural Science and Biotechnology* 77(6), 733–8. doi:10.1080/14620316.2002.11511565.

Andersson, N. E. and Nielsen, O. F. 2000. Energy consumption, light distribution and plant growth in greenhouse partly insulated with non-transparent material. *Gartenbauwissenschaft* 65(5), 190–4.

Arizov, A. and Niyazov, S. K. 1980. Rock-bed as a heat storage material for greenhouse applications. *Applied Solar Energy* 16, 430–7.

Baeza, E. and López, J. C. 2012. Light transmission through greenhouse covers. *Acta Horticulturae* 956, 425–40. doi:10.17660/ActaHortic.2012.956.50.

Baeza, E. J., Medrano, E., Sánchez-Guerrero, M. C., Sánchez-González, M. J., Porras, M. E., Giménez, M. and Lorenzo, P. 2017. An alternative to conventional fossil fuel heating systems: Water filled passive NIR absorbing polyethylene sleeves. *Acta Horticulturae* 1170, 765–72. doi:10.17660/ActaHortic.2017.1170.97.

Bailey, B. J. 1981. The reduction of thermal radiation in glasshouses by thermal screens. *Journal of Agricultural Engineering Research* 26(3), 215–24. doi:10.1016/0021-8634(81)90106-2.

Baille, M., Baille, A. and Tchamitchian, M. 1990. A simple model for the estimation of greenhouse transmission: Influence of structures and internal equipment. *Acta Horticulturae* 281, 35–46. doi:10.17660/ActaHortic.1990.281.3.

Baille, A., Kittas, C. and Katsoulas, N. 2001. Influence of whitening on greenhouse microclimate and crop energy partitioning. *Agricultural and Forest Meteorology* 107(4), 293–306. doi:10.1016/S0168-1923(01)00216-7.

Baille, A., López, J. C., Bonachela, S., González-Real, M. M. and Montero, J. I. 2006. Night energy balance in a heated low-cost plastic greenhouse. *Agricultural and Forest Meteorology* 137(1–2), 107–18. doi:10.1016/j.agrformet.2006.03.008.

Bakker, J. C. and Van Holsteijn, G. P. A. 1995. Screens. In: Bakker, J. C., Bot, G. P. A., Challa, H. and Van de Braak, N. J. (Eds), *Greenhouse Climate Control: An Integrated Approach*. Wageningen, The Netherlands: Wageningen Press, pp. 185-95.

Bakker, J. C. and Van Uffelen, J. A. M. 1988. The effects of diurnal temperature regimes on growth and yield of glasshouse sweet pepper. *Netherlands Journal of Agricultural Sciences* 36, 201-8.

Bakker, J. C., De Zwart, H. F. and Campen, J. B. 2006. Greenhouse cooling and heat recovery using fine wire heat exchangers in a closed pot plant greenhouse: Design of an energy producing greenhouse. *Acta Horticulturae* 719, 263-70. doi:10.17660/ActaHortic.2006.719.29.

Bakker, J. C., Adams, S. R., Boulard, T. and Montero, J. I. 2008. Innovative technologies for an efficient use of energy. *Acta Horticulturae* 801, 49-62. doi:10.17660/ActaHortic.2008.801.1.

Balendonck, J. and Janssen, J. J. 2015. The roaming climate measuring box greenhouse climate control based upon a wireless sensor network. *Proceedings of the 4th International Symposium on Sensor Science (I3S2015)*, 13-15 July 2015, Basel, Switzerland.

Balendonck, J., Sapounas, A. A., Kempkes, F., Van Os, E. A., Van der Schoor, R., Van Tuijl, B. A. J. and Keizer, L. C. P. 2014. Using a wireless sensor network to determine climate heterogeneity of a greenhouse environment. *Acta Horticulturae* 1037, 539-46. doi:10.17660/ActaHortic.2014.1037.67.

Blanchard, M. G. and Runkle, E. S. 2010. Influence of NIR-reflecting shading paint on greenhouse environment, plant temperature, and growth and flowering of bedding plants. *Transactions of the ASABE* 53(3), 939-44. doi:10.13031/2013.30076.

Blom, T. J. and Ingratta, F. J. 1981. The use of low intensity infrared for greenhouse heating in Southern Ontario. *Acta Horticulturae* 115, 205-16. doi:10.17660/ActaHortic.1981.115.23.

Bot, G. P. A. 1983. Greenhouse climate: From physical processes to a dynamic model. Ph.D. thesis, Wageningen, The Netherlands: University of Wageningen, 240pp.

Bot, G. P. A. 1992. Greenhouse growing: Outlook 2000. *Acta Horticulturae* 304, 43-47. doi:10.17660/ActaHortic.1992.304.3

Bot, G. P. A. 2001. Developments in indoor sustainable plant production with emphasis on energy saving. *Computers and Electronics in Agriculture* 30(1-3), 151-65. doi:10.1016/S0168-1699(00)00162-9.

Bot, G., Van de Braak, N., Challa, H., Hemming, S., Rieswijk, T., Van Straten, G. and Verlodt, I. 2005. The solar greenhouse: State of the art in energy saving and sustainable energy supply. *Acta Horticulturae* 691, 501-8. doi:10.17660/ActaHortic.2005.691.59.

Boulard, T., Razafinjohany, E. and Baille, J. 1990. Thermal performance and model of two types of greenhouses with solar energy storage. *Acta Horticulturae* 263, 121-30. doi:10.17660/ActaHortic.1990.263.11.

Boulard, T., Meneses, J. F., Mermier, M. and Papadakis, G. 1996. The mechanisms involved in the natural ventilation of greenhouses. *Agricultural and Forest Meteorology* 79(1-2), 61-77. doi:10.1016/0168-1923(95)02266-X.

Breuer, J. J. G. and Knies, P. 1995. Ventilation and cooling. In: Bakker, J. C., Bot, G. P. A., Challa, H. and Van de Braak, N. J. (Eds), *Greenhouse Climate Control - An Integrated Approach*. Wageningen, The Netherlands: Wageningen Academic Publishers, pp. 179-85.

Buchholz, M., Jochum, P. and Zaragoza, G. 2005. Concept for water, heat and food supply from a closed greenhouse – The Watergy project. *Acta Horticulturae* 691, 509–16. doi:10.17660/ActaHortic.2005.691.60.

Buwalda, F., Rijsdijk, A. A., Vogelezang, J. V. M., Hattendorf, A. and Batta, L. G. G. 1999. An energy efficient heating strategy for cut rose production based on crop tolerance to temperature fluctuations. *Acta Horticulturae* 507, 117–26. doi:10.17660/ActaHortic.1999.507.13.

Campen, J. B. and Bot, G. P. A. 2002. Dehumidification in greenhouses by condensation on finned pipes. *Biosystems Engineering* 82(2), 177–85. doi:10.1006/bioe.2002.0058.

Campen, J. B. and De Zwart, H. F. 2005. Practical measurements of a soil-based energy storage system and validation of a dynamic simulation model. Wageningen University and Research, Agrotechnology and Food innovations, report 058 (in Dutch). Available at: http://edepot.wur.nl/28400.

Campen, J. B., Bot, G. P. A. and De Zwart, H. F. 2003. Dehumidification of greenhouses at northern latitudes. *Biosystems Engineering* 86, 487–93.

Castilla, N., Hernández, J., Quesada, F. M., Morales, M. I., Guillén, A., Soriano, M. T., Escobar, I., Anton, A. and Montero, J. I. 2001. Comparison of asymmetrical greenhouse types in the Mediterranean area of Spain. *Acta Horticulturae* 559, 183–6. doi:10.17660/ActaHortic.2001.559.26.

Chandra, P. and Albright, L. D. 1980. Analytical determination of the effect on greenhouse heating requirements of using night curtains. *Transactions of the American Society for Agricultural Engineers* 23(4), 994–1000.

Chiapale, J.-P., Van Bavel, C. H. M. and Sadler, E. J. 1983. Comparison of calculated and measured performance of a fluid-roof and a standard greenhouse. *Energy in Agriculture* 2, 75–89. doi:10.1016/0167-5826(83)90008-6.

Cockshull, K. E., Graves, C. J. and Cave, C. R. J. 1992. The influence of shading on yield of glasshouse tomatoes. *Journal of Horticultural Science* 67(1), 11–24. doi:10.1080/00221589.1992.11516215.

Cohen, S. and Fuchs, M. 1999. Measuring and predicting radiometric properties of reflective shade nets and thermal screens. *Journal of Agricultural Engineering Research* 73(3), 245–55. doi:10.1006/jaer.1999.0410.

Corrado, C., Leow, S. W., Osborn, M., Carbone, I., Hellier, K., Short, M., Alers, G. and Carter, S. A. 2016. Power generation study of luminescent solar concentrator greenhouse. *Journal of Renewable and Sustainable Energy* 8(4), 043502. doi:10.1063/1.4958735.

Courtois, N., Petit, V., Thiéry, D., Grisey, A., Grasselly, D. and Pommier, F. 2008. Reversible climatisation of greenhouses in France using aquifer thermal energy storage: Pre-feasibility study. *Acta Horticulturae* 801, 819–26. doi:10.17660/ActaHortic.2008.801.96.

Cuce, E., Harjunowibowo, D. and Cuce, P. M. 2016. Renewable and sustainable energy saving strategies for greenhouse systems: A comprehensive review. *Renewable and Sustainable Energy Reviews* 64, 34–59. doi:10.1016/j.rser.2016.05.077.

De Gelder, A. and Dieleman, J. A. 2012. Validating the concept of the next generation greenhouse cultivation: An experiment with tomato. *Acta Horticulturae* 952, 545–50. doi:10.17660/ActaHortic.2012.952.69.

De Gelder, A. and Kipp, J. A. 2005. Perspectiefstudie voor (semi-)gesloten kassystemen. Wageningen University & Research, report PPO 41616095. Available at: http://edepot.wur.nl/272097.

De Gelder, A., Dieleman, J. A., Bot, G. P. A. and Marcelis, L. F. M. 2012a. An overview of climate and crop yield in closed greenhouses. *Journal of Horticultural Science and Biotechnology* 87(3), 193–202. doi:10.1080/14620316.2012.11512852.

De Gelder, A., Poot, E. H., Dieleman, J. A. and De Zwart, H. F. 2012b. A concept for reduced energy demand of greenhouses: The next generation greenhouse cultivation in the Netherlands. *Acta Horticulturae* 952, 539–44. doi:10.17660/ActaHortic.2012.952.68.

De Gelder, A., Warmenhoven, M. G. and Dieleman, J. A. 2014. Sustainable gerbera production realised with the next generation greenhouse cultivation. *Acta Horticulturae* 1037, 701–8. doi:10.17660/ActaHortic.2014.1037.90.

De Gelder, A., Van der Burg, R. and Zwinkels, J. 2017. Sweet pepper with reduced energy and good quality. Wageningen University and Research, Bleiswijk, report GTB-1434 (in Dutch). Available at: http://edepot.wur.nl/413253.

De Halleux, D. and Gauthier, L. 1998. Energy consumption due to dehumidification of greenhouses under northern latitudes. *Journal of Agricultural Engineering Research* 69(1), 35–42. doi:10.1006/jaer.1997.0221.

De Jong, T. 1990. Natural ventilation of large multi-span greenhouses. Ph.D. thesis, Wageningen, The Netherlands: University of Wageningen.

De Koning, A. N. M. 1994. Development and dry matter distribution in glasshouse tomato: A quantitative approach. Ph.D. thesis, Wageningen, The Netherlands: University of Wageningen.

De la Faille, B., Campen, J. B. and Oversloot, H. 2009. Emission and transmission of heat radiation through screens. TNO rapport 034-DTM-2009-04659.

De Pascale, S. and Maggio, A. 2005. Sustainable protected cultivation at a Mediterranean climate: Perspectives and challenges. *Acta Horticulturae* 691, 29–42. doi:10.17660/ActaHortic.2005.691.1.

De Pascale, S. and Stanghellini, C. 2011. High temperature control in Mediterranean greenhouse production: The constraints and the options. *Acta Horticulturae* 893, 103–16. doi:10.17660/ActaHortic.2011.893.6.

De Zwart, H. F. 1996. Analyzing energy saving options in greenhouse cultivation using a simulation model. Ph.D. thesis, Wageningen, The Netherlands: University of Wageningen, 236pp.

De Zwart, H. F. 2010. Performance of three demo-greenhouses in the Innovation and Demo Centrum. Wageningen University and Research, Wageningen, report GTB-1030 (in Dutch). Available at: http://library.wur.nl/WebQuery/wurpubs/fulltext/151912.

De Zwart, H. F. 2011. The Sunergy Greenhouse – One year of measurements in a next generation greenhouse. *Acta Horticulturae* 893, 351–8. doi:10.17660/ActaHortic.2011.893.31.

De Zwart, H. F. 2014. Energy conserving dehumidification of greenhouses. *Acta Horticulturae* 1037, 203–10. doi:10.17660/ActaHortic.2014.1037.21.

De Zwart, H. F. and Ruijs, M. N. A. 2009. Geothermal energy source as opportunity for Turkish greenhouse horticulture and the Dutch commercial sector. Wageningen University and Research, Wageningen, report 225. Available at: http://library.wur.nl/WebQuery/wurpubs/fulltext/301388.

De Zwart, H. F. and Speentjes, B. 2013. The next generation semi-closed greenhouse – perspective of a dehumidification system based on a cooling surface and based on hygroscopic salts. Wagenigen University and Research, Wageningen, report GTB-1292 (in Dutch). Available at: http://edepot.wur.nl/297286.

De Zwart, H. F., Van Noort, F. R. and Bakker, J. C. 2008. Energy performance and cropping experience in an energy producing greenhouse: Final report of a 2 years practical experiment. Wageningen University and Research, Wageningen, report 215, 128pp (in Dutch). Available at: http://library.wur.nl/WebQuery/wurpubs/fulltext/5353.

Dieleman, J. A. and Hemming, S. 2011. Energy saving: From engineering to crop management. *Acta Horticulturae* 893, 65–73. doi:10.17660/ActaHortic.2011.893.2.

Dieleman, J. A. and Kempkes, F. L. K. 2006. Energy screens in tomato: Determining the optimal opening strategy. *Acta Horticulturae* 718, 599–606. doi:10.17660/ActaHortic.2006.718.70.

Dieleman, J. A., Raaijmakers, E. and Meinen, E. 2005. Temperature research in tomato. Effect of a temporary reduction in temperature (DROP) on growth, development and production. Wageningen University and Research, Wageningen, report 360, 42pp.

Dieleman, A., Janse, J., De Gelder, A., Kempkes, F. L. K., De Visser, P., Lagas, P., Meinen, E., Warmenhoven, M. and Elings, A. 2015. Artificial lighting of tomato with less electricity. Wageningen University and Research, Wageningen, report GTB-1338 (in Dutch). Available at: http://edepot.wur.nl/338578.

Dijke, V. 2016. A method for measuring the moisture transport properties of greenhouse screens. *Swerea IVF* 82-11 (personal communication).

Dueck, T. A., Grashoff, C., Broekhuijsen, G. and Marcelis, L. F. M. 2006. Efficiency of light energy used by leaves situated in different levels of a sweet pepper canopy. *Acta Horticulturae* 711, 201–6. doi:10.17660/ActaHortic.2006.711.25.

Dueck, T. A., Meinen, E., Steenhuizen, J. W., Muusers, R., Uenk, D. and Marcelis, L. F. M. 2007. Artificial lighting in tomato: Using every lighting hour efficiently. Wageningen University and Research, Wageningen, report 439 (in Dutch). Available at: http://library.wur.nl/WebQuery/wurpubs/fulltext/20530.

Dueck, T. A., Janse, J., Li, T., Kempkes, F. L. K. and Eveleens-Clark, B. A. 2012a. Influence of diffuse glass on the growth and production of tomato. *Acta Horticulturae* 956, 75–82. doi:10.17660/ActaHortic.2012.956.6.

Dueck, T. A., Janse, J., Eveleens, B. A., Kempkes, F. L. K. and Marcelis, L. F. M. 2012b. Growth of tomatoes under hybrid LED and HPS lighting. *Acta Horticulturae* 952, 335–42. doi:10.17660/ActaHortic.2012.952.42.

Elings, A., Kempkes, F. L. K., Kaarsemaker, R. C., Ruijs, M. N. A., Van de Braak, N. J. and Dueck, T. A. 2005. The energy balance and energy-saving measures in greenhouse tomato cultivation. *Acta Horticulturae* 691, 67–74. doi:10.17660/ActaHortic.2005.691.5.

Fridleifsson, I. B. 2001. Geothermal energy for the benefit of the people. *Renewable and Sustainable Energy Reviews* 5(3), 299–312. doi:10.1016/S1364-0321(01)00002-8.

Gale, J., Feuermann, D., Kopel, R. and Levi, S. 1996. Liquid radiation filter greenhouses (LRFGs) and their use of low quality hot and cold water, for heating and cooling. *Acta Horticulturae* 440, 93–8. doi:10.17660/ActaHortic.1996.440.17.

Garcıa-Alonso, Y., Espi, E., Salmeron, A., Fontecha, A., Gonzalez, A. and Lopez, J. 2006. New cool plastic films for greenhouse covering in tropical and subtropical areas. *Acta Horticulturae* 719, 131–8. doi:10.17660/ActaHortic.2006.719.12.

Gbiorczyk, K. 1997. Untersuchung der Tropfengeometrie zur Bewertung des Kondensationsverhaltens von Gewächshausfolien. Diploma thesis, Hannover, Germany: Inst. Hort. Agri. Eng., University of Hannover.

Gbiorczyk, K., Sonneveld, P. J., Bot, G. P. A. and Von Elsner, B. 2004. The effect of roof inclivation on the condensation behavior of plastic films used as greenhouse covering materials. *Acta Horticulturae* 633, 127–36. doi:10.17660/ActaHortic.2004.633.15.

Geoola, F., Peiper, U. M. and Geoola, F. 1994. Outdoor testing of the condensation characteristics of plastic film covering materials using a model greenhouse. *Journal of Agricultural Engineering Research* 57(3), 167–72. doi:10.1006/jaer.1994.1016.

Geoola, F., Kashti, Y. and Peiper, U. M. 1998. A model greenhouse for testing the role of condensation, dust and dirt on the solar radiation transmissivity of greenhouse cladding materials. *Journal of Agricultural Engineering Research* 71(4), 339–46. doi:10.1006/jaer.1998.0333.

Geoola, F., Kashti, Y., Levi, A. and Brickman, R. 2009. A study of the overall heat transfer coefficient of greenhouse cladding materials with thermal screens using the hot box method. *Polymer Testing* 28(5), 470–4. doi:10.1016/j.polymertesting.2009.02.006.

Giacomelli, G. A., Giniger, M. S., Krass, A. E. and Mears, D. R. 1985. Improved methods of greenhouse evaporative cooling. *Acta Horticulturae* 174, 49–56. doi:10.17660/ActaHortic.1985.174.4.

Gómez, C. and Mitchell, C. A. 2016. In search of an optimized supplemental lighting spectrum for greenhouse tomato production with intracanopy lighting. *Acta Horticulturae* 1134, 57–62. doi:10.17660/ActaHortic.2016.1134.8.

Goodhind, G. W. 1965. Air movement in glass houses. Shinfield Paper No. 7.

Grafiadellis, M. 1985. Development of a passive solar system for heating greenhouses. *Proceedings of Second Greek Conference on Renewable Energy Sources*, Salonika, pp. 418–23.

Grange, R. I. and Hand, D. W. 1987. A review of the effects of atmospheric humidity on the growth of horticultural crops. *Journal of Horticultural Science* 62(2), 125–34. doi:10.1080/14620316.1987.11515760.

Grimstad, S. O. 1993. The effect of a daily low temperature pulse on growth and development of greenhouse cucumber and tomato plants during propagation. *Scientia Horticulturae* 53(1–2), 53–62. doi:10.1016/0304-4238(93)90137-F.

Grisey, A., Grasselly, D., Rosso, L., D'Amaral, F. and Melamedoff, S. 2011. Using heat exchangers to cool and heat a closed tomato greenhouse: Application in the South of France. *Acta Horticulturae* 893, 405–11. doi:10.17660/ActaHortic.2011.893.38.

Grisey, A., Rosso, L., Auger, E., Tisiot, R. and d'Amaral, F. 2014. 'Sustainable energy greenhouse' Results of the project: Reduction of energy consumption and energy storage in aquifer. *Acta Horticulturae* 1037, 147–54. doi:10.17660/ActaHortic.2014.1037.14.

Gunnlaugsson, B. and Adalsteinsson, S. 2006. Interlight and plant density in year-round production of tomato at northern latitudes. *Acta Horticulturae* 711, 71–6. doi:10.17660/ActaHortic.2006.711.6.

Guo, X., Hao, X., Khosla, S., Kumar, K. G. S., Cao, R. and Bennett, N. 2016. Effect of LED interlighting combined with overhead HPS light on fruit yield and quality of year-round sweet pepper in commercial greenhouse. *Acta Horticulturae* 1134, 71–8. doi:10.17660/ActaHortic.2016.1134.10.

Hammam, M., El-Mansy, M. K., El-Bashir, S. M. and El-Shaarawy, M. G. 2007. Performance evaluation of thin-film solar concentrators for greenhouse applications. *Desalination* 209(1–3), 244–50. doi:10.1016/j.desal.2007.04.034.

Hao, X., Little, C., Zheng, J. M. and Cao, R. 2016. Far-red LEDs improve fruit production in greenhouse tomato grown under high-pressure sodium lighting. *Acta Horticulturae* 1134, 95–102. doi:10.17660/ActaHortic.2016.1134.13.

Hemming, S. 2011. Use of natural and artificial light in horticulture – Interaction of plant and technology. *Acta Horticulturae* 907, 25–35. doi:10.17660/ActaHortic.2011.907.1.

Hemming, S., Kempkes, F., Mohammadkhani, V., Stanghellini, C., Swinkels, G. J. and Holterman, H. J. 2006a. Anti-reflection-coating for horticulture glass – first practical experiences. Wageningen University and Research, Wageningen, report 130 (in Dutch). Available at: http://library.wur.nl/WebQuery/wurpubs/fulltext/33330.

Hemming, S., Kempkes, F., van der Braak, N., Dueck, T. and Marissen, N. 2006b. Filtering natural light at the greenhouse covering – Better greenhouse climate and higher production by filtering out NIR? *Acta Horticulturae* 711, 411–6. doi:10.17660/ActaHortic.2006.711.58.

Hemming, S., Dueck, T. A., Janse, J. and Van Noort, F. R. 2008. The effect of diffuse light on crops. *Acta Horticulturae* 801, 1293–300. doi:10.17660/ActaHortic.2008.801.158.

Hemming, S., Kempkes, F. L. K. and Mohammadkhani, V. 2011. New glass coatings for high insulating greenhouses without light losses – Energy saving crop production and economic potentials. *Acta Horticulturae* 893, 217–26. doi:10.17660/ActaHortic.2011.893.15.

Hemming, S., Kempkes, F. L. K. and Janse, J. 2012. New greenhouse concept with high insulating double glass with modern coatings and new climate control strategies – Modelling and first results from a cucumber experiment. *Acta Horticulturae* 952, 233–40. doi:10.17660/ActaHortic.2012.952.28.

Hemming, S., De Zwart, H. F., Swinkels, G. L. A. M. and Janssen, H. J. J. 2014a. Development of electricity producing greenhouses – Two case studies. *Acta Horticulturae* 1037, 129–36. doi:10.17660/ActaHortic.2014.1037.12.

Hemming, S., Mohammadkhani, V. and Van Ruijven, J. P. M. 2014b. Material technology of diffuse greenhouse covering materials – influence on light transmission, light scattering and light spectrum. *Acta Horticulturae* 1037, 883–95. doi:10.17660/ActaHortic.2014.1037.118.

Hemming, S., Balendonck, J., Dieleman, J. A., De Gelder, A., Kempkes, F. L. K., Swinkels, G. L. A. M., De Visser, P. H. B. and De Zwart, H. F. 2017a. Innovations in greenhouse systems – Energy conservation by system design, sensors and decision support systems. *Acta Horticulturae* 1170, 1–16. doi:10.17660/ActaHortic.2017.1170.1.

Hemming, S., Baeza, E., Mohammadkhani, V. and Van Breugel, B. 2017b. Energy saving screen materials – Measurement method of radiation exchange, air permeability and humidity transport and a calculation method for energy saving. Wageningen University and Research, Wageningen, report GTB-1431. Available at: http://edepot.wur.nl/409298.

Hernández, R. and Kubota, C. 2012. Tomato seedling growth and morphological responses to supplemental LED lighting red:blue ratios under varied daily solar light integrals. *Acta Horticulturae* 956, 187–94. doi:10.17660/ActaHortic.2012.956.19.

Heuvelink, E. and González-Real, M. M. 2008. Innovation in plant-greenhouse interactions and crop management. *Acta Horticulturae* 801, 63–74. doi:10.17660/ActaHortic.2008.801.2.

Higashide, T. and Heuvelink, E. 2009. Physiological and morphological changes over the past 50 years in yield components in tomato. *Journal of the American Society of Horticultural Science* 134, 460–5.

Hoes, H., De Smedt, J., Goen, K. and Wittemans, L. 2008. The GESKAS project, closed greenhouse as energy source and optimal growing environment. *Acta Horticulturae* 801, 1355–62. doi:10.17660/ActaHortic.2008.801.166.

Hogewoning, S. W., Trouwborst, G., Meinen, E. and Van Ieperen, W. 2012. Finding the optimal growth light spectrum for greenhouse crops. *Acta Horticulturae* 956, 357–63. doi:10.17660/ActaHortic.2012.956.41.

Hovi, T., Nakkila, J. and Tahvonen, R. 2004. Interlighting improves production of year- round cucumber. *Scientia Horticulturae* 102(3), 283–94. doi:10.1016/j.scienta.2004.04.003.

Hovi-Pekkanen, T. and Tahvonen, R. 2008. Effects of interlighting on yield and external fruit quality in year-round cultivated cucumber. *Scientia Horticulturae* 116(2), 152–61. doi:10.1016/j.scienta.2007.11.010.

Huang, K. T. and Hanan, J. J. 1976. Theoretical analysis of internal and external covers for greenhouse heat conservation. *Horticulture Science* 11(6), 582–3.

Hurd, R. G. and Graves, C. J. 1984. The influence of different temperature patterns having the same integral on the earliness and yield of tomatoes. *Acta Horticulturae* 148, 547–54. doi:10.17660/ActaHortic.1984.148.69.

Impron, I. 2011. A greenhouse crop production system for tropical lowland conditions. Ph.D. thesis, Wageningen, The Netherlands: University of Wageningen. Available at: http://edepot.wur.nl/176828.

Impron, I., Hemming, S. and Bot, G. P. A. 2008. Effects of cover properties, ventilation rate, and crop leaf area on tropical greenhouse climate. *Biosystems Engineering* 99(4), 553–64. doi:10.1016/j.biosystemseng.2007.11.003.

Janssen, E. and Hart, H. 2006. Light interception of lighting systems, venlo and wide span type greenhouses. *Acta Horticulturae* 711, 431–6. doi:10.17660/ActaHortic.2006.711.61.

Jirka, V., Kuceravy, V., Maly, M., Pech, F. and Pokorný, J. 1999. Energy flow in a greenhouse equipped with glass raster lenses. *Renewable Energy* 16(1–4), 660–4. doi:10.1016/S0960-1481(98)00247-X.

Jongschaap, R. E. E., Dueck, T. A., Marissen, N., Hemming, S. and Marcelis, L. F. M. 2006. Simulating seasonal patterns of increased greenhouse crop production by conversion of direct radiation into diffuse radiation. *Acta Horticulturae* 718, 315–22. doi:10.17660/ActaHortic.2006.718.36.

KasAlsEnergiebron. 2018. Available at: https://www.kasalsenergiebron.nl/besparen/co2 /and https://www.kasalsenergiebron.nl/besparen/wkk-restwarmte/.

Katsoulas, N., Baille, A. and Kittas, C. 2001. Effect of misting on transpiration and conductances of a greenhouse rose canopy. *Agricultural and Forest Meteorology* 106(3), 233–47. doi:10.1016/S0168-1923(00)00211-2.

Katsoulas, N., Sapounas, A., Zwart, F. de, Dieleman, J. A. and Stanghellini, C. 2015. Reducing ventilation requirements in semi-closed greenhouses increases water use efficiency. *Agricultural Water Management* 156, 90–9. doi:10.1016/j.agwat.2015.04.003.

Kempkes, F. L. K. and Hemming, S. 2012. Calculation of NIR effect on greenhouse climate in various conditions. *Acta Horticulturae* 927(927), 543–50. doi:10.17660/ActaHortic.2012.927.66.

Kempkes, F. L. K., Stanghellini, C., Hemming, S. and Dai, J. 2008. Cover materials excluding near infrared radiation: Effect on greenhouse climate and plant processes. *Acta Horticulturae* 797(797), 477–82. doi:10.17660/ActaHortic.2008.797.69.

Kempkes, F. L. K., Janse, J. and Hemming, S. 2014a. Greenhouse concept with high insulating double glass with coatings and new climate control strategies: From design to results from tomato experiments. *Acta Horticulturae* 1037, 83-92. doi:10.17660/ActaHortic.2014.1037.6.

Kempkes, F., De Zwart, F. and Janse, J. 2014b. The impacts of a concept greenhouse with highly insulating double glass and a new method for greenhouse dehumidification management on energy use. *Acta Horticulturae* 1041, 253-63. doi:10.17660/ActaHortic.2014.1041.30.

Kempkes, F. L. K., Swinkels, G. L. A. M. and Hemming, S. 2017a. Increase of light transmission of a Venlo-type greenhouse during winter by 10%: A design study. Presentation during ISHS Greensys2017, Beijing, China. *Acta Horticulturae* 1227, 133–40.

Kempkes, F. L. K., Janse, J. and Hemming, S. 2017b. Greenhouse concept with high insulating cover by combination of glass and film: Design and first experimental results. *Acta Horticulturae* 1170, 469–86. doi:10.17660/ActaHortic.2017.1170.58.

Kittas, C. and Baille, A. 1998. Determination of the spectral properties of several greenhouse cover materials and evaluation of specific parameters related to plant response. *Journal of Agricultural Engineering Research* 71(2), 193–202. doi:10.1006/jaer.1998.0310.

Kittas, C., Baille, A. and Giaglaras, P. 1999. Influence of covering material and shading on the spectral distribution of light in greenhouses. *Journal of Agricultural Engineering Research* 73(4), 341–51. doi:10.1006/jaer.1999.0420.

Kittas, C., Karamanis, M. and Katsoulas, N. 2005. Air temperature regime in a forced ventilated greenhouse with rose crop. *Energy and Buildings* 37(8), 807–12. doi:10.1016/j.enbuild.2004.10.009.

Knies, P. and Bakker, J. C. 2005. Geothermal heat as sustainable heat source for greenhouse horticulture. Wageningen University and Research, Agrotechnology & Food Innovations, report 322 (in Dutch). Available at: http://library.wur.nl/WebQuery/wurpubs/fulltext/92047.

Korecko, J., Jirka, V., Sourek, B. and Cerveny, J. 2010. Module greenhouse with high efficiency of transformation of solar energy, utilizing active and passive glass optical rasters. *Solar Energy* 84(10), 1794–808. doi:10.1016/j.solener.2010.07.004.

Kumar, K. G. S., Hao, X., Khosla, S., Guo, X. and Bennett, N. 2016. Comparison of HPS lighting and hybrid lighting with top HPS and intra-canopy LED lighting for high-wire mini-cucumber production. *Acta Horticulturae* 1134, 111–8. doi:10.17660/ActaHortic.2016.1134.15.

Kurata, K. 1991. Scale-model experiments of applying a fresnel prism to greenhouse covering. *Solar Energy* 46(1), 53–7. doi:10.1016/0038-092X(91)90106-7.

Kurata, K., Quan, Z. and Nunomura, O. 1991. Optimal shapes of parallel East-West oriented single-span tunnels with respect to direct light transmissivity. *Journal of Agricultural Engineering Research* 48, 89–100. doi:10.1016/0021-8634(91)80006-Z.

Kurklu, A. 1998. Energy storage applications in greenhouses by means of phase change materials (PCMs): A review. *Renewable Energy* 13(1), 89–103. doi:10.1016/S0960-1481(97)83337-X.

Labrie, C. W. and De Zwart, F. 2012. Energy efficient climate control for cut flower alstroemeria. *Acta Horticulturae* 927, 581–7. doi:10.17660/ActaHortic.2012.927.71.

Labrie, C. and Verkerke, W. 2014. Healthy harvest from the greenhouse. *Acta Horticulturae* 1027, 423–6. doi:10.17660/ActaHortic.2014.1017.51.

Lamnatou, C. and Chemisana, D. 2013. Solar radiation manipulations and their role in greenhouse claddings: Fresnel lenses, NIR- and UV-blocking materials. *Renewable and Sustainable Energy Reviews* 18, 271–87. doi:10.1016/j.rser.2012.09.041.

Li, Q. and Kubota, C. 2009. Effects of supplemental light quality on growth and phytochemicals of baby leaf lettuce. *Environmental and Experimental Botany* 67(1), 59–64. doi:10.1016/j.envexpbot.2009.06.011.

Li, Y. L., Stanghellini, C. and Challa, H. 2001. Effect of electrical conductivity and transpiration on production of greenhouse tomato (*Lycopersicon esculentum* L.). *Scientia Horticulturae* 88(1), 11–29. doi:10.1016/S0304-4238(00)00190-4.

Li, T., Heuvelink, E., Van Noort, F., Kromdijk, J. and Marcelis, L. F. M. 2014a. Responses of two Anthurium cultivars to high daily integrals of diffuse light. *Scientia Horticulturae* 179, 306–13. doi:10.1016/j.scienta.2014.09.039.

Li, T., Heuvelink, E., Dueck, T. A., Janse, J., Gort, G. and Marcelis, L. F. M. 2014b. Enhancement of crop photosynthesis by diffuse light: Quantifying the contributing factors. *Annals of Botany* 114(1), 145–56. doi:10.1093/aob/mcu071.

Marcelis, L. F. M., Broekhuijsen, A. G. M., Meinen, E., Nijs, E. M. F. M. and Raaphorst, M. G. M. 2006. Quantification of the growth response to light quality of greenhouse grown crops. *Acta Horticulturae* 711, 97–104. doi:10.17660/ActaHortic.2006.711.9.

Marcelis, L. F. M., Buwalda, F., Dieleman, J. A., Dueck, T. A., Elings, A., De Gelder, A., Hemming, S., Kempkes, F. L. K., Li, T., Van Noort, F., et al. 2014. Innovations in crop production: A matter of physiology and technology. *Acta Horticulturae* 1037, 39–45. doi:10.17660/ActaHortic.2014.1037.1.

Max, J. F. J., Schurr, U., Tantau, H. J., Mutwiwa, U. N., Hofmann, T. and Ulbrich, A. 2012a. Greenhouse cover technology. In: Janick, J. (Ed.), *Horticultural Reviews*, vol. 40 (1st edn.). doi:10.1002/9781118351871.ch7.

Max, J. F. J., Reisinger, G., Hofmann, T., Hinken, J., Tantau, H. J., Ulbrich, A., Lambrecht, S., Von Elsner, B. and Schurr, U. 2012b. Glass-film-combination: Opto-physical properties and energy saving potential of a novel greenhouse glazing material. *Energy and Buildings* 50, 298–307. doi:10.1016/j.enbuild.2012.03.051.

McCree, K. J. 1971. The action spectrum, absorptance and quantum yield of photosynthesis in crop plants. *Agricultural Meteorology* 9, 191–216. doi:10.1016/0002-1571(71)90022-7.

Meyer, J. 1984. The influence of thermal screens on energy consumption of greenhouse. *Gartenbauwissenschaft* 49(2), 74–6.

Miguel, A. F. 1998. Transport phenomena through porous screens and openings: From theory to greenhouse practice. Ph.D. thesis, Wageningen, The Netherlands: University of Wageningen, 239pp. doi:10.5424/sjar/2006044-204.

Miguel, A. F., Van de Braak, N. J. and Bot, G. P. A. 1997. Analysis of the airflow characteristics of greenhouse screening materials. *Journal of Agricultural Engineering Research* 67(2), 105–12. doi:10.1006/jaer.1997.0157.

Miguel, A. F., Silva, A. M., Van de Braak, N. J. and Bot, G. P. A. 1998. Modelling of transport phenomena near greenhouse screens. *Acta Horticulturae* 456, 371–4. doi:10.17660/ActaHortic.1998.456.44.

Mistriotis, A., Bot, G. P. A., Picuno, P. and Scarascia-Mugnozza, G. 1997. Analysis of the efficiency of greenhouse ventilation using computational fluid dynamics. *Agricultural and Forest Meteorology* 85(3–4), 217–28. doi:10.1016/S0168-1923(96)02400-8.

Montero, J. I., Anton, A., Beil, C. and Franquet, A. 1990. Cooling of greenhouses with compressed air fogging nozzles. *Acta Horticulturae* 281, 199–210. doi:10.17660/ActaHortic.1990.281.22.

Morris, L. G., Trickett, E. S., Vanstone, F. H. and Wells, D. A. 1958. The limitation of maximum temperature in a glasshouse by the use of a water film on the roof. *Journal of Agricultural Engineering Research* 3, 121–30.

Myster, J. and Moe, R. 1995. Effect of diurnal temperature alternations on plant morphology in some greenhouse crops – A mini review. *Scientia Horticulturae* 62(4), 205–15. doi:10.1016/0304-4238(95)00783-P.

Näkkilä, J., Hovi-Pekkanen, T. and Tahvonen, R. 2006. Interplanting ensures continuous tomato production. *Acta Horticulturae* 711(711), 255–60. doi:10.17660/ActaHortic.2006.711.33.

Nayak, S. and Tiwari, G. N. 2008. Energy and exergy analysis of photovoltaic/thermal integrated with a solar greenhouse. *Energy and Buildings* 40(11), 2015–21. doi:10.1016/j.enbuild.2008.05.007.

Nederlands Normalisatie Instituut NEN. 2018. Greenhouse glass – Determination of optical properties of greenhouse covering materials and screens. NEN2675:2018 en.

Nijskens, J., Deltour, J., Coutisse, S. and Nisen, A. 1984. Heat transfer through covering materials of greenhouses. *Agricultural and Forest Meteorology* 33(2–3), 193–214. doi:10.1016/0168-1923(84)90070-4.

Ntinas, G. K., Koukounaras, A. and Kotsopoulos, T. 2015. Effect of energy saving solar sleeves on characteristics of hydroponic tomatoes grown in a greenhouse. *Scientia Horticulturae* 194, 126–33. doi:10.1016/j.scienta.2015.08.013.

Olofsson, T., Majabacka, A. and Engblom, S. 2006. Evaluation of energy performance for greenhouses based on multivariate analysis methodology. *Acta Horticulturae* 718, 211–8. doi:10.17660/ActaHortic.2006.718.23.

Opdam, J. J. G., Schoonderbeek, G. G., Heller, E. M. B. and De Gelder, A. 2005. Closed greenhouse: A starting point for sustainable entrepreneurship in horticulture. *Acta Horticulturae* 691, 517–24. doi:10.17660/ActaHortic.2005.691.61.

Ouzounis, T., Fretté, X., Rosenqvist, E. and Ottosen, C. O. 2014. Spectral effects of supplementary lighting on the secondary metabolites in roses, chrysanthemums, and campanulas. *Journal of Plant Physiology* 171(16), 1491–9. doi:10.1016/j.jplph.2014.06.012.

Ouzounis, T., Heuvelink, E., Ji, Y., Schouten, H. J., Visser, R. G. F. and Marcelis, L. F. M. 2016. Blue and red LED lighting effects on plant biomass, stomatal conductance, and metabolite content in nine tomato genotypes. *Acta Horticulturae* 1134, 251–8. doi:10.17660/ActaHortic.2016.1134.34.

Öztürk, H. H. and Bascetincelik, A. 2003. Energy and exergy efficiency of a packed-bed heat storage unit for greenhouse heating. *Biosystems Engineering* 86(2), 231–45. doi:10.1016/S1537-5110(03)00134-X.

Papadakis, G., Frangoudakis, A. and Kyritsis, S. 1992. Mixed, forced and free convection heat transfer at greenhouse cover. *Journal of Agricultural Engineering Research* 51, 191–205. doi:10.1016/0021-8634(92)80037-S.

Papadakis, G., Briassoulis, D., Scarascia Mugnozza, G., Vox, G., Feuilloley, P. and Stoffers, J. A. 2000. Radiometric and thermal properties of, and testing methods for, greenhouse covering materials. *Journal of Agricultural Engineering Research* 77(1), 7–38. doi:10.1006/jaer.2000.0525.

Paradiso, R., Meinen, E., Snel, J. F. H., De Visser, P., Van Ieperen, W., Hogewoning, S. W. and Marcelis, L. F. M. 2011. Spectral dependence of photosynthesis and light absorptance in single leaves and canopy in rose. *Scientia Horticulturae* 127(4), 548–54. doi:10.1016/j.scienta.2010.11.017.

Pieters, J. G., Deltour, J. M. and Debruyckere, M. J. 1997. Light transmission through condensation on glass and polyethylene. *Agricultural and Forest Meteorology* 85(1-2), 51–62. doi:10.1016/S0168-1923(96)02393-3.

Philips. 2018a. Available at: https://www.assets.lighting.philips.com/is/content/PhilipsLighting/fp928152419230-pss-global.

Philips. 2018b. Available at: http://images.philips.com/is/content/PhilipsConsumer/PDFDownloads/Netherlands/Horticulture/ODLI20180806_001-UPD-nl_NL-PS_toplighting_1-2_vegetables_NL_442294405317_B_LR-05-2018.pdf.

Pollet, I. V. and Pieters, J. G. 2000. Condensation and radiation transmittance of greenhouse cladding materials, part 3: Results for glass plates and plastic films. *Journal of Agricultural Engineering Research* 77(4), 419–28. doi:10.1006/jaer.2000.0628.

Qian, T., Dieleman, J. A., Elings, A., De Gelder, A., Marcelis, L. F. M. and Van Kooten, O. 2011. Comparison of climate and production in closed, semi-closed and open greenhouses. *Acta Horticulturae* 893, 807–14. doi:10.17660/ActaHortic.2011.893.88.

Qian, T., Dieleman, J. A., Elings, A. and Marcelis, L. F. M. 2012. Leaf photosynthetic and morphological responses to elevated CO_2 concentration and altered fruit number in the semi-closed greenhouse. *Scientia Horticulturae* 145, 1–9. doi:10.1016/j.scienta.2012.07.015.

Rath, T. 1992. Einsatz wissensbasierter Systeme zur Modellierung und Darstellung von gartenbautechnischem Fachwissen am Beispiel des hybriden Expertensystems HORTEX. Schriftenreihe Gartenbautechnische Informationen, Heft 34. Hannover. Ph.D. thesis, Hannover, Germany: Inst. Hort. Agri. Eng., University of Hannover.

Reisinger, G., Müller, J. and Mühlbauer, W. 1987. Development of an air-conditioning system for greenhouses in hot and dry climates. *Solar and Wind Technology* 4(1), 17–20. doi:10.1016/0741-983X(87)90004-X.

Ruijs, M. N. A., Raaphorst, M. G. M. and Dijkxhoorn, Y. 2010. More possibilities of energy saving crop control: Economic perspectives. The Hague: Wageningen University and Research, LEI, report 2010-006, 56pp (in Dutch). Available at: http://library.wur.nl/WebQuery/wurpubs/fulltext/133463.

Runkle, E. S., Jaster, P., Heins, R. D. and Thill, C. 2002a. Environmental conditions under an experimental near infra-red reflecting greenhouse film. *Acta Horticulturae* 578, 181–5. doi:10.17660/ActaHortic.2002.578.21.

Runkle, E. S., Heins, R. D., Jaster, P. and Thill, C. 2002b. Plant responses under an experimental near infra-red reflecting greenhouse film. *Acta Horticulturae* 580, 137–43. doi:10.17660/ActaHortic.2002.580.16.

Runkle, E. S., Padhye, S. R., Blanchard, M. G. and Oh, W. 2011. Energy efficient greenhouse lighting of ornamentals. *Acta Horticulturae* 907, 53–9. doi:10.17660/ActaHortic.2011.907.4.

Sadler, E. J. and Van Bavel, C. H. M. 1984. Simulation and measurement of energy partition in a fluid-roof greenhouse. *Agricultural and Forest Meteorology* 33(1), 1–13. doi:10.1016/0168-1923(84)90089-3.

Särkkä, L. E., Hovi-Pekkanen, T., Kaukoranta, T., Tahvonen, R. and Huttunen, J. 2006. Greenhouse cooling in summer in Finland – Preliminary results of climate control and plant response. *Acta Horticulturae* 719, 439–46. doi:10.17660/ActaHortic.2006.719.49.

Sase, S., Takakura, T. and Nara, M. 1984. Wind tunnel testing on air flow and temperature distribution of a naturally ventilated greenhouse. *Acta Horticulturae* 148, 329–36. doi:10.17660/ActaHortic.1984.148.42.

Schultz, W. 1997. Verfahren zur Untersuchung und Bewertung des Kondensationsverhaltens von Gewächshausfolien. Schriftenreihe Gartenbautechnische

Informationen, Heft 41. Ph.D. thesis, Hannover, Germany: Inst. Hort. Agri. Eng., University of Hannover.

Sethi, V. P. and Sharma, S. K. 2007. Survey of cooling technologies for worldwide agricultural greenhouse applications. *Solar Energy* 81(12), 1447–59. doi:10.1016/j.solener.2007.03.004.

Sethi, V. P. and Sharma, S. K. 2008. Survey and evaluation of heating technologies for worldwide agricultural greenhouse applications. *Solar Energy* 82(9), 832–59. doi:10.1016/j.solener.2008.02.010.

Silva, A. M. and Rosa, R. 1987. Radiative heat loss inside a greenhouse. *Journal of Agricultural Engineering Research* 37, 155–62.

Sol, J. A. H. P., Dehm, V., Hecht, R., Würthner, F., Schenning, A. P. H. J. and Debije, M. G. 2018a. Temperature-responsive luminescent solar concentrators: Tuning energy transfer in a liquid crystalline matrix. *Angewandte Chemie* 57(4), 1030–3.

Sol, J. A. H. P., Timmermans, G. H., Van Breugel, A. J., Schenning, A. P. H. J. and Debije, M. G. 2018b. Multistate luminescent solar concentrator 'smart' windows. *Advanced Energy Materials* 8(12), 1702922. doi:10.1002/aenm.201702922.

Sonneveld, P. J., Swinkels, G. L. A. M., Campen, J., Van Tuijl, B. A. J., Janssen, H. J. J. and Bot, G. P. A. 2010a. Performance results of a solar greenhouse combining electrical and thermal energy production. *Biosystems Engineering* 106(1), 48–57. doi:10.1016/j.biosystemseng.2010.02.003.

Sonneveld, P. J., Swinkels, G. L. A. M., Bot, G. P. A. and Flamand, G. 2010b. Feasibility study for combining cooling and high grade energy production in a solar greenhouse. *Biosystems Engineering* 105(1), 51–8. doi:10.1016/j.biosystemseng.2009.09.012.

Sonneveld, P. J., Swinkels, G. L. A. M., Van Tuijl, B. A. J., Janssen, H. J. J. and Gieling, T. H. 2011a. A fresnel lenses based concentrated PV system in a greenhouse. *Acta Horticulturae* 893, 343–50. doi:10.17660/ActaHortic.2011.893.30.

Sonneveld, P. J., Swinkels, G. L. A. M., Van Tuijl, B. A. J., Janssen, H. J. J., Campen, J. and Bot, G. P. A. 2011b. Performance of a concentrated photovoltaic energy system with static linear Fresnel lenses. *Solar Energy* 85(3), 432–42. doi:10.1016/j.solener.2010.12.001.

Sonneveld, P. J., Swinkels, G. L. A. M., Van Tuijl, B. A. J., Janssen, H. and Bot, G. P. A. 2012a. Greenhouse with a CPV system based on NIR reflecting lamellae. *Acta Horticulturae* 927, 43–50. doi:10.17660/ActaHortic.2012.927.3.

Sonneveld, P. J., Swinkels, G. L. A. M., Van Tuijl, B. A. J., Janssen, H. and de Zwart, H. F. 2012b. Up scaling and test results of an advanced fresnel greenhouse. *Acta Horticulturae* 952, 531–7. doi:10.17660/ActaHortic.2012.952.67.

Sørensen, B. 1989. Experiments with energy storage in a high-latitude greenhouse. *Solar Energy* 42(4), 293–301. doi:10.1016/0038-092X(89)90031-5.

Soriano, T. 2002. Validación de un modelo de cálculo de transmisividad a radiación solar directa en invernadero mediante maquetas a escala y determinación del prototipo óptimo para la costa Mediterránea. Ph.D. thesis, Spain: University of Almería.

Soriano, T., Montero, J. I., Sánchez-Guerrero, M. C., Medrano, E., Antón, A. J., Hernández, J., Morales, M. I. and Castilla, N. 2004. A study of direct solar radiation transmission in asymmetrical multi-span greenhouses using scale models and simulation models. *Biosystems Engineering* 88(2), 243–53. doi:10.1016/j.biosystemseng.2004.03.006.

Souliotis, M., Tripanagnostopoulos, Y. and Kavga, A. 2006. The use of Fresnel lenses to reduce the ventilation needs of greenhouses. *Acta Horticulturae* 719, 107–14. doi:10.17660/ActaHortic.2006.719.9.

Stadler, C. 2012. Effects of lighting time and light intensity on growth, yield and quality of greenhouse tomato. Landbúnaðarháskóli Íslands Rit LbhÍ nr. 40. Available at: https://skemman.is/bitstream/1946/23208/1/Rit_Lbhi_40.pdf.

Stanghellini, C. 1987. Transpiration of greenhouse crops, an aid to climate management. Ph.D. thesis, Wageningen, The Netherlands: University of Wageningen, 150pp.

Stanghellini, C., Dai, J. and Kempkes, F. L. K. 2011. Effect of near-infrared-radiation reflective screen materials on ventilation requirement, crop transpiration and water use efficiency of a greenhouse rose crop. *Biosystems Engineering* 110(3), 261–71. doi:10.1016/j.biosystemseng.2011.08.002.

Stanghellini, C., Bruins, M., Mohammadkhani, V., Swinkels, G. J. and Sonneveld, P. J. 2012. Effect of condensation on light transmission and energy budget of seven greenhouse cover materials. *Acta Horticulturae* 952(952), 249–54. doi:10.17660/ActaHortic.2012.952.30.

Stoffers, J. A. 1967. Light transmission of flat materials of protected greenhouses. ITT publication 14, Wageningen: IMAG, 35pp (in Dutch).

Swinkels, G. L. A. M., Sonneveld, P. J. and Bot, G. P. A. 2001. Improvement of greenhouse insulation with restricted transmission loss through zigzag covering material. *Journal of Agricultural Engineering Research* 79(1), 91–7. doi:10.1006/jaer.2000.0676.

Swinkels, G. L. A. M., Kempkes, F. L. K. and Hemming, S. 2017. Using a ray-tracing model to design a greenhouse with maximized transmission during winter. *Acta Horticulturae* 1154, 179–84. doi:10.17660/ActaHortic.2017.1154.23.

Sysoeva, M. I., Markovskaya, E. F., Kharkina, T. G. and Sherudilo, E. G. 1999. Temperature drop, dry matter accumulation and cold resistance of young cucumber plants. *Plant Growth Regulation* 28(2), 89–94. doi:10.1023/A:1006243230411.

Takakura, T., Jordan, K. A. and Boyd, L. L. 1971. Dynamic simulation of plant growth and environment in the greenhouse. *Transactions of the American Society of Agricultural Engineers* 14, 964–71.

Tanaka, J., Ishii, Y. and Ogura, T. 1997. A model experiment on reduction of greenhouse cooling load in the daytime using shading material against infrared radiation as covering material. *Environmental Control in Biology* 35, 15–20.

Tantau, H. -J. 1998. Energy saving potential of greenhouse climate control. *Mathematics and Computers in Simulation* 48(1), 93–101. doi:10.1016/S0378-4754(98)00145-1.

Tantau, H. -J.,Meyer, J., Schmidt, U. and Bessler, B. 2011. Low energy greenhouse – A system approach. *Acta Horticulturae* 893, 75–84. doi:10.17660/ActaHortic.2011.893.3.

Tantau, H.-J., Hinken, J., Von Elsner, B., Max, J. F. J., Ulbrich, A., Schurr, U., Hofmann, T. and Reisinger, G. 2012. Solar transmittance of greenhouse covering materials. *Acta Horticulturae* 956, 441–8. doi:10.17660/ActaHortic.2012.956.51.

Teitel, M., Peiper, U. M. and Zvieli, Y. 1996. Shading screens for frost protection. *Agricultural and Forest Meteorology* 81(3–4), 273–86. doi:10.1016/0168-1923(95)02321-6.

Teitel, M., Segal, I., Shklyar, A. and Barak, M. 1999. A comparison between pipe and air heating methods for greenhouses. *Journal of Agricultural Engineering Research* 72(3), 259–73. doi:10.1006/jaer.1998.0370.

Teitel, M., Barak, M. and Antler, A. 2009. Effect of cyclic heating and a thermal screen on the nocturnal heat loss and microclimate of a greenhouse. *Biosystems Engineering* 102(2), 162–70. doi:10.1016/j.biosystemseng.2008.11.013.

Tong, G., Christopher, D. M., Li, T. and Wang, T. 2013. Passive solar energy utilization: A review of cross-section building parameter selection for Chinese solar greenhouses. *Renewable and Sustainable Energy Reviews* 26, 540–8. doi:10.1016/j.rser.2013.06.026.

Tripanagnostopoulos, Y., Souliotis, M., Tonui, J. K. and Kavga, A. 2005. Irradiation aspects for energy balance in greenhouses. *Acta Horticulturae* 691, 733–40. doi:10.17660/ActaHortic.2005.691.90.

Trouwborst, G., Schapendonk, A. H. C. M., Rappoldt, K., Pot, S., Hogewoning, S. W. and Van Ieperen, W. 2011. The effect of intracanopy lighting on cucumber fruit yield–Model analysis. *Scientia Horticulturae* 129(2), 273–8. doi:10.1016/j.scienta.2011.03.042.

Van de Braak, N. J. 1995. Heating equipment. In: Bakker, J. C., Bot, G. P. A., Challa, H. and Van de Braak, N. J. (Eds), *Greenhouse Climate Control - An Integrated Approach*. Wageningen, The Netherlands: Wageningen Academic Publishers, pp. 171–7.

Van de Braak, N. J. and Breuer, J. J. G. 1991. Ventilation of greenhouses. Wageningen: IMAG-DLO, IMAG-DLO report 91–14, 21pp (in Dutch).

Van der Knijff, A., Benninga, J., Reijnders, C. and Nienhuis, J. 2004. Energy in greenhouse horticulture in The Netherlands: Developments in the sector and individual enterprises until 2004. Den Haag: LEI, report 3.06.02, 77pp (in Dutch). Available at: http://edepot.wur.nl/33057.

Van der Velden, N. and Smit, P. 2017. Energy monitor of the Dutch greenhouse horticulture 2016. The Hague: Wageningen University and Research, LEI, report 2017-094. Available at: http://edepot.wur.nl/424123.

Van Ieperen, W. 2012. Plant morphological and developmental responses to light quality in a horticultural context. *Acta Horticulturae* 956, 131–9. doi:10.17660/ActaHortic.2012.956.12.

Van Ieperen, W., Savvides, A. and Fanourakis, D. 2012. Red and blue light effects during growth on hydraulic and stomatal conductance in leaves of young cucumber plants. *Acta Horticulturae* 956, 223–30. doi:10.17660/ActaHortic.2012.956.24.

Vänninen, I., Pinto, D., Nissinen, A., Johansen, N. S. and Shipp, L. 2012. Prospecting the use of artificial lighting for integrated pest management. *Acta Horticulturae* 956, 593–608. doi:10.17660/ActaHortic.2012.956.71.

Vanthoor, B. H. E. 2011. A model-based greenhouse design method. Ph.D. thesis, Wageningen, The Netherlands: University of Wageningen, 307pp.

Varga, D. and Horváth, I. 2014. Integration of renewable energy sources into the energy supply flows of greenhouses: Development and validation of a forecasting model. In: Horváth, I. and Rusák, Z. (Eds), *Proceedings of TMCE 2014*, 19–23 May 2014, Budapest, Hungary.

Verheul, M. J. 2012. Effects of plant density, leaf removal and light intensity on tomato quality and yield. *Acta Horticulturae* 956, 365–72. doi:10.17660/ActaHortic.2012.956.42.

Verlodt, I. and Verschaeren, P. 1997. New interference film for climate control. *Plasticulture* 115, 27–35.

Victoria-Garcia, N., Kempkes, F. L. K., Van Weel, P., Stanghellini, C., Dueck, T. A. and Bruins, M. 2012. Effect of a diffuse glass greenhouse cover on rose production and quality. *Acta Horticulturae* 952, 241–8. doi:10.17660/ActaHortic.2012.952.29.

Villeneuve, J., De Halleux, D., Gosselin, A. and Amar, D. 2005. Concept of dynamic liquid foam insulation for greenhouse insulation and the assessment of its energy consumption and agronomic performances. *Acta Horticulturae* 691, 605–10. doi:10.17660/ActaHortic.2005.691.73.

Vogelezang, J. V. M. and Van Weel, P. A. 1988. Application of bench an floor heating systems. Aalsmeer: Proefstation voor de Bloemisterij, report 48 (in Dutch).

Von Elsner, B. and Xie, J. 2003. Effects of interference pigments in shading paint for greenhouses. *Proceedings of the Thirty-first Agricultural Plastics Congress*, 16–19 August 2003 in Grand Rapids, Michigan, USA, pp. 6–16.

Von Zabeltitz, C. 1987. Energy conservation and renewable energies for greenhouse heating. REU Technical Series 3, Roma: FAO and ENEA.

Von Zabeltitz, C. 2011. *Integrated Greenhouse Systems For Mild Climates*. Berlin: Springer.

Vos, J., Evers, J. B., Buck-Sorlin, G. H., Andrieu, B., Chelle, M. and De Visser, P. H. B. 2010. Functional-structural plant modelling: A new versatile tool in crop science. *Journal of Experimental Botany* 61(8), 2101–15. doi:10.1093/jxb/erp345.

Waaijenberg, I. D. 1984. Research on plastic greenhouse cladding materials. *Acta Horticulturae* 154, 57–64. doi:10.17660/ActaHortic.1984.154.4.

Waaijenberg, D. 2006. Design, construction and maintenance of greenhouse structures. *Acta Horticulturae* 710, 31–42. doi:10.17660/ActaHortic.2006.710.1.

Waaijenberg, D., Hemming, S. and Campen, J. B. 2005. The solar greenhouse: A highly insulated greenhouse design with an inflated roof system with PVDF or ETFE membranes. *Acta Horticulturae* 691, 561–8. doi:10.17660/ActaHortic.2005.691.67.

Wang, S. and Boulard, T. 2000. Measurement and prediction of solar radiation distribution in full scale greenhouse tunnel. *Agronomie* 20(1), 41–50. doi:10.1051/agro:2000107.

Willits, D. H. 2003. Cooling fan ventilated greenhouse: A modeling study. *Biosystems Engineering* 84(3), 315–29. doi:10.1016/S1537-5110(02)00270-2.

Wong, B., McClung, L., Snijders, A., McClenahan, D. and Thornton, J. 2011. The application of aquifer thermal energy storage in the Canadian greenhouse industry. *Acta Horticulturae* 893, 437–44. doi:10.17660/ActaHortic.2011.893.42.

Yano, A., Kadowaki, M., Furue, A., Tamaki, N., Tanaka, T., Hiraki, E., Kato, Y., Ishizu, F. and Noda, S. 2010. Shading and electrical features of a photovoltaic array mounted inside the roof of an east-west oriented greenhouse. *Biosystems Engineering* 106(4), 367–77. doi:10.1016/j.biosystemseng.2010.04.007.

Yildiz, I. and Stombaugh, D. P. 2008. Simulated effects of various environmental management practices on energy consumption in open and confined greenhouse systems. *Acta Horticulturae* 801, 1317–24. doi:10.17660/ActaHortic.2008.801.161.

Zhang, Y., Gauthier, L., De Halleux, D., Dansereau, B. and Gosselin, A. 1996. Effect of covering materials on energy consumption and greenhouse microclimate. *Agricultural and Forest Meteorology* 82(1–4), 227–44. doi:10.1016/0168-1923(96)02332-5.

Chapter 16

Assessing the environmental impact of greenhouse cultivation

Assumpció Antón, Erica Montemayor and Nancy Peña, Institute of Agrifood Research and Technology (IRTA), Spain

1 Introduction

Life cycle assessment (LCA) evaluates the environmental burdens associated with the life cycle of a product, process or activity by identifying and quantifying the flows of materials, energy and emissions, and ultimately their sustainability level and environmental impact. LCA can also evaluate and recommend implementation of strategies that reduce the environmental footprint of products, processes and services. It is now commonplace that policies, actions and environmental quantifications are planned taking into account the life cycle perspective, or in other words, accounting for the entire production chain. It is pertinent to focus not only on just one part of the chain, but on the chain as a whole. A clear example of this is waste management, a stage often overlooked in terms of a product's life cycle. Focusing on just one part of the chain could otherwise lead to the transfer of environmental problems to other points upstream or downstream of the analysed process.

Another great advantage of environmental quantification within the LCA framework is its ability to provide multiple responses to various environmental problems; it is not focussed simply on the calculation of carbon footprint or

http://dx.doi.org/10.19103/AS.2019.0052.18

water footprint, but rather it aims to address the diversity of environmental problems holistically. This helps to avoid the transfer of problems from one impact category to another. Finally, LCA methodology aims for objectivity and transparency. The LCA tool is supported by ISO regulations (ISO-14040, 2006; ISO-14044, 2006), and in accordance with ISO, LCA is an iterative process, divided into four steps.

1.1 Goal and scope of the study

This phase includes defining the purpose of the study and setting the system boundaries, which will depend on the final goal of the assessment. System boundaries define the unit processes that should be included in the product system. Normally, all processes from raw materials extraction to farm (greenhouse) gate, as well as material waste disposal, are considered. As the focus of the current chapter is greenhouse cultivation, the study will stop at the farm gate. However, when performing a consumer-focussed case study, processing, transport, commercialization and further steps in the chain should be included.

This phase also defines the functional unit (FU). The FU describes the main function of the process, and it is used as a mathematical reference to standardize all the different inputs and outputs of the system (flows). Since the most important function of horticultural greenhouse crops is the production of vegetables and flowers, the FU chosen is usually the yield of the horticultural product.

System boundaries separate the analysed system from the rest of the technosphere. At the same time, the system boundaries also define the boundary between the analysed system and the ecosphere. The inclusion of the crop field as part of the ecosphere or technosphere is one of the issues, which has been widely discussed in agricultural LCAs (Notarnicola et al., 2017; Rosenbaum et al., 2015). In the case of greenhouse production, there is a general agreement to consider the area covered by the greenhouse crop as part of the technosphere, which means that emissions going out of the greenhouse are taken into account, but not those produced into the greenhouse/technosphere.

1.2 Life cycle inventory

In this phase, the inventory is prepared through the collection of inputs and outputs of the different processes involved in the activity. Data collection on consumption (materials and energy) and emission modelling should be conducted across the different stages of the products life cycle.

Life cycle inventory (LCI) data preparation comprises primary data and the collection of specific data related to the core or foreground process. This data usually comes from companies, farms, growers and ultimately actors interested in the environmental assessment of their systems. Data for background processes are regarded as secondary data, which can be obtained from databases.

In greenhouse crop production, foreground processes include those related to greenhouse infrastructure, such as material and dimensions, greenhouse management, climate control systems, crop cultivation, labour operations, watering, fertilizer application and plant protection products and the relevant emissions released (Fig. 1).

The background processes could be split into upstream and downstream processes. The former include the inflow of raw materials and energy wares needed for the production of the final product, such as (1) manufacture of greenhouse infrastructure, including structure and covering material; (2) production of seeds, cuttings or plants for the cultivation; (3) production of agrochemicals, fertilizers and pesticides used in the cultivation; (4) production processes of energy wares used in the greenhouse; (5) manufacture of auxiliary products used such as watering pipes, heating or thermal systems and shading screens; (6) production of packaging materials; and (7) transport of all materials utilized from the manufacturer(s) to the greenhouse field. The downstream

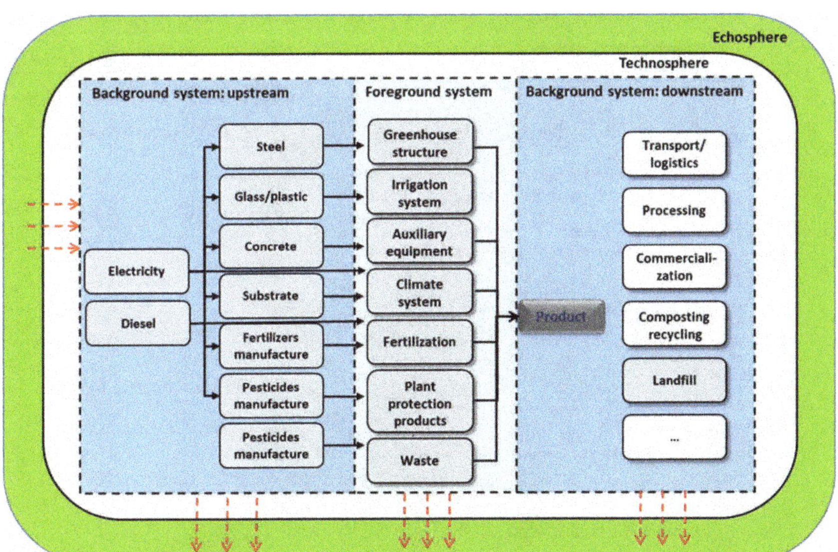

Figure 1 Scheme of the greenhouse crop production life cycle, including foreground system greenhouse cultivation and background upstream and downstream processes. Grey boxes are mandatory processes that must be included for assessments ending at the farm gate/greenhouse, and white boxes are optional and depend on the final goal and limits of the assessment defined in the goal and scope phase. Red dotted lines represent elementary flows and resources from nature and emissions to nature.

processes include (1) transport of the waste to the corresponding treatment plant and (2) waste management, landfill, incineration, composting or recycling of the different materials after use. For the products that will be reused through a transformation process, such as the ones for recycling or composting, there are different criteria to allocate burdens of these processes. The allocation criteria approach used will provide different environmental results. Our proposal is to follow the cut-off allocation procedure of Ekvall and Tillman (1997), for which recycling processes will be included as part of the recycled element. Therefore, in case of use of recycled materials such as recycled steel, the recycling processes shall be included as part of the greenhouse infrastructure, but not considered as waste treatment.

1.3 Life cycle impact assessment

In this phase, results of the inventory, emissions and resources are classified according to their impact categories and converted through corresponding characterization factors (CFs) to common impact units to make them comparable. Figure 2 outlines the classification and characterization elements of the life cycle impact assessment (LCIA), using the global warming category

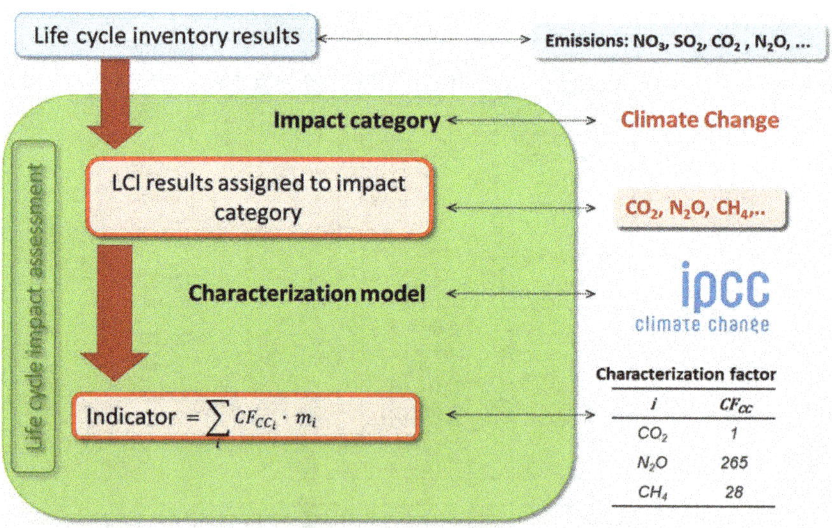

Figure 2 Diagram of LCIA phase. Adapted from ISO 14040 (2006) for climate change impact category. From life cycle inventory results, different intervention emission outputs (e.g. CO_2, CH_4, N_2O etc.) are selected based on their relevance and meaningfulness to the category chosen (e.g. global warming). For the given impact category (e.g. global warming), a characterization method comprises a category indicator (e.g. kg CO_2-eq.), a characterization model (e.g. IPCC) and a characterization factor (e.g. 265 kg N_2O per kg CO_2) derived from the model.

as an example. In the LCI, a list of interventions are recorded and quantified including different inputs and outputs of the processes. From that list, different emission outputs (e.g. CO_2, CH_4, N_2O) are selected based on their relevance and meaningfulness to the category chosen (e.g. global warming). For a given impact category (e.g. global warming), a characterization method comprises a category indicator (e.g. kg CO_2-eq.), a characterization model (e.g. IPCC, 2013) and a CF (e.g. 265 kg N_2O per kg CO_2) derived from the model. By means of CFs, also called equivalent factors, the addition of the different interventions is possible in order to provide a total value. Different LCIA models have been developed, such as the recommended by ILCD (EU-JRC-IES, 2011), ReCiPe (Goedkoop et al., 2009), IMPACT World+, updated form of Impact 2002 (Jolliet et al., 2003) among others. Table 1 provides the main contributing substances for each environmental impact category.

In the case of global impacts such as climate change and ozone depletion, we have common international characterization or equivalent factors, which only depend on the flow itself independently of location, where emissions take place, and to the conditions that surround it. In accordance with ILCD

Table 1 Main contributing substances to different impact categories

Impact category	Main contributing substances
Abiotic depletion	Coal, hard Gas, natural Oil crude Metals
Air acidification	Ammonia Nitrogen oxides Sulphur dioxide
Climate change	Carbon dioxide fossil Dinitrogen monoxide Methane
Ecotoxicity	Heavy metals Pesticides
Eutrophication	Ammonia Nitrogen oxides Chemical oxygen demand (COD) Phosphate Phosphorus
Particulate matter	Ammonia Nitrogen oxides Particulates, <2.5 mm Sulphur dioxide
Photochemical oxidation	Carbon monoxide fossil Sulphur dioxide

assessment framework recommendations, models and consequent CFs for such global impact categories are, in general, considered as recommendable and satisfactory (EU-JRC-IES, 2011). However, there are other impact categories, such as eutrophication, air acidification, land use and water use, that strongly depend on local conditions. These categories require site-specific CFs that depend on the activity's location. However, most site-specific CFs are still under development, and logically, agricultural activities are strongly related to regionalized impact categories.

The ultimate purpose of LCIA is to account for the damage in the three areas of protection: (1) human health, (2) ecosystems and (3) natural resources (EU-JRC-IES, 2010). Due to the complexity in the aggregation of the different environmental impacts into these damage categories, there are still no agreed equivalent factors at damage level. Therefore, midpoint indicators in contrast to endpoint indicators are often used in LCA studies.

1.4 Interpretation

This is the phase of an LCA that combines the results of the inventory analysis with the evaluation of the impact. The results of this interpretation are presented in the form of conclusions and recommendations to assist in decision-making. This step helps to identify and quantify what stage of the product's life cycle generates the most environmental impacts. It allows opportunities to identify improvements in the studied system. Interpretation is also a critical review of the quality of the data and the limitations of the analysis.

2 Greenhouse structures

Greenhouse structural components, including the framework, cladding materials and non-structural accessories, may account for nearly 30% of the total greenhouse cultivation systems in environmental impact categories, such as resource depletion and global warming (Antón, 2004; Martínez-Blanco et al., 2011; Russo and Scarascia-Mugnozza, 2004; Torrellas et al., 2012). Therefore, a fair comparison between greenhouse and open-field production involves infrastructure inclusion due to their high environmental impact and their role in producing higher yields. Despite this, the importance of including agricultural capital goods in environmental assessments is sometimes neglected (PAS2050.1, 2012).

In greenhouse infrastructure, there is great variability in the types of materials used (glass cover, plastic cover etc.), as well as greenhouse geometry (single span, multi-span, arched roof, flat roof, etc.). Moreover, greenhouse size or the scale factor varies greatly, depending on the features of the geographic area. This scale factor indicates that the larger the covered

Table 2 Results for climate change impact category expressed as kg CO_2-eq. m^{-2} and year for different types of greenhouse structures

Structure	Kg CO_2-eq m^{-2} $year^{-1}$	CV (%)	Main contributor
Glasshouse	2.9	17.5	Glass
Plastic multi-tunnel greenhouse	1.1	21.6	Steel
Steel *parral* greenhouse	0.64	28	Plastic
Wooden *parral* greenhouse	0.50	32.4	Plastic
Low tunnel	0.50	11	Plastic

CV: coefficient of variation.
Adapted from Antón et al. (2014b).

area, the lower the amount of materials required per area, and therefore, the lower the environmental load per area and production unit. Consequently, given the significance of infrastructure contribution and the difficulty in defining a 'representative' greenhouse as a single-data source, Antón et al. (2014b) provide a modelling approach for the greenhouse inventory stage to provide a more comprehensive way to assess the amount of materials used in greenhouse structures. Four main greenhouse structures are included: a glass greenhouse, a multi-tunnel greenhouse, a local Mediterranean model known as the *parral* greenhouse and a low-tunnel greenhouse. The glass greenhouse had a considerable environmental impact (2.94 kg CO_2-eq. m^{-2} $year^{-1}$) compared to the other types of greenhouses due to the high amount of glass and steel required to hold the glazing. Multi-tunnel structures showed average contributions between 0.97 and 1.19 kg CO_2-eq. m^{-2} $year^{-1}$ depending on the number of vent openings. Average values for the *parral* greenhouse were slightly lower depending on the use of wooden (0.50 kg CO_2-eq. m^{-2} $year^{-1}$) or steel (0.64 kg CO_2-eq. m^{-2} $year^{-1}$) structures. Finally, low-tunnel greenhouses showed average values between 0.45 and 0.53 kg CO_2-eq. m^{-2} $year^{-1}$ depending on the hoop material. In addition, Boulard et al. (2011) developed specific datasets for glass and plastic greenhouses in France, including not only the greenhouse frames but also all the equipment used; in which case, values are always twice the values shown in Table 2.

3 Greenhouse irrigation systems

An irrigation system includes all the equipment and processes involved in the irrigation of greenhouse crops. The watering system starts with extraction of water from the well, channel or rain reservoir tank, which provides the water from the source to water tanks and fertilizer tanks. Pumps and injectors supply water and fertilizers to a main pipe and later to secondary pipes, which finally provide water to the crop usually using trough dripper systems. The irrigation system could be an open-loop system, without recirculation of drainage water or a closed irrigation system, where leached water is collected and reused, after the corresponding cleaning and filtering processes.

In comparison to infrastructure impact contributions, the auxiliary equipment involved at the irrigation system represents between 15% and 40% of the total for infrastructure. In order to reduce infrastructure and equipment impacts, durable materials with a long lifespan are recommended, in addition to the use of recycled materials.

Electricity consumption included the electricity required to operate the irrigation system, which can range between 0.10 kWh m^{-3} for superficial water and 0.50 kWh m^{-3} for groundwater (Corominas, 2010). Mainly, in the case of groundwater use, a correct hydraulic design could help to save water and energy.

Regarding impacts in relation to water consumption, it is important to highlight the importance of location, 1 L of water consumption in an arid region is not equivalent to the consumption value in a humid region. Currently, the method AWARE (Boulay et al., 2018) has been developed with the aim to geographically weigh the value of water and is recommended as an assessment indicator for water use (Jolliet et al., 2018). AWARE is thus used as a water use midpoint indicator representing the relative Available WAter REmaining per area in a watershed, after the demand of humans and aquatic ecosystems have been met. It assesses the potential of water deprivation from either humans or ecosystems, building on the assumption that the less water remaining available per area, the more likely another user will be deprived (Boulay et al., 2018). Table 3 provides the values for the CFs of AWARE in a few relevant countries in relation to greenhouse production. According to this table, it should be understood that due to supply and demand of water in the different countries, there are different equivalence values of water. For instance, the consumption of 1 L of

Table 3 Characterization factors used to compare 1 m^3 of water according to location

Country	Characterization factor	Country	Characterization factor
Algeria	66.56	Japan	0.51
Bulgaria	26.41	South Korea	1.28
Canada	9.65	Mexico	34.25
China	45.53	Morocco	86.79
Egypt	98.51	Spain	79.13
France	8.29	The Netherlands	1.51
Greece	70.11	Tunisia	68.33
Israel	83.68	Turkey	57.39
Italy	48.50	USA	36.49

Values are provided by country but the reader is encouraged to consult specific values for each watershed (WULCA).

water in the Netherlands represents 1.51 L equivalents of water at world scale, while 1 L of water consumed in Spain represents 79.1 L equivalent, it means 52 times higher impact in Spain than in the Netherlands. In addition, it should be pointed out that the importance of using specific values for watersheds instead of those for countries. Mainly in large countries, with important geographical contrasts, averaged countries values could be misleading.

The origin or source of water is very important to consider in environmental impact accounting. Nowadays, the recommended AWARE method takes into account the use of blue water regardless of its origin. Although the scientific community has not yet reached a sufficient consensus on this topic, we could assume that the quantity of rainwater used for watering is green water, and hence is not accounted for in the current methods as a non-renewable resource. Therefore, it should be subtracted from the water used for irrigation and calculate the total available water remaining. For instance, it was estimated that greenhouses in Almeria, except flat *parral* structures, collected rainwater by greenhouse gutters and then rainwater was stored in irrigation ponds. Therefore, the estimated amount of rainwater collected by greenhouses was calculated as a weighted average, taking into account (1) the total greenhouse area in Almeria province (46 676 ha) after subtracting the flat *parral* greenhouse area (12 103 ha); (2) an average annual precipitation of 320 L m^{-2} (Mojonera, 2001–11 and Adra, 2003–11, weather stations); and (3) a 0.8 roof coefficient of rain collection (Farreny, 2011), resulting in 21% of total rainwater collected. Assuming all greenhouses in the area are collecting water and that there are enough irrigation ponds to store the water, an annual average estimated value of 66.7 L m^{-2} rainwater could be collected. This value is in accordance with data suggested by Pérez Parra et al. (2002) and means around 15% of total water consumption for a tomato crop.

4 Greenhouse climate systems

Depending on each production system, the climate control system can include the heating system, cogeneration system, distribution equipment, thermal water use, natural ventilation, CO_2 enrichment system, roof cooling and crop lighting. For those systems that include heating or lighting, these are by far the most environmentally contributing aspect. It is quite clear that this stage is clearly dominated by energy consumption. Therefore, impacts will mostly depend on the energy source. Accordingly, the use of renewable energy could help to considerably decrease environmental impacts.

In the case of cogeneration and combined heat and power (CHP), special attention should be paid to their multifunctionality, in that they generate three products: (1) heat for the greenhouses; (2) CO_2 to be used as CO_2 enrichment; and (3) electricity, which could be used in the own greenhouse (e.g. lighting)

Table 4 Results according to two different accounting criteria, system expansion and energy allocation, for cogeneration use in tomato greenhouse crop production in the Netherlands

Impact category (units)	System expansion	Energy allocation
Abiotic depletion (kg Sb-eq.)	5.6	15
Air acidification (kg SO_2-eq.)	1.2	2.9
Global warming (kg CO_2-eq.)	780	2000
Photochemical oxidant formation (kg C_2H_4-eq.)	0.19	0.21
Eutrophication (kg PO_4^--eq.)	−1.1	0.72

Results are expressed for different impact categories per 1 ton of tomato produced (Torrellas et al., 2012).

or is discharged to the public grid. Two approaches could be applied to deal with the multifunctionality of CHP or cogeneration plants: (1) system expansion approach, where the system boundaries are expanded to include the amount of electricity produced as an avoided product in the system, by subtracting it from the impacts of the current production mix process, and hence the electricity generated would also substitute part of the current production mix; and (2) allocation approach, following physical (energy) criteria of the energy inputs used (i.e. natural gas), the environmental loads of the inputs and outputs are allocated between the two products obtained, the crop and electricity.

As seen in Table 4, the results can differ considerably between the two types of approaches. Nevertheless, there is uncertainty surrounding the electricity that is substituted in the system expansion approach; electricity is accounted as an average, which may not reflect the real substituted one (Blonk et al., 2010). Additionally, this approach may involve aspects more related to consequential LCAs; therefore, adherence to allocation criteria in attributional LCAs would be advisable in order to gain insight into the amount of natural gas utilized for heating the greenhouse.

5 Use of substrates

Crops can be cultivated directly in the soil or in substrates. While in Northern countries the use of substrates is quite common, soilless crops represent around 20% of total crops in the Mediterranean area. The most popular substrates in hydroponics are stonewool, perlite, coconut fibre and peat. They are usually important contributors to the total environmental impact for a variety of material-dependent reasons, including destruction of natural resources (e.g. peat and quarries), high manufacturing and processing energy consumption (e.g. stonewool), transport from long distances (e.g. coconut) and their relatively short lifespan (e.g. stonewool for one crop and one season). Reducing the amount of substrate and substituting the classical substrates with local products, preferably coming from waste management could result in cleaner alternatives (Torrellas et al., 2012). However the rationale for using waste stream materials as growing media is often based on the assumption that reuse has less environmental impact than disposal. For an accurate understanding of the environmental impact of any growing medium, consideration should be given to its entire life cycle. Accounting impacts taking into account plant nutrient, watering efficiency as well as pollution potential during plant production (Barret et al., 2016). As far as these authors aware, this deep environmental assessment of growing media is still missing.

6 Fertilization

Fertilization involves important contributions to a range of impact categories, due to high energy and resource use during the manufacturing stage, and emissions produced during field application. Abiotic depletion and climate change are important indicators related to energy consumption and mineral fertilizer manufacturing. Mineral and organic fertilizer application emissions, such as NH_3, CH_4, NO_x and NO_3^-, are important contributors to air acidification, particulate matter and eutrophication impact categories. In heated greenhouses, fertilizer impacts are not as relevant as non-heated greenhouses due to impact domination by energy consumption contributions, more sophisticated closed systems and high yields, which means a major efficiency. Although these environmental values could be of high variability because of the different crops and fertilizer types used, seasonality, type of soilless crop and other several factors, their environmental impact could range between 1% and 10% of the total impact in heated greenhouses and between 10% and 50% in unheated greenhouses depending on the impact category.

Databases provide datasets for the manufacturing processes for the most common mineral fertilizers (e.g. calcium ammonium nitrate) or global averages for N, P or K fertilizers, but more often than not, only the total amount of N, P and K is known. In this case, the AGRIBALYSE® methodology (Koch and Salou, 2016) is recommended in order to calculate regionalized average N, P and K values based on the local consumption.

Accounting for fertilizer emissions is a controversial subject in both greenhouse and open-field production systems due to high pedoclimatic and practice-dependent variability and a lack of harmonized estimation models and measurement techniques. Therefore, a consensus and in-depth subject knowledge is required. Emissions from fertilizer application include N_2, N_2O, NO_x and NH_3 to air, and NO_3, PO_4 and K to soil where it could travel along surface and groundwater courses and eventually end up in the sea. Emissions from agriculture are highly variable depending on climate and soil type, but also on farming practices and many other inter-related factors. However, common or default values are usually used to account for these emissions (e.g. Tier 1), when regionalized values would be more suitable (e.g. Tier 2 or 3). For example, IPCC (2006) guidelines provide Tier 1 default values for N_2O air emission factors, which are based on the work by Bouwman et al. (2002), where they modelled N_2O air emission variability depending on soil, crops, climate and so on. Lately, Cayuela et al. (2017) provided emission factors for Mediterranean open-field conditions, thus addressing an important variability. However, regionalized values for greenhouses and substrates should also be developed. Similarly, default emission factors for ammonia emissions are currently being used (EEA, 2013) and are dependent on the soil pH. However, updated information regarding substrates and closed buildings such as greenhouses is required. In open-field agriculture, emissions of NH_3 and NO_x from fertilizers vary greatly depending on the method of application, where injection usually results in low emission quantities and broad spraying results in very high emission quantities. In greenhouses, there are three common methods of application, which may also result in important variations in emissions quantities. Thus, emission factors specific to greenhouses should be developed. Common greenhouse application methods include (1) fertilizers that can either be added during the pre-plant phase as a granular or slow- or controlled release fertilizer to the potting media; (2) fertilizers that can be applied to the plant's growing media via the irrigation system using a water-soluble fertilizer known as fertigation; or (3) fertilizers that can be applied as foliar sprays directly to the plant foliage.

For closed systems, lixiviates are agreed to be equal to zero, provided that they are collected and reused. In the case of open systems, surplus and potential lixiviates depend on the nutrient balance between inputs and outputs and eventually on characteristics of the soil and watering. Then, as a general rule, the balance between applied nutrients (N and P) and plant uptake, soil or substrate retention and air emissions for N need to be calculated. Nutrient uptake differs strongly among crops and is affected by the growth stage, climate conditions and ion composition of the nutrient solution. A dynamic adjustment between fertilization and growth stage could be the best way to save nutrients.

7 Plant protection products

Plant protection products (PPPs) contribute mainly to the toxicity impact category due to its application. For other impact categories such as climate change, manufacturing and transport usually contribute around 3–5% of the total contribution, due to the small quantities used in relation to other contributors such as fertilizers or greenhouse materials.

The assessment of impacts from toxic emissions is crucial to the relevance of LCA for several products. There are large differences among existing inventory emissions modelling as well as characterization impact methodologies. Regarding impact assessment methods, a consensus to characterize toxicity impacts was reached and the USEtox model was recommended (Rosenbaum et al., 2008). However, there is more work to be done to further refine and reduce uncertainty of the CFs used, mainly regarding the inclusion of site-specific characteristics.

Despite this, from an agricultural perspective, a good estimation of inventory data is even more relevant than the impact assessment stage. The proportion of PPPs emitted to the different media and environmental compartments in the ecosphere is required for environmental impact accounting. However, the total amount of PPP applied to the agricultural field is usually the only known input. This is a controversial issue, as the scientific community has not yet reached an agreement on the most appropriate model. Currently, an accounting project called Operationalization Life Cycle Assessment of Pesticide emissions (OLCA-PEST, 2017–20) is being conducted with the aim to provide a consensus on this matter. Although greenhouse crops are not included in this project, their basic principles can be followed, which suggest ways to deal with pesticide emissions inventory accounting (Peña et al., 2019; Rosenbaum et al., 2015).

Assuming that the best available technologies are applied, pesticide application is carried out with the greenhouse vents closed. Based on the review performed by Antón (2004), it can be assumed that air emissions due to the drift fraction exiting the greenhouse is 5% of the average amount applied, independent of the machinery used. When estimating secondary distribution, we should account for volatilization occurring from the plant and soil, which depends on the vapour pressure of the PPP. In this case, it is recommended to follow the consensus method explained by Peña et al. (2019), while taking into account the geographical ratio between surface water and soil of the surrounding area to account for the emissions that could potentially reach water (Fig. 3).

With regard to emissions to soil, we should account for the secondary emissions leached through the soil because the greenhouse is assumed to be part of the technosphere. However, greenhouse soils are usually covered with mulch or directly sealed. Therefore, in cases where crops are grown directly on the soil, emissions to soil should be accounted, following the consensus principles mentioned previously (Peña et al., 2019).

It is difficult to provide averages and standard deviations for pesticide emissions because pesticide application is strongly dependent on climate conditions, pests and farm practices. Nevertheless, it is worth mentioning that the use of mulching or hydroponic systems has decreased herbicide use, and insecticide use is decreasing thanks to biological control. As a result, fungicide use is more common than insecticide use.

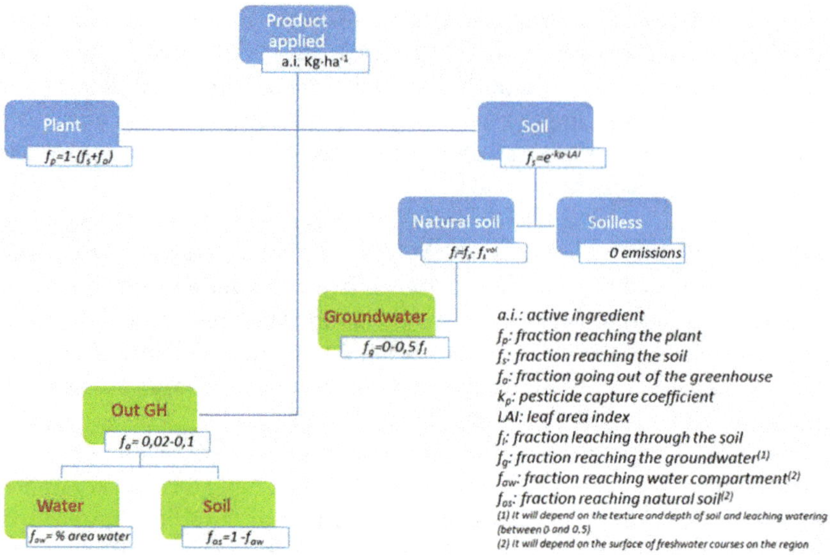

Figure 3 Scheme for accounting emission fractions of plant protection products applied (in green environmental compartments). Adapted from Antón et al. (2004) and Juraske et al. (2007).

8 Waste management

All products, elements and devices that take part in the production system have a specific life cycle. At the end of their life cycle, different waste management options can be applied, including reuse, recycling, composting, incineration or delivery to landfill. Waste management also includes the transport of waste materials from their source to the relevant waste management plant (e.g. landfill, compost plant, recycling plant).

According to the European circular economy policies, the highest priority should be the avoidance and reduction of waste generation, and recycling of waste should only be used as a last resort. Although recycling and composting of materials seems like a reasonable alternative, the results of environmental accounting may say otherwise. Therefore, it is advisable to use LCA tools to select the best option among those studied and neglect options when this recovery entails major environmental impacts.

Different approaches could be used to account for recycling processes, but they still require more development. Approaches may include (1) addition of recycling as a phase within the greenhouse production system; (2) allocation between both, our greenhouse production system and the new one, which is using the recycled materials; (3) assignment of recycling and composting processes to the new recycled material, when it replaces a non-recycled material made from raw materials. Then apply a cut-off at the point where the recyclable material leaves the greenhouse system. This last approach also referred to as the 'cut-off' approach (Ekvall and Tillman, 1997) would be highly recommended. In accordance with this method, each product should only be assigned the environmental impacts directly caused by the use of that product. Hence, each system receives the burdens for which it is directly responsible. Under this method, there is no uncertainty in the case of the extraction of raw materials, production processes or transport, because these were all directly assigned to the system. In the case of waste disposal, landfill and incineration, such treatment is fully attributable to the system being studied; while for waste, which is recycled or reused, its burdens are considered to be attributed to the system that will use it as a material source.

A separate issue exists with regard to transport from the waste source to the waste treatment plants. In cases where recycling companies go to the greenhouse and collect the materials, this transport will be part of the recycled process. In cases where transport is carried out by the grower or transported to final disposal sites such as landfills, transport should be included as part of the process system.

9 Conclusion and future trends

Throughout Sections 2 through 8 we have seen the importance of each process involved in greenhouse cultivation in terms of their environmental contributions, as well as different methodological criteria requirements.

From an environmental point of view, two completely different production systems need to be assessed: those from Northern European countries, where energy saving or renewable energy use is prioritized; and those from Southern areas (e.g. Mediterranean basin), where climate conditions are warmer, hence requiring lower energy input. The example in Fig. 4 reflects the potential contributions of each life cycle stage in an unheated, passive greenhouse cultivation system. The main aspects that require improvement include structure, auxiliary equipment (which in the figure also includes the substrate perlite) and fertilizers. Fertilizer impacts are important because of manufacturing processes and its emissions during its application.

As water consumption is usually high, fertilizer consumption applied during irrigation is also high. Therefore, an efficient balance of both fertilizers and water is recommended. In addition, the implementation of a closed-loop irrigation and water recovery system could reduce leaching emissions completely and contribute to water and nutrient conservation. Additionally, the incorporation of sensors and other technological devices could contribute to improved efficiency.

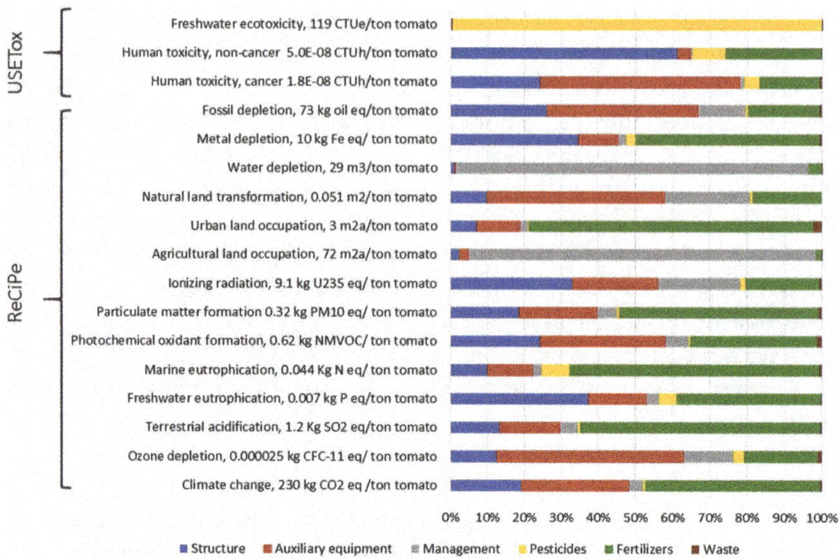

Figure 4 Example of total impact values per ton of tomato crop and stage contributions to impact assessment midpoint categories. Structure corresponds to a parral greenhouse; auxiliar equipment corresponds to perlite substrate and irrigation system; management includes the energy consumption to pump water for irrigation and the different labour operations. Adapted from Antón et al. (2014a).

In both Northern and Southern European systems, the structure becomes a major burden. Similarly, most of the substrates used are also major contributors due to the energy consumption within the manufacturing process.

On the other hand, the higher yields in greenhouse cultivation mean a more efficient use of resources, mainly land and water use. Besides that, a better climate control also means an improved efficiency for the use of other resources such as fertilizers. Being aware of the increasing population and food demands in the coming years, production systems that use resources more efficiently could be an excellent advancement, especially if we focus on reducing all the aforementioned aspects, either directly or indirectly, and reduce fossil energy consumption and adapt to renewable energies. This would greatly reduce contributions related to greenhouse energy consumption. Finally, taking advantage of circular economy principles could contribute to an improved environmental behaviour, while preserving and improving the socio-economic aspects related to the production system.

As this chapter is devoted to environmental impact quantification, it is important to examine the challenges and limitations that still exist, in order to refine the LCA methodology.

First, it is important to highlight the potential variability among production systems, not only for different farmers and regions but also from year to year. Thus, environmental quantification represents a range rather than absolute values, therefore guidelines on how to manage or average this variability should be developed.

Environmental quantification is provided as a function of the FU, which is usually yield. Although this could be appropriate as a reference or analysis flow, there is a common agreement among agricultural LCA practitioners that the real function of agriculture is to provide food as nutrients. Thus, some studies (Castañe and Antón, 2017) suggest the use of nutritional indexes, such as the Nutrient Rich Food Index NRF9.3 as the FU (Drewnowski and Fulgoni, 2008; Fulgoni et al., 2009).

Other methodological improvements still requiring consensus are allocation for multifunctional systems and emissions modelling. As seen in Section 4, LCA results could change substantially when different allocation criteria are applied, that is physical allocation or system expansion. Emissions modelling for both fertilizers and PPPs would benefit from regionalized and/or spatialized emission factors for greenhouse production.

In addition, there is a need for better modelling development for impact categories mainly related to agricultural processes, such as land use, biodiversity, eutrophication and its relationship with N and P cycles and water use. These aspects are being addressed in several current projects (Agronomics, 2018–22; Nutri2Cycle, 2018–22; OLCA-PEST, 2017–20; Organic-PLUS, 2018–22).

The fact that there is a consensus standard for some aspects of LCA facilitates analyses and adds credibility. There are several efforts from the European Commission (e.g. ILCD, EU-JRC, 2010, 2011; PEF/OEF, 2018) and international bodies (e.g. UNEP-SETAC Life Cycle Initiative) to harmonize models and rules. However, to our knowledge, no detailed rules have been agreed upon for environmental accounting of greenhouse cultivation.

Conclusively, the environmental requirements of greenhouse cultivation can be summarized in the following points:

1. Need for a consensus on inventory preparation, emissions modelling and impact characterization methods. For instance, building greenhouse product category rules in the LCA framework and environmental PEF initiatives.
2. Ensuring comprehensive and specific databases for environmental assessment of greenhouse products, including those for organic greenhouse production.
3. Advancing the development of regionalized CF, for greenhouse cultivation (e.g. land use or soil ecosystems) as well as differentiating degrees of intensiveness.

4 Incorporating eco-design tools and concepts as a way of improving greenhouse cultivation.
5 Adapting principles of circular economy into greenhouse cultivation, mainly for aspects related to the reduction of natural resource use, such as optimizing fertilizers and water, the use of innovative technologies and reutilization of waste.

10 Where to look for further information

As the other chapters in this book encompass technologic and agronomic improvement recommendations, this chapter mainly focuses on the aspects related to life cycle assessment (LCA) studies.

The LCA methodology has been developed and applied in the agricultural and food sectors to quantify environmental impacts and assist in decision-making. LCA in the agri-food sector has rapidly developed over the last two decades. LCA Food conferences appeared as a result of this rapid development, highlighting the importance of exchanging experiences, discussing issues and sharing methodologies, databases and tools. The first LCA Food conference took place in Brussels in 1996, followed by four more conferences: Brussels (1998), Göteborg (2001), Denmark (2003) and Göteborg (2007). It was not until the Zürich conference in 2008 that LCA Food finally gained momentum. Momentum that would be reinforced in the following conferences in Bari in 2008 and Saint-Malo in 2010. In 2012 the conference took place outside of Europe for the first time, in San Francisco, USA. A biannual cycle was then initiated that intends to alternate the location between Europe and a different continent. The last conference took place in 2016 in Dublin, and the next one will take place on 17–19 October 2018 in Bangkok.

With regard to impact assessment method improvements, the importance of Society of Environmental Toxicology and Chemistry (SETAC) conferences should be highlighted. SETAC Europe is a geographic unit (GU) of the SETAC, established to promote and undertake activities of SETAC in Europe and to support activities of SETAC in the Middle East and in Africa, SETAC's newest GU. There are annual meetings where aspects are mainly related to environment, toxicity and chemistry, as well as sessions on LCA. In 2018, the annual meeting took place in Rome and in 2019, it will be in Helsinki. The SETAC framework includes an LCA case study symposium, where the next one will take place on 24–27 September 2018 in Vienna.

It is also worth mentioning the Life Cycle Management (LCM) conferences. They are established as one of the leading conference series worldwide in the field of environmental, economic and social sustainability. The unique feature of LCM is developing practical solutions for the implementation of life cycle approaches into strategic and operational decision-making. Next, the 9th

International Life Cycle Management conference will take place at Poznań, Poland, in 2019.

Regarding specific journals in the area of LCA, *the International Journal of Life Cycle Assessment* (IJLCA, 2018), *Journal of Cleaner Production* (JCP, 2018) and *Environmental Science and Technology* (EST, 2018), among others, are recommendable.

Moreover, the following web pages are recommended in relation to LCA and key environmental issues:

UNEP-life cycle Initiative: https://www.lifecycleinitiative.org/
SETAC: https://www.setac.org/default.aspx
WWF: https://www.worldwildlife.org/
UN-SDG: http://www.un.org/sustainabledevelopment/sustainable-development-goals/
IPCC: http://www.ipcc.ch/
EU-JRC: http://ec.europa.eu/environment/eussd/smgp/ef_pilots.htm
LCAFood conferences: https://www6.inra.fr/lcafoodconferencearchives/

11 References

Agronomics, C. 2018–22. Efficient Carbon, Nitrogen and Phosphorus cycling in the European Agri-food System and related up- and down-stream processes to mitigate emissions. *H2020 ref* 773649.

Antón, A. 2004. Utilización del Análisis del Ciclo de Vida en la Evaluación del Impacto ambiental del cultivo bajo invernadero Mediterráneo. Tesis doctoral, Universitat Politècnica de Catalunya.

Antón, A., Castells, F., Montero, J. and Huijbregts, M. 2004. Comparison of toxicological impacts of integrated and chemical pest management in Mediterranean greenhouses. *Chemosphere*, 54, 1225–35.

Antón, A., Torrellas, M., Nuñez, M., Sevigné, E., Amores, M., Muñoz, P. and Montero, J. 2014a. Improvement of agricultural life cycle assessment studies through spatial differentiation and new impact categories: Case study on greenhouse tomato production. *Environmental Science & Technology*, 48, 9454–62.

Antón, A., Torrellas, M., Raya, V. and Montero, J. I. 2014b. Modelling the amount of materials to improve inventory datasets of greenhouse infrastructures. *International Journal of Life Cycle Assessment*, 19, 29–41

Barret, G. E., Alexander, P. D., Robinson, J. S. and Bragg, N. C. 2016. Achieving environmentally sustainable growing media for soilless plant cultivation systems – A review. *Scientia Horticulturae*, 212, 220–34.

Blonk, H., Kool, A., Luske, B., Ponsioen, T. and Scholten, J. 2010. Methodology for assessing carbon footprints of horticultural products. A study of methodological issues and solutions for the development of the Dutch carbon footprint protocol for horticultural products. Gouda, The Netherlands: Blonk Milieu Advies.

Boulard, T., Raeppel, C., Brun, R., Lecompte, F., Hayer, F., Carmassi, G. and Gaillard, G. 2011. Environmental impact of greenhouse tomato production in France. *Agronomy for Sustainable Development*, 31, 757–77.

Boulay, A., Bare, J., Benini, L., Berge, M., Lathuilliere, M., Manzardo, A., Margni, M., Motoshita, M., Núñez, M., Oki, T., Pastor, A., Ridoutt, B., Worbe, S. and Pfister, S. 2018. The WULCA consensus characterization model for water scarcity footprints: Assessing impacts of water consumption based on Available WAter REmaining (AWARE). *International Journal of Life Cycle Assessment*, 23, 368–78.

Bouwman, A., Boumans, L. and Batjes, N. 2002. Emissions of N_2O and NO from fertilized fields: Summary of available measurement data. *Global Biogeochemical Cycles*, 16, 6-1–6-13.

Castañe, S. and Antón, A. 2017. Assessment of the nutritional quality and environmental impact of two food diets: A Mediterranean and a vegan diet. *Journal of Cleaner Production*, 167, 929–37.

Cayuela, M., Aguilera, E., Sanz-Cobena, A., Adams, D., Abalos, D., Barton, L., Ryals, R., Silver, W., Alfaro, M., Pappa, V., Smith, P., Garnier, J., Billen, G., Bouwman, L., Bondeau, A. and Lassaletta, L. 2017. Direct nitrous oxide emissions in Mediterranean climate cropping systems: Emission factors based on a meta-analysis of available measurement data. *Agriculture, Ecosystems and Environment*, 238, 25–35.

Corominas, J. 2010. Agua y energía en el riego, en la época de la sostenibilidad. *Ingeniería del Agua*, 17, 219–33.

Drewnowski, A. and Fulgoni, V. 2008. Nutrient profiling of foods: Creating a nutrient rich food index. *Nutrition Reviews*, 66, 23–39.

EEA. 2013. EMEP/EEA air pollutant emission inventory guidebook 2013. Technical guidance to prepare national emission inventories EEA European Environmental Agency.

Ekvall, T. and Tillman, A. 1997. Open-loop recycling, criteria for allocation procedures. *International Journal of Life Cycle Assessment*, 2, 155–62.

EST. 2018. *Environmental Science and Technology* (Online). Available: https://pubs.acs.org/journal/esthag (Accessed 28 June 2018).

EU-JRC-IES. 2010. *International Reference Life Cycle Data System (ILCD) Handbook – General Guide for Life Cycle Assessment – Detailed Guidance*. First edition March 2010. EUR 24708 EN. Luxembourg: Publications Office of the European Union.

EU-JRC-IES. 2011. *International Reference Life Cycle Data System (ILCD) Handbook – Recommendations for Life Cycle Impact Assessment in the European Context*. European Commission-Joint Research Centre-Institute for Environment and Sustainability.

Farreny, R. 2011. Roof selection for rainwater harvesting: Quantity and quality assessments in Spain. *Water Research*, 45, 3245–54.

Fulgoni, V., Keast, D. and Drewnowski, A. 2009. Development and validation of the nutrient-rich foods index: A tool to measure nutritional quality of foods. *Journal of Nutrition*, 139, 1549–54.

Goedkoop, M., Heijungs, R., Huijbregts, M., De Schryver, A., Struijs, J. and van Zelm, R. 2009. ReCiPe 2008 A life cycle impact assessment method which comprises harmonised category indicators at the midpoint and the endpoint level. PRé Consultants, Amersfoort, Netherlands; CML, University of Leiden, The Netherlands; RUN, Radboud University, Nijmegen, The Netherlands; RIVM, Bilthoven, The Netherlands.

IJLCA. 2018. *International Journal of Life Cycle Assessment* (Online). Springer. Available: (https://www.springer.com/environment/journal/11367) (Accessed 28 June 2018).

IPCC. 2006. *Guidelines for National Greenhouse Gas Inventories. Volume 4: Agriculture, Forestry and Other Land Use*. Kanagawa, Japan: IGES.

IPCC. 2013. *Climate Change 2013: The Physical Science Basis. Contribution of Working Group I to the Fifth Assessment Report of the Intergovernmental Panel on Climate Change* (Stocker, T. F., Qin, D., Plattner, G.-K., Tignor, M., Allen, S. K., Boschung, J., Nauels, A., Xia, Y., Bex, V. and Midgley, P. M. (Eds)). Cambridge, UK and New York, NY: Cambridge University Press, 1535pp.

ISO-14040. 2006. *Environmental Management - Life Cycle Assessment - Principles and Framework*. Geneva: International Organisation for Standardisation ISO.

ISO-14044. 2006. *Environmental Management - Life Cycle Assessment - Requirements and Guidelines*. Geneva: International Organisation for Standardisation ISO.

JCP. 2018. *Journal of Cleaner Production* (Online). Available: https://www.journals.elsevier.com/journal-of-cleaner-production (Accessed 28 June 2018).

Jolliet, O., Margni, M., Charles, R., Humbert, S., Payet, J., Rebitzer, G. and Rosenbaum, R. 2003. IMPACT 2002+: A new life cycle impact assessment methodology. *International Journal Life Cycle Assessment*, 8, 324–30.

Jolliet, O., Antón, A., Boulay, A.-M., Cherubini, F., Fantke, P., Levasseur, A., Mckone, T. E., Michelsen, O., Milà I Canals, L., Motoshita, M., Pfister, S., Verones, F., Vigon, B. and Frischknecht, R. 2018. Global guidance on environmental life cycle impact assessment indicators: Impacts of climate change, fine particulate matter formation, water consumption and land use. *International Journal of Life Cycle Assessment*, 23, 2189–2207.

Juraske, R., Antón, A., Castells, F. and Huijbregts, M. 2007. Human intake fractions of pesticides via greenhouse tomato consumption: Comparing model estimates with measurements for Captan. *Chemosphere*, 67, 1102–7

Koch, P. and Salou, T. 2016. AGRIBALYSE®: Rapport Méthodologique - Version 1.3. November 2016. Angers, France: Ed ADEME, 332p.

Martínez-Blanco, J., Muñoz, P., Antón, A. and Rieradevall, J. 2011. Assessment of tomato Mediterranean production in open-field and standard multi-tunnel greenhouse, with compost or mineral fertilizers, from an agricultural and environmental standpoint. *Journal of Cleaner Production*, 19, 985–97.

Notarnicola, B., Sala, S., Antón, A., Mclaren, S. J., Saouter, E. and Sonesson, U. 2017. The role of life cycle assessment in supporting sustainable agri-food systems: A review of the challenges. *Journal of Cleaner Production*, 140(Part 2), 399–409

NUTRI2CYCLE. 2018–22. Closing the transition towards a more carbon and nutrient efficient agriculture in Europe. *H2020 ref* 773682.

OLCA-PEST. 2017–20. *Operationalising Life Cycle Assessment of Pesticides (OLCA-Pest)* (Online). Available: http://www.qsa.man.dtu.dk/research/research-projects/olca-pest (Accessed 28 June 2018).

ORGANIC-PLUS. 2018–22. Pathways to phase-out contentious inputs from organic agriculture in Europe. *H2020 ref 774340*.

PAS2050.1. 2012. Assessment of life cycle greenhouse gas emissions from horticultural products. Supplementary requirements for the cradle to gate stages of GHG assessments of horticultural products undertaken in accordance with PAS 2050. British Standard Institution.

Peña, N., Antón, A. and Fantke, P. 2019. Towards a consensus to estimate global emission fractions of pesticides: Definition of application scenarios. *Journal of Cleaner Production*, 209, 1332–41.

Pérez Parra, J., López, J. C. and Fernández, M. D. 2002. La agricultura del Sureste: Situación actual y tendencias de las estructuras de produccción en la horticultura almeriense. In: García Álvarez-Coque, J. M. C. (Ed.), *La Agricultura Mediterránea en el siglo XXI*. Almería: Instituto Cajamar.

Rosenbaum, R. K., Bachmann, T. M., Swirsky Gold, L., Huijbregts, M. A. J., Jolliet, O., Juraske, R., Koehler, A., Henrik, F., Larsen, H. F., Matthew Macleod, M., Manuele Margni, M., Mckone, T. E., Payet, J., Schuhmacher, M., Meent, D. and Hauschild, M. Z. 2008. USEtox - the UNEP-SETAC toxicity model: Recommended characterisation factors for human toxicity and freshwater ecotoxicity in life cycle impact assessment. *International Journal of Life Cycle Assessment*, 13, 532–46.

Rosenbaum, R., Anton, A., Bengoa, X., Bjørn, A., Brain, R., Bulle, C., Cosme, N., Dijkman, T. J., Fantke, P., Felix, M., Geoghegan, T. S., Gottesbüren, B., Hammer, C., Humbert, S., Jolliet, O., Juraske, R., Lewis, F., Maxime, D., Nemecek, T., Payet, J., Räsänen, K., Roux, P., Schau, E. M., Sourisseau, S., Van Zelm, R., Von Streit, B. and Wallman, M. 2015. The Glasgow consensus on the delineation between pesticide emission inventory and impact assessment for LCA. *International Journal of Life Cycle Assessment*, 20, 765–76.

Russo, G. and Scarascia-Mugnozza, G. 2004. LCA methodology applied to various typology of greenhouses. In: van Straten, G., Bot, G. P. A., van Meurs, W. T. M. and Marcelis, L. M. F. (Eds), *International Conference on Sustainable Greenhouse Systems - GREENSYS 2004*. Leuven, Belgium: ISHS. *Acta Horticulturae*, pp. 837–44.

Torrellas, M., Antón, A., Ruijs, M., García Victoria, N., Stanghellini, C. and Montero, J. 2012. Environmental and economic assessment of protected crops in four European scenarios. *International Journal of Cleaner Production*, 28, 45–55.

WULCA. Available: www.wulca-waterlca.org/ (Accessed 28 June 2018).

Index

CPSIA information can be obtained
at www.ICGtesting.com
Printed in the USA
BVHW010815161219
566725BV00013B/7/P

9 781786 762801